Introduction to Earth Science

INTRODUCTION TO EARTH SCIENCE

ADAPTED BY LAURA NESER

PDF AND EPUB FREE ONLINE AT: HTTPS://DOI.ORG/10.21061/INTROEARTHSCIENCE

VIRGINIA TECH DEPARTMENT OF GEOSCIENCES IN ASSOCIATION WITH VIRGINIA TECH PUBLISHING
BLACKSBURG, VA

Attributions: This work includes content from multiple sources reproduced under the terms of Creative Commons licenses, Public Domain, and Fair Use. Specifically: Chapters 1-16 are adapted from *An Introduction to Geology* (CC BY-NC-SA) by Chris Johnson, Matthew D. Affolter, Paul Inkenbrandt, and Cam Mosher. Chapter 17 is adapted from Section 22.1 of Chapter 22 "The Origin of Earth and the Solar System" by Karla Panchuk in *Physical Geology*, 2nd edition (CC BY) by Steven Earle, with Sections 7.1, 7.2, 7.3, and 7.4 of Chapter 7 "Other Worlds: An Introduction to the Solar System" from OpenStax *Astronomy*, 2nd edition (CC BY). And, figures are from a variety of sources; references at the end of each chapter describe the terms of reuse for each figure. Version notes located at the end of the book describe author changes made to these materials by chapter.

Suggested citation: Neser, Laura (2023). Introduction to Earth Science. Blacksburg: Virginia Tech Department of Geosciences. https://doi.org/10.21061/introearthscience. Licensed with CC BY-NC-SA 4.0. https://creativecommons.org/licenses/by-nc-sa/4.0.

Publisher: This work is published by the Virginia Tech Department of Geosciences in association with Virginia Tech Publishing, a division of the University Libraries at Virginia Tech.

Virginia Tech Department of Geosciences
4044 Derring Hall, 926 West Campus Drive, Blacksburg, VA 24061
https://geos.vt.edu

Virginia Tech Publishing University Libraries at Virginia Tech
560 Drillfield Drive Blacksburg, VA 24061 USA
https://publishing.vt.edu publishing@vt.edu

Accessibility statement: Virginia Tech Publishing is committed to making its publications accessible in accordance with the Americans with Disabilities Act of 1990. The Pressbooks (HTML) and ePub versions of this text are tagged structurally and include alternative text, which allows for machine readability.

Publication cataloging information:
Neser, Laura, author
Introduction to Earth Science / Laura Neser
Pages cm
ISBN 978-1-957213-34-7 (PDF)
ISBN 978-1-957213-35-4 (ePub)
ISBN 978-1-957213-33-0 (Pressbooks) https://pressbooks.lib.vt.edu/introearthscience
ISBN 978-1-957213-36-1 (Print)
URI (Universal Resource Identifier): http://hdl.handle.net/10919/112740
DOI: https://doi.org/10.21061/introearthscience

1. Earth sciences--Textbooks.
2. Geology--Textbooks. Title
   QE26.3

Cover image: Toby Elliott via Unsplash, Unsplash license
Cover design: Kindred Grey

CONTENTS

INTRODUCTION

Introduction to Earth Science is designed to provide a comprehensive introduction to Earth Science that can be freely accessed online, read offline, printed, or purchased as a print-on-demand book. It is intended for a typical 1000-level university introductory course in the Geosciences, although its contents could be applied to many other related courses. The textbook is an amalgamation of existing open-source textbooks (An Introduction to Geology from Salt Lake Community College, Physical Geology from BCcampus, and Astronomy by OpenStax). It has been customized to match the Pathways General Education Curriculum at Virginia Tech with a focus on Student Learning Outcomes (SLOs) of Pathways Concept 4: Reasoning in the Natural Sciences.

The sequence of chapters in this book may differ from a typical commercial publisher's introductory Earth Science textbook. Selected concepts in the book have been reorganized to summarize elementary or foundation topics prior to discussion of more complex topics in Earth Science. A foundational knowledge of minerals and rocks is essential in an introductory Earth Science course. Oftentimes all three rock classifications are bundled into a single "rocks" chapter but because they vary extensively in the way they form and their overall features, this textbook gives the three major rock types their own dedicated chapter. A similar issue occurs with mass wasting which is usually included as a small section within an overall "water" chapter but in the geosciences, mass wasting is an important surficial process that moves material across the surface of Earth. Therefore, mass wasting is dedicated to a single chapter. Geologic time is commonly paired with Earth history but both topics are uniquely important in a foundational knowledge of Earth and as such, each topic has its own chapter.

This textbook includes a chapter dedicated to the origin of the universe and our solar system. Many existing introductory Earth Science books lack a chapter dedicated to astronomy and the solar system. Understanding our solar system is important for a variety of reasons: it contains the only known example of a habitable planet, the only star observable close-up, and the only planets we can visit with modern technology such as satellites, probes, and landers. Knowledge of Earth's place within the solar system is essential to understand the origin of planets, along with the conditions that allow life to exist on Earth.

This open-source textbook includes various important features designed to enhance the student learning experience in introductory Earth Science courses. These include a multitude of high-quality figures and images within each chapter that help to clarify key concepts and are optimized for viewing online. Self-test assessment questions are embedded in each online chapter that help students focus their learning. QR codes are provided for each assessment to allow students using print or PDF versions to easily access the quiz from an internet-capable device of their choice. Selected graphics and tables have been replaced or updated to enhance quality and clarity.

Features of this book

- Example-rich narrative
- Graphic elements which illustrate and reinforce concepts
- Linked online glossary (glossary appears at the end for PDF)
- Embedded navigation and image alt-text for screen readers
- QR Online interactive self-test quiz questions
- Free online and in PDF, and in-print at vendor cost of production
- Open license, Creative Commons BY NC-SA 4.0 permits customization and sharing
- Instructor community portal enables sharing of ancillary resources
- Register your Use form allows instructors to opt in to receive book updates
- Errata and report-an-error/share-a-suggestion forms promote currency

ABOUT THE AUTHOR

Adapted by

Laura Neser, Ph.D., is an Instructor in the Department of Geosciences at Virginia Tech. Dr. Neser earned her B.S. in Geosciences at Virginia Tech in the spring of 2008 and completed her Ph.D. in Geological Sciences At the University of North Carolina at Chapel Hill (UNC) in 2014. Her doctoral research focused on the structural geology, sedimentology, and stratigraphy of formations that were deposited along the flanks of the Beartooth Mountains as they rose during late Paleocene-Eocene time. Dr. Neser has worked as an athletic tutor and online instructor at The University of North Carolina (Chapel Hill, NC), in temporary positions as an Adjunct Instructor at Chowan University (Murfreesboro, NC) and Full-Time Lecturer at Indiana State University (Terre Haute, IN), and as a Professor at Seminole State College (Sanford, FL) before starting as an Instructor at Virginia Tech in the fall of 2021.

Although she is currently focused on teaching online sections of Introduction to Earth Science, Earth Resources, Society and the Environment, and Climate History, her teaching background is significantly broader and includes Environmental Science, Astronomy, Environmental Ethics, Earth History, Structural Geology, and Field Geology.

Incorporating original works by

Chris Johnson, Matthew D. Affolter, Paul Inkenbrandt, and Cam Mosher (2017). *An Introduction to Geology,* **(CC BY NC SA 4.0) Salt Lake City Community College. Retrieved from https://slcc.pressbooks.pub/introgeology.**

An Introduction to Geology (2017) was edited and reviewed by

Johnathan Barnes – Salt Lake Community College
Jack Bloom – Rio Tinto Kennecott Copper (retired)
Deanna Brandau – Utah Museum of Natural History
Gereld Bryant – Dixie State University
Gregg Beukelman – Utah Geological Survey
Peter Davis – Pacific Lutheran University
Renee Faatz – Snow College
Gabriel Filippelli – Indiana University-Purdue University Indianapolis
Michelle Cooper Fleck – Utah State University
Colby Ford – Dixie State University
Michael Hylland – Utah Geological Survey
J. Lucy Jordan – Utah Geological Survey
Mike Kass – Salt Lake Community College
Ben Laabs – North Dakota State University
Tom Lachmar – Utah State University
Johnny MacLean – Southern Utah University
Erich Peterson – University of Utah
Tiffany Rivera – Westminster College
Leif Tapanila – Idaho State University

With Special Thanks to Jason Pickavance – Salt Lake Community College, Director of Educational Initiatives; and R. Adam Dastrup – Salt Lake Community College, Geoscience Dept. Chair, Director of Opengeography.org

Karla Panchuk (2019) "Starting with a Big Bang" in Steven Earle (Editor) *Physical Geology*, 2nd edition. (CC BY 4.0) BC Campus. Retrieved from https://opentextbc.ca/physicalgeology2ed/chapter/22-1-starting-with-a-big-bang.

Andrew Fraknoi, David Morrison, and Sidney C. Wolff (2016). *Astronomy*. (CC BY 4.0). OpenStax. Retrieved from https://openstax.org/details/books/astronomy.

Contributing authors to Astronomy (2016) included

John Beck, Stanford University

Susan D. Benecchi, Planetary Science Institute

John Bochanski, Rider University

Howard Bond, Pennsylvania State University, Emeritus, Space Telescope Science Institute

Jennifer Carson, Occidental College

Bryan Dunne, University of Illinois at Urbana-Champaign

Martin Elvis, Harvard-Smithsonian Center for Astrophysics

Debra Fischer, Yale University

Heidi Hammel, Association of Universities for Research in Astronomy

Tori Hoehler, NASA Ames Research Center

Douglas Ingram, Texas Christian University

Steven Kawaler, Iowa State University

Lloyd Knox, University of California, Davis

Todd Young, Wayne State College

Mark Krumholz, Australian National University

James Lowenthal, Smith College

Siobahn Morgan, University of Northern Iowa

Daniel Perley, California Institute of Technology

Claire Raftery, National Solar Observatory

Deborah Scherrer, retired, Stanford University

Phillip Scherrer, Stanford University

Sanjoy Som, Blue Marble Space Institute of Science, NASA Ames Research Center

Wes Tobin, Indiana University East

William H. Waller, retired, Tufts University, Rockport (MA) Public Schools

ACKNOWLEDGMENTS

This publication was made possible in part by funding and publishing support provided by the Open Education Initiative of the University Libraries at Virginia Tech.

Editorial Team

Kindred Grey, Graphic Design and Editorial Assistance

Kindred Grey is the OER and Graphic Design Specialist in the University Libraries at Virginia Tech. She received her B.S. in Statistics and Psychology from Virginia Tech in 2020 and has worked in University Libraries since then, assisting to publish open textbooks. Kindred's design abilities are demonstrated in the visual elements of this book: cover design, interior formatting, and creation of new figures. She prioritizes the accessibility of textbooks without compromising design. She provided project coordination and editorial assistance, including tracking and updating the over eight hundred figures and tables which were reviewed, revised, or replaced through the course of the project. Kindred's contributions have resulted in a text that is accessible to a wider range of readers and uses visual content to illustrate and more clearly convey conceptual information.

Anita Walz, Project Manager

Anita Walz is Associate Professor, and Assistant Director of Open Education and Scholarly Communication Librarian in the University Libraries at Virginia Tech. She received her MS in Library and Information Science from the University of Illinois at Urbana-Champaign and has worked in university, government, school, and international libraries for over 20 years. She is the founder of the Open Education Initiative at Virginia Tech and the managing editor of several open textbooks adapted or created at Virginia Tech, many of which may be found here: https://vtechworks.lib.vt.edu/handle/10919/70959. She has provided overall planning, project coordination, coaching, problem-solving, and oversight for this adapted version.

INSTRUCTOR RESOURCES

How to Adopt This Book

This is an open textbook. That means that this book is freely available and you are welcome to use, adapt, and share this book with attribution according to the Creative Commons NonCommercial ShareAlike 4.0 (CC BY-NC-SA 4.0) license https://creativecommons.org/licenses/by-nc-sa/4.0. (Many, but not all images, illustrations, etc., in this book are licensed under Creative Commons licenses.)

Instructors reviewing, adopting, or adapting this textbook are encouraged to register at https://bit.ly/interest_intro_earth_science. This assists the Open Education Initiative at Virginia Tech in assessing the impact of the book and allows us to more easily alert instructors of additional resources, features and opportunities.

Finding Additional Resources for Your Course

The main landing page for the book is https://doi.org/10.21061/introearthscience.

This page includes:

- Links to multiple electronic versions of the textbook (PDF, ePub, HTML)

- Links to the instructor resource-sharing portal https://www.oercommons.org/groups/introduction-to-earth-science-instructor-group/12785

- Link to errata document (report errors at https://bit.ly/report_error_intro_earth_science)

Sharing Resources You've Created

Have you created any supplementary materials for use with this book such as presentation slides, activities, test items, or a question bank? If so, please consider sharing your materials related to this open textbook. Please tell us about resources you wish to share by using this form: https://bit.ly/interest_intro_earth_science or by directly sharing resources under an open license to the public-facing instructor sharing portal https://www.oercommons.org/groups/introduction-to-earth-science-instructor-group/12785.

Making Your Own Version of This Book

The Creative Commons Attribution NonCommercial-ShareAlike 4.0 license https://creativecommons.org/licenses/by-nc-sa/4.0/legalcode on this book allows customization and redistribution which is NonCommercial, that is "not primarily intended for or directed towards commercial advantage or monetary compensation."

Best practices for attribution are provided at https://wiki.creativecommons.org/wiki/ Best_practices_for_attribution.

This book is hosted in PDF and ePub in VTechWorks http://hdl.handle.net/10919/112740 in HTML in Pressbooks https://pressbooks.lib.vt.edu/introearthscience. Note that the Pressbooks platforms offers customization/remixing.

UNDERSTANDING SCIENCE

By the end of this chapter, students should be able to:

- Contrast **objective** versus **subjective** observations, and **quantitative** versus **qualitative** observations.
- Identify a **pseudoscience** based on its lack of falsifiability.
- Contrast the methods used by Aristotle and Galileo to describe the natural environment.
- Explain the **scientific method** and apply it to a problem or question.
- Describe the foundations of modern geology, such as the **principle of uniformitarianism**.
- Contrast **uniformitarianism** with **catastrophism**.
- Explain why studying geology is important.
- Identify how Earth materials are transformed by **rock cycle** processes.
- Describe the steps involved in a reputable scientific study.
- Explain rhetorical arguments used by science deniers.

1.1 What is Science?

Figure 1.1: This is Grand Canyon of the Yellowstone in Yellowstone National Park. An objective statement about this would be: "The picture is of a waterfall." A subjective statement would be: "The picture is beautiful." or "The waterfall is there because of erosion."

Scientists seek to understand the fundamental principles that explain natural patterns and processes. Science is more than just a body of knowledge, science provides a means to evaluate and create new knowledge without bias. Scientists use **objective** evidence over **subjective** evidence, to reach sound and logical conclusions.

An **objective observation** is without personal bias and the same by all individuals. Humans are biased by nature, so they cannot be completely **objective**; the goal is to be as unbiased as possible. A **subjective observation** is based on a person's feelings and beliefs and is unique to that individual.

Another way scientists avoid bias is by using **quantitative** over **qualitative** measurements whenever possible. A **quantitative** measurement is expressed with a specific numerical value. **Qualitative** observations are general or relative descriptions. For example, describing a rock as red or heavy is a **qualitative observation**. Determining a rock's color by measuring wavelengths of reflected light or its density by measuring the proportions of minerals it contains is **quantitative**. Numerical values are more precise than general descriptions, and they can be analyzed using statistical calculations. This is why **quantitative** measurements are much more useful to scientists than **qualitative** observations.

Establishing truth in science is difficult because all scientific claims are **falsifiable**, which means any initial **hypothesis** may be tested and proven false. Only after exhaustively eliminating false results, competing ideas, and possible variations does a **hypothesis** become regarded as a reliable scientific **theory**. This meticulous scrutiny reveals weaknesses or flaws in a **hypothesis** and is the strength that supports all scientific ideas and procedures. In fact, proving current ideas are wrong has been the **driving force** behind many scientific careers.

Figure 1.2: Canyons like this, carved in the deposit left by the May 18th, 1980 eruption of Mt. St. Helens is sometimes used by purveyors of pseudoscience as evidence for the Earth being very young. In reality, the unconsolidated and unlithified volcanic deposit is carved much more easily than other canyons like the Grand Canyon.

Falsifiability separates science from **pseudoscience**. Scientists are wary of explanations of natural phenomena that discourage or avoid falsifiability. An explanation that cannot be tested or does not meet scientific standards is not considered science, but **pseudoscience**. **Pseudoscience** is a collection of ideas that may appear scientific but does not use the **scientific method**. Astrology is an example of **pseudoscience**. It is a belief system that attributes the movement of celestial bodies to influencing human behavior. Astrologers rely on celestial observations, but their conclusions are not based on experimental evidence and their statements are not **falsifiable**. This is not to be confused with astronomy which is the scientific study of celestial bodies and the cosmos.

Figure 1.3: Geologists share information by publishing, attending conferences, and even going on field trips, such as this trip to the Lake Owyhee Volcanic Field in Oregon by the Bureau of Land Management in 2019.

Science is also a social process. Scientists share their ideas with peers at conferences, seeking guidance and feedback. Research papers and data submitted for publication are rigorously reviewed by qualified peers, scientists who are experts in the same field. The scientific review process aims to weed out misinformation, invalid research results, and wild speculation. Thus, it is slow, cautious, and conservative. Scientists tend to wait until a **hypothesis** is supported by overwhelming amount of evidence from many independent researchers before accepting it as scientific **theory**.

Take this quiz to check your comprehension of this section.

If you are using an offline version of this text, access the quiz for section 1.1 via the QR code.

An interactive H5P element has been excluded from this version of the text. You can view it online here:
https://pressbooks.lib.vt.edu/introearthscience/?p=66#h5p-1

1.2 The Scientific Method

Modern science is based on the **scientific method**, a procedure that follows these steps:

- Formulate a question or observe a problem
- Apply **objective** experimentation and **observation**
- Analyze collected data and Interpret results
- Devise an evidence-based **theory**
- Submit findings to **peer review** and/or publication

This has a long history in human thought but was first fully formed by Ibn al-Haytham over 1,000 years ago. At the forefront of the **scientific method** are conclusions based on **objective** evidence, not opinion or hearsay.

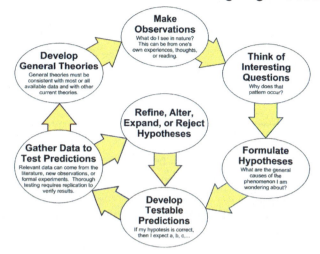

Figure 1.4: Diagram of the cyclical nature of the scientific method.

Step One: Observation, Problem, or Research Question

The procedure begins with identifying a problem or research question, such as a geological phenomenon that is not well explained in the scientific community's collective knowledge. This step usually involves reviewing the scientific literature to understand previous studies that may be related to the question.

Step Two: Hypothesis

Once the problem or question is well defined, the scientist proposes a possible answer, a **hypothesis**, before conducting an **experiment** or field work. This **hypothesis** must be specific, **falsifiable**, and should be based on other scientific work. Geologists often develop multiple working **hypotheses** because they usually cannot impose strict experimental controls or have limited opportunities to visit a field location.

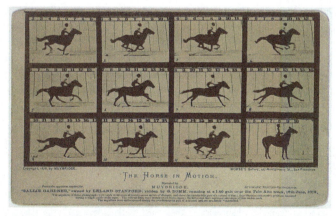

Figure 1.5: A famous hypothesis: Leland Stanford wanted to know if a horse lifted all 4 legs off the ground during a gallop, since the legs are too fast for the human eye to perceive it. These series of photographs by Eadweard Muybridge proved the horse, in fact, does have all four legs off the ground during the gallop.

Step Three: Experiment and Hypothesis Revision

The next step is developing an **experiment** that either supports or refutes the **hypothesis**. Many people mistakenly think experiments are only done in a lab; however, an **experiment** can consist of observing natural processes in the field. Regardless of what form an **experiment** takes, it always includes the systematic gathering of **objective** data. This data is interpreted to determine whether it contradicts or supports the **hypothesis**, which may be revised and tested again. When a **hypothesis** holds up under experimentation, it is ready to be shared with other experts in the field.

Step Four: Peer Review, Publication, and Replication

Scientists share the results of their research by publishing articles in scientific journals, such as *Science* and *Nature*. Reputable journals and publishing houses will not publish an experimental study until they have determined its methods are scientifically rigorous and the conclusions are supported by evidence. Before an article is published, it undergoes a rigorous **peer review** by scientific experts who scrutinize the methods, results, and discussion. Once an article is published, other scientists may attempt to replicate the results. This replication is necessary to confirm the reliability of the study's reported results. A **hypothesis** that seemed compelling in one study might be proven false in studies conducted by other scientists. New technology can be applied to published studies, which can aid in confirming or rejecting once-accepted ideas and/or **hypotheses**.

Figure 1.6: An experiment at the University of Queensland has been going since 1927. A petroleum product called pitch, which is highly viscous, drips out of a funnel about once per decade.

Step Five: Theory Development

In casual conversation, the word theory implies guesswork or speculation. In the language of science, an explanation or conclusion made in a *theory* carries much more weight because it is supported by experimental verification and widely accepted by the scientific community. After a **hypothesis** has been repeatedly tested for falsifiability through documented and independent studies, it eventually becomes accepted as a scientific **theory**.

While a **hypothesis** provides a tentative explanation *before* an **experiment**, a **theory** is the best explanation *after* being confirmed by multiple independent experiments. Confirmation of a **theory** may take years, or even longer. For example, the continental drift hypothesis first proposed by Alfred Wegener in 1912 was initially dismissed. After decades of additional evidence collection by other scientists using more advanced technology, Wegener's **hypothesis** was accepted and revised as the theory of **plate tectonics**.

Figure 1.7: Wegener later in his life, ca. 1924-1930.

The theory of evolution by natural selection is another example. Originating from the work of Charles Darwin in the mid-19th century, the theory of evolution has withstood generations of scientific testing for falsifiability. While it has been updated and revised to accommodate knowledge gained by using modern technologies, the theory of evolution continues to be supported by the latest evidence.

Take this quiz to check your comprehension of this section.

If you are using an offline version of this text, access the quiz for section 1.2 via the QR code.

 An interactive H5P element has been excluded from this version of the text. You can view it online here:
https://pressbooks.lib.vt.edu/introearthscience/?p=66#h5p-2

1.3 Early Scientific Thought

Figure 1.8: Fresco by Raphael of Plato (left) and Aristotle (right).

Western scientific thought began in the ancient city of Athens, Greece. Athens was governed as a democracy, which encouraged individuals to think independently, at a time when most civilizations were ruled by monarchies or military conquerors. Foremost among the early philosopher/scientists to use empirical thinking was Aristotle, born in 384 BCE. Empiricism emphasizes the value of evidence gained from experimentation and **observation**. Aristotle studied under Plato and tutored Alexander the Great. Alexander would later conquer the Persian Empire, and in the process spread Greek culture as far east as India.

Aristotle applied an empirical method of analysis called **deductive reasoning**, which applies known principles of thought to establish new ideas or predict new outcomes. **Deductive reasoning** starts with generalized principles and logically extends them to new ideas or specific conclusions. If the initial principle is valid, then it is highly likely the conclusion is also valid. An example of **deductive reasoning** is if A=B, and B=C, then A=C. Another example is if all birds have feathers, and a sparrow is a bird, then a sparrow must also have feathers. The problem with **deductive reasoning** is if the initial principle is flawed, the conclusion will inherit that flaw. Here is an example of a flawed initial principle leading to the wrong conclusion; if all animals that fly are birds, and bats also fly, then bats must also be birds.

This type of empirical thinking contrasts with **inductive reasoning**, which begins from new observations and attempts to discern underlying generalized principles. A conclusion made through **inductive reasoning** comes from analyzing measurable evidence, rather making a logical connection. For example, to determine whether bats are birds a scientist might list various characteristics observed in birds–the presence of feathers, a toothless beak, hollow bones, lack of forelegs, and externally laid eggs. Next, the scientist would check whether bats share the same characteristics, and if they do not, draw the conclusion that bats are not birds.

Both types of reasoning are important in science because they emphasize the two most important aspects of science: **observation** and inference. Scientists test existing principles to see if they accurately infer or predict their observations. They also analyze new observations to determine if the inferred underlying principles still support them.

Greek culture was spread by Alexander and then absorbed by the Romans, who help further extend Greek knowledge into Europe through their vast infrastructure of roads, bridges, and aqueducts. After the fall of the Roman Empire in 476 CE, scientific progress in Europe stalled. Scientific thinkers of medieval time had such high regard for Aristotle's wisdom and knowledge they faithfully followed his logical approach to understanding nature for centuries. By contrast, science in the Middle East flourished and grew between 800 and 1450 CE, along with culture and the arts.

Near the end of the medieval **period**, empirical experimentation became more common in Europe. During the Renaissance, which lasted from the 14th through 17th centuries, artistic and scientific thought experienced a great awakening. European scholars began to criticize the traditional Aristotelian approach and by the end of the Renaissance **period**, empiricism was poised to become a key component of the scientific revolution that would arise in the 17th century.

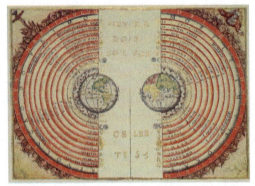

Figure 1.10: Geocentric drawing by Bartolomeu Velho in 1568.

Figure 1.9: Drawing of Avicenna (Ibn Sina). He is among the first to link mountains to earthquakes and erosion.

An early example of how Renaissance scientists began to apply a modern empirical approach is their study of the **solar system**. In the second century, the Greek astronomer Claudius Ptolemy observed the Sun, Moon, and stars moving across the sky. Applying Aristotelian logic to his astronomical calculations, he deductively reasoned all celestial bodies orbited around the Earth, which was located at the center of the universe. Ptolemy was a highly regarded mathematician, and his mathematical calculations were widely accepted by the scientific community. The view of the cosmos with Earth at its center is called the geocentric model. This geocentric model persisted until the Renaissance **period**, when some revolutionary thinkers challenged the centuries-old **hypothesis**.

By contrast, early Renaissance scholars such as astronomer Nicolaus Copernicus (1473-1543) proposed an alternative explanation for the perceived movement of the Sun, Moon, and stars. Sometime between 1507 and 1515, he provided credible mathematical proof for a radically new model of the cosmos, one in which the Earth and other planets orbited around a centrally located Sun. After the invention of the telescope in 1608, scientists used their enhanced astronomical observations to support this heliocentric, Sun-centered, model.

Two scientists, Johannes Kepler and Galileo Galilei, are credited with jumpstarting the scientific revolution. They accomplished this by building on Copernicus work and challenging long-established ideas about nature and science.

Johannes Kepler (1571-1630) was a German mathematician and astronomer who expanded on the heliocentric model—improving Copernicus' original calculations and describing planetary motion as elliptical paths. Galileo Galilei (1564 – 1642) was an Italian astronomer who used the newly developed telescope to observe the four largest moons of Jupiter. This was the first piece of direct evidence to contradict the geocentric model, since moons orbiting Jupiter could not also be orbiting Earth.

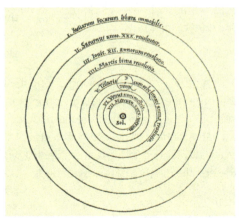

Figure 1.12: Copernicus' heliocentric model.

Figure 1.11: Galileo's first mention of moons of Jupiter.

Galileo strongly supported the heliocentric model and attacked the geocentric model, arguing for a more scientific approach to determine the credibility of an idea. Because of this he found himself at odds with prevailing scientific views and the Catholic Church. In 1633 he was found guilty of heresy and placed under house arrest, where he would remain until his death in 1642.

Galileo is regarded as the first modern scientist because he conducted experiments that would prove or disprove **falsifiable** ideas and based his conclusions on mathematical analysis of quantifiable evidence—a radical departure from the deductive thinking of Greek philosophers such as Aristotle. His methods marked the beginning of a major shift in how scientists studied the natural world, with an increasing number of them relying on evidence and experimentation to form their **hypotheses**. It was during this revolutionary time that geologists such as James Hutton and Nicolas Steno also made great advances in their scientific fields of study.

Take this quiz to check your comprehension of this section.

If you are using an offline version of this text, access the quiz for section 1.3 via the QR code.

 An interactive H5P element has been excluded from this version of the text. You can view it online here:
https://pressbooks.lib.vt.edu/introearthscience/?p=66#h5p-3

1.4 Foundations of Modern Geology

As part of the scientific revolution in Europe, modern geologic principles developed in the 17th and 18th centuries. One major contributor was Nicolaus Steno (1638-1686), a Danish priest who studied anatomy and geology. Steno was the first to propose the Earth's surface could change over time. He suggested sedimentary rocks, such as **sandstone** and **shale**, originally formed in horizontal layers with the oldest on the bottom and progressively younger layers on top.

Figure 1.13: Illustration by Steno showing a comparison between fossil and modern shark teeth.

In the 18th century, Scottish naturalist James Hutton (1726–1797) studied **rivers** and coastlines and compared the sediments they left behind to exposed **sedimentary** rock strata. He hypothesized the ancient rocks must have been formed by processes like those producing the features in the oceans and **streams**. Hutton also proposed the Earth was much older than previously thought. Modern geologic processes operate slowly. Hutton realized if these processes formed rocks, then the Earth must be very old, possibly hundreds of millions of years old.

Hutton's idea is called the **principle of uniformitarianism** and states that natural processes operate the same now as in the past, i.e. the laws of nature are uniform across space and time. Geologist often state "the present is the key to the past," meaning they can understand ancient rocks by studying modern geologic processes.

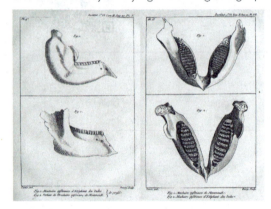

Figure 1.14: Cuvier's comparison of modern elephant and mammoth jaw bones.

Prior to the acceptance of **uniformitarianism**, scientists such as German geologist Abraham Gottlob Werner (1750-1817) and French anatomist Georges Cuvier (1769-1832) thought rocks and landforms were formed by great catastrophic events. Cuvier championed this view, known as **catastrophism**, and stated, "The thread of operation is broken; nature has changed course, and none of the agents she employs today would have been sufficient to produce her former works." He meant processes that operate today did not operate in the past. Known as the father of **vertebrate** paleontology, Cuvier made significant contributions to the study of ancient life and taught at Paris's Museum of Natural History. Based on his study of large **vertebrate fossils**, he was the first to suggest species could go **extinct**. However, he thought new species were introduced by special creation after catastrophic floods.

Hutton's ideas about **uniformitarianism** and Earth's age were not well received by the scientific community of his time. His ideas were falling into obscurity when Charles Lyell, a British lawyer and geologist (1797-1875), wrote the *Principles of Geology* in the early 1830s and later, *Elements of Geology*. Lyell's books promoted Hutton's **principle of uniformitarianism**, his studies of rocks and the processes that formed them, and the idea that Earth was possibly over 300 million years old. Lyell and his three-volume *Principles of Geology* had a lasting influence on the geologic community and public at large, who eventually accepted **uniformitarianism** and millionfold age for the Earth. The **principle of uniformitarianism** became so widely accepted, that geologists regarded cata-

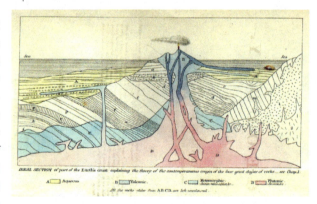

Figure 1.15: Inside cover of Lyell's Elements of Geology.

strophic change as heresy. This made it harder for ideas like the sudden demise of the dinosaurs by asteroid impact to gain traction.

A contemporary of Lyell, Charles Darwin (1809-1882) took *Principles of Geology* on his five-year trip on the HMS Beagle. Darwin used **uniformitarianism** and deep geologic time to develop his initial ideas about evolution. Lyell was one of the first to publish a reference to Darwin's idea of evolution.

The next big advancement, and perhaps the largest in the history of geology, is the **theory** of **plate tectonics** and continental drift. Dogmatic acceptance of **uniformitarianism** inhibited the progress of this idea, mainly because of the permanency placed on the continents and their positions. Ironically, slow and steady movement of plates would fit well into a uniformitarianism model. However, much time passed and a great deal of scientific resistance had to be overcome before the idea took hold. This happened for several reasons. Firstly, the movement was so slow it was overlooked. Secondly, the best evidence was hidden under the ocean. Finally, the accepted theories were anchored by a large amount of inertia. Instead of being bias free, scientists resisted and ridiculed the emerging idea of **plate tectonics**. This example of dogmatic thinking is still to this day a tarnish on the geoscience community.

Figure 1.16: J. Tuzo Wilson.

Plate tectonics is most commonly attributed to Alfred Wegener, the first scientist to compile a large data set supporting the idea of continents shifting places over time. He was mostly ignored and ridiculed for his ideas, but later workers like Marie Tharp, Bruce Heezen, Harry Hess, Laurence Morley, Frederick Vine, Drummond Matthews, Kiyoo Wadati, Hugo Benioff, Robert Coats, and J. Tuzo Wilson benefited from advances in sub-sea technologies. They discovered, described, and analyzed new features like the **mid-ocean ridge**, alignment of earthquakes, and **magnetic striping**. Gradually these scientists introduced a paradigm shift that revolutionized geology into the science we know today.

Take this quiz to check your comprehension of this section.

If you are using an offline version of this text, access the quiz for section 1.4 via the QR code.

An interactive H5P element has been excluded from this version of the text. You can view it online here:
https://pressbooks.lib.vt.edu/introearthscience/?p=66#h5p-4

1.5 The Study of Geology

Geologists apply the **scientific method** to learn about Earth's materials and processes. Geology plays an important role in society; its principles are essential to locating, extracting, and managing **natural resources**; evaluating environmental impacts of using or extracting these resources; as well as understanding and mitigating the effects of natural hazards.

Geology often applies information from physics and chemistry to the natural world, like understanding the physical forces in a **landslide** or the chemical interaction between water and rocks. The term comes from the Greek word *geo*, meaning Earth, and *logos*, meaning to think or reckon with.

Figure 1.17: Girls into Geoscience inaugural Irish Fieldtrip.

1.5.1 Why Study Geology?

Figure 1.18: Hoover Dam provides hydroelectric energy and stores water for southern Nevada.

Geology plays a key role in how we use **natural resources**—any naturally occurring material that can be extracted from the Earth for economic gain. Our developed modern society, like all societies before it, is dependent on geologic resources. Geologists are involved in extracting **fossil fuels**, such as **coal** and **petroleum**; metals such as copper, aluminum, and iron; and water resources in **streams** and underground **reservoirs** inside **soil** and rocks. They can help conserve our planet's finite supply of **nonrenewable** resources, like **petroleum**, which are fixed in quantity and depleted by consumption. Geologists can also help manage **renewable** resources that can be replaced or regenerated, such as solar or wind energy, and timber.

Resource extraction and usage impacts our environment, which can negatively affect human health. For example, burning **fossil fuels** releases chemicals into the air that are unhealthy for humans, especially children. **Mining** activities can release toxic heavy metals, such as lead and mercury, into the **soil** and waterways. Our choices will have an effect on Earth's environment for the foreseeable future. Understanding the remaining quantity, extractability, and renewability of geologic resources will help us better sustainably manage those resources.

Figure 1.19: Coal power plant in Helper, Utah.

Figure 1.20: Buildings toppled from liquefaction during a 7.5 magnitude earthquake in Japan.

Geologists also study natural hazards created by geologic processes. Natural hazards are phenomena that are potentially dangerous to human life or property. No place on Earth is completely free of natural hazards, so one of the best ways people can protect themselves is by understanding geology. Geology can teach people about the natural hazards in an area and how to prepare for them. Geologic hazards include **landslides**, earthquakes, **tsunamis**, floods, **volcanic** eruptions, and sea-level rise.

Finally, geology is where other scientific disciplines intersect in the concept known as **Earth System Science**. In science, a **system** is a group of interactive objects and processes. **Earth System Science** views the entire planet as a combination of systems that interact with each other via complex relationships. This geology textbook provides an introduction to science in general and will often reference other scientific disciplines.

Earth System Science includes five basic systems (or spheres), the **Geosphere** (the solid body of the Earth), the **Atmosphere** (the gas envelope surrounding the Earth), the **Hydrosphere** (water in all its forms at and near the surface of the Earth), the **Cryosphere** (frozen water part of Earth), and the **Biosphere** (life on Earth in all its forms and interactions, including humankind).

Rather than viewing geology as an isolated **system**, earth system scientists study how geologic processes shape not only the world, but all the spheres it contains. They study how these multidisciplinary spheres relate, interact, and change in response to natural cycles and human-driven forces. They use elements from physics, chemistry, biology, meteorology, environmental science, zoology, hydrology, and many other sciences.

Figure 1.21: Oregon's Crater Lake was formed about 7700 years ago after the eruption of Mount Mazama.

1.5.2 Rock Cycle

The most fundamental view of Earth materials is the **rock cycle**, which describes the major materials that comprise the Earth, the processes that form them, and how they relate to each other. It usually begins with hot molten liquid rock called **magma** or **lava**. **Magma** forms under the Earth's surface in the **crust** or **mantle**. **Lava** is molten rock that erupts onto the Earth's surface. When **magma** or **lava** cools, it solidifies by a process called **crystallization** in which **minerals** grow within the **magma** or **lava**. The rocks resulting rocks are **igneous** rocks. *Ignis* is Latin for fire.

Igneous rocks, as well as other types of rocks, on Earth's surface are exposed to **weathering** and **erosion**, which produces **sediments**. **Weathering** is the physical and chemical breakdown of rocks into smaller fragments. **Erosion** is the removal of those fragments from their original location. The broken-down and transported fragments or grains are considered **sediments**, such as gravel, sand, silt, and clay. These **sediments** may be transported by **streams** and **rivers**, ocean currents, **glaciers**, and wind.

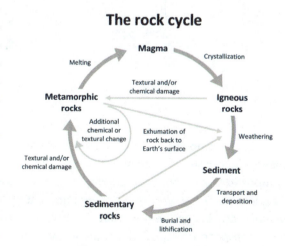

Figure 1.22: Rock cycle showing the five materials (such as igneous rocks and sediment) and the processes by which one changes into another (such as weathering).

Figure 1.23: Mississippian raindrop impressions over wave ripples from Nova Scotia.

Sediments come to rest in a process known as **deposition**. As the deposited **sediments** accumulate—often under water, such as in a shallow **marine** environment—the older **sediments** get buried by the new deposits. The deposits are compacted by the weight of the overlying **sediments** and individual grains are cemented together by **minerals** in **groundwater**. These processes of **compaction** and **cementation** are called **lithification**. Lithified **sediments** are considered a **sedimentary rock**, such as **sandstone** and **shale**. Other sedimentary rocks are made by direct chemical **precipitation** of **minerals** rather than eroded **sediments**, and are known as **chemical sedimentary** rocks.

Pre-existing rocks may be transformed into a **metamorphic rock**; *meta-* means change and *-morphos* means form or shape. When rocks are subjected to extreme increases in **temperature** or pressure, the **mineral** crystals are enlarged or altered into entirely new **minerals** with similar chemical make up. High temperatures and pressures occur in rocks buried deep within the Earth's **crust** or that come into contact with hot **magma** or **lava**. If the **temperature** and pressure conditions melt the rocks to create **magma** and **lava**, the **rock cycle** begins anew with the creation of new rocks.

Figure 1.24: Metamorphic rock in Georgian Bay, Ontario.

1.5.3 Plate Tectonics and Layers of Earth

The theory of **plate tectonics** is the fundamental unifying principle of geology and the **rock cycle**. **Plate tectonics** describes how Earth's layers move relative to each other, focusing on the **tectonic** or lithospheric **plates** of the outer layer. **Tectonic plates** float, collide, slide past each other, and split apart on an underlying mobile layer called the **asthenosphere**. Major landforms are created at the **plate** boundaries, and rocks within the **tectonic plates** move through the **rock cycle**. **Plate tectonics** is discussed in more detail in chapter 2.

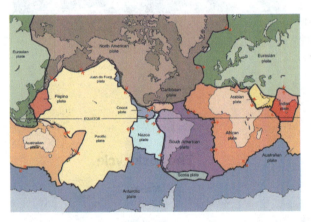

Figure 1.25: Map of the major plates and their motions along boundaries.

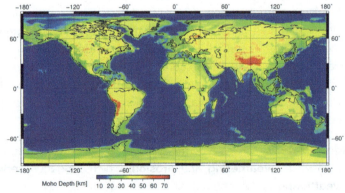

Figure 1.26: The global map of the depth of the moho.

Earth's three main geological layers can be categorized by chemical **composition** or the chemical makeup: **crust**, **mantle**, and **core**. The **crust** is the outermost layer and **composed** of mostly silicon, oxygen, aluminum, iron, and magnesium. There are two types, **continental crust** and **oceanic crust**. **Continental crust** is about 50 km (30 mi) thick, **composed** of low-density **igneous** and sedimentary rocks, **Oceanic crust** is approximately 10 km (6 mi) thick and made of high-density **igneous basalt**-type rocks. **Oceanic crust** makes up most of the **ocean floor**, covering about 70% of the planet. **Tectonic plates** are made of **crust** and a portion the upper **mantle**, forming a rigid physical layer called the **lithosphere**.

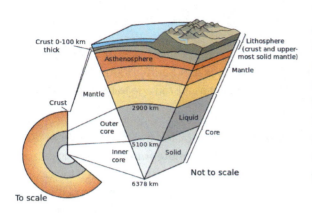

The **mantle**, the largest chemical layer by volume, lies below the **crust** and extends down to about 2,900 km (1,800 mi) below the Earth's surface. The mostly solid **mantle** is made of **peridotite**, a high-density **composed** of silica, iron, and magnesium. The upper part of mantle is very hot and flexible, which allows the overlying **tectonic plates** to float and move about on it. Under the **mantle** is the Earth's **core**, which is 3,500 km (2,200 mi) thick and made of iron and nickel. The **core** consists of two parts, a liquid **outer core** and solid **inner core**. Rotations within the solid and liquid **metallic core** generate Earth's magnetic field (see figure 1.27).

Figure 1.27: The layers of the Earth. Physical layers include lithosphere and asthenosphere; chemical layers are crust, mantle, and core.

1.5.4 Geologic Time and Deep Time

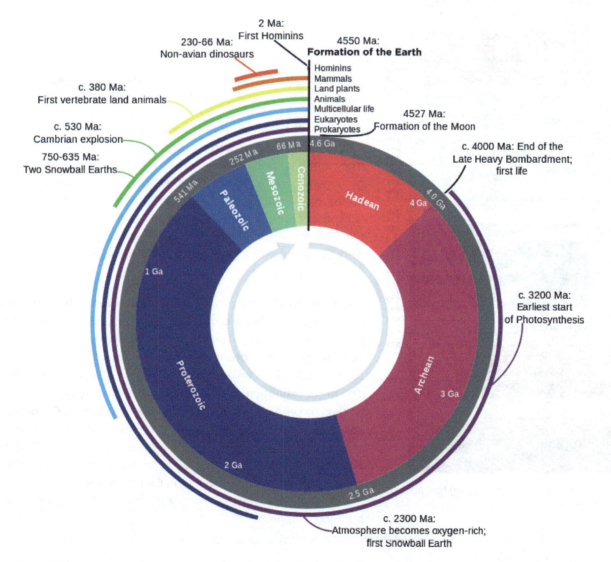

Figure 1.28: Geologic time on Earth, represented circularly, to show the individual time divisions and important events. Ga=billion years ago, Ma=million years ago.

"The result, therefore, of our present enquiry is, that we find no vestige of a beginning; no prospect of an end." (James Hutton, 1788)

One of the early pioneers of geology, James Hutton, wrote this about the age of the Earth after many years of geological study. Although he wasn't exactly correct—there is a beginning and will be an end to planet Earth—Hutton was expressing the difficulty humans have in perceiving the vastness of geological time. Hutton did not assign an age to the Earth, although he was the first to suggest the planet was very old.

Today we know Earth is approximately 4.54 ± 0.05 billion years old. This age was first calculated by Caltech professor Clair Patterson in 1956, who measured the half-lives of lead **isotopes** to radiometrically date a **meteorite** recovered in Arizona. Studying geologic time, also known as deep time, can help us overcome a perspective of Earth that is limited to our short lifetimes. Compared to the geologic scale, the human lifespan is very short, and we struggle to comprehend the depth of geologic time and slowness of geologic processes. For example, the study of earthquakes only goes back about 100 years; however, there is geologic evidence of large earthquakes occurring thousands of years ago. And scientific evidence indicates earthquakes will continue for many centuries into the future.

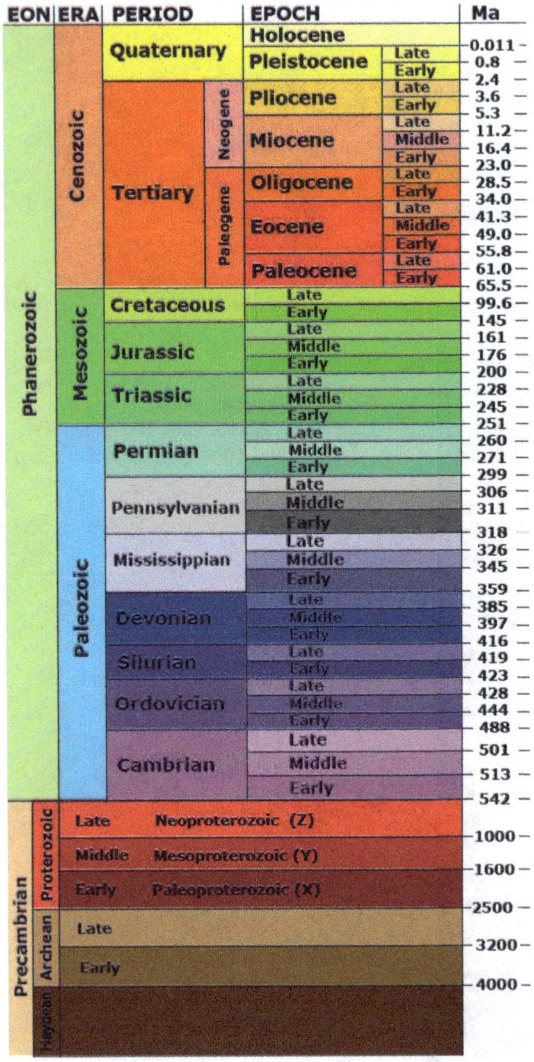

EON	ERA	PERIOD	EPOCH		Ma
Phanerozoic	Cenozoic	Quaternary	Holocene		0.011
			Pleistocene	Late	0.8
				Early	2.4
		Tertiary (Neogene)	Pliocene	Late	3.6
				Early	5.3
			Miocene	Late	11.2
				Middle	16.4
				Early	23.0
		Tertiary (Paleogene)	Oligocene	Late	28.5
				Early	34.0
			Eocene	Late	41.3
				Middle	49.0
				Early	55.8
			Paleocene	Late	61.0
				Early	65.5
	Mesozoic	Cretaceous	Late		99.6
			Early		145
		Jurassic	Late		161
			Middle		176
			Early		200
		Triassic	Late		228
			Middle		245
			Early		251
	Paleozoic	Permian	Late		260
			Middle		271
			Early		299
		Pennsylvanian	Late		306
			Middle		311
			Early		318
		Mississippian	Late		326
			Middle		345
			Early		359
		Devonian	Late		385
			Middle		397
			Early		416
		Silurian	Late		419
			Early		423
		Ordovician	Late		428
			Middle		444
			Early		488
		Cambrian	Late		501
			Middle		513
			Early		542
Precambrian	Proterozoic	Late	Neoproterozoic (Z)		1000
		Middle	Mesoproterozoic (Y)		1600
		Early	Paleoproterozoic (X)		2500
	Archean	Late			3200
		Early			4000
	Hadean				

Figure 1.29: Geologic time scale showing time period names and ages.

Eons are the largest divisions of time, and from oldest to youngest are named **Hadean, Archean, Proterozoic,** and **Phanerozoic.** The three oldest **eons** are sometimes collectively referred to as **Precambrian** time.

Life first appeared more than 3,800 million of years ago (Ma). From 3,500 Ma to 542 Ma, or 88% of geologic time, the predominant life forms were single-celled organisms such as bacteria. More complex organisms appeared only more recently, during the current **Phanerozoic** Eon, which includes the last 542 million years or 12% of geologic time.

The name **Phanerozoic** comes from *phaneros*, which means visible, and *zoic*, meaning life. This **eon** marks the proliferation of multicellular animals with hard body parts, such as shells, which are preserved in the geological record as **fossils.** Land-dwelling animals have existed for 360 million years, or 8% of geologic time. The demise of the dinosaurs and subsequent rise of mammals occurred around 65 Ma, or 1.5% of geologic time. Our human ancestors belonging to the genus *Homo* have existed since approximately 2.2 Ma—0.05% of geological time or just 1/2,000th the total age of Earth.

The **Phanerozoic** Eon is divided into three **eras**: **Paleozoic, Mesozoic,** and **Cenozoic. Paleozoic** means *ancient life*, and organisms of this era included invertebrate animals, fish, amphibians, and reptiles. The **Mesozoic** (*middle life*) is popularly known as the Age of Reptiles and is characterized by the abundance of dinosaurs, many of which evolved into birds. The **mass extinction** of the dinosaurs and other apex predator reptiles marked the end of the **Mesozoic** and beginning of the **Cenozoic. Cenozoic** means *new life* and is also called the Age of Mammals, during which mammals evolved to become the predominant land-dwelling animals. **Fossils** of early humans, or hominids, appear in the rock record only during the last few million years of the **Cenozoic**. The geologic time scale, geologic time, and geologic history are discussed in more detail in chapter 7 and chapter 8.

1.5.5 The Geologist's Tools

In its simplest form, a geologist's tool may be a rock hammer used for sampling a fresh surface of a rock. A basic tool set for fieldwork might also include:

- Magnifying lens for looking at mineralogical details
- Compass for measuring the orientation of geologic features
- Map for documenting the local distribution of rocks and **minerals**
- Magnet for identifying magnetic **minerals** like magnetite
- Dilute **solution** of hydrochloric acid to identify **carbonate**-containing **minerals** like **calcite** or **limestone**.

Figure 1.30: Iconic Archaeopteryx lithographica fossil from Germany.

In the laboratory, geologists use optical microscopes to closely examine rocks and **soil** for mineral **composition** and **grain size**. Laser and mass spectrometers precisely measure the chemical **composition** and geological age of **minerals**. **Seismographs** record and locate earthquake activity, or when used in conjunction with ground penetrating radar, locate objects buried beneath the surface of the earth. Scientists apply computer simulations to turn their collected data into testable, theoretical models. Hydrogeologists drill wells to sample and analyze underground water quality and availability. Geochemists use scanning electron microscopes to analyze **minerals** at the atomic level, via x-rays. Other geologists use gas chromatography to analyze liquids and gases trapped in **glacial** ice or rocks.

Technology provides new tools for scientific **observation**, which leads to new evidence that helps scientists revise and even refute old ideas. Because the ultimate technology will never be discovered, the ultimate **observation** will never be made. And this is the beauty of science—it is ever-advancing and always discovering something new.

Take this quiz to check your comprehension of this section.

If you are using an offline version of this text, access the quiz for section 1.5 via the QR code.

 An interactive H5P element has been excluded from this version of the text. You can view it online here: https://pressbooks.lib.vt.edu/introearthscience/?p=66#h5p-5

1.6 Science Denial and Evaluating Sources

 One or more interactive elements has been excluded from this version of the text. You can view them online here:
https://www.youtube.com/watch?v=8MqTOEospfo

Video 1.1: Science in America.

If you are using an offline version of this text, access this YouTube video via the QR code.

Introductory science courses usually deal with accepted scientific **theory** and do not include opposing ideas, even though these alternate ideas may be credible. This makes it easier for students to understand the complex material. Advanced students will encounter more controversies as they continue to study their discipline.

Some groups of people argue that some established scientific theories are wrong, not based on their scientific merit but rather on the ideology of the group. This section focuses on how to identify evidence based information and differentiate it from **pseudoscience**.

1.6.1 Science Denial

Figure 1.31: Anti-evolution league at the infamous Tennessee v. Scopes trial.

Figure 1.32: 2017 March for Science in Washington, DC. This and other similar marches were in response to funding cuts and anti-science rhetoric.

Science denial happens when people argue that established scientific theories are wrong, not based on scientific merit but rather on **subjective** ideology—such as for social, political, or economic reasons. Organizations and people use **science denial** as a rhetorical argument against issues or ideas they oppose. Three examples of **science denial** versus science are: 1) teaching evolution in public schools, 2) linking tobacco smoke to cancer, and 3) linking human activity to climate change. Among these, denial of climate change is strongly connected with geology. A **climate** denier specifically denies or doubts the **objective** conclusions of geologists and climate scientists.

Science denial generally uses three false arguments. The first argument tries to undermine the credibility of the scientific conclusion by claiming the research methods are flawed or the **theory** is not universally accepted—the science is unsettled. The notion that scientific ideas are not absolute creates doubt for non-scientists; however, a lack of universal truths should not be confused with scientific uncertainty. Because science is based on falsfiabiity, scientists avoid claiming universal truths and use language that conveys uncertainty. This allows scientific ideas to change and evolve as more evidence is uncovered.

The second argument claims the researchers are not **objective** and motivated by an ideology or economic agenda. This is an *ad hominem* argument in which a person's character is attacked instead of the merit of their argument. They claim results have been manipulated so researchers can justify asking for more funding. They claim that because the researchers are funded by a federal grant, they are using their results to lobby for expanded government regulation.

The third argument is to demand a balanced view, equal time in media coverage and educational curricula, to engender the false illusion of two equally valid arguments. Science deniers frequently demand equal coverage of their proposals, even when there is little scientific evidence supporting their ideology. For example, science deniers might demand religious explanations be taught as an alternative to the well-established **theory** of evolution. Or that all possible causes of **climate** change be discussed as equally probable, regardless of the body of evidence. Conclusions derived using the **scientific method** should not be confused with those based on ideologies.

Furthermore, conclusions about nature derived from ideologies have no place in science research and education. For example, it would be inappropriate to teach the flat Earth model in a modern geology course because this idea has been disproved by the **scientific method**. The formation of new conclusions based on the scientific method is the only way to change scientific conclusions. The fact that scientists avoid universal truths and change their ideas as more evidence is uncovered shouldn't be seen as meaning that the science is unsettled. Unfortunately, widespread scientific illiteracy allows these arguments to be used to suppress scientific knowledge and spread misinformation.

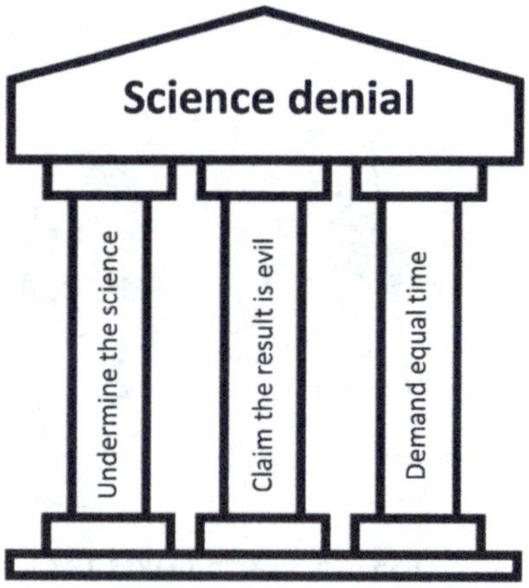

Figure 1.33: Three false rhetorical arguments of science denial.

In a classic case of **science denial**, beginning in the 1960s and for the next three decades, the tobacco industry and their scientists used rhetorical arguments to deny a connection between tobacco usage and cancer. Once it became clear scientific studies overwhelmingly found that using tobacco dramatically increased a person's likelihood of getting cancer, their next strategy was to create a sense of doubt about on the science. The tobacco industry suggested the results were not yet fully understood and more study was needed. They used this doubt to lobby for delaying legislative action that would warn consumers of the potential health hazards. This same tactic is currently being employed by those who deny the significance of human involvement in **climate** change.

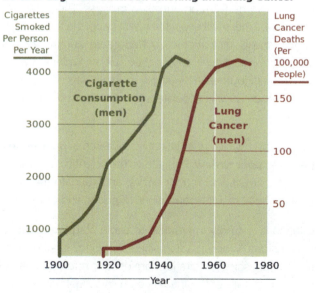

Figure 1.34: The lag time between cancer after smoking, plus the ethics of running human trials, delayed the government in taking action against tobacco.

1.6.2 Evaluating Sources of Information

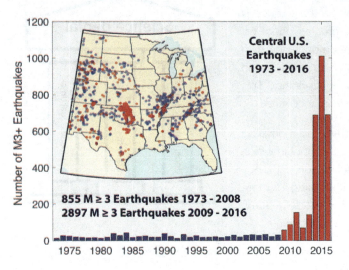

Figure 1.35: This graph shows earthquake data. To call this data induced, due to fracking, would be an interpretation.

In the age of the internet, information is plentiful. Geologists, scientists, or anyone exploring scientific inquiry must discern valid sources of information from **pseudoscience** and misinformation. This evaluation is especially important in scientific research because scientific knowledge is respected for its reliability. Textbooks such as this one can aid this complex and crucial task. At its roots, quality information comes from the **scientific method**, beginning with the empirical thinking of Aristotle. The application of the **scientific method** helps produce unbiased results. A valid inference or interpretation is based on **objective** evidence or data. Credible data and inferences are clearly labeled, separated, and differentiated. Anyone looking over the data can understand how the author's conclusion was derived or come to an alternative conclusion. Scientific procedures are clearly defined so the investigation can be replicated to confirm the original results or expanded further to produce new results. These measures make a scientific inquiry valid and its use as a source reputable. Of course, substandard work occasionally slips through and retractions are published from time to time. An infamous article linking the MMR vaccine to autism appeared in the highly reputable journal *Lancet* in 1998. Journalists discovered the author had multiple conflicts of interest and fabricated data, and the article was retracted in 2010.

In addition to methodology, data, and results, the authors of a study should be investigated. When looking into any research, the author(s) should be investigated. An author's credibility is based on multiple factors, such as having a degree in a relevant topic or being funded from an unbiased source.

The same rigor should be applied to evaluating the publisher, ensuring the results reported come from an unbiased process. The publisher should be easy to discover. Good publishers will show the latest papers in the journal and make their contact information and identification clear. Reputable journals show their **peer review** style. Some journal are predatory, where they use unexplained and unnecessary fees to submit and access journals. Reputable journals have recognizable editorial boards. Often, a reliable journal will associate with a trade, association, or recognized open source initiative.

Figure 1.36: Logo for The Geological Society of America, one of the leading geoscience organizations. They also publish GSA Bulletin, a reputable geology journal.

One of the hallmarks of scientific research is **peer review**. Research should be transparent to **peer review**. This allows the scientific community to reproduce experimental results, correct and retract errors, and validate theories. This allows reproduction of experimental results, corrections of errors, and proper justification of the research to experts.

Citation is not only imperative to avoid plagiarism, but also allows readers to investigate an author's line of thought and conclusions. When reading scientific works, it is important to confirm the citations are from reputable scientific research. Most often, scientific citations are used to reference paraphrasing rather than quotes. The number of times a work is cited is said to measure of the influence an investigation has within the scientific community, although this technique is inherently biased.

Take this quiz to check your comprehension of this section.

If you are using an offline version of this text, access the quiz for section 1.6 via the QR code.

  An interactive H5P element has been excluded from this version of the text. You can view it online here:
https://pressbooks.lib.vt.edu/introearthscience/?p=66#h5p-6

Summary

Science is a process, with no beginning and no end. Science is never finished because a full truth can never be known. However, science and the **scientific method** are the best way to understand the universe we live in. Scientists draw conclusions based on **objective** evidence; they consolidate these conclusions into unifying models. Geologists likewise understand studying the Earth is an ongoing process, beginning with James Hutton who declared the Earth has "…no vestige of a beginning, no prospect of an end." Geologists explore the 4.5 billion-year history of Earth, its resources, and its many hazards. From a larger viewpoint, geology can teach people how to develop credible conclusions, as well as identify and stop misinformation.

Take this quiz to check your comprehension of this chapter.

If you are using an offline version of this text, access the quiz for chapter 1 via the QR code.

  An interactive H5P element has been excluded from this version of the text. You can view it online here:
https://pressbooks.lib.vt.edu/introearthscience/?p=66#h5p-7

Text References

1. Adams, F.D., 1954, The birth and development of the geological sciences.
2. Alfe, D., Gillan, M.J., and Price, G.D., 2002, Composition and temperature of the Earth's core constrained by combining ab initio calculations and seismic data: Earth Planet. Sci. Lett., v. 195, no. 1, p. 91–98.
3. Alkin, M.C., 2004, Evaluation Roots: Tracing theorists' views and influences: SAGE.
4. Beckwith, C., 2013, How western Europe developed a full scientific method: Berfrois.
5. Birch, F., 1952, Elasticity and constitution of the Earth's interior: J. Geophys. Res., v. 57, no. 2, p. 227–286., doi: 10.1029/JZ057i002p00227.
6. Bocking, S., 2004, Nature's experts: science, politics, and the environment: Rutgers University Press.

7. Chamberlin, T.C., 1890, The method of multiple working hypotheses: Science, v. 15, no. 366, p. 92–96.

8. Cohen, H.F., 2010, How modern science came into the world: Four civilizations, one 17th-century breakthrough: Amsterdam University Press.

9. Darwin, C., 1846, Geological Observations on South America: Being the Third Part of the Geology of the Voyage of the Beagle, Under the Command of Capt. Fitzroy, R.N. During the Years 1832 to 1836: Smith, Elder and Company.

10. Drake, S., 1990, Galileo: Pioneer Scientist: University of Toronto Press.

11. Engdahl, E.R., Flynn, E.A., and Masse, R.P., 1974, Differential PkiKP travel times and the radius of the core: Geophysical J Royal Astro Soc, v. 40, p. 457–463.

12. Everitt, A., 2016, The Rise of Athens: The Story of the World's Greatest Civilization.

13. Goldstein, B.R., 2002, Copernicus and the origin of his heliocentric system: Journal for the History of Astronomy, v. 33, p. 219–235.

14. Goldsworthy, A.K., 2011, The complete Roman army: Thames & Hudson.

15. Hans Wedepohl, K., 1995, The composition of the continental crust: Geochim. Cosmochim. Acta, v. 59, no. 7, p. 1217–1232.

16. Heilbron, J.L., 2012, Galileo: Oxford, Oxford University Press, 528 p.

17. Hogendijk, J.P., and Sabra, A.I., 2003, The Enterprise of Science in Islam: New Perspectives: MIT Press.

18. Jakosky, B.M., Grebowsky, J.M., Luhmann, J.G., Connerney, J., Eparvier, F., Ergun, R., Halekas, J., Larson, D., Mahaffy, P., McFadden, J., Mitchell, D.F., Schneider, N., Zurek, R., Bougher, S., and others, 2015, MAVEN observations of the response of Mars to an interplanetary coronal mass ejection: Science, v. 350, no. 6261, p. aad0210.

19. Kerferd, G.B., 1959, The Biography of Aristotle Ingemar Düring: Aristotle in the Ancient Biographical Tradition. (Studia Graeca et Latina Gothoburgensia v.) Pp. 490; 1 plate. Gothenburg: Institute of Classical Studies, 1957. Paper, Kr. 32: Classical Rev., v. 9, no. 02, p. 128–130.

20. Kolbert, E., 2014, The sixth extinction: an unnatural history: New York, Henry Holt and Co., 336 p.

21. Krimsky, S., 2013, Do financial conflicts of interest bias research? An inquiry into the "funding effect" hypothesis: Sci. Technol. Human Values, v. 38, no. 4, p. 566–587.

22. Lehmann, I., 1936, P', Publ: Bur. Centr. Seism. Internat. Serie A, v. 14, p. 87–115.

23. Marshall, J., 2010, A short history of Greek philosophy: Andrews UK Limited.

24. Martin, C., 2014, Subverting Aristotle: Religion, History, and Philosophy in Early Modern Science: Baltimore: Johns Hopkins University Press.

25. Mayr, E., 1942, Systematics and the Origin of Species, from the Viewpoint of a Zoologist: Harvard University Press.

26. Montgomery, K., 2003, Siccar Point and teaching the history of geology: J. Geosci. Educ.

27. Mooney, W.D., Laske, G., and Masters, T.G., 1998, CRUST 5.1: A global crustal model: J. Geophys. Res. [Solid Earth], v. 103, no. B1, p. 727–747.

28. Moustafa, K., 2016, Aberration of the Citation: Account. Res., v. 23, no. 4, p. 230–244.

29. National Center for Science Education, 2016, Climate change denial: Online, http://ncse.com/climate/denial, accessed April 2016.

30. Oreskes, N., Conway, E., and Cain, S., 2010, Merchants of doubt: how a handful of scientists obscured the truth on issues from tobacco smoke to global warming: Bloomsbury Press, 368 p.

31. Paradowski, R.J., 2012, Nicolas Steno: Danish anatomist and geologist: Great Lives from History: Scientists & Science, p. 830–832.

32. Patterson, C., 1956, Age of meteorites and the earth: Geochim. Cosmochim. Acta, v. 10, no. 4, p. 230–237.

33. Popper, K., 2002, Conjectures and Refutations: The Growth of Scientific Knowledge: London; New York, Routledge, 608 p.

34. Porter, R., 1976, Charles Lyell and the Principles of the History of Geology: Br. J. Hist. Sci., v. 9, no. 02, p. 91–103.

35. Railsback, B.L., 1990, T. C. Chamberlin's "Method of Multiple Working Hypotheses": An encapsulation for modern students: Online, http://www.gly.uga.edu/railsback/railsback_chamberlin.html, accessed December 2016.

36. Railsback, B.L., 2004, T. C. Chamberlin's "Method of Multiple Working Hypotheses": An encapsulation for modern students: Houston Geological Society Bulletin, v. 47, no. 2, p. 68–69.

37. Rappaport, R., 1994, James Hutton and the History of Geology. Dennis R. Dean: Isis, v. 85, no. 3, p. 524–525.

38. Repcheck, J., 2007, Copernicus' secret: How the scientific revolution began: Simon and Schuster.

39. Repcheck, J., 2009, The Man Who Found Time: James Hutton and the Discovery of the Earth's Antiquity: New York: Basic Books.

40. Sabra, A.I. and Others, 1989, The optics of Ibn al-Haytham: Books I-III: On direct vision: Warburg Institute, University of London.

41. Saliba, G., 2007, Islamic science and the making of the European renaissance: MIT Press.

42. Shermer, M., 2011, What Is Pseudoscience? Scientific American.

43. Snow, C.E. (Ed.), 2016, Science literacy: concepts, contexts, and consequences: Washington, DC, National Academies Press (US).

44. Spier, R., 2002, The history of the peer-review process: Trends Biotechnol., v. 20, no. 8, p. 357–358.

45. Van Reybrouck, D., 2012, From Primitives to Primates: A History of Ethnographic and Primatological Analogies in the Study of Prehistory: Sidestone Press.

46. Waters, C.N., Zalasiewicz, J., Summerhayes, C., Barnosky, A.D., Poirier, C., Galuszka, A., Cearreta, A., Edgeworth, M., Ellis, E.C., Ellis, M., Jeandel, C., Leinfelder, R., McNeill, J.R., Richter, D.D., and others, 2016, The Anthropocene is functionally and stratigraphically distinct from the Holocene: Science, v. 351, no. 6269, p. aad2622.

47. de Wijs, G.A., Kresse, G., Vočadlo, L., Dobson, D., Alfè, D., Gillan, M.J., and Price, G.D., 1998, The viscosity of liquid iron at the physical conditions of the Earth's core: Nature, v. 392, no. 6678, p. 805–807., doi: 10.1038/33905.

48. Wyhe, J.V., 2008, Darwin: Andre Deutsch, 72 p.

49. Wyllie, P.J., 1970, Ultramafic rocks and the upper mantle, in Morgan, B.A., editor, Fiftieth anniversary symposia: Mineralogy and petrology of the Upper Mantle; Sulfides; Mineralogy and geochemistry of non-marine evaporites: Washington, DC, Mineralogical Society of America, p. 3–32.

50. Zalasiewicz, J., Williams, M., Smith, A., Barry, T.L., Coe, A.L., Bown, P.R., Brenchley, P., Cantrill, D., Gale, A., Gibbard, P., and Others, 2008, Are we now living in the Anthropocene? GSA Today, v. 18, no. 2, p. 4.

Figure References

Figure 1.1: This is Grand Canyon of the Yellowstone in Yellowstone National Park. Grastel. 2021. CC BY-SA 4.0. https://commons.wikimedia.org/wiki/File:Grand_Canyon_of_yellowstone.jpg

Figure 1.2: Canyons like this, carved in the deposit left by the May 18th, 1980 eruption of Mt. St. Helens is sometimes used by purveyors of pseudoscience as evidence for the Earth being very young. Richard Droker. 2011. CC BY-NC-ND 2.0. https://flic.kr/p/2cNFV8D

Figure 1.3: Geologists share information by publishing, attending conferences, and even going on field trips, such as this trip to the Lake Owyhee Volcanic Field in Oregon by the Bureau of Land Management in 2019. Bureau of Land Management Oregon and Washington. 2019. CC BY 2.0. https://flic.kr/p/RCtkjX

Figure 1.4: Diagram of the cyclical nature of the scientific method. ArchonMagnus. 2015. CC BY-SA 4.0. https://commons.wikimedia.org/wiki/File:The_Scientific_Method_as_an_Ongoing_Process.svg

Figure 1.5: A famous hypothesis: Leland Stanford wanted to know if a horse lifted all 4 legs off the ground during a gallop, since the legs are too fast for the human eye to perceive it. Eadweard Muybridge. 1878. Public domain. https://commons.wikimedia.org/wiki/File:Eadweard_Muybridge-Sallie_Gardner_1878.jpg

Figure 1.6: An experiment at the University of Queensland has been going since 1927. John Mainstone; adapted by Amada44. 2012. CC BY-SA 3.0. https://commons.wikimedia.org/wiki/File:University_of_Queensland_Pitch_drop_experiment-white_bg.jpg

Figure 1.7: Wegener later in his life, ca. 1924-1930. Author unknown. ca.1924-1930. Public domain. https://commons.wikimedia.org/wiki/File:Alfred_Wegener_ca.1924-30.jpg

Figure 1.8: Fresco by Raphael of Plato (left) and Aristotle (right). Raphael. 1509. Public domain. https://commons.wikimedia.org/wiki/File:Sanzio_01_Plato_Aristotle.jpg

Figure 1.9: Drawing of Avicenna (Ibn Sina). Unknown author. Unknown date. Public domain. https://commons.wikimedia.org/wiki/File:Avicenna-miniatur.jpg

Figure 1.10: Geocentric drawing by Bartolomeu Velho in 1568. Bartolomeu Velho. Original work, 1568. Photo taken in 2008. Public domain. https://en.wikipedia.org/wiki/File:Bartolomeu_Velho_1568.jpg

Figure 1.11: Galileo's first mention of moons of Jupiter. Galileo Galilei. ca. 1609 and 1610. Public domain. https://commons.wikimedia.org/wiki/File:Galileo_manuscript.png

Figure 1.12: Copernicus' heliocentric model. Copernicus; adapted by Professor marginalia. 1543; 2010. Public domain. https://commons.wikimedia.org/wiki/File:Copernican_heliocentrism_diagram-2.jpg

Figure 1.13: Illustration by Steno showing a comparison between fossil and modern shark teeth. Steensen, Niels. 1669 AD. Public domain. https://commons.wikimedia.org/wiki/File:Steensen_-_Elementorum_myologiae_specimen,_1669_-_4715289.tif

Figure 1.14: Cuvier's comparison of modern elephant and mammoth jaw bones. Georges Cuvier. 1796. Public domain. https://commons.wikimedia.org/wiki/File:Cuvier_elephant_jaw.jpg

Figure 1.15: Inside cover of Lyell's Elements of Geology. Charles Lyell. 1857. Public domain. https://commons.wikimedia.org/wiki/File:Lyell_Principles_frontispiece.jpg

Figure 1.16: J. Tuzo Wilson. Stephen Morris. 1992. CC BY-SA 3.0. https://commons.wikimedia.org/wiki/File:John_Tuzo_Wilson_in_1992.jpg

Figure 1.17: Girls into Geoscience inaugural Irish Fieldtrip. Aileen Doran. 2019. CC BY 4.0. https://commons.wikimedia.org/wiki/File:Exceptional_folds_during_the_Girls_into_Geoscience_inaugural_Irish_Fieldtrip.jpg

Figure 1.18: Hoover Dam provides hydroelectric energy and stores water for southern Nevada. Ubergirl. 2012. CC BY-SA 3.0. https://commons.wikimedia.org/wiki/File:Hoover_Dam,_Colorado_River.JPG

Figure 1.19: Coal power plant in Helper, Utah. David Jolley. 2007. CC BY-SA 3.0. https://commons.wikimedia.org/wiki/File:Castle_Gate_Power_Plant,_Utah_2007.jpg

Figure 1.20: Buildings toppled from liquefaction during a 7.5 magnitude earthquake in Japan. Ungtss. 1964. Public domain. https://commons.wikimedia.org/wiki/File:Liquefaction_at_Niigata.JPG

Figure 1.21: Oregon's Crater Lake was formed about 7700 years ago after the eruption of Mount Mazama. Zainubrazvi. 2006. CC BY-SA 3.0. https://en.wikipedia.org/wiki/File:Crater_lake_oregon.jpg

Figure 1.22: Rock cycle showing the five materials (such as igneous rocks and sediment) and the processes by which one changes into another (such as weathering). Kindred Grey. 2022. CC BY 4.0.

Figure 1.23: Mississippian raindrop impressions over wave ripples from Nova Scotia. Rygel, M.C.. 2006. CC BY-SA 3.0. https://commons.wikimedia.org/wiki/File:Raindrop_impressions_mcr1.jpg

Figure 1.24: Metamorphic rock in Georgian Bay, Ontario. P199. 2013. CC BY-SA 4.0. https://commons.wikimedia.org/wiki/File:Metamorphic_rock_Georgian_Bay.jpg

Figure 1.25: Map of the major plates and their motions along boundaries. Scott Nash via USGS. 1996. Public domain. https://commons.wikimedia.org/wiki/File:Plates_tect2_en.svg

Figure 1.26: The global map of the depth of the moho. AllenMcC. 2013. CC BY-SA 3.0. https://commons.wikimedia.org/wiki/File:Mohomap.png

Figure 1.27: The layers of the Earth. Drlauraguertin. 2015. CC BY-SA 3.0. https://wiki.seg.org/wiki/File:Earthlayers.png#file

Figure 1.28: Geologic time on Earth, represented circularly, to show the individual time divisions and important events. Woudloper; adapted by Hardwigg. 2010. Public domain. https://commons.wikimedia.org/wiki/File:Geologic_Clock_with_events_and_periods.svg

Figure 1.29: Geologic time scale showing time period names and ages. USGS. 2009. Public domain. https://commons.wikimedia.org/wiki/File:Geologic_time_scale.jpg

Figure 1.30: Iconic Archaeopteryx lithographica fossil from Germany. H. Raab. 2009. CC BY-SA 3.0. https://commons.wikimedia.org/wiki/File:Archaeopteryx_lithographica_(Berlin_specimen).jpg

Figure 1.31: Anti-evolution league at the infamous Tennessee v. Scopes trial. Mike Licht. 1925. CC BY 2.0. https://commons.wikimedia.org/wiki/File:Anti-EvolutionLeague.jpg

Figure 1.32: 2017 March for Science in Washington, DC. Becker1999. 2017. CC BY 2.0. https://commons.wikimedia.org/wiki/File:March_for_Science,_Washington,_DC_(34168985286).jpg

Figure 1.33: Three false rhetorical arguments of science denial. Kindred Grey. 2022. CC BY 4.0. Includes columns by Med MB from Noun Project (Noun Project license).

Figure 1.34: The lag time between cancer after smoking, plus the ethics of running human trials, delayed the government in taking action against tobacco. Sakurambo. 2007. Public domain. https://commons.wikimedia.org/wiki/File:Cancer_smoking_lung_cancer_correlation_from_NIH.svg

Figure 1.35: This graph shows earthquake data. USGS. 2019. Public domain. https://en.wikipedia.org/wiki/File:Cumulative_induced_seismicity.png

Figure 1.36: Logo for The Geological Society of America, one of the leading geoscience organizations. GSA. Retrieved 2022. https://www.geosociety.org/GSA/About/Who_We_Are/Society_Documents/GSA/About/logo_usage.aspx

2.

PLATE TECTONICS

Learning Objectives

By the end of this chapter, students should be able to:

- Describe how the ideas behind **plate tectonics** started with Alfred Wegener's hypothesis of continental drift.
- Describe the physical and chemical layers of the Earth and how they affect **plate** movement.
- Explain how movement at the three types of **plate** boundaries causes earthquakes, **volcanoes**, and mountain building.
- Identify **convergent** boundaries, including **subduction** and collisions, as places where **plates** come together.
- Identify **divergent** boundaries, including **rifts** and **mid-ocean ridges**, as places where **plates** separate.
- Explain **transform** boundaries as places where adjacent **plates shear** past each other.
- Describe the **Wilson Cycle**, beginning with **continental rifting**, ocean **basin** creation, **plate subduction**, and ending with ocean **basin** closure.
- Explain how the tracks of **hotspots**, places that have continually rising **magma**, is used to calculate **plate** motion.

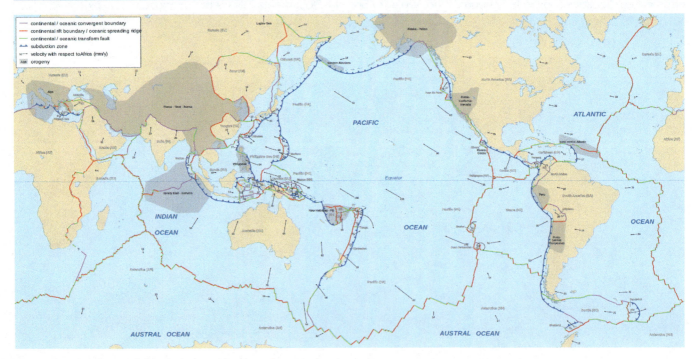

Figure 2.1: Detailed map of all known plates, their boundaries, and movements.

Revolution is a word usually reserved for significant political or social changes. Several of these idea revolutions forced scientists to re-examine their entire field, triggering a paradigm shift that shook up their conventionally held knowledge. Charles Darwin's book on evolution, *On the Origin of Species*, published in 1859; Gregor Mendel's discovery of the genetic principles of inheritance in 1866; and James Watson, Francis Crick, and Rosalind Franklin's model for the structure of DNA in 1953 did that for biology. Albert Einstein's relativity and quantum mechanics concepts in the early twentieth century did the same for Newtonian physics.

The concept of **plate tectonics** was just as revolutionary for geology. The **theory** of **plate tectonics** attributes the movement of **massive** sections of the Earth's outer layers with creating earthquakes, mountains, and **volcanoes**. Many earth processes make more sense when viewed through the lens of **plate tectonics**. Because it is so important in understanding how the world works, **plate tectonics** is the first topic of discussion in this textbook.

2.1 Alfred Wegener's Continental Drift Hypothesis

Alfred Wegener (1880-1930) was a German scientist who specialized in meteorology and climatology. His knack for questioning accepted ideas started in 1910 when he disagreed with the explanation that the Bering Land Bridge was formed by **isostasy**, and that similar land bridges once connected the continents. After reviewing the scientific literature, he published a **hypothesis** stating the continents were originally connected, and then drifted apart. While he did not have the precise mechanism worked out, his **hypothesis** was backed up by a long list of evidence.

2.1.1 Early Evidence for Continental Drift Hypothesis

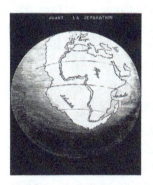 

Wegener's first piece of evidence was that the coastlines of some continents fit together like pieces of a jigsaw puzzle. People noticed the similarities in the coastlines of South America and Africa on the first world maps, and some suggested the continents had been ripped apart. Antonio Snider-Pellegrini did preliminary work on **continental** separation and matching **fossils** in 1858.

Figure 2.2: Wegener later in his life, ca. 1924-1930.

Figure 2.3: Snider-Pellegrini's map showing the continental fit and separation, 1858.

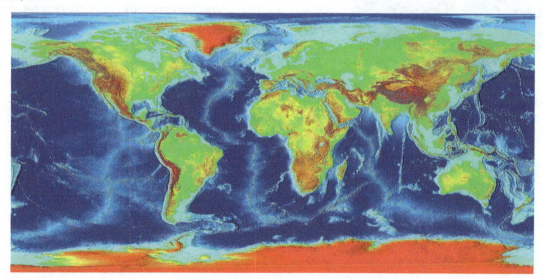

Figure 2.4: Map of world elevations. Note the light blue, which are continental shelves flooded by shallow ocean water. These show the true shapes of the continents.

What Wegener did differently was synthesize a large amount of data in one place. He used true edges of the continents, based on the shapes of the **continental** shelves. This resulted in a better fit than previous efforts that traced the existing coastlines.

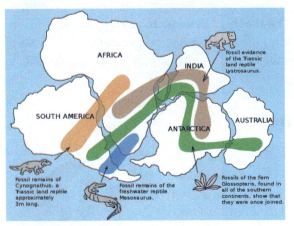

Figure 2.5: Image showing fossils that connect the continents of Gondwana (the southern continents of Pangea).

Wegener also compiled evidence by comparing similar rocks, mountains, **fossils**, and **glacial formations** across oceans. For example, the **fossils** of the primitive aquatic reptile *Mesosaurus* were found on the separate coastlines of Africa and South America. **Fossils** of another reptile, *Lystrosaurus*, were found on Africa, India, and Antarctica. He pointed out these were land-dwelling creatures could not have swum across an entire ocean.

Opponents of **continental** drift insisted trans-**oceanic** land bridges allowed animals and plants to move between continents. The land bridges eventually eroded away, leaving the continents permanently separated. The problem with this **hypothesis** is the improbability of a land bridge being tall and long enough to stretch across a broad, deep ocean.

More support for **continental** drift came from the puzzling evidence that **glaciers** once existed in normally very warm areas in southern Africa, India, Australia, and Arabia. These **climate anomalies** could not be explained by land bridges. Wegener found similar evidence when he discovered tropical plant **fossils** in the frozen region of the Arctic Circle. As Wegener collected more data, he realized the explanation that best fit all the **climate**, rock, and **fossil** observations involved moving continents.

2.1.2 Proposed Mechanism for Continental Drift

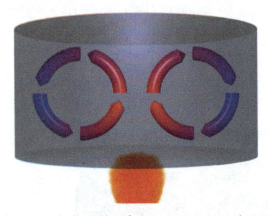

Figure 2.6: The basic idea of convection: an uneven heat source in a fluid causes rising material next to the heat and sinking material far from the heat.

Wegener's work was considered a fringe science **theory** for his entire life. One of the biggest flaws in his **hypothesis** was an inability to provide a mechanism for how the continents moved. Obviously, the continents did not appear to move, and changing the conservative minds of the scientific community would require exceptional evidence that supported a credible mechanism. Other pro-continental drift followers used expansion, contraction, or even the moon's origin to explain how the continents moved. Wegener used centrifugal forces and **precession**, but this model was proven wrong. He also speculated about seafloor spreading, with hints of **convection**, but could not substantiate these proposals. As it turns out, current scientific knowledge reveals **convection** is one the major forces in driving **plate** movements, along with gravity and density.

2.1.3 Development of Plate Tectonic Theory

Wegener died in 1930 on an expedition in Greenland. Poorly respected in his lifetime, Wegener and his ideas about moving continents seemed destined to be lost in history as fringe science. However, in the 1950s, evidence started to trickle in that made **continental** drift a more viable idea. By the 1960s, scientists had amassed enough evidence to support the missing mechanism—namely, seafloor spreading—for Wegener's **hypothesis** of **continental** drift to be accepted as the **theory** of **plate tectonics**. Ongoing GPS and earthquake data analyses continue to support this **theory**. The next section provides the pieces of evidence that helped **transform** one man's wild notion into a scientific **theory**.

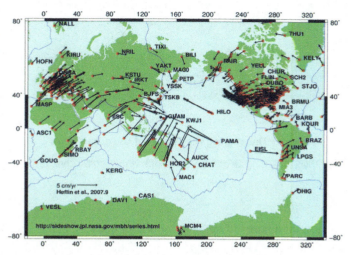

Figure 2.7: GPS measurements of plate motions.

Mapping of the Ocean Floors

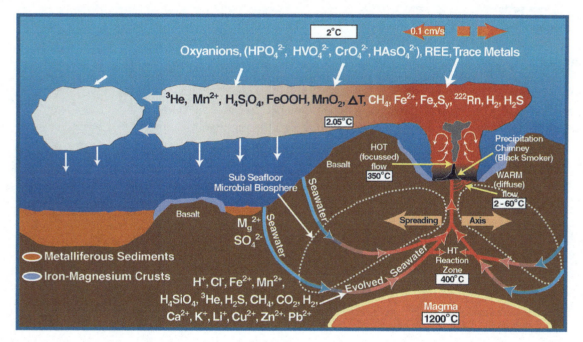

Figure 2.8: The complex chemistry around mid-ocean ridges.

In 1947 researchers started using an adaptation of **SONAR** to map a region in the middle of the Atlantic Ocean with poorly-understood topographic and thermal properties. Using this information, Bruce Heezen and Marie Tharp created the first detailed map of the **ocean floor** to reveal the Mid-Atlantic Ridge, a basaltic mountain range that spanned the length of the Atlantic Ocean, with rock chemistry and dimensions unlike the mountains found on the continents. Initially scientists thought the ridge was part of a mechanism that explained the expanding Earth or ocean-**basin** growth **hypotheses**. In 1959, Harry Hess proposed the **hypothesis** of seafloor spreading—that the **mid-ocean ridges** represented **tectonic plate** factories, where new **oceanic plate** was issuing from these long **volcanic** ridges. Scientists later included **transform faults** perpendicular to the ridges to better account for varying rates of movement between the newly formed **plates**. When earthquake **epicenters** were discovered along the ridges, the idea that earthquakes were linked to **plate** movement took hold.

One or more interactive elements has been excluded from this version of the text. You can view them online here:
https://www.youtube.com/watch?v=TgfYjSOOTWw

Video 2.1: Uncovering the secrets of the ocean floor.

If you are using an offline version of this text, access this YouTube video via the QR code.

Seafloor **sediment**, measured by dredging and drilling, provided another clue. Scientists once believed **sediment** accumulated on the ocean floors over a very long time in a static environment. When some studies showed less **sediment** than expected, these results were initially used to argue against **continental** movement. With more time, researchers discovered these thinner **sediment** layers were located close to **mid-ocean ridges**, indicating the ridges were younger than the surrounding **ocean floor**. This finding supported the idea that the sea floor was not fixed in one place.

Paleomagnetism

The seafloor was also mapped magnetically. Scientists had long known of strange magnetic **anomalies** that formed a striped pattern of symmetrical rows on both sides of mid-oceanic ridges. What made these features unusual was the north and south magnetic poles within each stripe was reversed in alternating rows. By 1963, Harry Hess and other scientists used these magnetic reversal patterns to support their model for seafloor spreading (see also Lawrence W. Morley).

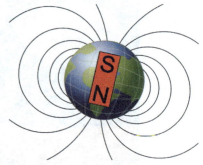

Figure 2.9: The magnetic field of Earth, simplified as a bar magnet.

Paleomagnetism is the study of magnetic fields frozen within rocks, basically a fossilized compass. In fact, the first hard evidence to support **plate** motion came from **paleomagnetism**.

Igneous rocks containing magnetic **minerals** like magnetite typically provide the most useful data. In their liquid state as **magma** or **lava**, the magnetic poles of the **minerals** align themselves with the Earth's magnetic field. When the rock cools and solidifies, this alignment is frozen into place, creating a permanent paleomagnetic record that includes magnetic inclination related to global **latitude**, and declination related to magnetic north.

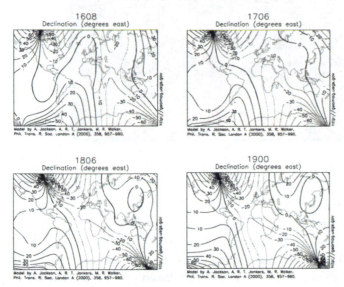

Figure 2.10: How the magnetic poles have moved over 400 years.

Scientists had noticed for some time the alignment of magnetic north in many rocks was nowhere close to the earth's current magnetic north. Some explained this away are part of the normal movement of earth's magnetic north pole. Eventually, scientists realized adding the idea of **continental** movement explained the data better than pole movement alone.

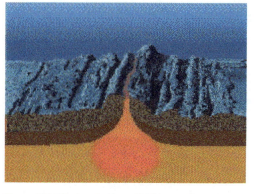

Figure 2.11: The iron in the solidifying rock preserves the current magnetic polarity as new oceanic plates form at mid ocean ridges.

Wadati–Benioff Zones

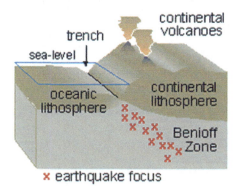

Figure 2.12: The Wadati-Benioff zone, showing earthquakes following the subducting slab down.

Around the same time **mid-ocean ridges** were being investigated, other scientists linked the creation of ocean trenches and island arcs to **seismic** activity and **tectonic plate** movement. Several independent research groups recognized earthquake **epicenters** traced the shapes of **oceanic plates** sinking into the mantle. These deep earthquake zones congregated in planes that started near the surface around ocean trenches and angled beneath the continents and island arcs. Today these earthquake zones called Wadati-Benioff zones.

Based on the mounting evidence, the **theory** of **plate tectonics** continued to take shape. J. Tuzo Wilson was the first scientist to put the entire picture together by proposing that the opening and closing of the ocean basins. Before long, scientists proposed other models showing **plates** moving with respect to each other, with clear boundaries between them. Others started piecing together complicated histories of **tectonic plate** movement. The **plate tectonic** revolution had taken hold.

Figure 2.13: J. Tuzo Wilson.

Complete this interactive activity to check your understanding.

If you are using an offline version of this text, access this interactive activity via the QR code.

 An interactive H5P element has been excluded from this version of the text. You can view it online here:
https://pressbooks.lib.vt.edu/introearthscience/?p=139#h5p-8

Take this quiz to check your comprehension of this section.

If you are using an offline version of this text, access the quiz for section 2.1 via the QR code.

 An interactive H5P element has been excluded from this version of the text. You can view it online here:
https://pressbooks.lib.vt.edu/introearthscience/?p=139#h5p-9

2.2 Layers of the Earth

In order to understand the details of **plate tectonics**, it is essential to first understand the layers of the earth. First-hand information about what is below the surface is very limited; most of what we know is pieced together from hypothetical models, and analyzing **seismic wave** data and **meteorite** materials. In general, the Earth can be divided into layers based on chemical **composition** and physical characteristics.

2.2.1 Chemical Layers

Certainly the earth is composed of a countless combination of **elements**. Regardless of what **elements** are involved two major factors—**temperature** and pressure—are responsible for creating three distinct chemical layers.

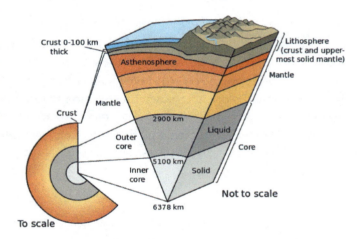

Figure 2.14: The layers of the Earth. Physical layers include lithosphere and asthenosphere; chemical layers are crust, mantle, and core.

Crust

The outermost chemical layer and the one we currently reside on, is the **crust**. There are two types of **crust**. **Continental crust** has a relatively low density and **composition** similar to **granite**. **Oceanic crust** has a relatively high density, especially when cold and old, and **composition** similar to **basalt**. The surface levels of **crust** are relatively **brittle**. The deeper parts of the **crust** are subjected to higher temperatures and pressure, which makes them more **ductile**. **Ductile** materials are like soft plastics or putty, they move under force. **Brittle** materials are like solid glass or pottery, they break under force, especially when it is applied quickly. Earthquakes, generally occur in the upper **crust** and are caused by the rapid movement of relatively **brittle** materials.

The base of the **crust** is characterized by a large increase in **seismic** velocity, which measures how fast earthquake waves travel through solid matter. Called the Mohorovičić Discontinuity, or **Moho** for short, this zone was discovered by Andrija Mohorovičić (pronounced mo-ho-ro-vee-cheech; audio pronunciation) in 1909 after studying earthquake wave paths in his native Croatia. The change in wave direction and speed is caused by dramatic chemical differences of the **crust** and **mantle**. Underneath the oceans, the **Moho** is found roughly 5 km below the **ocean floor**. Under the continents, it is located about 30-40 km below the surface. Near certain large mountain-building events known as orogenies, the **continental Moho** depth is doubled.

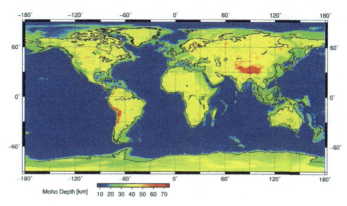

Figure 2.15: The global map of the depth of the Moho.

Mantle

The **mantle** sits below the **crust** and above the **core**. It is the largest chemical layer by volume, extending from the base of the **crust** to a depth of about 2900 km. Most of what we know about the **mantle** comes from **seismic wave** analysis, though information is gathered by studying **ophiolites** and **xenoliths**. **Ophiolites** are pieces of **mantle** that have risen through the **crust** until they are exposed as part of the **ocean floor**. **Xenoliths** are carried within **magma** and brought to the Earth's surface by volcanic eruptions. Most **xenoliths** are made of **peridotite**, an **ultramafic** class of **igneous rock** (see section 4.2 for explanation). Because of this, scientists hypothesize most of the **mantle** is made of **peridotite**.

Figure 2.16: This mantle xenolith containing olivine (green) is chemically weathering by hydrolysis and oxidation into the pseudo-mineral iddingsite, which is a complex of water, clay, and iron oxides. The more altered side of the rock has been exposed to the environment longer.

Core

Figure 2.17: A polished fragment of the iron-rich Toluca Meteorite, with octahedral Widmanstätten pattern.

The **core** of the Earth, which has both liquid and solid layers, and consists mostly of iron, nickel, and possibly some oxygen. Scientists looking at **seismic** data first discovered this innermost chemical layer in 1906. Through a union of hypothetical modeling, astronomical insight, and hard **seismic** data, they concluded the **core** is mostly **metallic** iron. Scientists studying **meteorites**, which typically contain more iron than surface rocks, have proposed the earth was formed from meteoric material. They believe the liquid component of the **core** was created as the iron and nickel sank into the center of the planet, where it was liquefied by intense pressure.

2.2.2 Physical Layers

The Earth can also be broken down into five distinct physical layers based on how each layer responds to **stress**. While there is some overlap in the chemical and physical designations of layers, specifically the **core–mantle** boundary, there are significant differences between the two systems.

Lithosphere

Lithos is Greek for stone, and the **lithosphere** is the outermost physical layer of the Earth. It is grouped into two types: **oceanic** and **continental**. **Oceanic lithosphere** is thin and relatively rigid. It ranges in thickness from nearly zero in new **plates** found around **mid-ocean ridges**, to an average of 140 km in most other locations. **Continental lithosphere** is generally thicker and considerably more plastic, especially at the deeper levels. Its thickness ranges from 40 to 280 km. The **lithosphere** is not continuous. It is broken into segments called **plates**. A **plate boundary** is where two **plates** meet and move relative to each other. **Plate** boundaries are where we see **plate tectonics** in action—mountain building, triggering earthquakes, and generating **volcanic** activity.

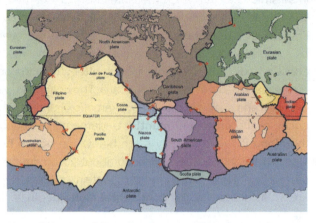

Figure 2.18: Map of the major plates and their motions along boundaries.

Asthenosphere

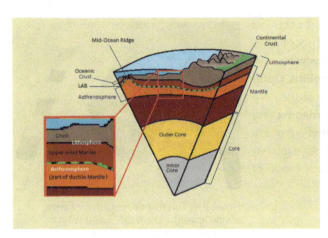

Figure 2.19: The lithosphere–asthenosphere boundary changes with certain tectonic situations.

The **asthenosphere** is the layer below the **lithosphere**. *Astheno-* means lacking strength, and the most distinctive property of the **asthenosphere** is movement. Because it is mechanically weak, this layer moves and flows due to **convection** currents created by heat coming from the earth's **core** cause. Unlike the **lithosphere** that consists of multiple **plates**, the **asthenosphere** is relatively unbroken. Scientists have determined this by analyzing **seismic** waves that pass through the layer. The depth of at which the **asthenosphere** is found is **temperature**-dependent. It tends to lie closer to the earth's surface around **mid-ocean ridges** and much deeper underneath mountains and the centers of lithospheric **plates**.

Mesosphere

The **mesosphere**, sometimes known as the lower **mantle**, is more rigid and immobile than the **asthenosphere**. Located at a depth of approximately 410 and 660 km below the earth's surface, the **mesosphere** is subjected to very high pressures and temperatures. These extreme conditions create a transition zone in the upper **mesosphere** where **minerals** continuously change into various forms, or pseudomorphs. Scientists identify this zone by changes in **seismic** velocity and sometimes physical barriers to movement. Below this transitional zone, the **mesosphere** is relatively uniform until it reaches the **core**.

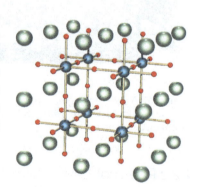

Figure 2.20: General perovskite structure. Perovskite silicates (i.e., Bridgmenite, (Mg,Fe)SiO3) are thought to be the main component of the lower mantle, making it the most common mineral in or on Earth.

Inner and Outer Core

Figure 2.21: Lehmann in 1932.

The **outer core** is the only entirely liquid layer within the Earth. It starts at a depth of 2,890 km and extends to 5,150 km, making it about 2,300 km thick. In 1936, the Danish geophysicist Inge Lehmann analyzed **seismic** data and was the first to prove a solid **inner core** existed within a liquid **outer core**. The solid **inner core** is about 1,220 km thick, and the **outer core** is about 2,300 km thick.

It seems like a contradiction that the hottest part of the Earth is solid, as the **minerals** making up the **core** should be liquified or vaporized at this **temperature**. Immense pressure keeps the **minerals** of the **inner core** in a solid phase. The **inner core** grows slowly from the lower **outer core** solidifying as heat escapes the interior of the Earth and is dispersed to the outer layers.

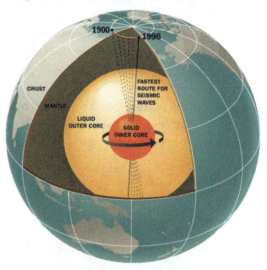

Figure 2.22: The outer core's spin causes our protective magnetic field.

The earth's liquid **outer core** is critically important in maintaining a breathable **atmosphere** and other environmental conditions favorable for life. Scientists believe the earth's magnetic field is generated by the circulation of molten iron and nickel within the **outer core**. If the **outer core** were to stop circulating or become solid, the loss of the magnetic field would result in Earth getting stripped of life-supporting gases and water. This is what happened, and continues to happen, on Mars.

Complete this interactive activity to check your understanding.

If you are using an offline version of this text, access this interactive activity via the QR code.

 An interactive H5P element has been excluded from this version of the text. You can view it online here:
https://pressbooks.lib.vt.edu/introearthscience/?p=139#h5p-10

2.2.3 Plate Tectonic Boundaries

At passive margins the **plates** don't move—the **continental lithosphere** transitions into oceanic lithosphere and forms **plates** made of both types. A **tectonic plate** may be made of both **oceanic** and **continental lithosphere** connected by a **passive margin**. North and South America's eastern coastlines are examples of passive margins. Active margins are places where the **oceanic** and **continental** lithospheric **tectonic plates** meet and move relative to each other, such as the western coasts of North and South America. This movement is caused by frictional drag created between the **plates** and differences in **plate** densities. The majority of mountain-building events, earthquake activity and active **volcanism** on the Earth's surface can be attributed to **tectonic plate** movement at active margins.

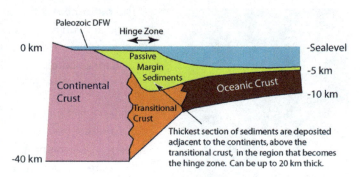

Figure 2.23: Passive margin.

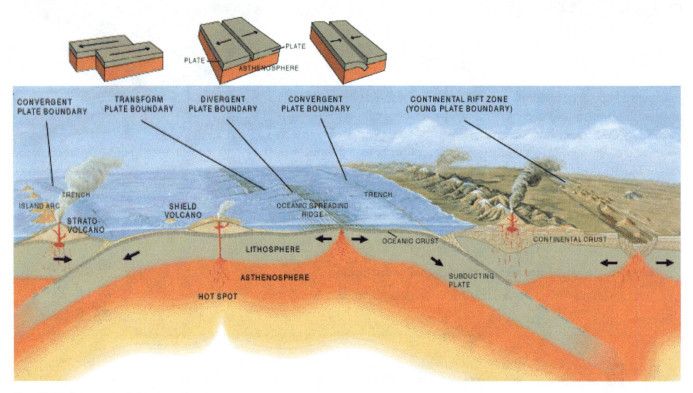

Figure 2.24: Schematic of plate boundary types.

In a simplified model, there are three categories of **tectonic plate** boundaries. **Convergent** boundaries are places where **plates** move toward each other. At **divergent** boundaries, the **plates** move apart. At **transform** boundaries, the **plates** slide past each other.

Take this quiz to check your comprehension of this section.

If you are using an offline version of this text, access the quiz for section 2.2 via the QR code.

 An interactive H5P element has been excluded from this version of the text. You can view it online here:
https://pressbooks.lib.vt.edu/introearthscience/?p=139#h5p-11

2.3 Convergent Boundaries

Convergent boundaries, also called destructive boundaries, are places where two or more **plates** move toward each other. **Convergent** boundary movement is divided into two types, **subduction** and **collision**, depending on the density of the involved **plates**. **Continental lithosphere** is of lower density and thus more buoyant than the underlying **asthenosphere**. **Oceanic lithosphere** is more dense than **continental lithosphere**, and, when old and cold, may even be more dense than **asthenosphere**.

When **plates** of different densities converge, the higher density plate is pushed beneath the more buoyant plate in a process called **subduction**. When **continental plates** converge without **subduction** occurring, this process is called **collision**.

Figure 2.25: Geologic provinces with the Shield (orange) and Platform (pink) comprising the craton, the stable interior of continents.

2.3.1. Subduction

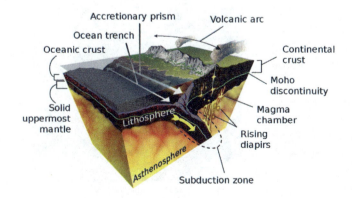

Figure 2.26: Diagram of ocean–continent subduction.

Subduction occurs when a dense **oceanic plate** meets a more buoyant **plate**, like a **continental plate** or warmer/ younger **oceanic plate**, and descends into the **mantle**. The worldwide average rate of **oceanic plate subduction** is 25 miles per million years, about a half-inch per year. As an **oceanic plate** descends, it pulls the **ocean floor** down into a **trench**. These trenches can be more than twice as deep as the average depth of the adjacent ocean **basin**, which is usually three to four km. The Mariana Trench, for example, approaches a staggering 11 km.

Within the **trench**, **ocean floor sediments** are scraped together and compressed between the **subducting** and overriding **plates**. This feature is called the **accretionary wedge**, mélange, or accretionary prism. Fragments of **continental** material, including microcontinents, riding atop the **subducting plate** may become sutured to the **accretionary wedge** and accumulate into a large area of land called a **terrane**. Vast portions of California are comprised of accreted terranes.

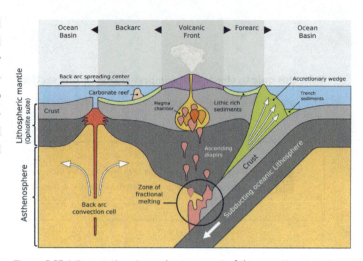

Figure 2.27: Microcontinents can become part of the accretionary prism of a subduction zone.

When the **subducting oceanic plate**, or **slab**, sinks into the **mantle**, the immense heat and pressure pushes volatile materials like water and carbon dioxide into an area below the **continental plate** and above the descending **plate** called the **mantle wedge**. The **volatiles** are released mostly by hydrated **minerals** that revert to non-hydrated **minerals** in these higher **temperature** and pressure conditions. When mixed with asthenospheric material above the plate, the volatile lower the melting point of the **mantle wedge**, and through a process called **flux melting** it becomes liquid **magma**. The molten **magma** is more buoyant than the lithospheric plate above it and migrates to the Earth's surface where it emerges as **volcanism**. The resulting **volcanoes** frequently appear as curved mountain chains, volcanic arcs, due to the curvature of the earth. Both **oceanic** and **continental** plates can contain volcanic arcs.

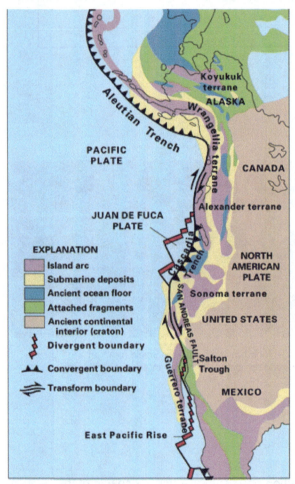

Figure 2.28: Accreted terranes of western North America. Everything that is not the "Ancient continental interior (craton)" has been smeared onto the side of the continent by accretion from subduction.

How **subduction** is initiated is still a matter of scientific debate. It is generally accepted that **subduction** zones start as passive margins, where **oceanic** and **continental plates** come together, and then gravity initiates **subduction** and converts the **passive margin** into an active one. One **hypothesis** is gravity pulls the denser oceanic plate down or the plate can start to flow ductility at a low angle. Scientists seeking to answer this question have collected evidence that suggests a new **subduction** zone is forming off the **coast** of Portugal. Some scientists have proposed large earthquakes like the 1755 Lisbon earthquake may even have something to do with this process of creating a **subduction** zone, although the evidence is not definitive. Another **hypothesis** proposes **subduction** happens at **transform** boundaries involving **plates** of different densities.

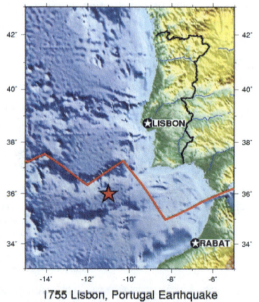

1755 Lisbon, Portugal Earthquake

Figure 2.29: Location of the large (Mw 8.5-9.0) 1755 Lisbon earthquake.

Some plate boundaries look like they should be active, but show no evidence of **subduction**. The **oceanic** lithospheric **plates** on either side of the Atlantic Ocean for example, are denser than the underlying **asthenosphere** and are not **subducting** beneath the **continental plates**. One **hypothesis** is the **bond** holding the **oceanic** and **continental plates** together is stronger than the downwards force created by the difference in **plate** densities.

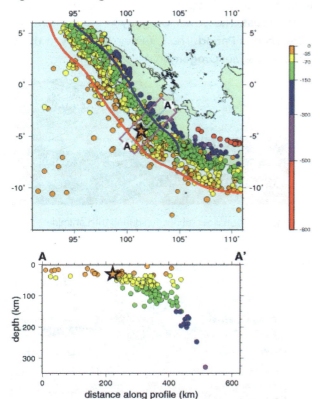

Figure 2.30: Earthquakes along the Sunda megathrust subduction zone, along the island of Sumatra, showing the 2006 Mw 9.1-9.3 Indian Ocean earthquake as a star.

Subduction zones are known for having the largest earthquakes and **tsunamis**; they are the only places with **fault** surfaces large enough to create **magnitude**-9 earthquakes. These **subduction**-zone earthquakes not only are very large, but also are very deep. When a **subducting slab** becomes stuck and cannot descend, a massive amount of energy builds up between the stuck **plates**. If this energy is not gradually dispersed, it may force the **plates** to suddenly release along several hundred kilometers of the **subduction** zone. Because **subduction**-zone **faults** are located on the **ocean floor**, this massive amount of movement can generate giant **tsunamis** such as those that followed the 2004 Indian Ocean Earthquake and 2011 Tōhoku Earthquake in Japan.

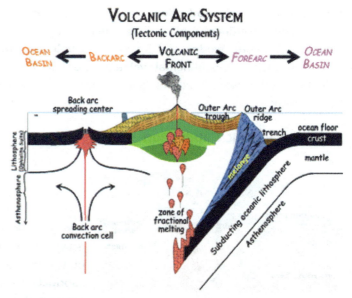

Figure 2.31: Various parts of a subduction zone. This subduction zone is ocean–ocean subduction, though the same features can apply to continent–ocean subduction.

All **subduction** zones have a **forearc basin**, a feature of the overriding plate found between the **volcanic arc** and **oceanic trench**. The **forearc basin** experiences a lot of **faulting** and **deformation** activity, particularly within the **accretionary wedge**.

In some **subduction** zones, **tensional** forces working on the **continental plate** create a backarc **basin** on the interior side of the **volcanic arc**. Some scientists have proposed a **subduction** mechanism called **oceanic slab** rollback creates **extension faults** in the overriding plates. In this model, the descending **oceanic slab** does not slide directly under the overriding plate but instead rolls back, pulling the overlying plate seaward. The **continental plate** behind the **volcanic arc** gets stretched like pizza dough until the surface cracks and collapses to form a backarc **basin**. If the **extension** activity is extensive and deep enough, a backarc **basin** can develop into a **continental rifting** zone. These **continental divergent** boundaries may be less symmetrical than their **mid-ocean ridge** counterparts.

In places where numerous young buoyant **oceanic plates** are converging and **subducting** at a relatively high velocity, they may force the overlying **continental plate** to buckle and crack. This is called **back-arc faulting**. Extensional **back-arc faults** pull rocks and chunks of plates apart. **Compressional back-arc faults**, also known as **thrust faults**, push them together.

The dual spines of the Andes Mountain range include a example of **compressional** thrust **faulting**. The western spine is part of a **volcanic arc**. Thrust faults have deformed the non-volcanic eastern spine, pushing rocks and pieces of **continental plate** on top of each other.

There are two styles of **thrust fault deformation**: **thin-skinned faults** that occur in superficial rocks lying on top of the **continental plate** and **thick-skinned faults** that reach deeper into the **crust**. The Sevier **Orogeny** in the western U.S. is a notable **thin-skinned** type of **deformation** created during the **Cretaceous Period**. The Laramide **Orogeny**, a **thick-skinned** type of **deformation**, occurred near the end of and slightly after the Sevier **Orogeny** in the same region.

Flat-**slab**, or shallow, **subduction** caused the Laramide **Orogeny**. When the descending **slab subducts** at a low angle, there is more contact between the **slab** and the overlying **continental plate** than in a typical **subduction** zone. The shallowly-**subducting slab** pushes against the overriding plate and creates an area of **deformation** on the overriding plate many kilometers away from the **subduction** zone.

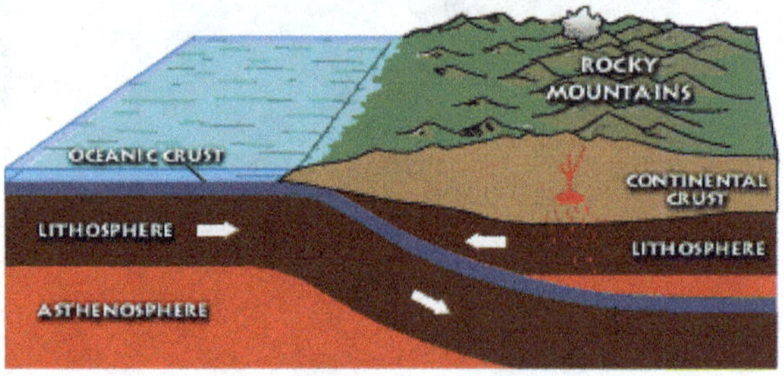

Figure 2.32: Shallow subduction during the Laramide orogeny.

Oceanic–Continental Subduction

Oceanic-continental subduction occurs when an **oceanic plate** dives below a **continental plate**. This **convergent** boundary has a **trench** and **mantle wedge** and frequently, a **volcanic arc**. Well-known examples of **continental volcanic arcs** are the Cascade Mountains in the Pacific Northwest and western Andes Mountains in South America.

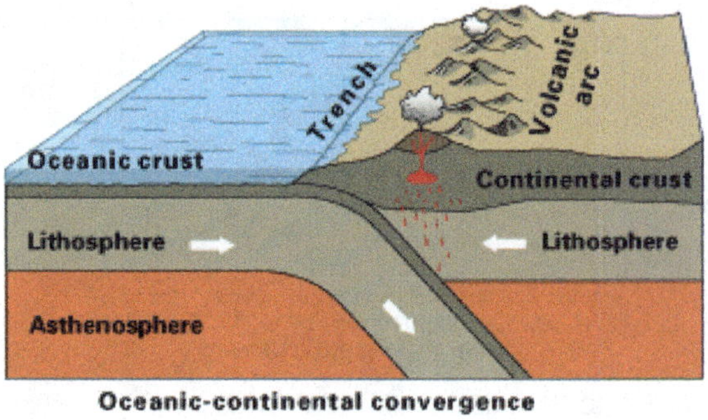

Oceanic–Oceanic Subduction

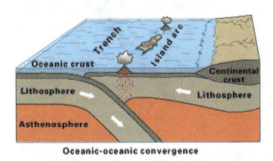

Oceanic-oceanic convergence

Figure 2.34: Subduction of an oceanic plate beneath another oceanic plate, forming a trench and an island arc.

Oceanic-continental convergence

Figure 2.33: Subduction of an oceanic plate beneath a continental plate, forming a trench and volcanic arc.

The boundaries of **oceanic-oceanic subduction** zones show very different activity from those involving **oceanic–continental plates**. Since both **plates** are made of **oceanic lithosphere**, it is usually the older **plate** that **subducts** because it is colder and denser. The **volcanism** on the overlying **oceanic plate** may remain hidden underwater.. If the **volcanoes** rise high enough the reach the ocean surface, the chain of **volcanism** forms an **island arc**. Examples of these island arcs include the Aleutian Islands in the northern Pacific Ocean, Lesser Antilles in the Caribbean Sea, and numerous island chains scattered throughout the western Pacific Ocean.

2.3.2. Collisions

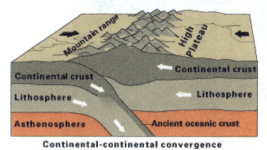

Figure 2.35: Two continental plates colliding.

When **continental plates** converge, during the closing of an ocean **basin** for example, **subduction** is not possible between the equally buoyant **plates**. Instead of one **plate** descending beneath another, the two masses of **continental lithosphere** slam together in a process known as **collision**. Without **subduction**, there is no **magma formation** and no **volcanism**. **Collision** zones are characterized by tall, non-**volcanic** mountains; a broad zone of frequent, large earthquakes; and very little **volcanism**.

When **oceanic crust** connected by a **passive margin** to **continental crust** completely **subducts** beneath a **continent**, an ocean **basin** closes, and **continental collision** begins. Eventually, as ocean basins close, continents join together to form a **massive** accumulation of continents called a **supercontinent**, a process that has taken place in ~500 million year old cycles over earth's history.

The process of **collision** created **Pangea**, the supercontinent envisioned by Wegener as the key component of his **continental** drift **hypothesis**. Geologists now have evidence that **continental plates** have been continuously converging into **supercontinents** and splitting into smaller **basin**-separated continents throughout Earth's existence, calling this process the **supercontinent** cycle, a process that takes place in approximately 500 million years. For example, they estimate **Pangea** began separating 200 million years ago. **Pangea** was preceded by an earlier **supercontinents**, one of which being **Rodinia**, which existed 1.1 billion years ago and started breaking apart 800 million to 600 million years ago.

Figure 2.36: A reconstruction of Pangaea, showing approximate positions of modern continents.

Figure 2.37: The tectonics of the Zagros Mountains. Note the Persian Gulf foreland basin.

A foreland **basin** is a feature that develops near mountain belts, as the combined mass of the mountains forms a depression in the lithospheric **plate**. While foreland basins may occur at **subduction** zones, they are most commonly found at **collision** boundaries. The Persian Gulf is possibly the best modern example, created entirely by the weight of the nearby Zagros Mountains.

If **continental** and **oceanic** **lithosphere** are fused on the same **plate**, it can partially subduct but its buoyancy prevents it from fully descending. In very rare cases, part of a **continental plate** may become trapped beneath a descending **oceanic plate** in a process called **obduction**. When a portion of the **continental crust** is driven down into the **subduction** zone, due to its buoyancy it returns to the surface relatively quickly.

As pieces of the **continental lithosphere** break loose and migrate upward through the **obduction** zone, they bring along bits of the **mantle** and **ocean floor** and amend them on top of the **continental plate**. Rocks composed of this **mantle** and ocean-floor material are called **ophiolites** and they provide valuable information about the **composition** of the **mantle**.

Figure 2.38: Pillow lavas, which only form under water, from an ophiolite in the Apennine Mountains of central Italy.

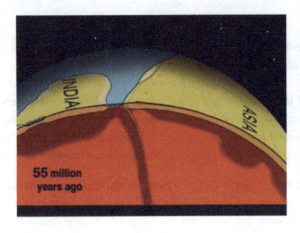 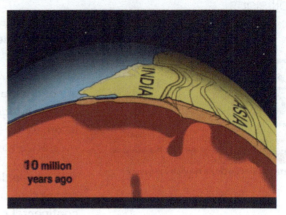

Figure 2.39: India crashing into Asia.

The area of **collision**-zone **deformation** and **seismic** activity usually covers a broader area because **continental lithosphere** is plastic and malleable. Unlike **subduction**-zone earthquakes, which tend to be located along a narrow swath near the **convergent** boundary, **collision**-zone earthquakes may occur hundreds of kilometers from the boundary between the **plates**.

The Eurasian **continent** has many examples of **collision**-zone deformations covering vast areas. The Pyrenees mountains begin in the Iberian Peninsula and cross into France. Also, there are the Alps stretching from Italy to central Europe; the Zagros mountains from Arabia to Iran; and Himalaya mountains from the Indian subcontinent to central Asia.

Take this quiz to check your comprehension of this section.

If you are using an offline version of this text, access the quiz for section 2.3 via the QR code.

 An interactive H5P element has been excluded from this version of the text. You can view it online here:
https://pressbooks.lib.vt.edu/introearthscience/?p=139#h5p-12

2.4 Divergent Boundaries

At **divergent** boundaries, sometimes called constructive boundaries, lithospheric **plates** move away from each other. There are two types of **divergent** boundaries, categorized by where they occur: **continental rift** zones and **mid-ocean ridges**. **Continental rift** zones occur in weak spots in the **continental** lithospheric **plate**. A **mid-ocean ridge** usually originates in a **continental plate** as a **rift** zone that expands to the point of splitting the **plate** apart, with seawater filling in the gap. The separate pieces continue to drift apart and become individual continents. This process is known as rift-to-drift.

2.4.1. Continental Rifting

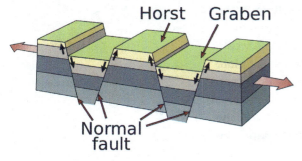

Figure 2.40: Faulting that occurs in divergent boundaries.

In places where the **continental plates** are very thick, they reflect so much heat back into the **mantle** it develops strong **convection** currents that push super-heated **mantle** material up against the overlying **plate**, softening it. **Tensional** forces created by this convective upwelling begin to pull the weakened **plate** apart. As it stretches, it becomes thinner and develops deep cracks called **extension** or normal **faults**. Eventually plate sections located between large faults drop into deep depressions known as **rift** valleys, which often contain keystone-shaped blocks of down-dropped **crust** known as **grabens**. The shoulders of these **grabens** are called **horsts**. If only one side of a section drops, it is called a **half-graben**. Depending on the conditions, **rifts** can grow into very large lakes and even oceans.

While seemingly occurring at random, **rifting** is dictated by two factors. **Rifting** does not occur in continents with older and more stable interiors, known as **cratons**. When **continental rifting** does occur, the break-up pattern resembles the seams of a soccer ball, also called a truncated icosahedron. This is the most common surface-**fracture** pattern to develop on an evenly expanding sphere because it uses the least amount of energy.

Using the soccer ball model, **rifting** tends to lengthen and expand along a particular seam while fizzling out in the other directions. These seams with little or no **tectonic** activity are called failed rift arms. A **failed rift arm** is still a weak spot in the **continental plate**; even without the presence of active **extension faults**, it may develop into a called an **aulacogen**. One example of a **failed rift arm** is the Mississippi Valley Embayment, a depression through which the upper end of the Mississippi River flows. Occasionally connected **rift** arms do develop concurrently, creating multiple boundaries of active **rifting**. In places where the **rift** arms do not fail, for example the Afar Triangle, three **divergent** boundaries can develop near each other forming a **triple junction**.

Figure 2.41: The Afar Triangle (center) has the Red Sea ridge (center to upper left), Gulf of Aden ridge (center to right), and East African Rift (center to lower left) form a triple junction that are about 120° apart.

Figure 2.42: NASA image of the Basin and Range horsts and grabens across central Nevada.

Rifts come in two types: narrow and broad. Narrow **rifts** are characterized by a high density of highly active **divergent** boundaries. The East African Rift Zone, where the **horn** of Africa is pulling away from the mainland, is an excellent example of an active narrow **rift**. Lake Baikal in Russia is another. Broad **rifts** also have numerous **fault** zones, but they are distributed over wide areas of **deformation**. The **Basin and Range** region located in the western United States is a type of broad **rift**. The Wasatch Fault, which also created the Wasatch Mountain Range in the state of Utah, forms the eastern **divergent** boundary of this broad **rift** (animation 1 and animation 2).

Rifts have earthquakes, although not of the **magnitude** and frequency of other boundaries. They may also exhibit **volcanism**. Unlike the flux-melted **magma** found in **subduction** zones, **rift**-zone **magma** is created by **decompression melting**. As the **continental plates** are pulled apart, they create a region of low pressure that melts the **lithosphere** and draws it upwards. When this molten **magma** reaches the weakened and **fault**-riddled **rift** zone, it migrates to surface by breaking through the **plate** or escaping via an open **fault**. Examples of young **rift volcanoes** dot the **Basin and Range** region in the United States. **Rift**-zone activity is responsible for generating some unique **volcanism**, such as the Ol Doinyo Lengai in Tanzania. This **volcano** erupts **lava** consisting largely of **carbonatite**, a relatively cold, liquid **carbonate mineral**.

2.4.2. Mid-ocean Ridges

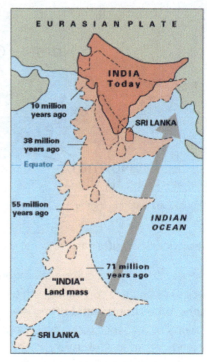

Figure 2.43: India colliding into Eurasia to create the modern day Himalayas.

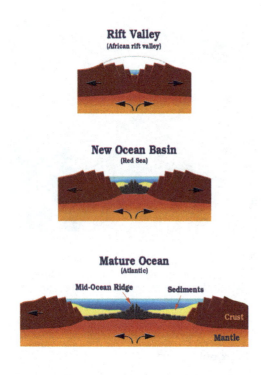

Figure 2.44: Progression from rift to mid-ocean ridge.

As **rifting** and **volcanic** activity progress, the **continental lithosphere** becomes more **mafic** (see chapter 4) and thinner, with the eventual result transforming the **plate** under the **rifting** area into **oceanic lithosphere**. This is the process that gives birth to a new ocean, much like the narrow Red Sea emerged with the movement of Arabia away from Africa. As the **oceanic lithosphere** continues to diverge, a **mid-ocean ridge** is formed.

Mid-ocean ridges, also known as **spreading centers**, have several distinctive features. They are the only places on earth that create new **oceanic lithosphere**. **Decompression melting** in the **rift** zone changes **asthenosphere** material into new **lithosphere**, which oozes up through cracks in **oceanic plate**. The amount of new **lithosphere** being created at **mid-ocean ridges** is highly significant. These undersea **rift volcanoes** produce more **lava** than all other types of **volcanism** combined. Despite this, most mid-oceanic ridge **volcanism** remains unmapped because the volcanoes are located deep on the **ocean floor**.

In rare cases, such as a few locations in Iceland, **rift** zones display the type of **volcanism**, spreading, and ridge **formation** found on the **ocean floor**.

The ridge feature is created by the accumulation of hot **lithosphere** material, which is lighter than the dense underlying **asthenosphere**. This chunk of isostatically buoyant **lithosphere** sits partially submerged and partially exposed on the **asthenosphere**, like an ice cube floating in a glass of water.

As the ridge continues to spread, the **lithosphere** material is pulled away from the area of **volcanism** and becomes colder and denser. As it continues to spread and cool, the **lithosphere** settles into wide swathes of relatively featureless topography called **abyssal** plains with lower topography.

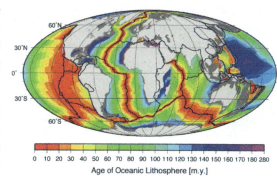

Figure 2.45: Age of oceanic lithosphere, in millions of years. Notice the differences in the Atlantic Ocean along the coasts of the continents.

This model of ridge **formation** suggests the sections of **lithosphere** furthest away from the **mid-ocean ridges** will be the oldest. Scientists have tested this idea by comparing the age of rocks located in various locations on the **ocean floor**. Rocks found near ridges are younger than those found far away from any ridges. **Sediment** accumulation patterns also confirm the idea of sea-floor spreading. **Sediment** layers tend to be thinner near **mid-ocean ridges**, indicating it has had less time to build up.

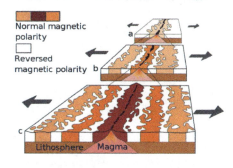

Figure 2.46: A time progression (with "a" being youngest and "c" being oldest) showing a spreading center getting wider while recording changes in the magnetic field of the Earth.

As mentioned in the section on **paleomagnetism** and the development of **plate tectonic theory**, scientists noticed **mid-ocean ridges** contained unique magnetic **anomalies** that show up as symmetrical striping on both sides of the ridge. The Vine-Matthews-Morley hypothesis proposes these alternating reversals are created by the earth's magnetic field being imprinted into **magma** after it emerges from the ridge. Very hot **magma** has no magnetic field. As the **oceanic plates** get pulled apart, the **magma** cools below the Curie point, the **temperature** below which a magnetic field gets locked into magnetic **minerals**. The alternating magnetic reversals in the rocks reflects the periodic swapping of earth's magnetic north and south poles. This paleomagnetic pattern provides a great historical record of ocean-floor movement, and is used to reconstruct past **tectonic** activity and determine rates of ridge spreading.

 One or more interactive elements has been excluded from this version of the text. You can view them online here: https://www.youtube.com/watch?v=6o1HawAOTEI

Video 2.2: Pangea breakup and formation of the northern Atlantic Ocean.

If you are using an offline version of this text, access this YouTube video via the QR code.

Thanks to their distinctive geology, **mid-ocean ridges** are home to some of the most unique ecosystems ever discovered. The ridges are often studded with **hydrothermal** vents, deep fissures that allow seawater to circulate through the upper portions of the **oceanic plate** and interact with hot rock. The super-heated seawater rises back up to the surface of the **plate**, carrying **dissolved** gasses and **minerals**, and small particulates. The resulting emitted **hydrothermal** water looks like black underwater smoke.

Scientists had known about these geothermal areas on the **ocean floor** for some time. However, it was not until 1977, when scientists piloting a deep sub-mergence vehicle, the Alvin, discovered a thriving community of organisms clustered around these **hydrothermal** vents. These unique organisms, which include 10-foot-long tube worms taller than people, live in the complete dark-ness of the **ocean floor** deprived of oxygen and sunlight. They use geothermal energy provided by the vents and a process called bacterial **chemosynthesis** to feed on sulfur compounds. Before this discovery, scientists believed life on earth could not exist without photosynthesis, a process that requires sunlight. Some scientists suggest this type of environment could have been the origin of life on Earth, and perhaps even extraterrestrial life elsewhere in the galaxy, such as on Jupiter's moon Europa.

Figure 2.47: Black smoker hydrothermal vent with a colony of giant (6'+) tube worms.

Take this quiz to check your comprehension of this section.

If you are using an offline version of this text, access the quiz for section 2.4 via the QR code.

An interactive H5P element has been excluded from this version of the text. You can view it online here: https://pressbooks.lib.vt.edu/introearthscience/?p=139#h5p-13

2.5 Transform Boundaries

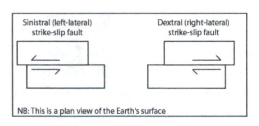

Figure 2.48: The two types of transform/strike-slip faults.

A **transform** boundary, sometimes called a **strike-slip** or conservative boundary, is where the lithospheric plates slide past each other in the hori-zontal plane. This movement is described based on the perspective of an observer standing on one of the plates, looking across the boundary at the opposing plate. **Dextral**, also known as **right-lateral**, movement describes the opposing plate moving to the right. **Sinistral**, also known as **left lateral**, movement describe the opposing plate moving to the left.

Most **transform** boundaries are found on the **ocean floor**, around **mid-ocean ridges**. These boundaries form **aseismic fracture** zones, filled with earthquake-free **transform faults**, to accommodate different rates of spreading occurring at the ridge.

Some **transform** boundaries produce significant **seismic** activity, primarily as earth-quakes, with very little mountain-building or **volcanism**. This type of **transform** boundary may contain a single **fault** or series of **faults**, which develop in places where **plate tectonic stresses** are transferred to the surface. As with other types of active boundaries, if the **plates** are unable to **shear** past each other the **tectonic** forces will continue to build up. If the built up energy between the **plates** is suddenly released, the result is an earth-quake.

In the eyes of humanity, the most significant **transform faults** occur within **continental plates**, and have a **shearing** motion that frequently produces moderate-to-large **magnitude** earthquakes. Notable examples include the San Andreas Fault in California, Northern and Eastern Anatolian Faults in Turkey, Altyn Tagh Fault in central Asia, and Alpine Fault in New Zealand.

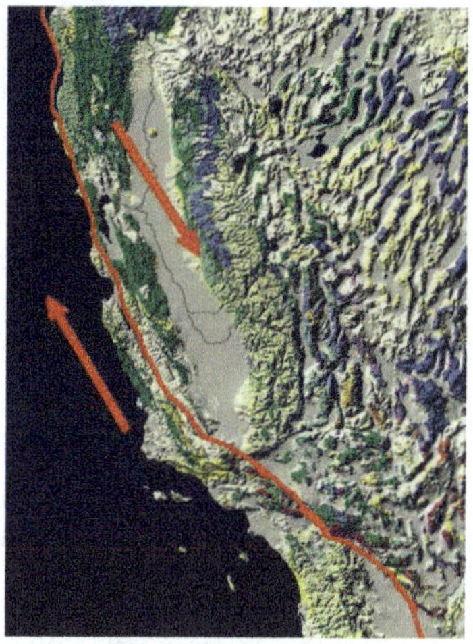

Figure 2.49: Map of the San Andreas fault, showing relative motion.

2.5.1 Transpression and Transtension

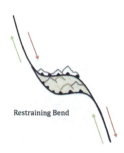

Restraining Bend

Figure 2.50: A transpressional strike-slip fault, causing uplift called a restraining bend.

Bends along **transform faults** may create **compressional** or **extensional** forces that cause secondary **faulting** zones. **Transpression** occurs where there is a component of **compression** in addition to the **shearing** motion. These forces build up around the area of the bend, where the opposing plates are restricted from sliding past each other. As the forces continue to build up, they create mountains in the restraining bend around the **fault**. The Big Bend area, located in the southern part of the San Andreas Fault includes a large area of **transpression** where many mountains have been built, moved, and even rotated.

Transtension zones require a **fault** that includes a releasing bend, where the plates are being pulled apart by **extensional** forces. Depressions and sometimes **volcanism** develop in the releasing bend, along the fault. The Dead Sea found between Israel and Jordan, and the Salton Sea of California are examples of basins formed by transtensional forces.

Releasing Bend

2.5.2 Piercing Points

Figure 2.52: Wallace (dry) Creek on the Cariso Plain, California. Note as the creek flows from the northern mountainous part of the image, it takes a sharp right (as viewed from the flow of water), then a sharp left. This is caused by the San Andreas Fault cutting roughly perpendicular to the creek, and shifting the location of the creek over time. The fault can be seen about halfway down, trending left to right, as a change in the topography.

When a geological feature is cut by a **fault**, it is called a **piercing point**. Piercing points are very useful for recreating past fault movement, especially along **transform** boundaries. **Transform** faults are unique because their horizontal motion keeps a geological feature relatively intact, preserving the record of what happened. Other types of faults—normal and reverse—tend to be more destructive, obscuring or destroying these features. The best type of **piercing point** includes unique patterns that are used to match the parts of a geological feature separated by fault movement. Detailed studies of piercing points show the San Andreas Fault has experienced over 225 km of movement in the last 20 million years, and this movement occurred at three different fault traces.

Figure 2.51: A transtensional strike-slip fault, causing a restraining bend. In the center of the fault, a depression with extension would be found.

Video 2.3: Video of the origin of the San Andreas fault. As the mid-ocean ridge subducts, the relative motion between the remaining plates become transform, forming the fault system. Note that because the motion of the plates is not exactly parallel to the fault, it causes divergent motion in the interior of North America.

If you are using an offline version of this text, access this YouTube video via the QR code.

Take this quiz to check your comprehension of this section.

If you are using an offline version of this text, access the quiz for section 2.5 via the QR code.

2.6 The Wilson Cycle

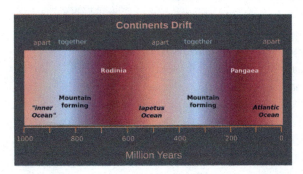

Figure 2.53: Diagram of the Wilson Cycle, showing rifting and collision phases.

The **Wilson Cycle** is named for J. Tuzo Wilson who first described it in 1966, and it outlines the ongoing origin and breakup of **supercontinents**, such as **Pangea** and **Rodinia**. Scientists have determined this cycle has been operating for at least three billion years and possibly earlier.

There are a number of **hypotheses** about how the **Wilson Cycle** works. One mechanism proposes that **rifting** happens because **continental plates** reflect the heat much better than **oceanic plates**. When continents congregate together, they reflect more of the Earth's heat back into the **mantle**, generating more vigorous **convection** currents that then start the **continental rifting** process. Some geologists believe **mantle** plumes are remnants of these **periods** of increased **mantle temperature** and **convection** upwelling, and study them for clues about the origin of **continental rifting**.

The mechanism behind how **supercontinents** are created is still largely a mystery. There are three schools of thought about what continues to drive the continents further apart and eventually bring them together. The ridge-push **hypothesis** suggests after the initial **rifting** event, **plates** continue to be pushed apart by mid-ocean **spreading centers** and their underlying **convection** currents. Slab-pull proposes the **plates** are pulled apart by descending slabs in the **subduction** zones of the **oceanic–continental** margins. A third idea, gravitational sliding, attributes the movement to gravitational forces pulling the lithospheric **plates** down from the elevated **mid-ocean ridges** and across the underlying **asthenosphere**. Current evidence seems to support **slab** pull more than ridge push or gravitational sliding.

2.7 Hotspots

The **Wilson Cycle** provides a broad overview of **tectonic plate** movement. To analyze **plate** movement more precisely, scientists study **hotspots**. First postulated by J. Tuzo Wilson in 1963, a **hotspot** is an area in the lithospheric **plate** where molten **magma** breaks through and creates a **volcanic** center, islands in the ocean and mountains on land. As the **plate** moves across the **hotspot**, the **volcano** center becomes **extinct** because it is no longer over an active **magma** source. Instead, the **magma** emerges through another area in the **plate** to create a new active **volcano**. Over time, the combination of moving **plate** and stationary **hotspot** creates a chain of islands or mountains. The classic definition of **hotspots** states they do not move, although recent evidence suggests that there may be exceptions.

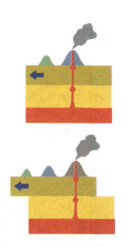

Figure 2.54: Diagram showing a non-moving source of magma (mantle plume) and a moving overriding plate.

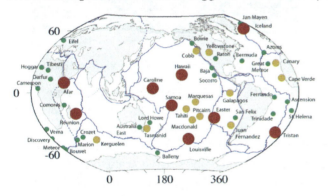

Figure 2.55: Map of world hotspots. Larger circles indicate more active hotspots.

Hotspots are the only types of **volcanism** not associated with **subduction** or **rifting** zones at **plate** boundaries; they seem totally disconnected from any **plate tectonics** processes, such as earthquakes. However, there are relationships between **hotspots** and **plate tectonics**. There are several **hotspots**, current and former, that are believed to have begun at the time of **rifting**. Also, scientists use the age of **volcanic** eruptions and shape of the chain to quantify the rate and direction of **plate** movement relative to the **hotspot**.

Scientists are divided over how **magma** is generated in **hotspots**. Some suggest that **hotspots** originate from super-heated material from as deep as the **core** that reaches the Earth's **crust** as a **mantle plume**. Others argue the molten material that feeds **hotspots** is sourced from the **mantle**. Of course, it is difficult to collect data from these deep-Earth features due to the extremely high pressure and **temperature**.

How **hotspots** are initiated is another highly debated subject. The prevailing mechanism has **hotspots** starting in **divergent** boundaries during **supercontinent rifting**. Scientists have identified a number of current and past **hotspots** believed to have begun this way. **Subducting** slabs have also been named as causing **mantle** plumes and hot-spot **volcanism**. Some geologists have suggested another geological process not involving **plate tectonics** may be involved, such as a large space objects crashing into Earth. Regardless of how they are formed, dozens are on the Earth. Some well-known examples include the Tahiti Islands, Afar Triangle, Easter Island, Iceland, Galapagos Islands, and Samoan Islands. The United States is home to two of the largest and best-studied **hotspots**: Hawaii and Yellowstone.

2.7.1 Hawaiian Hotspot

The active **volcanoes** in Hawaii represent one of the most active **hotspot** sites on earth. Scientific evidence indicates the Hawaiian **hotspot** is at least 80 million years old. Geologists believe it is actually much older; however any rocks with proof of this have been **subducted** under the **ocean floor**. The big island of Hawaii sits atop a large **mantle plume** that marks the active **hotspot**. The Kilauea **volcano** is the main **vent** for this **hotspot** and has been actively erupting since 1983.

This enormous **volcanic** island chain, much of which is underwater, stretches across the Pacific for almost 6,000 km. The **seamount** chain's most striking feature is a sharp 60-degree bend located at the midpoint, which marks a significant change in **plate** movement direction that occurred 50 million years ago. The change in direction has been more often linked to a **plate** reconfiguration, but also to other things like plume migration.

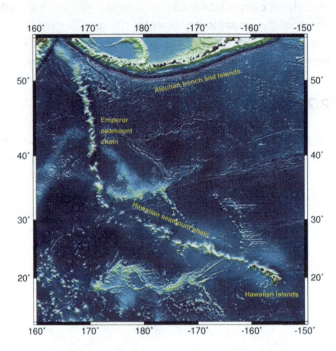

Figure 2.56: The Hawaii–Emperor seamount and island chain.

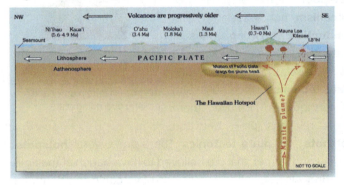

Figure 2.57: Diagram of the Hawaiian hotspot and islands that it formed.

In an attempt to map the Hawaiian **mantle plume** as far down as the lower **mantle**, scientists have used **tomography**, a type of three-dimensional **seismic** imaging. This information—along with other evidence gathered from rock ages, vegetation types, and island size—indicate the oldest islands in the chain are located the furthest away from the active **hotspot**.

2.7.2 Yellowstone Hotspot

Like the Hawaiian version, the Yellowstone **hotspot** is formed by **magma** rising through the **lithosphere**. What makes it different is this **hotspot** is located under a thick, **continental plate**. Hawaii sits on a thin **oceanic plate**, which is easily breached by **magma** coming to the surface. At Yellowstone, the thick **continental plate** presents a much more difficult barrier for **magma** to penetrate. When it does emerge, the eruptions are generally much more violent. Thankfully they are also less frequent.

Over 15 million years of eruptions by this **hotspot** have carved a curved path across the western United States. It has been suggested the Yellowstone **hotspot** is connected to the much older Columbia River **flood basalts** and even to 70 million-year-old **volcanism** found in the Yukon region of Canada.

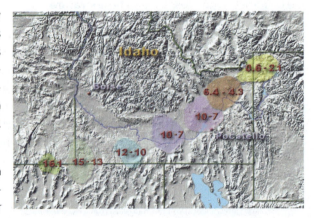

Figure 2.58: The track of the Yellowstone hotspot, which shows the age of different eruptions in millions of years ago.

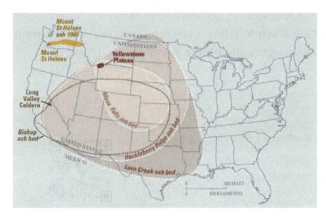

Figure 2.59: Several prominent ash beds found in North America, including three Yellowstone eruptions shaded pink (Mesa Falls, Huckleberry Ridge, and Lava Creek), the Bisho Tuff ash bed (brown dashed line), and the modern May 18th, 1980 ash fall (yellow).

The most recent major eruption of this **hotspot** created the Yellowstone Caldera and Lava Creek **tuff formation** approximately 631,000 years ago. The eruption threw 1,000 cubic kilometers of **ash** and **magma** into the **atmosphere**, some of which was found as far away as Mississippi. Should the **hotspot** erupt again, scientists predict it will be another massive event. This would be a calamity reaching far beyond the western United States. These super **volcanic** eruptions fill the earth's **atmosphere** with so much gas and **ash**, they block sunlight from reaching the earth. Not only would this drastically alter climates and environments around the globe, it could affect worldwide food production.

Take this quiz to check your comprehension of this section.

If you are using an offline version of this text, access the quiz for sections 2.6 and 2.7 via the QR code.

 An interactive H5P element has been excluded from this version of the text. You can view it online here:
https://pressbooks.lib.vt.edu/introearthscience/?p=139#h5p-15

Summary

 One or more interactive elements has been excluded from this version of the text. You can view them online here:
https://www.youtube.com/watch?v=6wJBOk9xjto&t=2s

Video 2.4: Plate tectonics.

If you are using an offline version of this text, access this YouTube video via the QR code.

Plate tectonics is a unifying **theory**; it explains nearly all of the major geologic processes on Earth. Since its early inception in the 1950s and 1960s, geologists have been guided by this revolutionary perception of the world. The **theory** of **plate tectonics** states the surface layer of the Earth is broken into a network of solid, relatively **brittle plates**. Underneath the **plates**

is a much hotter and more **ductile** layer that contains zones of convective upwelling generated by the interior heat of Earth. These **convection** currents move the surface **plates** around—bringing them together, pulling them apart, and **shearing** them side-by-side. Earthquakes and **volcanoes** form at the boundaries where the **plates** interact, with the exception of **volcanic hotspots**, which are not caused by **plate** movement.

Take this quiz to check your comprehension of this chapter.

If you are using an offline version of this text, access the quiz for chapter 2 via the QR code.

 An interactive H5P element has been excluded from this version of the text. You can view it online here:
https://pressbooks.lib.vt.edu/introearthscience/?p=139#h5p-16

URLs Linked Within This Chapter

Audio pronunciation: https://www.merriam-webster.com/dictionary/Mohorovicic%20discontinuity

Animation 1: Basin and Range Structures. How do they form? [Video: 0:52] https://youtu.be/TvvWqAdNV84

Animation 2: Basin & Range: Extension, Erosion, Sedimentation. [Video: 0:31] https://www.youtube.com/watch?v=7DxcAMmNeZk

Text References

1. Aitta, A., 2006, Iron melting curve with a tricritical point: J. Stat. Mech., v. 2006, no. 12, p. P12015.
2. Alfe, D., Gillan, M.J., and Price, G.D., 2002, Composition and temperature of the Earth's core constrained by combining ab initio calculations and seismic data: Earth Planet. Sci. Lett., v. 195, no. 1, p. 91–98.
3. Atwater, T., 1970, Implications of Plate Tectonics for the Cenozoic Tectonic Evolution of Western North America: Geol. Soc. Am. Bull., v. 81, no. 12, p. 3513–3536., doi: 10.1130/0016-7606(1970)81[3513:IOPTFT]2.0.CO;2.
4. Bacon, F., and Montagu, B., 1848, The Works of Francis Bacon, Lord Chancellor of England: With a Life of the Author: The Works of Francis Bacon, Lord Chancellor of England: With a Life of the Author, Parry & McMillan, The Works of Francis Bacon, Lord Chancellor of England: With a Life of the Author.
5. Benioff, H., 1949, Seismic evidence for the fault origin of oceanic deeps: Geological Society of America Bulletin, v. 60, no. 12, p. 1837–1856., doi: 10.1130/0016-7606(1949)60[1837:SEFTFO]2.0.CO;2.
6. Birch, F., 1952, Elasticity and constitution of the Earth's interior: J. Geophys. Res., v. 57, no. 2, p. 227–286., doi: 10.1029/JZ057i002p00227.
7. Birch, F., 1964, Density and composition of mantle and core: J. Geophys. Res., v. 69, no. 20, p. 4377–4388.
8. Bott, M.H.P., 1993, Modelling the plate-driving mechanism: Journal of the Geological Society, v. 150, no. 5, p. 941–951., doi: 10.1144/gsjgs.150.5.0941.
9. Coats, R.R., 1962, Magma type and crustal structure in the Aleutian Arc, *in* The Crust of the Pacific Basin: American Geophysical Union, p. 92–109., doi: 10.1029/GM006p0092.
10. Conrad, C.P., and Lithgow-Bertelloni, C., 2002, How mantle slabs drive plate tectonics: Science (New York, N.Y.), v. 298, no. 5591, p. 207–209., doi: 10.1126/science.1074161.
11. Corliss, J.B., Dymond, J.G., Gordon, L.I., Edmond, J.M., von Heezen, R.P., Ballard, R.D., Green, K., Williams, D.L., Bainbridge, A., Crane, K., and van Andel, T.H., 1979, Submarine thermal springs on the Galapagos Rift: Science, v. 203, p.

107321083.

12. Davis, E.E., and Lister, C.R.B., 1974, Fundamentals of ridge crest topography: Earth Planet. Sci. Lett., v. 21, no. 4, p. 405–413.

13. Dawson, J.B., Pinkerton, H., Norton, G.E., and Pyle, D.M., 1990, Physicochemical properties of alkali carbonatite lavas: Data from the 1988 eruption of Oldoinyo Lengai, Tanzania: Geology, v. 18, no. 3, p. 260–263.

14. Drake, E.T., 1976, Alfred Wegener's reconstruction of Pangea: Geology, v. 4, no. 1, p. 41–44., DOI: 10.1130/0091-7613(1976)4<41:AWROP>2.0.CO;2

15. Engdahl, E.R., Flynn, E.A., and Masse, R.P., 1974, Differential PkiKP travel times and the radius of the core: Geophysical J Royal Astro Soc, v. 40, p. 457–463.

16. Ewing, M., Ewing, J.I., and Talwani, M., 1964, Sediment distribution in the oceans: The Mid-Atlantic Ridge: Geol. Soc. Am. Bull., v. 75, no. 1, p. 17–36., doi: 10.1130/0016-7606(1964)75[17:SDITOT]2.0.CO;2.

17. Ewing, M., Houtz, R., and Ewing, J., 1969, South Pacific sediment distribution: J. Geophys. Res., v. 74, no. 10, p. 2477–2493., doi: 10.1029/JB074i010p02477.

18. Fernandez, L.M., and Careaga, J., 1968, The thickness of the crust in central United States and La Paz, Bolivia, from the spectrum of longitudinal seismic waves: Bull. Seismol. Soc. Am., v. 58, no. 2, p. 711–741.

19. Fluegel, von H.W., 1980, Wegener-Ampferer-Schwinner. Ein Beitrag zur Geschichte der Geologie in Österreich: Mitt. Oesterr. Geol. Ges., v. 73, p. 237–254.

20. Forsyth, D.W., 1975, The Early Structural Evolution and Anisotropy of the Oceanic Upper Mantle: Geophys. J. Int., v. 43, no. 1, p. 103–162., doi: 10.1111/j.1365-246X.1975.tb00630.x.

21. Frankel, H., 1982, The Development, Reception, and Acceptance of the Vine-Matthews-Morley Hypothesis: Hist. Stud. Phys. Biol. Sci., v. 13, no. 1, p. 1–39.

22. Fukao, Y., and Obayashi, M., 2013, Subducted slabs stagnant above, penetrating through, and trapped below the 660 km discontinuity: J. Geophys. Res. [Solid Earth], v. 118, no. 11, p. 2013JB010466.

23. Hagstrum, J.T., 2005, Antipodal hotspots and bipolar catastrophes: Were oceanic large-body impacts the cause? Earth Planet. Sci. Lett., v. 236, no. 1–2, p. 13–27.

24. Hanks, T.C., and Anderson, D.L., 1969, The early thermal history of the earth: Phys. Earth Planet. Inter., v. 2, no. 1, p. 19–29.

25. Heezen, B.C., and Tharp, M., 1965, Tectonic Fabric of the Atlantic and Indian Oceans and Continental Drift: Philosophical Transactions of the Royal Society of London A: Mathematical, Physical and Engineering Sciences, v. 258, no. 1088, p. 90–106., doi: 10.1098/rsta.1965.0024.

26. Heller, P.L., Bowdler, S.S., Chambers, H.P., Coogan, J.C., Hagen, E.S., Shuster, M.W., Winslow, N.S., and Lawton, T.F., 1986, Time of initial thrusting in the Sevier orogenic belt, Idaho-Wyoming and Utah: Geology, v. 14, no. 5, p. 388–391.

27. Herak, D., and Herak, M., 2007, Andrija Mohorovičić (1857-1936)—On the occasion of the 150th anniversary of his birth: Seismol. Res. Lett., v. 78, no. 6, p. 671–674.

28. Hess, H.H., 1962, History of ocean basins: Petrologic studies, v. 4, p. 599–620.

29. Hutson, P., Middleton, J., and Miller, D., 2003, Collision Zones: Online, http://www.geosci.usyd.edu.au/users/prey/ACSGT/EReports/eR.2003/GroupD/Report1/web%20pages/contents.html, accessed June 2017.

30. Isacks, B., Oliver, J., and Sykes, L.R., 1968, Seismology and the new global tectonics: J. Geophys. Res., v. 73, no. 18, p. 5855–5899.

31. Ito, E., and Takahashi, E., 1989, Postspinel transformations in the system Mg2SiO4-Fe2SiO4 and some geophysical implications: J. Geophys. Res. [Solid Earth], v. 94, no. B8, p. 10637–10646.

32. Jacoby, W.R., 1981, Modern concepts of Earth dynamics anticipated by Alfred Wegener in 1912: Geology, v. 9, no. 1, p. 25–27., doi: <a href="https://doi.org/10.1130/0091-7613(1981)92.0.CO;2">10.1130/0091-7613(1981)9<25:MCOEDA>2.0.CO;2.

33. Jakosky, B.M., Grebowsky, J.M., Luhmann, J.G., Connerney, J., Eparvier, F., Ergun, R., Halekas, J., Larson, D., Mahaffy, P., McFadden, J., Mitchell, D.F., Schneider, N., Zurek, R., Bougher, S., and others, 2015, MAVEN observations of the response of Mars to an interplanetary coronal mass ejection: Science, v. 350, no. 6261, p. aad0210.

34. James, D.E., Fouch, M.J., Carlson, R.W., and Roth, J.B., 2011, Slab fragmentation, edge flow and the origin of the Yellowstone hotspot track: Earth Planet. Sci. Lett., v. 311, no. 1–2, p. 124–135.

35. Ji, Y., and Nataf, H.-C., 1998, Detection of mantle plumes in the lower mantle by diffraction tomography: Hawaii: Earth

Planet. Sci. Lett., v. 159, no. 3–4, p. 99–115.

36. Johnston, S.T., Jane Wynne, P., Francis, D., Hart, C.J.R., Enkin, R.J., and Engebretson, D.C., 1996, Yellowstone in Yukon: The Late Cretaceous Carmacks Group: Geology, v. 24, no. 11, p. 997–1000.

37. Kearey, P., Klepeis, K.A., and Vine, F.J., 2009, Global Tectonics: Oxford; Chichester, West Sussex; Hoboken, NJ, Wiley-Blackwell, 496 p.

38. Le Pichon, X., 1968, Sea-floor spreading and continental drift: J. Geophys. Res., v. 73, no. 12, p. 3661–3697.

39. Lehmann, I., 1936, P', Publ: Bur. Centr. Seism. Internat. Serie A, v. 14, p. 87–115.

40. Mantovani, R., 1889, Les fractures de l'écorce terrestre et la théorie de Laplace: Bull. Soc. Sc. et Arts Réunion, p. 41–53.

41. Mason, R.G., 1958, A magnetic survey off the west coast of the United-States between latitudes 32-degrees-N and 36-degrees-N longitudes 121-degrees-W and 128-degrees-W: Geophysical Journal of the Royal Astronomical Society, v. 1, no. 4, p. 320.

42. Mason, R.G., and Raff, A.D., 1961, Magnetic Survey Off the West Coast of North America, 32° N. Latitude to 42° N. Latitude: Geological Society of America Bulletin, v. 72, no. 8, p. 1259–1265., doi: 10.1130/0016-7606(1961)72[1259:MSOTWC]2.0.CO;2.

43. McCollom, T.M., 1999, Methanogenesis as a potential source of chemical energy for primary biomass production by autotrophic organisms in hydrothermal systems on Europa: J. Geophys. Res., v. 104, no. E12, p. 30729–30742., doi: 10.1029/1999JE001126.

44. McKenzie, D.P., and Parker, R.L., 1967, The North Pacific: an Example of Tectonics on a Sphere: Nature, v. 216, p. 1276–1280., doi: 10.1038/2161276a0.

45. Miller, A.R., Densmore, C.D., Degens, E.T., Hathaway, J.C., Manheim, F.T., McFarlin, P.F., Pocklington, R., and Jokela, A., 1966, Hot brines and recent iron deposits in deeps of the Red Sea: Geochimica et Cosmochimica Acta, v. 30, no. 3, p. 341–359., doi: 10.1016/0016-7037(66)90007-X.

46. Morgan, W.J., 1968, Rises, trenches, great faults, and crustal blocks: J. Geophys. Res., v. 73, no. 6, p. 1959–1982., doi: 10.1029/JB073i006p01959.

47. Mueller, S., and Phillips, R.J., 1991, On the initiation of subduction: J. Geophys. Res. [Solid Earth], v. 96, no. B1, p. 651–665.

48. Oldham, R.D., 1906, The constitution of the interior of the Earth, as revealed by earthquakes: Q. J. Geol. Soc. London, v. 62, no. 1–4, p. 456–475.

49. Pasyanos, M.E., 2010, Lithospheric thickness modeled from long-period surface wave dispersion: Tectonophysics, v. 481, no. 1–4, p. 38–50.

50. Powell, R.E., and Weldon, R.J., 1992, Evolution of the San Andreas fault: Annu. Rev. Earth Planet. Sci., v. 20, p. 431.

51. Raff, A.D., and Mason, R.G., 1961, Magnetic Survey Off the West Coast of North America, 40 N. Latitude to 52 N. Latitude: Geological Society of America Bulletin, v. 72, no. 8, p. 1267–1270., doi: 10.1130/0016-7606(1961)72[1267:MSOTWC]2.0.CO;2.

52. Runcorn, S.K., 1965, Palaeomagnetic comparisons between Europe and North America: Philosophical Transactions of the Royal Society of London A: Mathematical, Physical and Engineering Sciences, v. 258, no. 1088, p. 1–11.

53. Saito, T., Ewing, M., and Burckle, L.H., 1966, Tertiary sediment from the mid-atlantic ridge: Science, v. 151, no. 3714, p. 1075–1079., doi: 10.1126/science.151.3714.1075.

54. Satake, K., and Atwater, B.F., 2007, Long-term perspectives on giant earthquakes and tsunamis at subduction zones*: Annu. Rev. Earth Planet. Sci., v. 35, p. 349–374.

55. Scheidegger, A.E., 1953, Examination of the physics of theories of orogenesis: Geol. Soc. Am. Bull., v. 64, no. 2, p. 127–150., doi: 10.1130/0016-7606(1953)64[127:EOTPOT]2.0.CO;2.

56. Simpson, G.G., 1943, Mammals and the nature of continents: Am. J. Sci., v. 241, no. 1, p. 1–31.

57. Starr, A.M., 2015, Ambient resonance of rock arches: Salt Lake City, Utah, University of Utah, 134 p.

58. Stern, R.J., 1998, A subduction primer for instructors of introductory geology courses and authors of introductory-geology textbooks: J. Geosci. Educ., v. 46, p. 221.

59. Stern, R.J., 2004, Subduction initiation: spontaneous and induced: Earth Planet. Sci. Lett., v. 226, no. 3–4, p. 275–292.

60. Stich, D., Mancilla, F. de L., Pondrelli, S., and Morales, J., 2007, Source analysis of the February 12th 2007, Mw 6.0 Horseshoe earthquake: Implications for the 1755 Lisbon earthquake: Geophys. Res. Lett., v. 34, no. 12, p. L12308.

61. Tatsumi, Y., 2005, The subduction factory: how it operates in the evolving Earth: GSA Today, v. 15, no. 7, p. 4.

62. Todo, Y., Kitazato, H., Hashimoto, J., and Gooday, A.J., 2005, Simple foraminifera flourish at the ocean's deepest point: Science, v. 307, no. 5710, p. 689., doi: 10.1126/science.1105407.

63. Tolstoy, I., and Ewing, M., 1949, North Atlantic hydrography and the Mid-Atlantic Ridge: Geol. Soc. Am. Bull., v. 60, no. 10, p. 1527–1540., doi: 10.1130/0016-7606(1949)60[1527:NAHATM]2.0.CO;2.

64. Vine, F.J., and Matthews, D.H., 1963, Magnetic anomalies over oceanic ridges: Nature, v. 199, no. 4897, p. 947–949.

65. Wächtershäuser, G., 1990, Evolution of the first metabolic cycles: Proc. Natl. Acad. Sci. U. S. A., v. 87, no. 1, p. 200–204.

66. Wadati, K., 1935, On the activity of deep-focus earthquakes in the Japan Islands and neighbourhoods: Geophys. Mag., v. 8, no. 3–4, p. 305–325.

67. Waszek, L., Irving, J., and Deuss, A., 2011, Reconciling the hemispherical structure of Earth's inner core with its super-rotation: Nat. Geosci., v. 4, no. 4, p. 264–267, doi: 10.1038/ngeo1083.

68. Wegener, A., 1912, Die Entstehung der Kontinente: Geol. Rundsch., v. 3, no. 4, p. 276–292., doi: 10.1007/BF02202896.

69. Wegener, A., 1920, Die entstehung der kontinente und ozeane: Рипол Классик.

70. Wells, H.G., Huxley, J., and Wells, G.P., 1931, The Science of Life: Philosophy, v. 6, no. 24, p. 506–507.

71. White, I.C., and Moreira, C., 1908, Commissão de estudos das minas de Carvão de Pedra do Brazil.

72. de Wijs, G.A., Kresse, G., Vočadlo, L., Dobson, D., Alfè, D., Gillan, M.J., and Price, G.D., 1998, The viscosity of liquid iron at the physical conditions of the Earth's core: Nature, v. 392, no. 6678, p. 805–807., doi: 10.1038/33905.

73. Wilson, J.T., 1966, Did the Atlantic close and then re-open? Nature.

74. Wilson, M., 1993, Plate-moving mechanisms: constraints and controversies: Journal of the Geological Society, v. 150, no. 5, p. 923–926., doi: 10.1144/gsjgs.150.5.0923.

75. Wyllie, P.J., 1970, Ultramafic rocks and the upper mantle, in Morgan, B.A., editor, Fiftieth anniversary symposia: Mineralogy and petrology of the Upper Mantle; Sulfides; Mineralogy and geochemistry of non-marine evaporites: Washington, DC, Mineralogical Society of America, p. 3–32.

76. Zhou, Z., 2004, The origin and early evolution of birds: discoveries, disputes, and perspectives from fossil evidence: Naturwissenschaften, v. 91, no. 10, p. 455–471.

Figure References

Figure 2.1: Detailed map of all known plates, their boundaries, and movements. Eric Gaba. 2006-10, updated 2015-09. CC BY-SA 2.5. https://commons.wikimedia.org/wiki/File:Tectonic_plates_boundaries_detailed-en.svg

Figure 2.2: Wegener later in his life, ca. 1924-1930. Author unknown. ca. 1924 and 1930. Public domain. https://commons.wikimedia.org/wiki/File:Alfred_Wegener_ca.1924-30.jpg

Figure 2.3: Snider-Pellegrini's map showing the continental fit and separation, 1858. Antonio Snider-Pellegrini. 1858. Public domain. https://commons.wikimedia.org/wiki/File:Antonio_Snider-Pellegrini_Opening_of_the_Atlantic.jpg

Figure 2.4: Map of world elevations. Tahaisik. 2013. CC BY-SA 3.0. https://commons.wikimedia.org/wiki/File:Tahacik.jpg

Figure 2.5: Image showing fossils that connect the continents of Gondwana (the southern continents of Pangea). Osvaldocangaspadilla. 2010. Public domain. https://commons.wikimedia.org/wiki/File:Snider-Pellegrini_Wegener_fossil_map.svg

Figure 2.6: Animation of the basic idea of convection: an uneven heat source in a fluid causes rising material next to the heat and sinking material far from the heat. Oni Lukos. 2006. CC BY-SA 3.0. https://commons.wikimedia.org/wiki/File:Convection.gif

Figure 2.7: GPS measurements of plate motions. NASA. Public domain. https://cddis.nasa.gov/docs/2009/HTS_0910.pdf

Figure 2.8: The complex chemistry around mid-ocean ridges. NOAA. https://oceanexplorer.noaa.gov/explorations/04fire/background/chemistry/media/chemistry_600.html

Figure 2.9: The magnetic field of Earth, simplified as a bar magnet. Zureks. 2012. Public domain. https://commons.wikimedia.org/wiki/File:Earth%27s_magnetic_field,_schematic.png

Figure 2.10: This animation shows how the magnetic poles have moved over 400 years. USGS. 2010. Public domain. https://commons.wikimedia.org/wiki/File:Earth_Magnetic_Field_Declination_from_1590_to_1990.gif

Figure 2.11: The iron in the solidifying rock preserves the current magnetic polarity as new oceanic plates form at mid ocean ridges. USGS. 2011. Public domain. https://commons.wikimedia.org/wiki/File:Mid-ocean_ridge_topography.gif

Figure 2.12: The Wadati-Benioff zone, showing earthquakes following the subducting slab down. USGS. 2013. Public domain. https://commons.wikimedia.org/wiki/File:Benioff_zone_earthquake_focus.jpg

Figure 2.13: J. Tuzo Wilson. Stephen Morris. 1992. CC BY-SA 3.0. https://commons.wikimedia.org/wiki/File:John_Tuzo_Wilson_in_1992.jpg

Figure 2.14: The layers of the Earth. Drlauraguertin. 2015. CC BY-SA 3.0. https://wiki.seg.org/wiki/File:Earthlayers.png#file

Figure 2.15: The global map of the depth of the Moho. AllenMcC. 2013. CC BY-SA 3.0. https://commons.wikimedia.org/wiki/File:Mohomap.png

Figure 2.16: This mantle xenolith containing olivine (green) is chemically weathering by hydrolysis and oxidation into the pseudo-mineral iddingsite, which is a complex of water, clay, and iron oxides. Matt Affolter. 2010. CC BY-SA 3.0. https://commons.wikimedia.org/wiki/File:Iddingsite.JPG

Figure 2.17: A polished fragment of the iron-rich Toluca Meteorite, with octahedral Widmanstätten pattern. H. Raab. 2005. CC BY-SA 3.0. https://commons.wikimedia.org/wiki/File:TolucaMeteorite.jpg

Figure 2.18: Map of the major plates and their motions along boundaries. Scott Nash via USGS. 1996. Public domain. https://commons.wikimedia.org/wiki/File:Plates_tect2_en.svg

Figure 2.19: The lithosphere–asthenosphere boundary changes with certain tectonic situations. Nealey Sims. 2015. CC BY-SA 3.0. https://commons.wikimedia.org/wiki/File:Earth%27s_Inner_Layers_denoting_the_LAB.png

Figure 2.20: General perovskite structure. Perovskite silicates (i.e., Bridgmenite, (Mg,Fe)SiO3) are thought to be the main component of the lower mantle, making it the most common mineral in or on Earth. Cadmium. 2006. Public domain. https://en.wikipedia.org/wiki/File:Perovskite.jpg

Figure 2.21: Lehmann in 1932. Even Neuhaus. 1932. Public domain. https://commons.wikimedia.org/wiki/File:Inge_Lehman.jpg

Figure 2.22: The outer core's spin causes our protective magnetic field. NASA. 2017. Public domain. https://www.nasa.gov/mission_pages/sunearth/news/gallery/earths-dynamiccore.html

Figure 2.23: Passive margin. Joshua Doubek. 2013. CC BY-SA 3.0. https://commons.wikimedia.org/wiki/File:Passive_Continental_Margin.jpg

Figure 2.24: Schematic of plate boundary types. NOAA via USGS. Public domain. https://oceanexplorer.noaa.gov/facts/plate-boundaries.html

Figure 2.25: Geologic provinces with the Shield (orange) and Platform (pink) comprising the craton, the stable interior of continents. USGS. 2005. Public domain. https://commons.wikimedia.org/wiki/File:World_geologic_provinces.jpg

Figure 2.26: Diagram of ocean–continent subduction. K. D. Schroeder. 2016. CC-BY-SA 4.0. https://commons.wikimedia.org/wiki/File:Subduction-en.svg

Figure 2.27: Microcontinents can become part of the accretionary prism of a subduction zone. MagentaGreen. 2014. CC BY-SA 3.0. https://commons.wikimedia.org/wiki/File:Volcanic_Arc_System_SVG_en.svg

Figure 2.28: Accreted terranes of western North America. Modified from illustration provided by Oceanus Magazine; original figure by Jack Cook, Woods Hole Oceanographic Institution; adapted by USGS. Used under fair use.

Figure 2.29: Location of the large (Mw 8.5-9.0) 1755 Lisbon earthquake. USGS. 2014. Public domain. https://commons.wikimedia.org/wiki/File:1755_Lisbon_Earthquake_Location.png

Figure 2.30: Earthquakes along the Sunda megathrust subduction zone, along the island of Sumatra, showing the 2006 Mw 9.1-9.3 Indian Ocean earthquake as a star. USGS. 2007. Public domain. https://commons.wikimedia.org/wiki/File:SundaMegathrustSeismicity.PNG

Figure 2.31: Various parts of a subduction zone. USGS. 2006. Public domain. https://commons.wikimedia.org/wiki/File:Volcanic_Arc_System.png

Figure 2.32: Shallow subduction during the Laramide orogeny. Melanie Moreno. 2006. Public domain. https://commons.wikimedia.org/wiki/File:Shallow_subduction_Laramide_orogeny.png

Figure 2.33: Subduction of an oceanic plate beneath a continental plate, forming a trench and volcanic arc. USGS. 1999. Public domain. https://commons.wikimedia.org/wiki/File:Oceanic-continental_convergence_Fig21oceancont.gif

Figure 2.34: Subduction of an oceanic plate beneath another oceanic plate, forming a trench and an island arc. USGS. 2005. Public domain. https://commons.wikimedia.org/wiki/File:Oceanic-oceanic_convergence_Fig21oceanocean.gif

Figure 2.35: Two continental plates colliding. USGS. 2005. Public domain. https://commons.wikimedia.org/wiki/File:Continental-continental_convergence_Fig21contcont.gif

Figure 2.36: A reconstruction of Pangaea, showing approximate positions of modern continents. Kieff. 2009. CC BY-SA 3.0. https://commons.wikimedia.org/wiki/File:Pangaea_continents.svg

Figure 2.37: The tectonics of the Zagros Mountains. Mikenorton. 2010. CC BY-SA 3.0. https://commons.wikimedia.org/wiki/File:ZagrosFTB.png

Figure 2.38: Pillow lavas, which only form under water, from an ophiolite in the Apennine Mountains of central Italy. Matt Affolter (Qfl247). 2010. CC BY-SA 3.0. https://commons.wikimedia.org/wiki/File:ItalyPillowBasalt.jpg

Figure 2.39: Animation of India crashing into Asia. Raynaldi rji. 2015. CC BY-SA 4.0. https://en.wikipedia.org/wiki/File:India-Eurasia_collision.gif

Figure 2.40: Faulting that occurs in divergent boundaries. USGS; adapted by Gregors. 2011. Public domain. https://commons.wikimedia.org/wiki/File:Fault-Horst-Graben.svg

Figure 2.41: The Afar Triangle (center) has the Red Sea ridge (center to upper left), Gulf of Aden ridge (center to right), and East African Rift (center to lower left) form a triple junction that are about 120° apart. Koba-chan. 2005. CC BY-SA 3.0. https://commons.wikimedia.org/wiki/File:Topographic30deg_N0E30.png

Figure 2.42: NASA image of the Basin and Range horsts and grabens across central Nevada. NASA. 2005. Public domain. https://commons.wikimedia.org/wiki/File:Basin_range_province.jpg

Figure 2.43: India colliding into Eurasia to create the modern day Himalayas. USGS. 2007. Public domain. https://commons.wikimedia.org/wiki/File:Himalaya-formation.gif

Figure 2.44: Progression from rift to mid-ocean ridge. Hannes Grobe, Alfred Wegener, Institute for Polar and Marine Research; adapted by Lichtspiel. 2011. CC BY-SA 2.5. https://commons.wikimedia.org/wiki/File:Ocean-birth.svg

Figure 2.45: Age of oceanic lithosphere, in millions of years. Muller, R.D., M. Sdrolias, C. Gaina, and W.R. Roest (2008) Age, spreading rates and spreading symmetry of the world's ocean crust, Geochem. Geophys. Geosyst., 9, Q04006, doi:10.1029/2007GC001743. CC BY-SA 3.0. https://commons.wikimedia.org/wiki/File:Age_of_oceanic_lithosphere.jpg

Figure 2.46: A time progression (with "a" being youngest and "c" being oldest) showing a spreading center getting wider while recording changes in the magnetic field of the Earth. Chmee2. 2012. Public domain. https://commons.wikimedia.org/wiki/File:Oceanic.Stripe.Magnetic.Anomalies.Scheme.svg

Figure 2.47: Black smoker hydrothermal vent with a colony of giant (6'+) tube worms. NOAA. 2006. Public domain. https://commons.wikimedia.org/wiki/File:Main_Endeavour_black_smoker.jpg

Figure 2.48: The two types of transform/strike-slip faults. Cferrero. 2003. CC BY-SA 3.0. https://commons.wikimedia.org/wiki/File:Strike_slip_fault.png

Figure 2.49: Map of the San Andreas fault, showing relative motion. Kate Barton, David Howell, and Joe Vigil via USGS. 2006. Public domain. https://commons.wikimedia.org/wiki/File:Sanandreas.jpg

Figure 2.50: A transpressional strike-slip fault, causing uplift called a restraining bend. GeoAsh. 2015. CC BY-SA 4.0. https://commons.wikimedia.org/wiki/File:Restraining_Bend.png

Figure 2.51: A transtensional strike-slip fault, causing a restraining bend. In the center of the fault, a depression with extension would be found. K. Martin. 2013. CC BY-SA 3.0. https://commons.wikimedia.org/wiki/File:Releasing_bend.png

Figure 2.52: Wallace (dry) Creek on the Cariso Plain, California. Robert E. Wallace via USGS. 2014. Public domain. https://commons.wikimedia.org/wiki/File:Wallace_Creek_offset_across_the_San_Andreas_Fault.png

Figure 2.53: Diagram of the Wilson Cycle, showing rifting and collision phases. Hannes Grobe. 2007. CC BY-SA 2.5. https://commons.wikimedia.org/wiki/File:Wilson-cycle_hg.png

Figure 2.54: Diagram showing a non-moving source of magma (mantle plume) and a moving overriding plate. Los688. 2008. Public domain. https://commons.wikimedia.org/wiki/File:Hotspot(geology)-1.svg

Figure 2.55: Map of world hotspots. Foulger. 2011. Public domain. https://commons.wikimedia.org/wiki/File:CourtHotspots.png

Figure 2.56: The Hawaii–Emperor seamount and island chain. Ingo Wölbern. 2008. Public domain. https://commons.wikimedia.org/wiki/File:Hawaii-Emperor_engl.png

Figure 2.57: Diagram of the Hawaiian hotspot and islands that it formed. Joel E. Robinson via USGS. 2006. Public domain. https://commons.wikimedia.org/wiki/File:Hawaii_hotspot_cross-sectional_diagram.jpg

Figure 2.58: The track of the Yellowstone hotspot, which shows the age of different eruptions in millions of years ago. Kelvin Case. 2013. CC BY 3.0. https://commons.wikimedia.org/wiki/File:HotspotsSRP_update2013.JPG

Figure 2.59: Several prominent ash beds found in North America, including three Yellowstone eruptions shaded pink (Mesa Falls, Huckleberry Ridge, and Lava Creek), the Bisho Tuff ash bed (brown dashed line), and the modern May 18th, 1980 ash fall (yellow). USGS. 2005. Public domain. https://commons.wikimedia.org/wiki/File:Yellowstone_volcano_-_ash_beds.svg

3.

MINERALS

Learning Objectives

By the end of this chapter, students should be able to:

- Define **mineral**.
- Describe the basic structure of the atom.
- Derive basic atomic information from the Periodic Table of Elements.
- Describe chemical **bonding** related to **minerals**.
- Describe the main ways **minerals** form.
- Describe the **silicon-oxygen tetrahedron** and how it forms common **silicate minerals**.
- List common non-**silicate minerals** in **oxide**, **sulfide**, **sulfate**, and **carbonate** groups.
- Identify **minerals** using physical properties and identification tables.

The term "minerals" as used in nutrition labels and pharmaceutical products is not the same as a **mineral** in a geological sense. In geology, the classic definition of a **mineral** is: 1) naturally occurring, 2) inorganic, 3) solid at room **temperature**, 4) regular crystal structure, and 5) defined chemical **composition**. Some natural substances technically should not be considered **minerals**, but are included by exception. For example, water and mercury are liquid at room **temperature**. Both are considered **minerals** because they were classified before the room-**temperature** rule was accepted as part of the definition. **Calcite** is quite often formed by organic processes, but is considered a **mineral** because it is widely found and geologically important. Because of these discrepancies, the International Mineralogical Association in 1985 amended the definition to: "A **mineral** is an **element** or chemical compound that is normally crystalline and that has been formed as a result of geological processes." This means that the **calcite** in the shell of a clam is not considered a **mineral**. But once that clam shell undergoes burial, **diagenesis**, or other geological processes, then the **calcite** is considered a **mineral**. Typically, substances like **coal**, pearl, opal, or **obsidian** that do not fit the definition of **mineral** are called mineraloids.

A rock is a substance that contains one or more **minerals** or mineraloids. As is discussed in later chapters, there are three types of rocks composed of **minerals**: **igneous** (rocks crystallizing from molten material), sedimentary (rocks **composed** of products of **mechanical weathering** (sand, gravel, etc.) and **chemical weathering** (things **precipitated** from **solution**), and **metamorphic** (rocks produced by alteration of other rocks by heat and pressure.

3.1 Chemistry of Minerals

Rocks are **composed** of **minerals** that have a specific chemical **composition**. To understand **mineral** chemistry, it is essential to examine the fundamental unit of all matter, the atom.

3.1.1 The Atom

One or more interactive elements has been excluded from this version of the text. You can view them online here:
https://www.youtube.com/watch?v=RF-1_JaND68

Video 3.1: Atomic orbitals.

If you are using an offline version of this text, access this YouTube video via the QR code.

Matter is made of atoms. Atoms consists of subatomic particles—protons, neutrons, and electrons. A simple model of the atom has a central nucleus composed of protons, which have positive charges, and neutrons which have no charge. A cloud of negatively charged electrons surrounds the nucleus, the number of electrons equaling the number of protons thus balancing the positive charge of the protons for a neutral atom. Protons and neutrons each have a mass number of 1. The mass of an electron is less than $1/1000^{th}$ that of a proton or neutron, meaning most of the atom's mass is in the nucleus.

3.1.2 Periodic Table of the Elements

Matter is composed of **elements** which are atoms that have a specific number of protons in the nucleus. This number of protons is called the Atomic Number for the **element**. For example, an oxygen atom has 8 protons and an iron atom has 26 protons. An **element** cannot be broken down chemically into a simpler form and retains unique chemical and physical properties. Each **element** behaves in a unique manner in nature. This uniqueness led scientists to develop a periodic table of the **elements**, a tabular arrangement of all known **elements** listed in order of their atomic number.

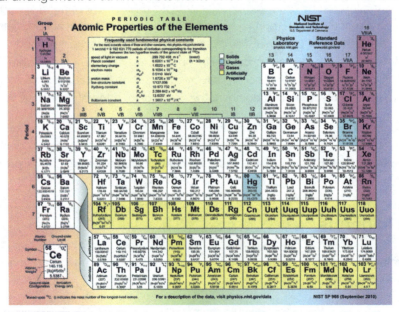

Figure 3.1: The periodic table of the elements.

The first arrangement of **elements** into a periodic table was done by Dmitri Mendeleev in 1869 using the **elements** known at the time. In the periodic table, each **element** has a chemical symbol, name, atomic number, and atomic mass. The chemical symbol is an abbreviation for the **element**, often derived from a Latin or Greek name for the substance. The atomic number is the number of protons in the nucleus. The atomic mass is the number of protons and neutrons in the nucleus, each with a mass number of one. Since the mass of electrons is so much less than the protons and neutrons, the atomic mass is effectively the number of protons plus neutrons.

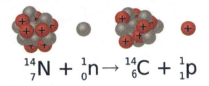

$$^{14}_{7}N + ^{1}_{0}n \rightarrow ^{14}_{6}C + ^{1}_{1}p$$

Figure 3.2: Formation of carbon-14 from nitrogen-14.

The atomic mass of natural **elements** represents an average mass of the atoms comprising that substance in nature and is usually not a whole number as seen on the periodic table, meaning that an **element** exists in nature with atoms having different numbers of neutrons. The differing number of neutrons affects the mass of an **element** in nature and the atomic mass number represents this average. This gives rise to the concept of **isotope**. **Isotopes** are forms of an **element** with the same number of protons but different numbers of neutrons. There are usually several **isotopes** for a particular **element**. For example, 98.9% of carbon atoms have 6 protons and 6 neutrons. This **isotope** of carbon is called carbon-12 (^{12}C). A few carbon atoms, carbon-13 (^{13}C), have 6 protons and 7 neutrons. A trace amount of carbon atoms, carbon-14 (^{14}C), has 6 protons and 8 neutrons.

Among the 118 known **elements**, the heaviest are fleeting human creations known only in high energy particle accelerators, and they decay rapidly. The heaviest naturally occurring **element** is uranium, atomic number 92. The eight most abundant elements in Earth's **continental crust** are shown in Table 1. These **elements** are found in the most common rock forming **minerals**.

Element	Symbol	Abundance %
Oxygen	O	47%
Silicon	Si	28%
Aluminum	Al	8%
Iron	Fe	5%
Calcium	Ca	4%
Sodium	Na	3%
Potassium	K	3%
Magnesium	Mg	2%

Table 3.1: Eight most abundant elements in the Earth's continental crust (% by weight). All other elements are less than 1%. (Source: USGS).

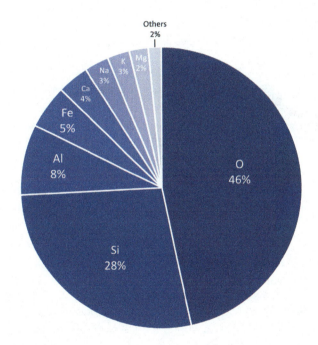

Figure 3.3: Element abundance pie chart for Earth's crust.

3.1.3 Chemical Bonding

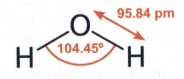

Figure 3.4: A model of a water molecule, showing the bonds between the hydrogen and oxygen.

Most substances on Earth are compounds containing multiple **elements**. Chemical **bonding** describes how these atoms attach with each other to form compounds, such as sodium and chlorine combining to form NaCl, common table salt. Compounds that are held together by chemical **bonds** are called molecules. Water is a compound of hydrogen and oxygen in which two hydrogen atoms are covalently **bonded** with one oxygen making the water molecule. The oxygen we breathe is formed when one oxygen atom covalently **bonds** with another oxygen atom to make the molecule O_2. The subscript 2 in the chemical formula indicates the molecule contains two atoms of oxygen.

Most **minerals** are also compounds of more than one **element**. The common **mineral calcite** has the chemical formula $CaCO_3$ indicating the molecule consists of one calcium, one carbon, and three oxygen atoms. In **calcite**, one carbon and three oxygen atoms are held together by covalent **bonds** to form a molecular **ion**, called **carbonate**, which has a negative charge. Calcium as an **ion** has a positive charge of plus two. The two oppositely charged ions attract each other and combine to form the **mineral calcite**, CaCO3. The name of the chemical compound is calcium **carbonate**, where calcium is Ca and **carbonate** refers to the molecular **ion** CO_3^{-2}.

The **mineral olivine** has the chemical formula $(Mg,Fe)_2SiO_4$, in which one silicon and four oxygen atoms are **bonded** with two atoms of either magnesium or iron. The comma between iron (Fe) and magnesium (Mg) indicates the two **elements** can occupy the same location in the crystal structure and substitute for one another.

Valence and Charge

The electrons around the atom's nucleus are located in shells representing different energy levels. The outermost shell is called the valence shell. Electrons in the valence shell are involved in chemical **bonding**. In 1913, Niels Bohr proposed a simple model of the atom that states atoms are more stable when their outermost shell is full. Atoms of most **elements** thus tend to gain or lose electrons so the outermost or valence shell is full. In Bohr's model, the innermost shell can have a maximum of two electrons and the second and third shells can have a maximum of eight electrons. When the innermost shell is the valence shell, as in the case of hydrogen and helium, it obeys the **octet rule** when it is full with two electrons. For **elements** in higher rows, the **octet rule** of eight electrons in the valence shell applies.

Figure 3.5: The carbon dioxide molecule. Since oxygen is -2 and carbon is +4, the two oxygens bond to the carbon to form a neutral molecule.

The rows in the periodic table present the **elements** in order of atomic number and the columns organize **elements** with similar characteristics, such as the same number of electrons in their valence shells. Columns are often labeled from left to right with Roman numerals I to VIII, and Arabic numerals 1 through 18. The **elements** in columns I and II have 1 and 2 electrons in their respective valence shells and the **elements** in columns VI and VII have 6 and 7 electrons in their respective valence shells.

In row 3 and column I, sodium (Na) has 11 protons in the nucleus and 11 electrons in three shells—2 electrons in the inner shell, 8 electrons in the second shell, and 1 electron in the valence shell. To maintain a full outer shell of 8 electrons per the **octet rule**, sodium readily gives up that 1 electron so there are 10 total electrons. With 11 positively charged protons in the nucleus and 10 negatively charged electrons in two shells, sodium when forming chemical **bonds** is an **ion** with an overall net charge of +1.

All **elements** in column I have a single electron in their valence shell and a valence of 1. These other column I **elements** also readily give up this single valence electron and thus become ions with a +1 charge. **Elements** in column II readily give up 2 electrons and end up as ions with a charge of +2. Note that elements in columns I and II which readily give up their valence electrons, often form bonds with **elements** in columns VI and VII which readily take up these electrons. **Elements** in columns 3 through 15 are usually involved in covalent **bonding**. The last column 18 (VIII) contains the noble gases. These **elements** are chemically inert because the valence shell is already full with 8 electrons, so they do not gain or lose electrons. An example is the noble gas helium which has 2 valence electrons in the first shell. Its valence shell is therefore full. All **elements** in column VIII possess full valence shells and do not form **bonds** with other **elements**.

As seen above, an atom with a net positive or negative charge as a result of gaining or losing electrons is called an **ion**. In general the **elements** on the left side of the table lose electrons and become positive ions, called **cations** because they are attracted to the cathode in an electrical device. The **elements** on the right side tend to gain electrons. These are called **anions** because they are attracted to the anode in an electrical device. The **elements** in the center of the periodic table, columns 3 through 15, do not consistently follow the **octet rule**. These are called transition **elements**. A common example is iron, which has a +2 or +3 charge depending on the **oxidation** state of the **element**. Oxidized Fe^{+3} carries a +3 charge and reduced Fe^{+2} is +2. These two different **oxidation** states of iron often impart dramatic colors to rocks containing their **minerals**—the oxidized form producing red colors and the reduced form producing green.

Ionic Bonding

Ionic **bonds**, also called electron-transfer **bonds**, are formed by the electrostatic attraction between atoms having opposite charges. Atoms of two opposite charges attract each other electrostatically and form an ionic **bond** in which the positive **ion** transfers its electron (or electrons) to the negative **ion** which takes them up. Through this transfer both atoms thus achieve a full valence shell. For example one atom of sodium (Na^{+1}) and one atom of chlorine (Cl^{-1}) form an ionic **bond** to make the compound sodium chloride (NaCl). This is also known as the **mineral halite** or common table salt. Another example is calcium (Ca^{+2}) and chlorine (Cl^{-1}) combining to make the compound calcium chloride ($CaCl_2$). The subscript 2 indicates two atoms of chlorine are ionically **bonded** to one atom of calcium.

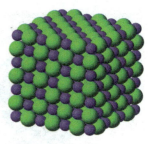

Figure 3.6: Cubic arrangement of Na and Cl in halite.

Covalent Bonding

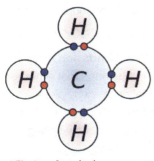

● Electron from hydrogen
● Electron from carbon

Figure 3.7: Methane molecule.

Ionic **bonds** are usually formed between a metal and a nonmetal. Another type, called a covalent or electron-sharing **bond**, commonly occurs between nonmetals. Covalent **bonds** share electrons between ions to complete their valence shells. For example, oxygen (atomic number 8) has 8 electrons—2 in the inner shell and 6 in the valence shell. Gases like oxygen often form diatomic molecules by sharing valence electrons. In the case of oxygen, two atoms attach to each other and share 2 electrons to fill their valence shells to become the common oxygen molecule we breathe (O_2). Methane (CH_4) is another covalently bonded gas. The carbon atom needs 4 electrons and each hydrogen needs 1. Each hydrogen shares its electron with the carbon to form a molecule as shown in the figure.

Take this quiz to check your comprehension of this section.

If you are using an offline version of this text, access the quiz for section 3.1 via the QR code.

 An interactive H5P element has been excluded from this version of the text. You can view it online here: https://pressbooks.lib.vt.edu/introearthscience/?p=234#h5p-17

3.2 Formation of Minerals

Minerals form when atoms **bond** together in a crystalline arrangement. Three main ways this occurs in nature are: 1) **precipitation** directly from an aqueous (water) **solution** with a **temperature** change, 2) **crystallization** from a **magma** with a **temperature** change, and 3) biological **precipitation** by the action of organisms.

3.2.1 Precipitation From Aqueous Solution

Figure 3.8: Calcium carbonate deposits from hard water.

Solutions consist of ions or molecules, known as solutes, **dissolved** in a medium or solvent. In nature this solvent is usually water. Many **minerals** can be dissolved in water, such as **halite** or table salt, which has the **composition** sodium chloride, NaCl. The Na^{+1} and Cl^{-1} ions separate and disperse into the **solution**.

Precipitation is the reverse process, in which ions in **solution** come together to form solid **minerals**. **Precipitation** is dependent on the concentration of ions in **solution** and other factors such as **temperature** and pressure. The point at which a solvent cannot hold any more solute is called **saturation**. **Precipitation** can occur when the **temperature** of the **solution** falls, when the solute evaporates, or with changing chemical conditions in the **solution**. An example of **precipitation** in our homes is when water evaporates and leaves behind a rind of **minerals** on faucets, shower heads, and drinking glasses.

In nature, changes in environmental conditions may cause the **minerals dissolved** in water to form **bonds** and grow into crystals or cement grains of **sediment** together. In Utah, deposits of **tufa** formed from **mineral**-rich springs that emerged into the **ice age** Lake Bonneville. Now exposed in dry valleys, this porous **tufa** was a natural insulation used by pioneers to build their homes with a natural protection against summer heat and winter cold. The **travertine terraces** at Mammoth Hot Springs in Yellowstone Park are another example formed by **calcite precipitation** at the edges of the shallow **spring**-fed ponds.

Another example of **precipitation** occurs in the Great Salt Lake, Utah, where the concentration of sodium chloride and other salts is nearly eight times greater than in the world's oceans. **Streams** carry salt ions into the lake from the surrounding mountains. With no other outlet, the water in the lake evaporates and the concentration of salt increases until **saturation** is reached and the **minerals precipitate** out as **sediments**. Similar salt deposits include **halite** and other precipitates, and occur in other lakes like Mono Lake in California and the Dead Sea.

Figure 3.9: The Bonneville Salt Flats of Utah.

3.2.2 Crystallization From Magma

Heat is energy that causes atoms in substances to vibrate. **Temperature** is a measure of the intensity of the vibration. If the vibrations are violent enough, chemical **bonds** are broken and the crystals melt releasing the ions into the melt. **Magma** is molten rock with freely moving ions. When **magma** is emplaced at depth or extruded onto the surface (then called **lava**), it starts to cool and **mineral** crystals can form.

Figure 3.10: Lava, magma at the Earth's surface.

3.2.3 Precipitation by Organisms

Figure 3.11: Ammonite shell made of calcium carbonate.

Many organisms build bones, shells, and body coverings by extracting ions from water and precipitating **minerals** biologically. The most common **mineral precipitated** by organisms is **calcite**, or calcium **carbonate** ($CaCO_3$). **Calcite** is often **precipitated** by organisms as a **polymorph** called aragonite. **Polymorphs** are crystals with the same chemical formula but different crystal structures. **Marine** invertebrates such as corals and clams **precipitate** aragonite or **calcite** for their shells and structures. Upon death, their hard parts accumulate on the **ocean floor** as **sediments**, and eventually may become the **sedimentary rock limestone**. Though **limestone** can form inorganically, the vast majority is formed by this biological process. Another example is **marine** organisms called radiolaria, which are zooplankton that **precipitate** silica for their microscopic external shells. When the organisms die, the shells accumulate on the **ocean floor** and can form the **sedimentary rock chert**. An example of biologic **precipitation** from the **vertebrate** world is bone, which is **composed** mostly of a type of apatite, a **mineral** in the **phosphate** group. The apatite found in bones contains calcium and water in its structure and is called hydroxycarbonate apatite, $Ca_5(PO_4)_3(OH)$. As mentioned above, such substances are not technically **minerals** until the organism dies and these hard parts become **fossils**.

Take this quiz to check your comprehension of this section.

If you are using an offline version of this text, access the quiz for section 3.2 via the QR code.

 An interactive H5P element has been excluded from this version of the text. You can view it online here: https://pressbooks.lib.vt.edu/introearthscience/?p=234#h5p-18

3.3 Silicate Minerals

Minerals are categorized based on their **composition** and structure. **Silicate minerals** are built around a molecular **ion** called the **silicon-oxygen tetrahedron**. A tetrahedron has a pyramid-like shape with four sides and four corners. **Silicate minerals** form the largest group of **minerals** on Earth, comprising the vast majority of the Earth's **mantle** and **crust**. Of the nearly four thousand known **minerals** on Earth, most are rare. There are only a few that make up most of the rocks likely to be encountered by surface dwelling creatures like us. These are generally called the rock-forming **minerals**.

Figure 3.12: A tetrahedra.

The **silicon-oxygen tetrahedron** (SiO₄) consists of a single silicon atom at the center and four oxygen atoms located at the four corners of the tetrahedron. Each oxygen **ion** has a -2 charge and the silicon **ion** has a +4 charge. The silicon **ion** shares one of its four valence electrons with each of the four oxygen ions in a covalent **bond** to create a symmetrical geometric four-sided pyramid figure. Only half of the oxygen's valence electrons are shared, giving the **silicon-oxygen tetrahedron** an ionic charge of -4. This **silicon-oxygen tetrahedron** forms bonds with many other combinations of ions to form the large group of **silicate minerals**.

Figure 3.13: Silicate tetrahedron.

The silicon **ion** is much smaller than the oxygen ions (see the figures) and fits into a small space in the center of the four large oxygen ions, seen if the top ball is removed (as shown in the figure to the right). Because only one of the valence electrons of the corner oxygens is shared, the **silicon-oxygen tetrahedron** has chemically active corners available to form **bonds** with other **silica tetrahedra** or other positively charged ions such as Al₊₃, Fe₊₂,₊₃, Mg₊₂, K₊₁, Na₊₁, and Ca₊₂. Depending on many factors, such as the original **magma** chemistry, **silica-oxygen tetrahedra** can combine with other tetrahedra in several different configurations. For example, tetrahedra can be isolated, attached in chains, sheets, or three dimensional structures. These combinations and others create the chemical structure in which positively charged ions can be inserted for unique chemical compositions forming **silicate mineral** groups.

3.3.1 The Dark Ferromagnesian Silicates

Olivine Family

Olivine is the primary **mineral** component in **mantle** rock such as **peridotite** and **basalt**. It is characteristically green when not weathered. The chemical formula is (Fe,Mg)₂SiO₄. As previously described, the comma between iron (Fe) and magnesium (Mg) indicates these two **elements** occur in a **solid solution**. Not to be confused with a liquid solution, a **solid solution** occurs when two or more **elements** have similar properties and can freely substitute for each other in the same location in the crystal structure.

Figure 3.14: Olivine crystals in basalt.

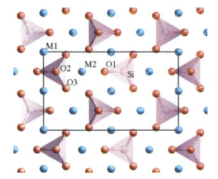

Figure 3.15: Tetrahedral structure of olivine.

Olivine is referred to as a **mineral** family because of the ability of iron and magnesium to substitute for each other. Iron and magnesium in the **olivine** family indicates a **solid solution** forming a compositional series within the **mineral** group which can form crystals of all iron as one end member and all mixtures of iron and magnesium in between to all magnesium at the other end member. Different **mineral** names are applied to compositions between these end members. In the **olivine** series of **minerals**, the iron and magnesium ions in the **solid solution** are about the same size and charge, so either atom can fit into the same location in the growing crystals. Within the cooling **magma**, the **mineral** crystals continue to grow until they solidify into **igneous rock**. The relative amounts of iron and magnesium in the parent **magma** determine which **minerals** in the series form. Other rarer **elements** with similar properties to iron or magnesium, like manganese (Mn), can substitute into the **olivine** crystalline structure in small amounts. Such ionic substitutions in **mineral** crystals give rise to the great variety of **minerals** and are often responsible for differences in color and other properties within a group or family of **minerals**. **Olivine** has a pure iron end-member (called fayalite) and a pure magnesium end-member (called forsterite). Chemically, **olivine** is mostly silica, iron, and magnesium and therefore is grouped among the dark-colored ferromagnesian (iron=ferro, magnesium=magnesian) or **mafic minerals**, a contraction of their chemical symbols Ma and Fe. **Mafic minerals** are also referred to as dark-colored ferromagnesian **minerals**. *Ferro* means iron and *magnesian* refers to magnesium. Ferromagnesian **silicates** tend to be more dense than non-ferromagnesian **silicates**. This difference in density ends up being important in controlling the behavior of the **igneous** rocks that are built from these **minerals**: whether a **tectonic plate subducts** or not is largely governed by the density of its rocks, which are in turn controlled by the density of the **minerals** that comprise them.

The crystal structure of **olivine** is built from independent **silica tetrahedra**. **Minerals** with independent tetrahedral structures are called neosilicates (or orthosilicates). In addition to **olivine**, other common neosilicate **minerals** include garnet, topaz, kyanite, and **zircon**.

Two other similar arrangements of tetrahedra are close in structure to the neosilicates and **grade** toward the next group of **minerals**, the pyroxenes. In a variation on independent tetrahedra called sorosilicates, there are **minerals** that share one oxygen between two tetrahedra, and include **minerals** like pistachio-green epidote, a gemstone. Another variation are the cyclosilicates, which as the name suggests, consist of tetrahedral rings, and include gemstones such as beryl, emerald, aquamarine, and tourmaline

Pyroxene Family

Figure 3.16: Crystals of diopside, a member of the pyroxene family.

Pyroxene is another family of dark ferromagnesian **minerals**, typically black or dark green in color. Members of the **pyroxene** family have a complex chemical **composition** that includes iron, magnesium, aluminum, and other **elements bonded** to polymerized **silica tetrahedra**. Polymers are chains, sheets, or three-dimensional structures, and are formed by multiple tetrahedra covalently **bonded** via their corner oxygen atoms. Pyroxenes are commonly found in **mafic igneous** rocks such as **peridotite**, **basalt**, and **gabbro**, as well as **metamorphic** rocks like eclogite and blue **schist**.

Pyroxenes are built from long, single chains of polymerized **silica tetrahedra** in which tetrahedra share two corner oxygens. The silica chains are **bonded** together into the crystal structures by metal cations. A common member of the **pyroxene** family is augite, itself containing several **solid solution** series with a complex chemical formula $(Ca,Na)(Mg,Fe,Al,Ti)(Si,Al)_2O_6$ that gives rise to a number of individual **mineral** names.

This single-chain crystalline structure **bonds** with many **elements**, which can also freely substitute for each other. The generalized chemical **composition** for **pyroxene** is $XZ(Al,Si)_2O_6$. X represents the ions Na, Ca, Mg, or Fe, and Z represents Mg, Fe, or Al. These ions have similar ionic sizes, which allows many possible substitutions among them. Although the **cations** may freely substitute for each other in the crystal, they carry different ionic charges that must be balanced out in the final crystalline structure. For example Na has a charge of +1, but Ca has charge of +2. If a Na^+ **ion** substitutes for a Ca^{+2} **ion**, it creates an unequal charge that must be balanced by other ionic substitutions elsewhere in the crystal. Note that ionic size is more important than ionic charge for substitutions to occur in **solid solution** series in crystals.

Figure 3.17: Single chain tetrahedral structure in pyroxene.

Amphibole Family

Figure 3.18: Elongated crystals of hornblende in orthoclase.

Amphibole minerals are built from polymerized double silica chains and they are also referred to as inosilicates. Imagine two **pyroxene** chains that connect together by sharing a third oxygen on each tetrahedra. Amphiboles are usually found in **igneous** and **metamorphic** rocks and typically have a long-bladed **crystal habit**. The most common **amphibole**, hornblende, is usually black; however, they come in a variety of colors depending on their chemical **composition**. The **metamorphic rock**, amphibolite, is primarily **composed** of **amphibole minerals**.

Figure 3.19: Hornblende crystals.

Amphiboles are **composed** of iron, magnesium, aluminum, and other **cations bonded** with **silica tetrahedra**. These dark ferromagnesian **minerals** are commonly found in **gabbro**, baslt, **diorite**, and often form the black specks in **granite**. Their chemical formula is very complex and generally written as $(RSi_4O_{11})_2$, where R represents many different **cations**. For example, it can also be written more exactly as $AX_2Z_5((Si,Al,Ti)_8O_{22})(OH,F,Cl,O)_2$. In this formula A may be Ca, Na, K, Pb, or blank; X equals Li, Na, Mg, Fe, Mn, or Ca; and Z is Li, Na, Mg, Fe, Mn, Zn, Co, Ni, Al, Cr, Mn, V, Ti, or Zr. The substitutions create a wide variety of colors such as green, black, colorless, white, yellow, blue, or brown. **Amphibole** crystals can also include hydroxide ions (OH^-)ʼ which occurs from an interaction between the growing **minerals** and water **dissolved** in **magma**.

3.3.2 Sheet Silicates

Figure 3.21: Sheet crystals of biotite mica.

Sheet **silicates** are built from tetrahedra which share all three of their bottom corner oxygens thus forming sheets of tetrahedra with their top corners available for **bonding** with other atoms. Micas and clays are common types of sheet **silicates**, also known as phyllosilicates. **Mica minerals** are usually found in **igneous** and **metamorphic** rocks, while clay **minerals** are more often found in sedimentary rocks. Two frequently found micas are dark-colored **biotite**, frequently found in **granite**, and light-colored **muscovite**, found in the **metamorphic rock** called **schist**.

Figure 3.20: Double chain structure.

Figure 3.22: Crystal of muscovite mica.

Chemically, sheet **silicates** usually contain silicon and oxygen in a 2:5 ratio (Si_4O_{10}). Micas contain mostly silica, aluminum, and potassium. **Biotite mica** has more iron and magnesium and is considered a ferromagnesian **silicate mineral**. **Muscovite** micas belong to the **felsic silicate minerals**. **Felsic** is a contraction formed from **feldspar**, the dominant **mineral** in **felsic** rocks.

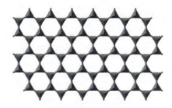

Figure 3.23: Sheet structure of mica.

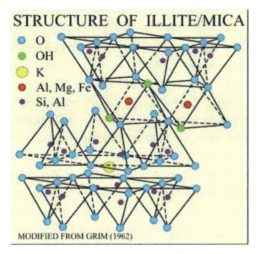

Figure 3.24: Crystal structure of a mica.

The illustration of the crystalline structure of **mica** shows the corner O atoms **bonded** with K, Al, Mg, Fe, and Si atoms, forming polymerized sheets of linked tetrahedra, with an octahedral layer of Fe, Mg, or Al, between them. The yellow potassium ions form Van der Waals **bonds** (attraction and repulsion between atoms, molecules, and surfaces) and hold the sheets together. Van der Waals **bonds** differ from covalent and ionic **bonds**, and exist here between the sandwiches, holding them together into a **stack** of sandwiches. The Van der Waals **bonds** are weak compared to the **bonds** within the sheets, allowing the sandwiches to be separated along the potassium layers. This gives **mica** its characteristic property of easily cleaving into sheets.

The mica "sandwich"

Figure 3.25: Mica "silica sandwich" structure. In this analogy, you may start with one "sandwich": the top bun is a silica sheet, with a "jam" of anions filling the sandwich. The bottom bun is another silica sheet. Then if you place this sandwich on top of an existing sandwich, you may use butter to hold the two sandwiches together—this "butter" would be the large potassium ions forming Van der Waals bonds that hold the two sandwiches' bottom and top buns (silica sheets) together.

Clays **minerals** occur in **sediments** formed by the **weathering** of rocks and are another family of **silicate minerals** with a tetrahedral sheet structure. Clay **minerals** form a complex family, and are an important component of many sedimentary rocks. Other sheet **silicates** include serpentine and chlorite, found in **metamorphic** rocks.

Clay **minerals** are **composed** of hydrous aluminum **silicates**. One type of clay, kaolinite, has a structure like an open-faced sandwich, with the bread being a single layer of **silicon-oxygen tetrahedra** and a layer of aluminum as the spread in an octahedral configuration with the top oxygens of the sheets.

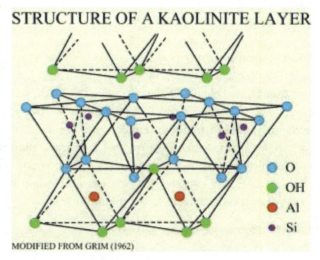

Figure 3.26: Structure of kaolinite.

3.3.3 Framework Silicates

Figure 3.27: Freely growing quartz crystals showing crystal faces.

Quartz and **feldspar** are the two most abundant **minerals** in the **continental crust**. In fact, **feldspar** itself is the single most abundant **mineral** in the Earth's **crust**. There are two types of **feldspar**, one containing potassium and abundant in **felsic** rocks of the **continental crust**, and the other with sodium and calcium abundant in the **mafic** rocks of **oceanic crust**. Together with **quartz**, these **minerals** are classified as framework **silicates**. They are built with a three-dimensional framework of **silica tetrahedra** in which all four corner oxygens are shared with adjacent tetrahedra. Within these frameworks in **feldspar** are holes and spaces into which other ions like aluminum, potassium, sodium, and calcium can fit giving rise to a variety of **mineral** compositions and **mineral** names.

Feldspars are usually found in **igneous** rocks, such as **granite**, **rhyolite**, and **basalt** as well as **metamorphic** rocks and **detrital** sedimentary rocks. **Detrital** sedimentary rocks are **composed** of mechanically weathered rock particles, like sand and gravel. **Quartz** is especially abundant in **detrital** sedimentary rocks because it is very resistant to disintegration by **weathering**. While **quartz** is the most abundant **mineral** on the Earth's surface, due to its durability, the **feldspar minerals** are the most abundant **minerals** in the Earth's **crust**, comprising roughly 50% of the total **minerals** that make up the **crust**.

Figure 3.29: Pink orthoclase crystals.

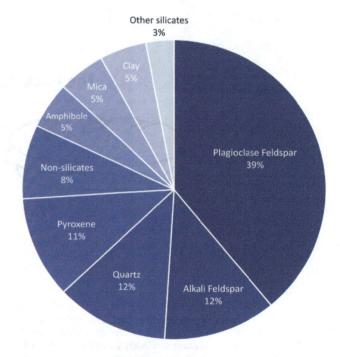

Figure 3.28: Mineral abundance pie chart in Earth's crust.

Quartz is **composed** of pure silica, SiO_2, with the tetrahedra arranged in a three dimensional framework. Impurities consisting of atoms within this framework give rise to many varieties of **quartz** among which are gemstones like amethyst, rose **quartz**, and citrine. **Feldspars** are mostly silica with aluminum, potassium, sodium, and calcium. Orthoclase **feldspar** ($KAlSi_3O_8$), also called potassium **feldspar** or **K-spar**, is made of silica, aluminum, and potassium. **Quartz** and orthoclase **feldspar** are **felsic minerals**. **Felsic** is the compositional term applied to **continental igneous minerals** and rocks that contain an abundance of silica. Another **feldspar** is **plagioclase** with the formula $(Ca,Na)AlSi_3O_8$, the **solid solution** (Ca,Na) indicating a series of **minerals**, one end of the series with calcium $CaAl_2Si_2O_8$, called anorthite, and the other end with sodium $NaAlSi_3O_8$, called albite. Note how the **mineral** accommodates the substitution of Ca^{++} and Na^+. **Minerals** in this solid solution series have different **mineral** names.

Note that aluminum, which has a similar ionic size to silicon, can substitute for silicon inside the tetrahedra. Because potassium ions are so much larger than sodium and calcium ions, which are very similar in size, the inability of the crystal lattice to accommodate both potassium and sodium/calcium gives rise to the two families of **feldspar**, orthoclase and **plagioclase** respectively. Framework **silicates** are called tectosilicates and include the alkali metal-rich feldspathoids and zeolites.

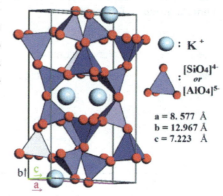

Figure 3.30: Crystal structure of feldspar.

Take this quiz to check your comprehension of this section.

If you are using an offline version of this text, access the quiz for section 3.3 via the QR code.

 An interactive H5P element has been excluded from this version of the text. You can view it online here:
https://pressbooks.lib.vt.edu/introearthscience/?p=234#h5p-19

3.4 Non-silicate Minerals

The crystal structure of non-**silicate minerals** (see table) does not contain **silica-oxygen tetrahedra**. Many non-**silicate minerals** are economically important and provide **metallic** resources such as copper, lead, and iron. They also include valuable non-**metallic** products such as salt, construction materials, and fertilizer.

Figure 3.31: Hanksite, Na22K(SO4)9(CO3)2Cl, one of the few minerals that is considered a carbonate and a sulfate.

Mineral group	Examples	Formula	Uses
Native elements	Gold, silver, copper	Au, Ag, Cu	Jewelry, coins, industry
Carbonates	Calcite, dolomite	$CaCO_3$, $CaMg(CO_3)_2$	Lime, Portland cement
Oxides	Hematite, magnetite, bauxite	Fe_2O_3, Fe_3O_4, a mixture of aluminum oxides	Ores of iron & aluminum, pigments
Halides	Halite, sylvite	NaCl, KCl	Table salt, fertilizer
Sulfides	Galena, chalcopyrite, cinnabar	PbS, $CuFeS_2$, HgS	Ores of lead, copper, mercury
Sulphates	Gypsum, epsom salts	$CaSo_4 \cdot 2H_2O$, $MgSO_4 \cdot 7H_2O$	Sheetrock, therapeutic soak
Phosphates	Apatite	$Ca_5(PO_4)_3(F,Cl,OH)$	Fertilizer, teeth, bones

Table 3.2: Common non-silicatemineral groups.

3.4.1 Carbonates

Figure 3.32: Calcite crystal in shape of rhomb. Note the double-refracted word "calcite" in the center of the figure due to birefringence.

Calcite ($CaCO_3$) and dolomite ($CaMg(CO_3)_2$) are the two most frequently occurring **carbonate minerals**, and usually occur in sedimentary rocks, such as **limestone** and dolostone rocks, respectively. Some **carbonate** rocks, such **calcite** and dolomite, are formed via evaporation and **precipitation**. However, most **carbonate**-rich rocks, such as **limestone**, are created by the **lithification** of fossilized **marine** organisms. These organisms, including those we can see and many microscopic organisms, have shells or exoskeletons consisting of calcium **carbonate** ($CaCO_3$). When these organisms die, their remains accumulate on the floor of the water body in which they live and the soft body parts decompose and **dissolve** away. The calcium **carbonate** hard parts become included in the **sediments**, eventually becoming the **sedimentary rock** called **limestone**. While **limestone** may contain large, easy to see **fossils**, most **limestones** contain the remains of microscopic creatures and thus originate from biological processes.

Figure 3.33: Limestone with small fossils.

Calcite crystals show an interesting property called birefringence, meaning they polarize light into two wave components vibrating at right angles to each other. As the two light waves pass through the crystal, they travel at different velocities and are separated by **refraction** into two different travel paths. In other words, the crystal produces a double image of objects viewed through it. Because they polarize light, **calcite** crystals are used in special petrographic microscopes for studying **minerals** and rocks.

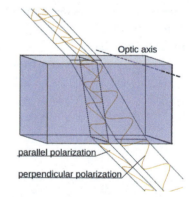

Figure 3.34: Bifringence in calcite crystals.

Many non-**silicate minerals** are referred to as salts. The term salts used here refers to compounds made by replacing the hydrogen in natural acids. The most abundant natural acid is **carbonic acid** that forms by the **solution** of carbon dioxide in water. **Carbonate minerals** are salts built around the **carbonate ion** (CO_3^{-2}) where calcium and/or magnesium replace the hydrogen in **carbonic acid** (H_2CO_3). **Calcite** and a closely related **polymorph** aragonite are secreted by organisms to form shells and physical structures like corals. Many such creatures draw both calcium and **carbonate** from **dissolved** bicarbonate ions (HCO_3^-) in ocean water. As seen in the **mineral** identification section below, **calcite** is easily **dissolved** in acid and thus effervesces in dilute hydrochloric acid (HCl). Small dropper bottles of dilute hydrochloric acid are often carried by geologists in the field as well as used in **mineral** identification labs.

Other salts include **halite** (NaCl) in which sodium replaces the hydrogen in hydrochloric acid and **gypsum** (Ca[SO$_4$] · 2 H$_2$O) in which calcium replaces the hydrogen in sulfuric acid. Note that some water molecules are also included in the **gypsum** crystal. Salts are often formed by evaporation and are called **evaporite minerals**.

The figure shows the crystal structure of **calcite** (CaCO$_3$). Like silicon, carbon has four valence electrons. The **carbonate** unit consists of carbon atoms (tiny white dots) covalently **bonded** to three oxygen atoms (red), one oxygen sharing two valence electrons with the carbon and the other two sharing one valence electron each with the carbon, thus creating triangular units with a charge of -2. The negatively charged **carbonate** unit forms an ionic **bond** with the Ca **ion** (blue), which as a charge of +2.

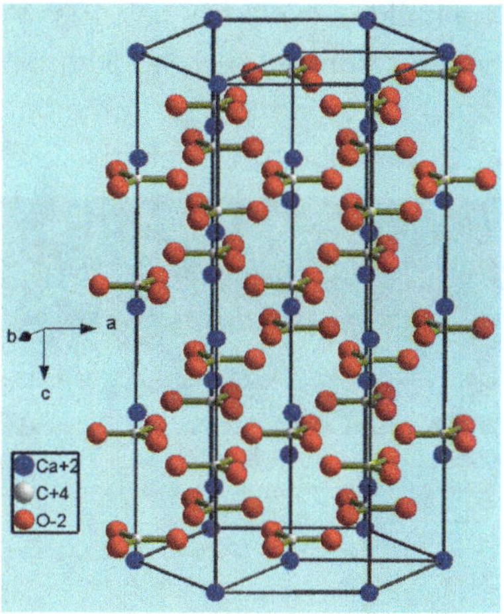

Figure 3.35: Crystal structure of calcite.

3.4.2 Oxides, Halides, and Sulfides

Figure 3.36: Limonite, a hydrated oxide of iron.

After **carbonates**, the next most common non-**silicate minerals** are the **oxides**, **halides**, and **sulfides**.

Oxides consist of metal ions covalently **bonded** with oxygen. The most familiar **oxide** is rust, which is a combination of iron **oxides** (Fe$_2$O$_3$) and hydrated **oxides**. Hydrated **oxides** form when iron is exposed to oxygen and water. Iron **oxides** are important for producing **metallic** iron. When iron **oxide** or **ore** is smelted, it produces carbon dioxide (CO$_2$) and **metallic** iron.

The red color in rocks is usually due to the presence of iron **oxides**. For example, the red **sandstone** cliffs in Zion National Park and throughout Southern Utah consist of white or colorless grains of **quartz** coated with iron **oxide** which serve as cementing agents holding the grains together.

Other iron **oxides** include limonite, magnetite, and hematite. Hematite occurs in many different crystal forms. The **massive** form shows no external structure. Botryoidal hematite shows large concentric blobs. Specular hematite looks like a mass of shiny **metallic** crystals. Oolitic hematite looks like a mass of dull red fish eggs. These different forms of hematite are **polymorphs** and all have the same formula, Fe$_2$O$_3$.

Other common **oxide minerals** include:

- ice (H$_2$O), an **oxide** of hydrogen
- **bauxite** (Al$_2$H$_2$O$_4$), hydrated **oxides** of aluminum, an **ore** for producing **metallic** aluminum
- corundum (Al$_2$O$_3$), which includes ruby and sapphire gemstones.

Figure 3.37: Oolitic hematite.

The **halides** consist of halogens in column VII, usually fluorine or chlorine, ionically **bonded** with sodium or other **cations**. These include **halite** or sodium chloride (NaCl), common table salt; sylvite or potassium chloride (KCl); and fluorite or calcium fluoride (CaF_2).

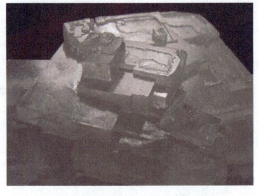

Figure 3.38: Halite crystal showing cubic habit.

Figure 3.39: Salt crystals at the Bonneville Salt Flats.

Halide minerals usually form from the evaporation of sea water or other isolated bodies of water. A well-known example of **halide mineral** deposits created by evaporation is the Bonneville Salt Flats, located west of the Great Salt Lake in Utah (see figure 3.39).

Figure 3.40: Fluorite. B shows fluorescence of fluorite under UV light.

Figure 3.41: Cubic crystals of pyrite.

Many important metal **ores** are **sulfides**, in which metals are **bonded** to sulfur. Significant examples include: galena (lead **sulfide**), sphalerite (zinc **sulfide**), pyrite (iron **sulfide**, sometimes called "fool's gold"), and chalcopyrite (iron-copper **sulfide**). **Sulfides** are well known for being important **ore minerals**. For example, galena is the main source of lead, sphalerite is the main source of zinc, and chalcopyrite is the main copper **ore mineral mined** in porphyry deposits like the Bingham **mine** (see chapter 16). The largest sources of nickel, antimony, molybdenum, arsenic, and mercury are also **sulfides**.

3.4.3 Sulfates

Sulfate minerals contain a metal **ion**, such as calcium, **bonded** to a **sulfate ion**. The **sulfate ion** is a combination of sulfur and oxygen (SO_4^{-2}). The **sulfate mineral gypsum** ($CaSO_4 \cdot 2H_2O$) is used in construction materials such as plaster and drywall. **Gypsum** is often formed from evaporating water and usually contains water molecules in its crystalline structure. The $2H_2O$ in the formula indicates the water molecules are whole H_2O. This is different from **minerals** like **amphibole**, which contain a hydroxide **ion** (OH^-) that is derived from water, but is missing a hydrogen ion (H^+). The calcium **sulfate** without water is a different **mineral** than **gypsum** called anhydrite ($CaSO_4$).

Figure 3.42: Gypsum crystal.

3.4.4 Phosphates

Phosphate minerals have a tetrahedral **phosphate** unit (PO_4^{-3}) combined with various **anions** and **cations**. In some cases arsenic or vanadium can substitute for phosphorus. **Phosphates** are an important ingredient of fertilizers as well as detergents, paint, and other products. The best known **phosphate mineral** is apatite, $Ca_5(PO_4)_3(F,Cl,OH)$, variations of which are found in teeth and bones. The gemstone turquoise [$CuAl_6(PO_4)_4(OH)_8 \cdot 4H2O$] is a copper-rich **phosphate mineral** that, like **gypsum**, contains water molecules.

Figure 3.43: Apatite crystal.

3.4.5 Native Element Minerals

Figure 3.44: Native sulfur deposited around a volcanic fumarole.

Native element minerals, usually metals, occur in nature in a pure or nearly pure state. Gold is an example of a **native element mineral**; it is not very reactive and rarely **bonds** with other **elements** so it is usually found in an isolated or pure state. The non-**metallic** and poorly-reactive **mineral** carbon is often found as a **native element**, such as graphite and diamonds. Mildly reactive metals like silver, copper, platinum, mercury, and sulfur sometimes occur as **native element minerals**. Reactive metals such as iron, lead, and aluminum almost always **bond** to other **elements** and are rarely found in a **native** state.

Figure 3.45: Native copper.

Take this quiz to check your comprehension of this section.

If you are using an offline version of this text, access the quiz for section 3.4 via the QR code.

An interactive H5P element has been excluded from this version of the text. You can view it online here:
https://pressbooks.lib.vt.edu/introearthscience/?p=234#h5p-20

3.5 Identifying Minerals

Geologists identify **minerals** by their physical properties. In the field, where geologists may have limited access to advanced technology and powerful machines, they can still identify **minerals** by testing several physical properties: **luster** and color, **streak**, **hardness**, **crystal habit**, **cleavage** and **fracture**, and some special properties. Only a few common **minerals** make up the majority of Earth's rocks and are usually seen as small grains in rocks. Of the several properties used for identifying **minerals**, it is good to consider which will be most useful for identifying them in small grains surrounded by other **minerals**.

Figure 3.46: The rover Curiosity drilled a hole in this rock from Mars, and confirmed the mineral hematite, as mapped from satellites.

3.5.1 Luster and Color

The first thing to notice about a **mineral** is its surface appearance, specifically **luster** and color. **Luster** describes how the **mineral** looks. **Metallic luster** looks like a shiny metal such as chrome, steel, silver, or gold. Submetallic **luster** has a duller appearance. Pewter, for example, shows submetallic **luster**.

Figure 3.47: 15 mm metallic hexagonal molybdenite crystal from Quebec.

Nonmetallic luster doesn't look like a metal and may be described as vitreous (glassy), earthy, silky, pearly, and other surface qualities. **Nonmetallic minerals** may be shiny, although their vitreous shine is different from **metallic luster**. See the table for descriptions and examples of **nonmetallic luster**.

Figure 3.48: Submetallic luster shown on an antique pewter plate.

Luster	Description	Image	Image description
Vitreous/glassy	Surface is shiny like glass		Quartz crystals showing vitreous luster
Earthy/dull	Dull, like dried mud or clay		Kaolin specimen showing dull or earthy luster
Silky	Soft shine like silk fabric		Gypsum specimen showing silky luster
Pearly	Like the inside of a clam shell or mother-of-pearl		Mica specimen showing pearly luster
Submetallic	Has the appearance of dull metal, like pewter. These minerals would usually still be considered metallic. Submetallic appearance can occur in metallic minerals because of weathering.		Sphalerite specimen showing submetallic luster

Table 3.3: Nonmetallic luster descriptions and examples.

Figure 3.49: Azurite is ALWAYS a dark blue color, and has been used for centuries for blue pigment.

Surface color may be helpful in identifying **minerals**, although it can be quite variable within the same **mineral** family. **Mineral** colors are affected by the main **elements** as well as impurities in the crystals. These impurities may be rare **elements**—like manganese, titanium, chromium, or lithium—even other molecules that are not normally part of the **mineral** formula. For example, the incorporation of water molecules gives **quartz**, which is normally clear, a milky color.

Some **minerals** predominantly show a single color. Malachite and azurite are green and blue, respectively, because of their copper content. Other **minerals** have a predictable range of colors due to elemental substitutions, usually via a **solid solution**. **Feldspars**, the most abundant **minerals** in the earth's **crust**, are complex, have **solid solution** series, and present several colors including pink, white, green, gray and others. Other **minerals** also come in several colors, influenced by trace amounts of several **elements**. The same **element** may show up as different colors, in different **minerals**. With notable exceptions, color is usually not a definitive property of **minerals**. For identifying many **minerals**, a more reliable indicator is **streak**, which is the color of the powdered **mineral**.

3.5.2 Streak

Streak examines the color of a powdered **mineral**, and can be seen when a **mineral** sample is scratched or scraped on an unglazed porcelain **streak plate**. A paper page in a field notebook may also be used for the **streak** of some **minerals**. **Minerals** that are harder than the **streak plate** will not show **streak**, but will scratch the porcelain. For these **minerals**, a **streak** test can be obtained by powdering the **mineral** with a hammer and smearing the powder across a **streak plate** or notebook paper.

Figure 3.50: Different minerals may have different streaks.

While **mineral** surface colors and appearances may vary, their **streak** colors can be diagnostically useful. An example of this property is seen in the iron-**oxide mineral** hematite. Hematite occurs in a variety of forms, colors and lusters, from shiny **metallic** silver to earthy red-brown, and different physical appearances. A hematite **streak** is consistently reddish brown, no matter what the original specimen looks like. Iron **sulfide** or pyrite, is a brassy **metallic** yellow. Commonly named fool's gold, pyrite has a characteristic black to greenish-black **streak**.

3.5.3 Hardness

Figure 3.51: Mohs hardness scale.

Hardness measures the ability of a **mineral** to scratch other substances. The Mohs Hardness Scale gives a number showing the relative scratch-resistance of **minerals** when compared to a standardized set of **minerals** of increasing hardness. The Mohs scale was developed by German geologist Fredrick Mohs in the early 20th century, although the idea of identifying **minerals** by **hardness** goes back thousands of years. Mohs **hardness** values are determined by the strength of a **mineral**'s atomic **bonds**.

The figure shows the **minerals** associated with specific **hardness** values, together with some common items readily available for use in field testing and **mineral** identification. The **hardness** values run from 1 to 10, with 10 being the hardest; however, the scale is not linear. Diamond defines a **hardness** of 10 and is actually about four times harder than corundum, which is 9. A steel pocketknife blade, which has a **hardness** value of 5.5, separates between hard and soft **minerals** on many **mineral** identification keys.

3.5.4 Crystal Habit

Minerals can be identified by **crystal habit**, how their crystals grow and appear in rocks. Crystal shapes are determined by the arrangement of the atoms within the crystal structure. For example, a cubic arrangement of atoms gives rise to a cubic-shaped **mineral** crystal. **Crystal habit** refers to typically observed shapes and characteristics; however, they can be affected by other **minerals** crystallizing in the same rock. When **minerals** are constrained so they do not develop their typical **crystal habit**, they are called **anhedral**. **Subhedral** crystals are partially formed shapes. For some **minerals** characteristic **crystal habit** is to grow crystal faces even when surrounded by other crystals in rock. An example is garnet. **Minerals** grown freely where the crystals are unconstrained and can take characteristic shapes often form crystal faces. A **euhedral** crystal has a perfectly formed, unconstrained shape. Some **minerals** crystallize in such tiny crystals, they do not show a specific **crystal habit** to the naked eye. Other **minerals**, like pyrite, can have an array of different crystal habits, including cubic, dodecahedral, octahedral, and **massive**. The table lists typical crystal habits of various **minerals**.

Habit	Image	Examples
Bladed Long and flat crystals	 Kyanite	Kyanite, amphibole, gypsum
Botryoidal/mammillary Blobby, circular crystals	 Malchite	Hematite, malachite, smithsonite
Coating/laminae/druse Crystals that are small and coat surfaces	 Quartz (var. amethyst) geode	Quartz, calcite, malachite, azurite
Cubic Cube-shaped crystals	 Calcite, Galena	Pyrite, galena, halite
Dodecahedral 12-sided polygon shapes	 Pyrite	Garnet, pyrite
Dendritic Branching crystals	 Manganese dendrites, scale in mm	Mn-oxides, copper, gold

Habit	Image	Examples
Equant Crystals that do not have a long direction	 Olivine	Olivine, garnet, pyroxene
Fibrous Thin, very long crystals	 Tremolite, a type of amphibole	Serpentine, amphibole, zeolite
Layered, sheets Stacked, very thin, flat crystals	 Muscovite	Mica (biotite, muscovite, etc.)
Lenticular/platy Crystals that are plate-like	 Orange wulfenite on calcite	Selenite roses, wulfenite, calcite
Hexagonal Crystals with six sides	 Hanksite	Quartz, hanksite, corundum
Massive/granular Crystals with no obvious shape, microscopic crystals	 Limonite, a hydrated oxide of iron	Limonite, pyrite, azurite, bornite

Habit	Image	Examples
Octahedral 4-sided double pyramid crystals	Fluorite	Diamond, fluorite, magnetite, pyrite
Prismatic/columnar Very long, cylindrical crystals	Tourmaline var. Elbaite with Quartz & Lepidolite on Cleavelandite	Tourmaline, beryl, barite
Radiating Crystals that grow from a point and fan out	Pyrophyllite	Pyrite "suns", pyrophyllite
Rhombohedral Crystals shaped like slanted cubes	Calcite	Calcite, dolomite
Tabular/blocky/stubby Sharp-sided crystals with no long direction	Diopside, a member of the pyroxene family	Feldspar, pyroxene, calcite
Tetrahedral 3-sided, pyramid-shaped crystals	Tetrahedrite	Magnetite, spinel, tetrahedrite

Table 3.4: Typical crystal habits of various minerals.

Figure 3.52: Gypsum with striations.

Another **crystal habit** that may be used to identify **minerals** is striations, which are dark and light parallel lines on a crystal face. Twinning is another, which occurs when the crystal structure replicates in mirror images along certain directions in the crystal.

Figure 3.53: Twinned staurolite found at Fairy Stone State Park, located in Patrick County, Virginia.

Striations and twinning are related properties in some **minerals** including **plagioclase feldspar**. Striations are optical lines on a **cleavage** surface. Because of twinning in the crystal, striations show up on one of the two cleavage faces of the **plagioclase** crystal.

3.5.5 Cleavage and Fracture

Figure 3.54: Striations on plagioclase.

Figure 3.55: Citrine, a variety of quartz showing conchoidal fracture.

Minerals often show characteristic patterns of breaking along specific cleavage planes or show characteristic **fracture** patterns. **Cleavage** planes are smooth, flat, parallel planes within the crystal. The cleavage planes may show as reflective surfaces on the crystal, as parallel cracks that penetrate into the crystal, or show on the edge or side of the crystal as a series of steps like rice **terraces**. **Cleavage** arises in crystals where the atomic **bonds** between atomic layers are weaker along some directions than others, meaning they will break preferentially along these planes. Because they develop on atomic surfaces in the crystal, cleavage planes are optically smooth and reflect light, although the actual break on the crystal may appear jagged or uneven. In such cleavages, the cleavage surface may appear like rice **terraces** on a mountainside that all reflect sunlight from a particular sun angle. Some **minerals** have a strong cleavage, some **minerals** only have weak cleavage or do not typically demonstrate cleavage.

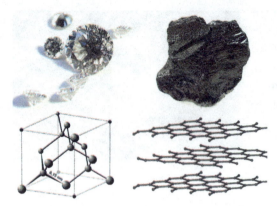

For example, **quartz** and **olivine** rarely show cleavage and typically break into **conchoidal fracture** patterns.

Graphite has its carbon atoms arranged into layers with relatively strong **bonds** within the layer and very weak **bonds** between the layers. Thus graphite cleaves readily between the layers and the layers slide easily over one another giving graphite its lubricating quality.

Figure 3.56: Graphite showing layers of carbon atoms separated by a gap with weak bonds holding the layers together.

Mineral fracture surfaces may be rough and uneven or they may be show **conchoidal fracture**. Uneven **fracture** patterns are described as irregular, splintery, fibrous. A conchoidal fracture has a smooth, curved surface like a shallow bowl or conch shell, often with curved ridges. Natural **volcanic** glass, called **obsidian**, breaks with this characteristic **conchoidal** pattern.

Figure 3.57: Cubic cleavage of galena; note how the cleavage surfaces show up as different but parallel layers in the crystal.

To work with **cleavage**, it is important to remember that cleavage is a result of **bonds** separating along planes of atoms in the crystal structure. On some **minerals**, cleavage planes may be confused with crystal faces. This will usually not be an issue for crystals of **minerals** that grew together within rocks. The act of breaking the rock to expose a fresh face will most likely break the crystals along cleavage planes. Some cleavage planes are parallel with crystal faces but many are not. Cleavage planes are smooth, flat, parallel planes within the crystal. The cleavage planes may show as parallel cracks that penetrate into the crystal (see **amphibole** below), or show on the edge or side of the crystal as a series of steps like rice **terraces**. For some **minerals** characteristic **crystal habit** is to grow crystal faces even when surrounded by other crystals in rock. An example is garnet. **Minerals** grown freely where the crystals are unconstrained and can take characteristic shapes often form crystal faces (see **quartz** below).

Figure 3.58: Freely growing quartz crystals showing crystal faces.

In some **minerals**, distinguishing cleavage planes from crystal faces may be challenging for the student. Understanding the nature of **cleavage** and referring to the number of cleavage planes and cleavage angles on identification keys should provide the student with enough information to distinguish cleavages from crystal faces. **Cleavage** planes may show as multiple parallel cracks or flat surfaces on the crystal. **Cleavage** planes may be expressed as a series of steps like terraced rice paddies. See the cleavage surfaces on galena above or **plagioclase** below. **Cleavage** planes arise from the tendency of **mineral** crystals to break along specific planes of weakness within the crystal favored by atomic arrangements. The number of cleavage planes, the quality of the cleavage surfaces, and the angles between them are diagnostic for many **minerals** and cleavage is one of the most useful properties for identifying **minerals**. Learning to recognize cleavage is an especially important and useful skill in studying **minerals**.

As an identification property of **minerals**, **cleavage** is usually given in terms of the quality of the cleavage (perfect, imperfect, or none), the number of cleavage surfaces, and the angles between the surfaces. The most common number of cleavage plane directions in the common rock-forming **minerals** are: one perfect cleavage (as in **mica**), two cleavage planes (as in **feldspar**, **pyroxene**, and **amphibole**), and three cleavage planes (as in **halite**, **calcite**, and galena). One perfect cleavage (as in **mica**) develops on the top and bottom of the **mineral** specimen with many parallel cracks showing on the sides but no angle of intersection. Two cleavage planes intersect at an angle. Common cleavage angles are 60°, 75°, 90°, and 120°. **Amphibole** has two cleavage planes at 60° and 120°. Galena and **halite** have three cleavage planes at 90° (cubic cleavage). **Calcite** cleaves readily in three directions producing a cleavage figure called a

Figure 3.59: Steps of cleavage along the same cleavage direction.

rhomb that looks like a cube squashed over toward one corner giving rise to the approximately 75° cleavage angles. **Pyroxene** has an imperfect cleavage with two planes at 90°.

Cleavages on common rock-forming minerals:

- **Quartz**—none (**conchoidal fracture**)
- **Olivine**—none (**conchoidal fracture**)
- **Mica**—1 perfect
- **Feldspar**—2 perfect at 90°
- **Pyroxene**—2 imperfect at 90°
- **Amphibole**—2 perfect at 60°/120°
- **Calcite**—3 perfect at approximately 75°
- **Halite**, galena, pyrite—3 perfect at 90°

3.5.6 Special Properties

Figure 3.60: Photomicrograph showing 120/60 degree cleavage within a grain of amphibole.

Special properties are unique and identifiable characteristics used to identify **minerals** or that allow some **minerals** to be used for special purposes. Ulexite has a fiber-optic property that can project images through the crystal like a high-definition television screen (see figure 3.61). A simple identifying special property is taste, such as the salty flavor of **halite** or common table salt (NaCl). Sylvite is potassium chloride (KCl) and has a more bitter taste.

Figure 3.61: A demonstration of ulexite's image projection.

Another property geologists may use to identify **minerals** is a property related to density called **specific gravity**. **Specific gravity** measures the weight of a **mineral** specimen relative to the weight of an equal volume of water. The value is expressed as a ratio between the **mineral** and water weights. To measure **specific gravity**, a **mineral** specimen is first weighed in grams then submerged in a graduated cylinder filled with pure water at room **temperature**. The rise in water level is noted using the cylinder's graduated scale. Since the weight of water at room **temperature** is 1 gram per cubic centimeter, the ratio of the two weight numbers gives the **specific gravity**. **Specific gravity** is easy to measure in the laboratory but is less useful for **mineral** identification in the field than other more easily observed properties, except in a few rare cases such as the very dense galena or **native** gold. The high density of these **minerals** gives rise to a **qualitative** property called "heft." Experienced geologists can roughly assess **specific gravity** by heft, a **subjective** quality of how heavy the specimen feels in one's hand relative to its size.

Figure 3.62: Native gold has one of the highest specific gravities.

A simple test for identifying **calcite** and dolomite is to drop a bit of dilute hydrochloric acid (10-15% HCl) on the specimen. If the acid drop effervesces or fizzes on the surface of the rock, the specimen is **calcite**. If it does not, the specimen is scratched to produce a small amount of powder and test with acid again. If the acid drop fizzes slowly on the powdered **mineral**, the specimen is dolomite. The difference between these two **minerals** can be seen in the video. Geologists who work with **carbonate** rocks carry a small dropper bottle of dilute HCl in their field kit. Vinegar, which contains acetic acid, can be used for this test and is used to distinguish non-**calcite fossils** from **limestone**. While acidic, vinegar produces less of a fizzing reaction because acetic acid is a weaker acid.

 One or more interactive elements has been excluded from this version of the text. You can view them online here:
https://www.youtube.com/watch?v=DX6ZMPbA09U

Video 3.2: Calcite and dolomite reacting with hydrochloric acid.

If you are using an offline version of this text, access this YouTube video via the QR code.

Some iron-**oxide minerals** are magnetic and are attracted to magnets. A common name for a naturally magnetic iron **oxide** is lodestone. Others include magnetite (Fe_3O_4) and ilmenite ($FeTiO_3$). Magnetite is strongly attracted to magnets and can be magnetized. Ilmenite and some types of hematite are weakly magnetic.

Figure 3.63: Paperclips attach to lodestone (magnetite).

Some **minerals** and mineraloids scatter light via a phenomenon called iridescence. This property occurs in labradorite (a variety of **plagioclase**) and opal. It is also seen in biologically created substances like pearls and seashells. Cut diamonds show iridescence and the jeweler's diamond cut is designed to maximize this property.

Figure 3.64: Iridescence on plagioclase, also showing striations on the cleavage surface.

Figure 3.65: Exsolution lamellae within potassium feldspar.

Striations on **mineral cleavage** faces are an optical property that can be used to separate **plagioclase feldspar** from potassium **feldspar** (**K-spar**). A process called twinning creates parallel zones in the crystal that are repeating mirror images. The actual cleavage angle in **plagioclase** is slightly different than 90° and the alternating mirror images in these twinned zones produce a series of parallel lines on one of **plagioclase**'s two cleavage faces. Light reflects off these twinned lines at slightly different angles which then appear as light and dark lines called striations on the cleavage surface. Potassium **feldspar** does not exhibit twinning or striations but may show linear features called exsolution lamellae, also known as perthitic **lineation** or simply perthite. Because sodium and potassium do not fit into the same **feldspar** crystal structure, the lines are created by small amounts of sodium **feldspar** (albite) separating from the dominant potassium **feldspar** (**K-spar**) within the crystal structure. The two different **feldspars** crystallize out into roughly parallel zones within the crystal, which are seen as these linear markings.

One of the most interesting special **mineral** properties is fluorescence. Certain **minerals**, or trace **elements** within them, give off visible light when exposed to ultraviolet radiation or black light. Many **mineral** exhibits have a fluorescence room equipped with black lights so this property can be observed. An even rarer optical property is phosphorescence. Phosphorescent **minerals** absorb light and then slowly release it, much like a glow-in-the-dark sticker.

Figure 3.66: Fluorite. B shows fluorescence of fluorite under UV light.

Take this quiz to check your comprehension of this section.

If you are using an offline version of this text, access the quiz for section 3.5 via the QR code.

 An interactive H5P element has been excluded from this version of the text. You can view it online here:
https://pressbooks.lib.vt.edu/introearthscience/?p=234#h5p-21

Summary

Minerals are the building blocks of rocks and essential to understanding geology. **Mineral** properties are determined by their atomic **bonds**. Most **minerals** begin in a fluid, and either crystallize out of cooling **magma** or **precipitate** as ions and molecules out of a **saturated solution**. The **silicates** are largest group of **minerals** on Earth, by number of varieties and relative quantity, making up a large portion of the **crust** and **mantle**. Based on the **silicon-oxygen tetrahedra**, the crystal structure of **silicates** reflects the fact that silicon and oxygen are the top two of Earth's most abundant **elements**. Non-**silicate minerals** are also economically important, and providing many types of construction and manufacturing materials. **Minerals** are identified by their unique physical properties, including **luster**, color, **streak**, **hardness**, **crystal habit**, **fracture**, **cleavage**, and special properties.

Take this quiz to check your comprehension of this chapter.

If you are using an offline version of this text, access the quiz for chapter 3 via the QR code.

 An interactive H5P element has been excluded from this version of the text. You can view it online here:
https://pressbooks.lib.vt.edu/introearthscience/?p=234#h5p-22

Text References

1. Clarke, F.W.H.S.W., 1927, The Composition of the Earth's Crust: Professional Paper, United States Geological Survey, Professional Paper.
2. Gordon, L.M., and Joester, D., 2011, Nanoscale chemical tomography of buried organic-inorganic interfaces in the chiton tooth: Nature, v. 469, no. 7329, p. 194–197.
3. Hans Wedepohl, K., 1995, The composition of the continental crust: Geochim. Cosmochim. Acta, v. 59, no. 7, p. 1217–1232.
4. Lambeck, K., 1986, Planetary evolution: banded iron formations: v. 320, no. 6063, p. 574–574.

5. Scerri, E.R., 2007, The Periodic Table: Its Story and Its Significance: Oxford University Press, USA.

6. Thomson, J.J., 1897, XL. Cathode Rays: Philosophical Magazine Series 5, v. 44, no. 269, p. 293–316.

7. Trenn, T.J., Geiger, H., Marsden, E., and Rutherford, E., 1974, The Geiger-Marsden Scattering Results and Rutherford's Atom, July 1912 to July 1913: The Shifting Significance of Scientific Evidence: Isis, v. 65, no. 1, p. 74–82.

Figure References

Figure 3.1: The periodic table of the elements. R.A. Dragoset, A. Musgrove, C.W. Clark, and W.C. Martin — NIST, Physical Measurement Laboratory. 2010. Public domain. https://commons.wikimedia.org/wiki/File:Periodic_Table_-_Atomic_Properties_of_the_Elements.png

Figure 3.2: Formation of carbon-14 from nitrogen-14. Sgbeer; adapted by NikNaks. 2011. CC BY-SA 3.0. https://commons.wikimedia.org/wiki/File:Carbon_14_formation_and_decay.svg

Figure 3.3: Element abundance pie chart for Earth's crust. Kindred Grey. 2022. CC BY 4.0.

Figure 3.4: A model of a water molecule, showing the bonds between the hydrogen and oxygen. Dan Craggs. 2009. Public domain. https://commons.wikimedia.org/wiki/File:H2O_2D_labelled.svg

Figure 3.5: The carbon dioxide molecule. Jynto. 2011. Public domain. https://commons.wikimedia.org/wiki/File:Carbon_dioxide_3D_ball.png

Figure 3.6: Cubic arrangement of Na and Cl in halite. Benjah-bmm27. 2006. Public domain. https://commons.wikimedia.org/wiki/File:Sodium-chloride-3D-ionic.png

Figure 3.7: Methane molecule. DynaBlast. 2006. CC BY-SA 2.5. https://commons.wikimedia.org/wiki/File:Covalent.svg

Figure 3.8: Calcium carbonate deposits from hard water. Bbypnda. 2014. CC BY-SA 3.0. https://commons.wikimedia.org/wiki/File:Hard_Water_Calcification.jpg

Figure 3.9: The Bonneville Salt Flats of Utah. Bureau of Land Management. 2015. CC BY 2.0. https://commons.wikimedia.org/wiki/File:Bonneville_Salt_Flats_(17423041595).jpg

Figure 3.10: Lava, magma at the Earth's surface. Hawaii Volcano Observatory (DAS). 2003. Public domain. https://commons.wikimedia.org/wiki/File:Pahoehoe_toe.jpg

Figure 3.11: Ammonite shell made of calcium carbonate. Dlloyd. 2006. CC BY-SA 3.0. https://commons.wikimedia.org/wiki/File:Ammonite_Asteroceras.jpg

Figure 3.12: Rotating animation of a tetrahedra. Kjell André. 2005. CC BY-SA 3.0. https://commons.wikimedia.org/wiki/File:Tetrahedron.gif

Figure 3.13: Silicate tetrahedron. Helgi. 2013. CC BY-SA 3.0. https://commons.wikimedia.org/wiki/File:Silicate_tetrahedron_%2B.svg

Figure 3.14: Olivine crystals in basalt. Vsmith. 2005. CC BY-SA 3.0. https://commons.wikimedia.org/wiki/File:Peridot_in_basalt.jpg

Figure 3.15: Tetrahedral structure of olivine. Matanya (usurped). 2005. Public domain. https://commons.wikimedia.org/wiki/File:Atomic_structure_of_olivine_1.png

Figure 3.16: Crystals of diopside, a member of the pyroxene family. Robert M. Lavinsky. 2010. CC BY-SA 3.0. https://commons.wikimedia.org/wiki/File:Diopside-172005.jpg

Figure 3.17: Single chain tetrahedral structure in pyroxene. Bubenik. 2012. CC BY-SA 3.0. https://en.wikipedia.org/wiki/File:Pyroxferroite-chain.png

Figure 3.18: Elongated crystals of hornblende in orthoclase. Dave Dyet. 2007. Public domain. https://commons.wikimedia.org/wiki/File:Orthoclase_Hornblende.jpg

Figure 3.19: Hornblende crystals. Saperaud~commonswiki. 2005. Public domain. https://commons.wikimedia.org/wiki/File:Amphibole.jpg

Figure 3.20: Double chain structure. Bubenik. 2012. CC BY-SA 3.0. https://commons.wikimedia.org/wiki/File:Jimthompsonite-chain.png

Figure 3.21: Sheet crystals of biotite mica. Fred Kruijen. 2005. CC BY-SA 3.0 NL. https://commons.wikimedia.org/wiki/File:Biotite_aggregate_-_Ochtendung,_Eifel,_Germany.jpg

Figure 3.22: Crystal of muscovite mica. Saperaud~commonswiki. 2005. Public domain. https://commons.wikimedia.org/wiki/File:MicaSheetUSGOV.jpg

Figure 3.23: Sheet structure of mica. Benjah-bmm27. 2007. Public domain. https://en.wikipedia.org/wiki/File:Silicate-sheet-3D-polyhedra.png

Figure 3.24: Crystal structure of a mica. Rosarinagazo. 2006. Public domain. https://commons.wikimedia.org/wiki/File:Ill-struc.jpg

Figure 3.25: Mica "silica sandwich" structure. Kindred Grey. 2022. CC BY 4.0. Includes Sandwich by Alex Muravev from Noun Project (Noun Project license).

Figure 3.26: Structure of kaolinite. USGS. Public domain. https://pubs.usgs.gov/of/2001/of01-041/htmldocs/clays/kaogr.htm

Figure 3.27: Freely growing quartz crystals showing crystal faces. JJ Harrison. 2009. CC BY-SA 2.5. https://en.wikipedia.org/wiki/File:Quartz,_Tibet.jpg

Figure 3.28: Mineral abundance pie chart in Earth's crust. Kindred Grey. 2022. CC BY 4.0.

Figure 3.29: Pink orthoclase crystals. Didier Descouens. 2009. CC BY-SA 4.0. https://commons.wikimedia.org/wiki/File:OrthoclaseBresil.jpg

Figure 3.30: Crystal structure of feldspar. Taisiya Skorina and Antoine Allanore (DOI:10.1039/C4GC02084G). 2015. CC BY 3.0. https://www.researchgate.net/figure/fig1_273641498

Figure 3.31: Hanksite, $Na_{22}K(SO_4)_9(CO_3)_2Cl$, one of the few minerals that is considered a carbonate and a sulfate. Matt Affolter(QFL247). 2009. CC BY-SA 3.0. https://commons.wikimedia.org/wiki/File:Hanksite.JPG

Figure 3.32: Calcite crystal in shape of rhomb. Note the double-refracted word "calcite" in the center of the figure due to birefringence. Alkivar. 2005. Public domain. https://commons.wikimedia.org/wiki/File:Calcite-HUGE.jpg

Figure 3.33: Limestone with small fossils. Jim Stuby. 2010. Public domain. https://commons.wikimedia.org/wiki/File:Limestone_etched_section_KopeFm_new.jpg

Figure 3.34: Bifringence in calcite crystals. Mikael Häggström. 2010. Public domain. https://commons.wikimedia.org/wiki/File:Positively_birefringent_material.svg

Figure 3.35: Crystal structure of calcite. Materialscientist. 2009. CC BY-SA 3.0. https://commons.wikimedia.org/wiki/File:Calcite.png

Figure 3.36: Limonite, a hydrated oxide of iron. USGS. 2005. Public domain. https://commons.wikimedia.org/wiki/File:LimoniteUSGOV.jpg

Figure 3.37: Oolitic hematite. Dave Dyet. 2007. Public domain. https://en.wikipedia.org/wiki/File:Hematite_-_oolitic_with_shale_Iron_Oxide_Clinton,_Oneida_County,_New_York.jpg

Figure 3.38: Halite crystal showing cubic habit. Saperaud~commonswiki. 2005. Public domain. https://en.wikipedia.org/wiki/File:ImgSalt.jpg

Figure 3.39: Salt crystals at the Bonneville Salt Flats. Michael. 2008. CC BY-SA 4.0. https://commons.wikimedia.org/wiki/File:Bonneville_salt_flats_pilot_peak.jpg

Figure 3.40: Fluorite. B shows fluorescence of fluorite under UV light. Didier Descouens. 2009. CC BY-SA 4.0. https://commons.wikimedia.org/wiki/File:FluoriteUV.jpg

Figure 3.41: Cubic crystals of pyrite. CarlesMillan. 2009. CC BY-SA 3.0. https://commons.wikimedia.org/wiki/File:2780M-pyrite1.jpg

Figure 3.42: Gypsum crystal. USGS. 2004. Public domain. https://commons.wikimedia.org/wiki/File:SeleniteGypsumUS-GOV.jpg

Figure 3.43: Apatite crystal. Didier Descouens. 2010. CC BY-SA 4.0. https://commons.wikimedia.org/wiki/File:Apatite_Canada.jpg

Figure 3.44: Native sulfur deposited around a volcanic fumarole. Brisk g. 2006. Public domain. https://commons.wikimedia.org/wiki/File:Fumarola_Vulcano.jpg

Figure 3.45: Native copper. Jonathan Zander (Digon3); adapted by Materialscientist. 2009. CC BY-SA 2.5. https://commons.wikimedia.org/wiki/File:NatCopper.jpg

Figure 3.46: The rover Curiosity drilled a hole in this rock from Mars, and confirmed the mineral hematite, as mapped from satellites. NASA/JPL-Caltech/MSSS. Public domain. https://www.nasa.gov/jpl/msl/pia19036/

Figure 3.47: 15 mm metallic hexagonal molybdenite crystal from Quebec. John Chapman. 2008. CC BY-SA 4.0. https://commons.wikimedia.org/wiki/File:Molly_Hill_molybdenite.JPG

Figure 3.48: Submetallic luster shown on an antique pewter plate. Unknown author. ca. 1770 and 1810. Public domain. https://commons.wikimedia.org/wiki/File:Pewter_Plate.jpg

Table 3.3: Nonmetallic luster descriptions and examples. Quartz Brésil by Didier Descouens, 2010 (CC BY-SA 4.0, https://commons.wikimedia.org/wiki/File:Quartz_Br%C3%A9sil.jpg). KaolinUSGOV by Saperaud~commonswiki, 2005 (Public domain, https://commons.wikimedia.org/wiki/File:KaolinUSGOV.jpg). Selenite Gips Marienglas by Ra'ike, 2006 (CC BY-SA 3.0, https://commons.wikimedia.org/wiki/File:Selenite_Gips_Marienglas.jpg). Mineral Mica GDFL006 by Luis Miguel Bugallo Sánchez, 2005 (CC BY-SA 3.0, https://commons.wikimedia.org/wiki/File:Mineral_Mica_GDFL006.JPG). Sphalerite4 by Andreas Früh, 2005 (CC BY-SA 3.0, https://commons.wikimedia.org/wiki/File:Sphalerite4.jpg).

Figure 3.49: Azurite is ALWAYS a dark blue color, and has been used for centuries for blue pigment. Graeme Churchard. 2013. CC BY 2.0. https://en.wikipedia.org/wiki/File:Azurite_in_siltstone,_Malbunka_mine_NT.jpg

Figure 3.50: Different minerals may have different streaks. Ra'ike. 2010. CC BY-SA 3.0. https://commons.wikimedia.org/wiki/File:Streak_plate_with_Pyrite_and_Rhodochrosite.jpg

Figure 3.51: Mohs hardness scale. NPS. Public domain (full license here). https://www.nps.gov/articles/mohs-hardness-scale.htm

Figure 3.63: Paperclips attach to lodestone (magnetite). Ryan Somma. 1980. CC BY-SA 2.0. https://commons.wikimedia.org/wiki/File:Magnetite_Lodestone.jpg

Figure 3.64: Iridescence on plagioclase, also showing striations on the cleavage surface. Mike Beauregard. 2011. CC BY 2.0. https://flic.kr/p/9xh4MS

Figure 3.65: Exsolution lamellae within potassium feldspar. Jstuby. 2009. Public domain. https://commons.wikimedia.org/wiki/File:Perthitic_feldspar_Dan_Patch_SD.jpg

Figure 3.66: Fluorite. B shows fluorescence of fluorite under UV light. Didier Descouens. 2009. CC BY-SA 4.0. https://commons.wikimedia.org/wiki/File:FluoriteUV.jpg

4.

IGNEOUS PROCESSES AND VOLCANOES

Learning Objectives

By the end of this chapter, students should be able to:

- Explain the origin of **magma** as it relates to **plate tectonics**.
- Describe how the **Bowen's Reaction Series** relates **mineral crystallization** and melting temperatures.
- Explain how cooling of **magma** leads to rock compositions and textures, and how these are used to classify **igneous** rocks.
- Analyze the features of common **igneous** landforms and how they relate to their origin.
- Describe how silica content affects **magma viscosity** and eruptive style of **volcanoes**.
- Describe **volcano** types, eruptive styles, **composition**, and their **plate tectonic** settings.
- Describe **volcanic** hazards.

Igneous rock is formed when liquid rock freezes into a solid rock. This molten material is called **magma** when it is in the ground and **lava** when it is on the surface. Only the Earth's **outer core** is liquid; the Earth's **mantle** and **crust** is naturally solid. However, there are a few minor pockets of **magma** that form near the surface where geologic processes cause melting. It is this **magma** that becomes the source for **volcanoes** and **igneous rocks**. This chapter will describe the classification of **igneous** rocks, the unique processes that form **magmas**, types of **volcanoes** and **volcanic** processes, **volcanic** hazards, and **igneous** landforms.

Figure 4.1: Lava flow in Hawai'i.

Lava cools quickly on the surface of the earth and forms tiny microscopic crystals. These are known as fine-grained **extrusive**, or **volcanic**, **igneous** rocks. **Extrusive** rocks are often **vesicular**, filled with holes from escaping gas bubbles. **Volcanism** is the process in which **lava** is erupted. Depending on the properties of the **lava** that is erupted, the **volcanism** can be drastically different, from smooth and gentle to dangerous and explosive. This leads to different types of **volcanoes** and different **volcanic** hazards.

In contrast, **magma** that cools slowly below the earth's surface forms larger crystals which can be seen with the naked eye. These are known as coarse-grained **intrusive**, or **plutonic**, **igneous** rocks. This relationship between cooling rates and grain sizes of the solidified **minerals** in **igneous** rocks is important for interpreting the rock's geologic history.

4.1 Classification of Igneous Rocks

Igneous rocks are classified based on **texture** and **composition**. **Texture** describes the physical characteristics of the **minerals**, such as grain size. This

Figure 4.2: Half Dome, an intrusive igneous batholith in Yosemite National Park.

relates to the cooling history of the molten **magma** from which it came. **Composition** refers to the rock's specific mineralogy and chemical **composition**. Cooling history is also related to changes that can occur to the **composition** of **igneous** rocks.

4.1.1 Texture

If **magma** cools slowly, deep within the **crust**, the resulting rock is called **intrusive** or **plutonic**. The slow cooling process allows crystals to grow large, giving **intrusive igneous rock** a coarse-grained or **phaneritic texture**. The individual crystals in **phaneritic texture** are readily visible to the unaided eye.

Figure 4.3: Granite is a classic coarse-grained (phaneritic) intrusive igneous rock. The different colors are unique minerals. The black colors are likely two or three different minerals.

Figure 4.4: Basalt is a classic fine-grained extrusive igneous rock.

When **lava** is extruded onto the surface, or intruded into shallow fissures near the surface and cools, the resulting **igneous rock** is called **extrusive** or **volcanic**. **Extrusive igneous** rocks have a fine-grained or **aphanitic texture**, in which the grains are too small to see with the unaided eye. The fine-grained **texture** indicates the quickly cooling **lava** did not have time to grow large crystals. These tiny crystals can be viewed under a petrographic microscope. In some cases, **extrusive lava** cools so rapidly it does not develop crystals at all. This non-crystalline material is not classified as **minerals**, but as **volcanic** glass. This is a common component of **volcanic ash** and rocks like **obsidian**.

Figure 4.5: Porphyritic texture.

Some **igneous** rocks have a mix of coarse-grained **minerals** surrounded by a matrix of fine-grained material in a **texture** called **porphyritic**. The large crystals are called **phenocrysts** and the fine-grained matrix is called the **groundmass** or matrix. **Porphyritic texture** indicates the **magma** body underwent a multi-stage cooling history, cooling slowly while deep under the surface and later rising to a shallower depth or the surface where it cooled more quickly.

Residual molten material expelled from **igneous** intrusions may form veins or masses containing very large crystals of **minerals** like **feldspar**, **quartz**, beryl, tourmaline, and **mica**. This **texture**, which indicates a very slow **crystallization**, is called **pegmatitic**. A rock that chiefly consists of **pegmatitic texture** is known as a **pegmatite**. To give an example of how large these crystals can get, transparent **cleavage** sheets of **pegmatitic muscovite mica** were used as windows during the Middle Ages.

Figure 4.6: Pegmatitic texture.

Figure 4.7: Scoria.

All **magmas** contain gases **dissolved** in solution called **volatiles**. As the **magma** rises to the surface, the drop in pressure causes the **dissolved volatiles** to come bubbling out of **solution**, like the fizz in an opened bottle of soda. The gas bubbles become trapped in the solidifying **lava** to create a **vesicular texture**, with the holes specifically called vesicles. The type of **volcanic rock** with common vesicles is called scoria.

An extreme version of scoria occurs when volatile-rich **lava** is very quickly quenched and becomes a meringue-like froth of glass called **pumice**. Some **pumice** is so full of vesicles that the density of the rock drops low enough that it will float.

Figure 4.8: Pumice.

Figure 4.9: Obsidian (volcanic glass). Note conchoidal fracture.

Lava that cools extremely quickly may not form crystals at all, even microscopic ones. The resulting rock is called **volcanic** glass. Obsidian is a rock consisting of **volcanic** glass. **Obsidian** as a glassy rock shows an excellent example of **conchoidal fracture** similar to the **mineral quartz** (see chapter 3).

When **volcanoes** erupt explosively, vast amounts of **lava**, rock, **ash**, and gases are thrown into the **atmosphere**. The solid parts, called **tephra**, settle back to earth and cool into rocks with **pyroclastic** textures. *Pyro*, meaning fire, refers to the **igneous** source of the **tephra** and *clastic* refers to the rock fragments. **Tephra** fragments are named based on size—**ash** (<2 mm), **lapilli** (2-64 mm), and **bombs** or blocks (>64 mm). **Pyroclastic texture** is usually recognized by the chaotic mix of crystals, angular glass shards, and rock fragments. Rock formed from large deposits of **tephra** fragments is called **tuff**. If the fragments accumulate while still hot, the heat may deform the crystals and weld the mass together, forming a welded **tuff**.

Figure 4.10: Welded tuff.

4.1.2 Composition

Composition refers to a rock's chemical and **mineral** make-up . For **igneous rock**, **composition** is divided into four groups: **felsic**, **intermediate**, **mafic**, and **ultramafic**. These groups refer to differing amounts of silica, iron, and magnesium found in the **minerals** that make up the rocks. It is important to realize these groups do not have sharp boundaries in nature, but rather lie on a continuous spectrum with many transitional compositions and names that refer to specific quantities of **minerals**. As an example, **granite** is a commonly-used term, but has a very specific definition which includes exact quantities of **minerals** like **feldspar** and **quartz**. Rocks labeled as '**granite**' in laymen applications can be several other rocks, including syenite, tonalite, and monzonite. To avoid these complications, the following figure presents a simplified version of **igneous rock** nomenclature focusing on the four main groups, which is adequate for an introductory student.

Felsic refers to a predominance of the light-colored (**felsic**) **miner-als fel**dspar and **si**lica in the form of **quartz**. These light-colored **minerals** have more silica as a proportion of their overall chemical formula. Minor amounts of dark-colored (**mafic**) **minerals** like **amphibole** and biotite **mica** may be present as well. **Felsic igneous** rocks are rich in silica (in the 65-75% range, meaning the rock would be 65-75% weight percent SiO^2) and poor in iron and magnesium.

Intermediate is a **composition** between **felsic** and **mafic**. It usually contains roughly-equal amounts of light and dark **minerals**, including light grains of **plagioclase feldspar** and dark **minerals** like amphibole. It is **intermediate** in silica in the 55-60% range.

Mafic refers to a abundance of ferromagnesian **minerals** (with magnesium and iron, chemical symbols **M**g and **Fe**) plus **plagioclase feldspar**. It is mostly made of dark **minerals** like **pyroxene** and **olivine**, which are rich in iron and magnesium and relatively poor in silica. **Mafic** rocks are low in silica, in the 45-50% range.

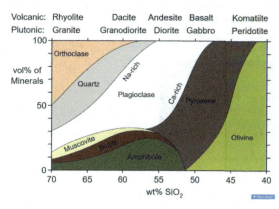

Figure 4.11: Mineral composition of common igneous rocks. Percentage of minerals is shown on the vertical axis. Percentage of silica is shown on the horizontal axis. Rock names at the top include a continuous spectrum of compositions grading from one into another.

Ultramafic refers to the extremely **mafic** rocks **composed** of mostly **olivine** and some **pyroxene** which have even more magnesium and iron and even less silica. These rocks are rare on the surface, but make up **peridotite**, the rock of the upper **mantle**. It is poor in silica, in the 40% or less range.

On the figure above, the top row has both **plutonic** and **volcanic igneous** rocks arranged in a continuous spectrum from **felsic** on the left to **intermediate**, **mafic**, and **ultramafic** toward the right. **Rhyolite** thus refers to the **volcanic** and **felsic igneous** rocks, and **granite** thus refer to **intrusive** and **felsic igneous** rocks. **Andesite** and **diorite** likewise refer to **extrusive** and **intrusive intermediate** rocks (with dacite and granodiorite applying to those rocks with **composition** between **felsic** and **intermediate**). **Basalt** and **gabbro** are the **extrusive** and **intrusive** names for **mafic igneous** rocks, and **peridotite** is **ultramafic**, with komatiite as the fine-grained **extrusive** equivalent. Komatiite is a rare rock because **volcanic** material that comes direct from the **mantle** is not common, although some examples can be found in ancient **Archean** rocks. Nature rarely has sharp boundaries and the classification and naming of rocks often imposes what appear to be sharp boundary names onto a continuous spectrum.

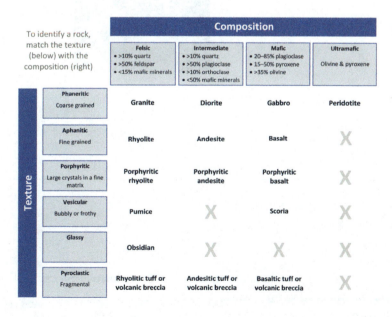

Figure 4.12: Igneous rock classification table with composition as vertical columns and texture as horizontal rows.

Aphanitic/phaneritic rock types

Felsic composition

 Granite

 Rhyolite

Granite is a course-crystalline felsic intrusive rock. The presence of quartz is a good indicator of granite. Granite commonly has large amounts of salmon pink potassium feldspar and white plagioclase crystals that have visible cleavage planes. Granite is a good approximation for the continental crust, both in density and composition.

Rhyolite is a fine-crystalline felsic extrusive rock. Rhyolite is commonly pink and will often have glassy quartz phenocrysts. Because felsic lavas are less mobile, it is less common than granite. Examples of rhyolite include several lava flows in Yellowstone National Park and the altered rhyolite that makes up the Grand Canyon of the Yellowstone.

Intermediate composition

 Diorite

Andesite

Diorite is a coarse-crystalline intermediate intrusive igneous rock. Diorite is identifiable by it's Dalmatian-like appearance of black hornblende and biotite and white plagioclase feldspar. It is found in its namesake, the Andes Mountains as well as the Henry and Abajo mountains of Utah.

Andesite is a fine crystalline intermediate extrusive rock. It is commonly grey and porphyritic. It can be found in the Andes Mountains and in some island arcs (see Chapter 2). It is the fine grained compositional equivalent of diorite.

Mafic composition

 Gabbro

 Vesicular basalt

Gabbro is a coarse-grained mafic igneous rock, made with mainly mafic minerals like pyroxene and only minor plagioclase. Because mafic lava is more mobile, it is less common than basalt. Gabbro is a major component of the lower oceanic crust.

Basalt is a fine-grained mafic igneous rock. It is commonly vesicular and aphanitic. When porphyritic, it often has either olivine or plagioclase phenocrysts. Basalt is the main rock which is formed at mid-ocean ridges, and is therefore the most common rock on the Earth's surface, making up the entirety of the ocean floor (except where covered by sediment).

Table 4.1: Aphanitic and phaneritic rock types with images.

4.1.3 Igneous Rock Bodies

Igneous rocks are common in the geologic record, but surprisingly, it is the **intrusive** rocks that are more common. **Extrusive** rocks, because of their small crystals and glass, are less durable. Plus, they are, by definition, exposed to the **elements** of **erosion** immediately. **Intrusive** rocks, forming underground with larger, stronger crystals, are more likely to last. Therefore, most landforms and rock groups that owe their origin to **igneous** rocks are **intrusive** bodies. A significant exception to this is active **volcanoes**, which are discussed in a later section on volcanism. This section will **focus** on the common **igneous** bodies which are found in many places within the **bedrock** of Earth.

When **magma** intrudes into a weakness like a crack or fissure and solidifies, the resulting cross-cutting feature is called a **dike** (sometimes spelled **dyke**). Because of this, **dikes** are often vertical or at an angle relative to the pre-existing rock layers that they intersect. **Dikes** are therefore discordant intrusions, not following any layering that was present. **Dikes** are important to geologists, not only for the study of **igneous** rocks themselves but also for dating rock sequences and interpreting the geologic history of an area. The **dike** is younger than the rocks it cuts across and, as discussed in the chapter on Geologic Time (chapter 7), may be used to assign actual numeric ages to sedimentary sequences, which are notoriously difficult to age date.

Figure 4.13: Dike of olivine gabbro cuts across Baffin Island in the Canadian Arctic.

Sills are another type of **intrusive** structure. A **sill** is a concordant intrusion that runs parallel to the sedimentary layers in the **country rock**. They are formed when **magma** exploits a weakness between these layers, shouldering them apart and squeezing between them. As with **dikes**, **sills** are younger than the surrounding layers and may be radioactively dated to study the age of sedimentary **strata**.

Figure 4.14: Igneous sill intruding between Paleozoic strata in Nova Scotia.

A **magma chamber** is a large underground **reservoir** of molten rock. The path of rising **magma** is called a **diapir**. The processes by which a **diapir** intrudes into the surrounding **native** or **country rock** are not well understood and are the subject of ongoing geological inquiry. For example, it is not known what happens to the pre-existing **country rock** as the **diapir** intrudes. One **theory** is the overriding rock gets shouldered aside, displaced by the increased volume of **magma**. Another is the **native** rock is melted and consumed into the rising **magma** or broken into pieces that settle into the **magma**, a process known as **stoping**. It has also been proposed that diapirs are not a real phenomenon, but just a series of **dikes** that blend into each other. The **dikes** may be intruding over millions of years, but since they may be made

Figure 4.15: Quartz monzonite in the Cretaceous of Montana, USA.

of similar material, they would be appearing to be formed at the same time. Regardless, when a **diapir** cools, it forms an mass of **intrusive** rock called a **pluton**. **Plutons** can have irregular shapes, but can often be somewhat round.

Figure 4.16: Half Dome in Yosemite National Park, California, is a part of the Sierra Nevada batholith which is mostly made of granite.

When many **plutons** merge together in an extensive single feature, it is called a **batholith**. **Batholiths** are found in the cores of many mountain ranges, including the **granite formations** of Yosemite National Park in the Sierra Nevada of California. They are typically more than 100 km^2 in area, associated with **subduction** zones, and mostly **felsic** in **composition**. A stock is a type of **pluton** with less surface exposure than a **batholith**, and may represent a narrower neck of material emerging from the top of a **batholith**. **Batholiths** and stocks are discordant intrusions that cut across and through surrounding **country rock**.

Laccoliths are blister-like, concordant intrusions of **magma** that form between sedimentary layers. The Henry Mountains of Utah are a famous topographic landform formed by this process. **Laccoliths** bulge upwards; a similar downward-bulging intrusion is called a **lopolith**.

Figure 4.17: The Henry Mountains in Utah are interpreted to be a laccolith, exposed by erosion of the overlying layers.

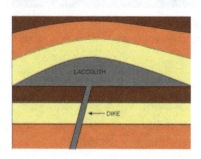

Figure 4.18: Laccolith forms as a blister in between sedimentary strata.

Complete this interactive activity to check your understanding. Click on the plus signs on the illustration for descriptions of several igneous features.

If you are using an offline version of this text, access this interactive activity via the QR code.

 An interactive H5P element has been excluded from this version of the text. You can view it online here:
https://pressbooks.lib.vt.edu/introearthscience/?p=331#h5p-23

Take this quiz to check your comprehension of this section.

If you are using an offline version of this text, access the quiz for section 4.1 via the QR code.

 An interactive H5P element has been excluded from this version of the text. You can view it online here:
https://pressbooks.lib.vt.edu/introearthscience/?p=331#h5p-24

4.2 Bowen's Reaction Series

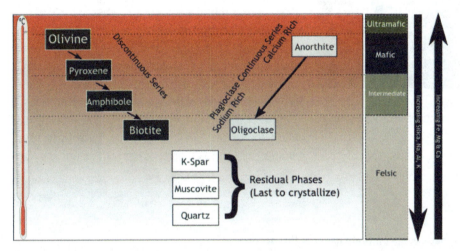

Figure 4.19: Bowen's Reaction Series. Higher temperature minerals shown at top (olivine) and lower temperature minerals shown at bottom (quartz).

Bowen's Reaction Series describes the **temperature** at which **minerals** crystallize when cooled, or melt when heated. The low end of the **temperature** scale where all **minerals** crystallize into solid rock, is approximately 700°C (1292°F). The upper end of the range where all **minerals** exist in a molten state, is approximately 1,250°C (2,282°F). These numbers reference **minerals** that crystallize at standard sea-level pressure, 1 bar. The values will be different for **minerals** located deep below the Earth's surface due to the increased pressure, which affects **crystallization** and melting temperatures. However, the order and relationships are maintained.

In the figure, the righthand column lists the four groups of **igneous rock** from top to bottom: **ultramafic**, **mafic**, **intermediate**, and **felsic**. The down-pointing arrow on the far right shows increasing amounts of silica, sodium, aluminum, and potassium as the **mineral composition** goes from **ultramafic** to **felsic**. The up-pointing arrow shows increasing ferromagnesian components, specifically iron, magnesium, and calcium. To the far left of the diagram is a **temperature** scale. **Minerals** near the top of diagram, such as olivine and anorthite (a type of plagioclase), crystallize at higher temperatures. Minerals near the bottom, such as **quartz** and **muscovite**, crystalize at lower temperatures.

Figure 4.20: Olivine, the first mineral to crystallize in a melt.

The most important aspect of **Bowen's Reaction Series** is to notice the relationships between **minerals** and **temperature**. Norman L. Bowen (1887-1956) was an early 20th Century geologist who studied igneous rocks. He noticed that in **igneous** rocks, certain **minerals** always occur together and these **mineral** assemblages exclude other **minerals**. Curious as to why, and with the **hypothesis** in mind that it had to do with the **temperature** at which the rocks cooled, he set about conducting experiments on **igneous** rocks in the early 1900s. He conducted experiments on **igneous rock**—grinding combinations of rocks into powder, sealing the powders into metal capsules, heating them to various temperatures, and then cooling them.

Figure 4.21: Norman L. Bowen.

Figure 4.22: Norman L. Bowen and his colleague working at the Carnegie Institution of Washington Geophysical Laboratory.

When he opened the quenched capsules, he found a glass surrounding **mineral** crystals that he could identify under his petrographic microscope. The results of many of these experiments, conducted at different temperatures over a **period** of several years, showed that the common **igneous minerals** crystallize from **magma** at different temperatures. He also saw that **minerals** occur together in rocks with others that crystallize within similar **temperature** ranges, and never crystallize with other **minerals**. This relationship can explain the main difference between **mafic** and **felsic igneous** rocks. **Mafic igneous** rocks contain more **mafic minerals**, and therefore, crystallize at higher temperatures than **felsic igneous** rocks. This is even seen in **lava** flows, with **felsic** lavas erupting hundreds of degrees cooler than their **mafic** counterparts. Bowen's work laid the foundation for understanding **igneous petrology** (the study of rocks) and resulted in his book, *The Evolution of the Igneous Rocks* in 1928.

Take this quiz to check your comprehension of this section.

If you are using an offline version of this text, access the quiz for section 4.2 via the QR code.

 An interactive H5P element has been excluded from this version of the text. You can view it online here:
https://pressbooks.lib.vt.edu/introearthscience/?p=331#h5p-25

4.3 Magma Generation

Magma and **lava** contain three components: melt, solids, and **volatiles**. The melt is made of ions from **minerals** that have liquefied. The solids are made of crystallized **minerals** floating in the liquid melt. These may be **minerals** that have already cooled **Volatiles** are gaseous components—such as water vapor, carbon dioxide, sulfur, and chlorine—**dissolved** in the **magma**. The presence and amount of these three components affect the physical behavior of the **magma** and will be discussed more below.

4.3.1 Geothermal Gradient

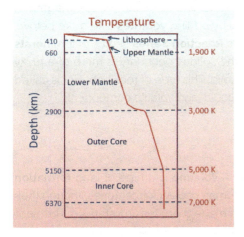

Figure 4.23: Geothermal gradient.

Below the surface, the **temperature** of the Earth rises. This heat is caused by residual heat left from the **formation** of Earth and ongoing **radioactive** decay. The rate at which **temperature** increases with depth is called the **geothermal gradient**. The average **geothermal gradient** in the upper 100 km (62 mi) of the **crust** is about 25°C per kilometer of depth. So for every kilometer of depth, the **temperature** increases by about 25°C.

The depth-**temperature** graph (see figure 4.23) illustrates the relationship between the **geothermal gradient** (geotherm, red line) and the start of rock melting (solidus, green line). The **geothermal gradient** changes with depth (which has a direct relationship to pressure) through the **crust** into upper **mantle**. The area to the left of the green line includes solid components; to the right is where liquid components start to form. The increasing **temperature** with depth makes the depth of about 125 kilometers (78 miles) where the natural **geothermal gradient** is closest to the solidus.

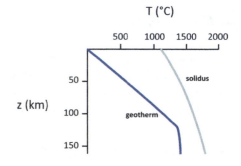

Figure 4.24: Pressure-temperature diagram showing temperature in degrees Celsius on the x-axis and depth below the surface in kilometers (km) on the y-axis. The red line is the geothermal gradient and the green solidus line represents the temperature and pressure regime at which melting begins. Rocks at pressures and temperatures left of the green line are solid. If pressure/temperature conditions change so that rocks pass to the right of the green line, then they will start to melt.

The **temperature** at 100 km (62 mi) deep is about 1,200°C (2,192°F). At bottom of the **crust**, 35 km (22 mi) deep, the pressure is about 10,000 bars. A bar is a measure of pressure, with 1 bar being normal atmospheric pressure at sea level. At these pressures and temperatures, the **crust** and **mantle** are solid. To a depth of 150 km (93 mi), the **geothermal gradient** line stays to the left of the solidus line. This relationship continues through the **mantle** to the **core–mantle** boundary, at 2,880 km (1,790 mi).

The solidus line slopes to the right because the melting **temperature** of any substance depends on pressure. The higher pressure created at greater depth increases the **temperature** needed to melt rock. In another example, at sea level with an atmospheric pressure close to 1 bar, water boils at 100°C. But if the pressure is lowered, as shown on the video below, water boils at a much lower **temperature**.

 One or more interactive elements has been excluded from this version of the text. You can view them online here: https://www.youtube.com/watch?v=Ks4VuXTTKmo

Video 4.1: Boiling water at room temperature.

If you are using an offline version of this text, access this YouTube video via the QR code.

There are three principal ways rock behavior crosses to the right of the green solidus line to create molten **magma**: 1) **decompression melting** caused by lowering the pressure, 2) **flux melting** caused by adding **volatiles** (see more below), and 3) heat-induced melting caused by increasing the **temperature**. The **Bowen's Reaction Series** shows that **minerals** melt at different temperatures. Since **magma** is a mixture of different **minerals**, the solidus boundary is more of a fuzzy zone rather than a well-defined line; some **minerals** are melted and some remain solid. This type of rock behavior is called **partial melting** and represents real-world **magmas**, which typically contain solid, liquid, and volatile components.

4.4 Volcanism

When **magma** emerges onto the Earth's surface, the molten rock is called **lava**. A **volcano** is a type of land **formation** created when **lava** solidifies into rock. **Volcanoes** have been an important part of human society for centuries, though their understanding has greatly increased as our understanding of **plate tectonics** has made them less mysterious. This section describes **volcano** location, type, hazards, and monitoring.

4.4.1. Distribution and Tectonics

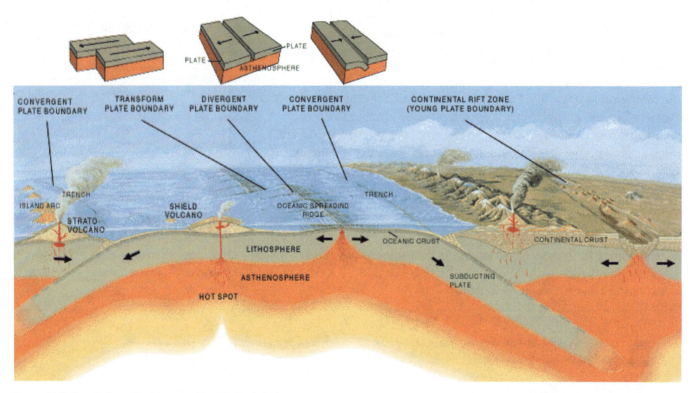

Figure 4.25: Association of volcanoes with plate boundaries.

Most **volcanoes** are **interplate volcanoes**. **Interplate volcanoes** are located at active **plate** boundaries created by **volcanism** at **mid-ocean ridges**, **subduction** zones, and **continental rifts**. The prefix "*inter-*" means between. Some volcanoes are **intraplate volcanoes**. The prefix "*intra-*" means within, and intraplate volcanoes are located within **tectonic plates**, far removed from plate boundaries. Many **intraplate volcanoes** are formed by **hotspots**.

Volcanoes at Mid-ocean Ridges

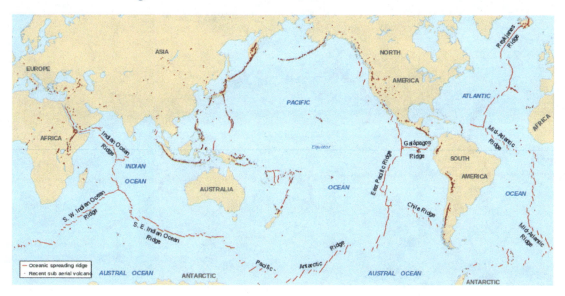

Figure 4.26: Map of spreading ridges throughout the world.

Most **volcanism** on Earth occurs on the **ocean floor** along **mid-ocean ridges**, a type of **divergent plate boundary** (see chapter 2). These **interplate volcanoes** are also the least observed and famous, since most of them are located under 3,000-4,500 m (10,000-15,000 ft) of ocean and the eruptions are slow, gentle, and oozing. One exception is the **interplate volcanoes** of Iceland. The diverging and thinning **oceanic plates** allow hot **mantle** rock to rise, releasing pressure and causing **decompression melting**. **Ultramafic mantle** rock, consisting largely of **peridotite**, partially melts and generates **magma** that is basaltic. Because of this, almost all **volcanoes** on the **ocean floor** are basaltic. In fact, most **oceanic lithosphere** is basaltic near the surface, with **phaneritic gabbro** and **ultramafic peridotite** underneath.

Figure 4.27: Pillow basalt on sea floor near Hawai'i.

Figure 4.28: Black smoker hydrothermal vent with a colony of giant (6'+) tube worms.

When basaltic **lava** erupts underwater it emerges in small explosions and/or forms pillow-shaped structures called pillow basalts. These seafloor eruptions enable entire underwater ecosystems to thrive in the deep ocean around **mid-ocean ridges**. This ecosystem exists around tall vents emitting black, hot **mineral**-rich water called deep-sea **hydrothermal** vents, also known as **black smokers**.

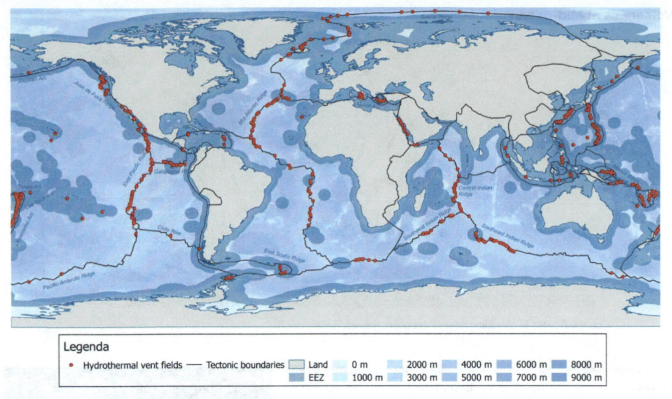

Figure 4.29: Distribution of hydrothermal vent fields.

Without sunlight to support photosynthesis, these organisms instead utilize a process called **chemosynthesis**. Certain bacteria are able to turn hydrogen **sulfide** (H_2S), a gas that smells like rotten eggs, into life-supporting nutrients and water. Larger organisms may eat these bacteria or absorb nutrients and water produced by bacteria living symbiotically inside their bodies. The videos show some of the ecosystems found around deep-sea **hydrothermal** vents.

 One or more interactive elements has been excluded from this version of the text. You can view them online here: https://www.youtube.com/watch?v=a5aQ4W9GbpU

Video 4.2: Updating the deep–diving submarine at 50-years-old.

If you are using an offline version of this text, access this YouTube video via the QR code.

One or more interactive elements has been excluded from this version of the text. You can view them online here: https://www.youtube.com/watch?v=dXOQFnU-49k

Video 4.3: Incredible views on board the deep-sea vessel.

If you are using an offline version of this text, access this YouTube video via the QR code.

Volcanoes at Subduction Zones

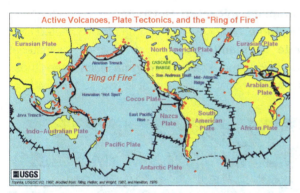

Figure 4.30: Distribution of volcanoes on the planet. Click here for an interactive map of volcano distributions.

The second most commonly found location for **volcanism** is adjacent to **subduction** zones, a type of **convergent plate boundary** (see chapter 2). The process of **subduction** expels water from hydrated **minerals** in the descending **slab**, which causes **flux melting** in the overlying **mantle** rock. Because **subduction volcanism** occurs in a **volcanic arc**, the thickened **crust** promotes **partial melting** and **magma** differentiation. These evolve the **mafic magma** from the mantle into more silica-rich **magma**. The Ring of Fire surrounding the Pacific Ocean, for example, is dominated by **subduction**-generated eruptions of mostly silica-rich **lava**; the **volcanoes** and **plutons** consist largely of **intermediate**-to-**felsic** rock such as **andesite, rhyolite, pumice**, and **tuff**.

Volcanoes at Continental Rifts

Some **volcanoes** are created at **continental rifts**, where crustal thinning is caused by diverging lithospheric **plates**, such as the East African Rift Basin in Africa. **Volcanism** caused by crustal thinning without **continental rifting** is found in the **Basin and Range** Province in North America. In this location, **volcanic** activity is produced by rising **magma** that stretches the overlying **crust**. Lower **crust** or upper **mantle** material rises through the thinned **crust**, releases pressure, and undergoes decompression-induced **partial melting**. This **magma** is less dense than the surrounding rock and continues to rise through the **crust** to the surface, erupting as basaltic **lava**. These eruptions usually result in **flood basalts**, **cinder** cones, and basaltic **lava** flows (see video). Relatively young **cinder** cones of basaltic **lava** can be found in south-central Utah, in the Black Rock Desert Volcanic Field, which is part of the zone of **Basin and Range** crustal **extension**.

Figure 4.31: Basaltic cinder cones of the Black Rock Desert near Beaver, Utah.

These Utah **cinder** cones and **lava** flows started erupting around 6 million years ago, with the last eruption occurring 720 years ago.

One or more interactive elements has been excluded from this version of the text. You can view them online here:
https://www.youtube.com/watch?v=4VgMe-JXOAM

Video 4.4: Basin and range volcanic processes.

If you are using an offline version of this text, access this YouTube video via the QR code.

Hotspots

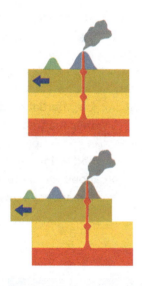

Figure 4.32: Diagram showing a non-moving source of magma (mantle plume) and a moving overriding plate.

Hotspots are the main source of **intraplate volcanism**. **Hotspots** occur when lithospheric **plates** glide over a hot **mantle plume**, an ascending column of solid heated rock originating from deep within the **mantle**. The **mantle plume** generates melts as material rises, with the **magma** rising even more. When the ascending **magma** reaches the lithospheric crust, it spreads out into a mushroom-shaped head that is tens to hundreds of kilometers across.

Since most **mantle plumes** are located beneath the **oceanic lithosphere**, the early stages of **intraplate volcanism** typically take place underwater. Over time, basaltic **volcanoes** may build up from the sea floor into islands, such as the Hawaiian Islands. Where a **hotspot** is found under a **continental plate**, contact with the hot **mafic** magma may cause the overlying **felsic** rock to melt and mix with the mafic material below, forming **intermediate magma**. Or the **felsic magma** may continue to rise, and cool into a granitic **batholith** or erupt as a **felsic volcano**. The Yellowstone caldera is an example of **hotspot volcanism** that resulted in an explosive eruption.

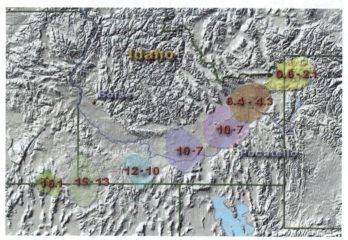

Figure 4.33: The track of the Yellowstone hotspot, which shows the age of different eruptions in millions of years ago.

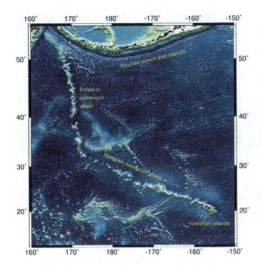

Figure 4.34: The Hawaiian–Emperor seamount and island chain.

A zone of actively erupting **volcanism** connected to a chain of **extinct volcanoes** indicates **intraplate volcanism** located over a **hotspot**. These **volcanic** chains are created by the overriding **oceanic plate** slowly moving over a **hotspot mantle plume**. These chains are seen on the seafloor and continents and include **volcanoes** that have been inactive for millions of years. The Hawaiian Islands on the Pacific Oceanic plate are the active end of a long **volcanic** chain that extends from the northwest Pacific Ocean to the Emperor Seamounts, all the way to the to the **subduction** zone beneath the Kamchatka Peninsula. The overriding North American **continental plate** moved across a **mantle plume hotspot** for several million years, creating a chain of **volcanic** calderas that extends from Southwestern Idaho to the presently active Yellowstone caldera in Wyoming.

The three-minute video (below) illustrates **hotspot volcanoes**.

 One or more interactive elements has been excluded from this version of the text. You can view them online here:
https://www.youtube.com/watch?v=AhSaE0omw9o

Video 4.5: What is a volcanic hotspot?

If you are using an offline version of this text, access this YouTube video via the QR code.

4.4.2 Volcano Features and Types

There are several different types of **volcanoes** based on their shape, eruption style, magmatic **composition**, and other aspects.

Complete this interactive activity to check your understanding.

If you are using an offline version of this text, access this interactive activity via the QR code.

 An interactive H5P element has been excluded from this version of the text. You can view it online here:
https://pressbooks.lib.vt.edu/introearthscience/?p=331#h5p-28

The figure shows the main features of a typical **stratovolcano**: 1) **magma chamber**, 2) upper layers of **lithosphere**, 3) the **conduit** or narrow pipe through which the **lava** erupts, 4) the base or edge of the **volcano**, 5) a **sill** of **magma** between layers of the **volcano**, 6) a **diapir** or feeder tube to the **sill**, 7) layers of **tephra** (**ash**) from previous eruptions, 8 & 9) layers of **lava** erupting from the **vent** and flowing down the sides of the **volcano**, 10) the crater at the top of the **volcano**, 11) layers of **lava** and **tephra** on (12), a **parasitic cone**. A **parasitic cone** is a small **volcano** located on the flank of a larger volcano such as Shastina on Mount Shasta. Kilauea sitting on the flank of Mauna Loa is not considered a **parasitic cone** because it has its own separate **magma chamber**, 13) the vents of the parasite and the main **volcano**, 14) the rim of the crater, 15) clouds of **ash** blown into the sky by the eruption; this settles back onto the **volcano** and surrounding land.

Figure 4.35: Mt. Shasta in Washington state with Shastina, its parasitic cone.

Figure 4.36: Oregon's Crater Lake was formed about 7700 years ago after the eruption of Mount Mazama.

The largest craters are called **calderas**, such as the Crater Lake Caldera in Oregon. Many **volcanic** features are produced by **viscosity**, a basic property of a **lava**. **Viscosity** is the resistance to flowing by a fluid. Low **viscosity magma** flows easily more like syrup, the basaltic **volcanism** that occurs in Hawai'i on **shield volcanoes**. High **viscosity** means a thick and sticky **magma**, typically **felsic** or **intermediate**, that flows slowly, similar to toothpaste.

Shield Volcano

Figure 4.37: Kilauea in Hawai'i.

The largest **volcanoes** are **shield volcanoes**. They are characterized by broad low-angle flanks, small vents at the top, and **mafic magma** chambers. The name comes from the side view, which resembles a medieval warrior's **shield**. They are typically associated with **hotspots**, **mid-ocean ridges**, or **continental rifts** with rising upper **mantle** material. The low-angle flanks are built up slowly from numerous low-**viscosity** basaltic **lava** flows that spread out over long distances. The basaltic **lava** erupts effusively, meaning the eruptions are small, localized, and predictable.

Typically, **shield volcano** eruptions are not much of a hazard to human life—although non-explosive eruptions of Kilauea (Hawai'i) in 2018 produced uncharacteristically large lavas that damaged roads and structures. Mauna Loa (see USGS page) and Kilauea (see USGS page) in Hawai'i are examples of **shield volcanoes**. **Shield volcanoes** are also found in Iceland, the Galapagos Islands, Northern California, Oregon, and the East African Rift.

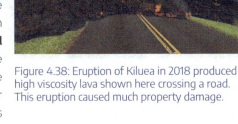

The largest **volcanic** edifice in the **Solar System** is Olympus Mons on Mars. This (possibly **extinct**) **shield volcano** covers an area the size of the state of Arizona. This may indicate the **volcano** erupted over a **hotspot** for millions of years, which means Mars had little, if any, **plate tectonic** activity.

Figure 4.38: Eruption of Kiluea in 2018 produced high viscosity lava shown here crossing a road. This eruption caused much property damage.

Figure 4.39: Olympus Mons, an enormous shield volcano on Mars, the largest volcano in the solar system, standing about two and a half times higher than Everest is above sea level.

Basaltic **lava** forms special landforms based on **magma temperature**, **composition**, and content of **dissolved** gases and water vapor. The two main types of basaltic **volcanic rock** have Hawaiian names—*pahoehoe* and *aa*. **Pahoehoe** might come from low-**viscosity** lava that flows easily into ropey strands.

Aa (sometimes spelled a'a or **'a'ā** and pronounced "ah-ah") is more

Figure 4.40: Ropey pahoehoe lava.

viscous and has a crumbly blocky appearance. The exact details of what forms the two types of flows are still up for debate. **Felsic** lavas have lower temperatures and more silica, and thus are higher **viscosity**. These also form **aa**-style flows.

Low-**viscosity**, fast-flowing basaltic **lava** tends to harden on the outside into a tube and continue to flow internally. Once **lava** flow sub-

Figure 4.41: Blocky a'a lava.

sides, the empty outer shell may be left as a **lava** tube. **Lava** tubes, with or without collapsed roofs, make famous caves in Hawai'i, Northern California, the Columbia River Basalt Plateau of Washington and Oregon, El Malpais National Monument in New Mexico, and Craters of the Moon National Monument in Idaho.

Figure 4.42: Volcanic fissure and flow, which could eventually form a lava tube.

Fissures are cracks that commonly originate from **shield**-style eruptions. **Lava** emerging from fissures is typically **mafic** and very fluid. The 2018 Kiluaea eruption included fissures associated with the **lava** flows. Some fissures are caused by the **volcanic seismic** activity rather than **lava** flows. Some fissures are influenced by **plate tectonics**, such as the common fissures located parallel to the **divergent** boundary in Iceland.

Figure 4.43: Devils Tower in Wyoming has columnar jointing.

Cooling **lava** can contract into columns with semi-hexagonal cross sections called columnar jointing. This feature forms the famous Devils Tower in Wyoming, possibly an ancient **volcanic vent** from which the surrounding layers of **lava** and **ash** have been removed by **erosion**. Another well-known exposed example of columnar jointing is the Giant's Causeway in Ireland.

Figure 4.44: Columnar jointing on Giant's Causeway in Ireland.

Stratovolcano

A **stratovolcano**, also called a **composite cone volcano**, has steep flanks, a symmetrical cone shape, distinct crater, and rises prominently above the surrounding landscape. The term composite refers to the alternating layers of **pyroclastic** fragments like **ash** and **bombs**, and solidified **lava** flows of varying **composition**. Examples include Mount Rainier in Washington state and Mount Fuji in Japan.

Figure 4.45: Mount Rainier towers over Tacoma, Washington.

Stratovolcanoes usually have **felsic** to **intermediate magma** chambers, but can even produce **mafic** lavas. Stratovolcanoes have **viscous lava** flows and **domes**, punctuated by explosive eruptions. This produces **volcanoes** with steep flanks.

Lava Domes

Lava domes are accumulations of silica-rich **volcanic rock**, such as **rhyolite** and

Figure 4.46: Mt. Fuji in Japan, a typical stratovolcano, symmetrical, increasing slope, visible crater at the top.

obsidian. Too **viscous** to flow easily, the **felsic lava** tends to pile up near the **vent** in blocky masses. **Lava domes** often form in a **vent** within the collapsed crater of a **stratovolcano**, and grow by internal expansion. As the **dome** expands, the outer surface cools, hardens, and shatters, and spills loose fragments down the sides. Mount Saint Helens has a good example of a **lava dome** inside of a collapsed **stratovolcano** crater. Examples of stand-alone **lava domes** are Chaiten in Chile and Mammoth Mountain in California.

Figure 4.47: Lava domes have started the rebuilding process at Mount St. Helens, Washington.

Caldera

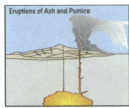

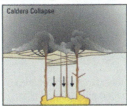

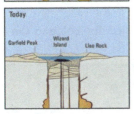

Figure 4.48: Timeline of events at Mount Mazama.

Calderas are steep-walled, **basin**-shaped depressions formed by the collapse of a **volcanic** edifice into an empty **magma chamber**. Calderas are generally very large, with diameters of up to 25 km (15.5 mi). The term **caldera** specifically refers to a **volcanic vent**; however, it is frequently used to describe a **volcano** type. **Caldera volcanoes** are typically formed by eruptions of high-**viscosity felsic lava** having high **volatiles** content.

Crater Lake, Yellowstone, and the Long Valley Caldera are good examples of this type of **volcanism**. The **caldera** at Crater Lake National Park in Oregon was created about 6,800 years ago when Mount Mazama, a **composite volcano**, erupted in a huge explosive blast. The **volcano** ejected large amounts of **volcanic ash** and rapidly drained the **magma chamber**, causing the top to collapse into a large depression that later filled with water. Wizard Island in the middle of the lake is a later resurgent **lava dome** that formed within the **caldera basin**.

Figure 4.49: Wizard Island sits in the caldera at Crater Lake.

The Yellowstone **volcanic system** erupted three times in the recent geologic past—2.1, 1.3, and 0.64 million years ago—leaving behind three **caldera** basins. Each eruption created large **rhyolite lava** flows as well as **pyroclastic** flows that solidified into **tuff formations**. These extra-large eruptions rapidly emptied the **magma chamber**, causing the roof to collapse and form a **caldera**. The youngest of the three calderas contains most of Yellowstone National Park, as well as two resurgent **lava domes**. The calderas are difficult to see today due to the amount of time since their eruptions and subsequent **erosion** and **glaciation**.

Yellowstone **volcanism** started about 17-million years ago as a **hotspot** under the North American lithospheric **plate** near the Oregon/Nevada border. As the **plate** moved to the southwest over the stationary **hotspot**, it left behind a track of past **volcanic** activities. Idaho's Snake **River** Plain was created from **volcanism** that produced a series of calderas and **lava** flows. The **plate** eventually arrived at its current location in northwestern Wyoming, where **hotspot volcanism** formed the Yellowstone calderas.

Figure 4.50: Map of calderas and related rocks around Yellowstone.

The Long Valley Caldera near Mammoth, California, is the result of a large **volcanic** eruption that occurred 760,000 years ago. The explosive eruption dumped enormous amounts of **ash** across the United States, in a manner similar to the Yellowstone eruptions. The Bishop Tuff deposit near Bishop, California, is made of **ash** from this eruption. The current **caldera basin** is 17 km by 32 km (10 mi by 20 mi), large enough to contain the town of Mammoth Lakes, major ski resort, airport, major highway, resurgent **dome**, and several hot springs.

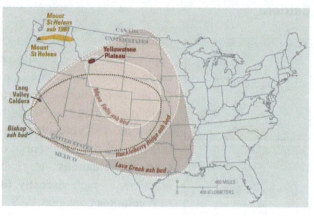

Figure 4.51: Several prominent ash beds found in North America, including three Yellowstone eruptions shaded pink (Mesa Falls, Huckleberry Ridge, and Lava Creek), the Bisho Tuff ash bed (brown dashed line), and the modern May 18th, 1980 ash fall (yellow).

Cinder Cone

Figure 4.52: Sunset Crater, Arizona is a cinder cone.

Cinder cones are small **volcanoes** with steep sides, and made of **pyroclastic** fragments that have been ejected from a pronounced central **vent**. The small fragments are called **cinders** and the largest are **volcanic bombs**. The eruptions are usually short-lived events, typically consisting of **mafic** lavas with a high content of **volatiles**. Hot **lava** is ejected into the air, cooling and solidifying into fragments that accumulate on the flank of the **volcano**. **Cinder** cones are found throughout western North America.

Figure 4.54: Lava from Parícutin covered the local church and destroyed the town of San Juan, Mexico.

A recent and striking example of a **cinder cone** is the eruption near the village of Parícutin, Mexico that started in 1943. The **cinder cone** started explosively shooting **cinders** out of the **vent** in the middle of a farmer's field. The **volcanism** quickly built up the cone to a height of over 90 m (300 ft) within a week, and 365 m (1,200 ft) within the first 8 months. After the initial explosive eruption of gases and **cinders**, basaltic **lava** poured out from the base of the cone. This is a common order of events for **cinder** cones: violent eruption, cone and crater **formation**, low-**viscosity lava** flow from the base. The **cinder cone** is not strong enough to support a column of **lava** rising to the top of the crater, so the **lava** breaks through and emerges near the bottom of the **volcano**. During nine years of eruption activity, the ashfall covered about 260 km² (100 mi²) and destroyed the nearby town of San Juan.

Figure 4.53: Soon after the birth of Parícutin in 1943.

Flood Basalts

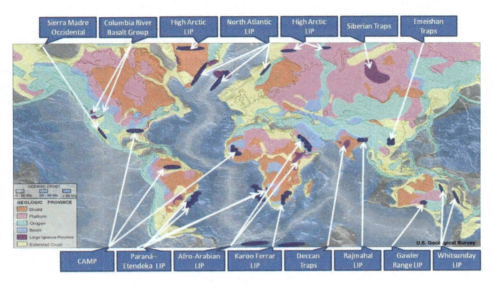

Figure 4.55: World map of flood basalts. Note the largest is the Siberian Traps.

A rare **volcanic** eruption type, unobserved in modern times, is the **flood basalt**. **Flood basalts** are some of the largest and lowest **viscosity** types of eruptions known. They are not known from any eruption in human history, so the exact mechanisms of eruption are still mysterious. Some famous examples include the Columbia River Flood Basalts in Washington, Oregon, and Idaho, the Deccan Traps, which cover about 1/3 of the country of India, and the Siberian Traps, which may have been involved in the Earth's largest **mass extinction** (see chapter 8).

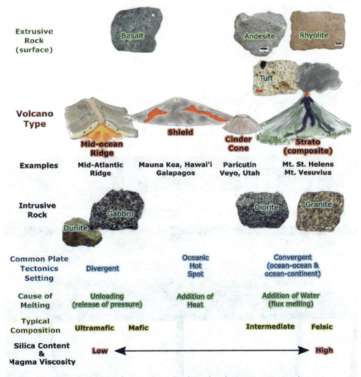

Figure 4.56: Igneous rock types and related volcano types. Mid-ocean ridges and shield volcanoes represent more mafic compositions, and strato (composite) volcanoes generally represent a more intermediate or felsic composition and a convergent plate tectonic boundary. Note that there are exceptions to this generalized layout of volcano types and igneous rock composition.

4.4.3 Volcanic Hazards and Monitoring

While the most obvious **volcanic** hazard is **lava**, the dangers posed by **volcanoes** go far beyond **lava** flows. For example, on May 18, 1980, Mount Saint Helens (Washington, United States) erupted with an explosion and **landslide** that removed the upper 400 m (1,300 ft) of the mountain. The initial explosion was immediately followed by a lateral blast, which produced a **pyroclastic flow** that covered nearly 600 km² (230 mi²) of forest with hot **ash** and debris. The pyroclastic flow moved at speeds of 80-130 kph (50-80 mph), flattening trees and ejecting clouds of ash into the air. The USGS video provides an account of this explosive eruption that killed 57 people.

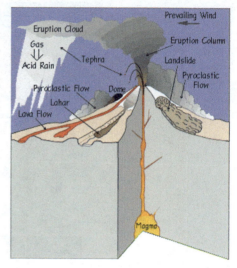

Figure 4.57: General diagram of volcanic hazards.

 One or more interactive elements has been excluded from this version of the text. You can view them online here:
https://www.youtube.com/watch?v=Ec30uU0G56U

Video 4.6: Mout St. Helens.

If you are using an offline version of this text, access this YouTube video via the QR code.

Figure 4.58: Human remains from the 79 CE eruption of Vesuvius.

In 79 AD, Mount Vesuvius, located near Naples, Italy, violently erupted sending a **pyroclastic flow** over the Roman countryside, including the cities of Herculaneum and Pompeii. The buried towns were discovered in an archeological expedition in the 18th century. Pompeii famously contains the remains (**casts**) of people suffocated by ash and covered by 10 feet (3 m) of **ash**, **pumice lapilli**, and collapsed roofs.

Before   **After**

Figure 4.59: Mount St. Helens, the day before the May 18th, 1980 eruption (left) and 4 months after the major eruption (right).

Figure 4.60: Image from the May 18, 1980, eruption of Mt. Saint Helens, Washington.

Pyroclastic Flows

The most dangerous **volcanic** hazard are **pyroclastic** flows (video). These flows are a mix of **lava** blocks, **pumice**, **ash**, and hot gases between 200°C-700°C (400°F-1,300°F). The turbulent cloud of **ash** and gas races down the steep flanks at high speeds up to 193 kph (120 mph) into the valleys around composite **volcanoes**. Most explosive, silica-rich, high **viscosity magma volcanoes** such as composite cones usually have **pyroclastic** flows. The rock **tuff** and welded **tuff** is often formed from these **pyroclastic** flows.

Figure 4.61: The material coming down from the eruption column is a pyroclastic flow.

There are numerous examples of deadly **pyroclastic** flows. In 2014, the Mount Ontake **pyroclastic flow** in Japan killed 47 people. The flow was caused by **magma** heating **groundwater** into steam, which then rapidly ejected with **ash** and **volcanic bombs**. Some were killed by inhalation of toxic gases and hot ash, while others were struck by volcanic bombs. Two short videos below document eye-witness video of **pyroclastic** flows. In the early 1990s, Mount Unzen erupted several times with **pyroclastic** flows. The **pyroclastic flow** shown in this famous short video killed 41 people. In 1902, on the Caribbean Island Martinique, Mount Pelee erupted with a violent **pyroclastic flow** that destroyed the entire town of St. Pierre and killing 28,000 people in moments.

Figure 4.62: The remains of St. Pierre.

 One or more interactive elements has been excluded from this version of the text. You can view them online here: https://www.youtube.com/watch?v=Cvjwt9nnwXY

Video 4.7: Dome collapse and pyroclastic flow at Unzen Volcano.

If you are using an offline version of this text, access this YouTube video via the QR code.

Landslides and Landslide–Generated Tsunamis

The steep and unstable flanks of a **volcano** can lead to slope failure and dangerous **landslides**. These **landslides** can be triggered by **magma** movement, explosive eruptions, large earthquakes, and/or heavy rainfall. During the 1980 Mount St. Helens eruption, the entire north flank of the **volcano** collapsed and released a huge **landslide** that moved at speeds of 160-290 kph (100-180 mph).

If enough **landslide** material reaches the ocean, it may cause a **tsunami**. In 1792, a **landslide** caused by the Mount Unzen eruption reached the Ariaka Sea, generating a **tsunami** that killed 15,000 people (see USGS page). When Mount Krakatau in Indonesia erupted in 1883, it generated ocean waves that towered 40 m (131 ft) above sea level. The **tsunami** killed 36,000 people and destroyed 165 villages.

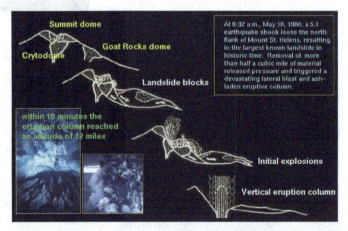

Figure 4.63: Sequence of events for Mount St. Helens, May 18th, 1980. Note that an earthquake caused a landslide, which caused the "uncorking" of the mountain and started the eruption.

Tephra

Figure 4.64: Aman sweeps ash from an eruption of Kelud, Indonesia.

Volcanoes, especially composite **volcanoes**, eject large amounts of **tephra** (ejected rock materials), most notably **ash** (**tephra** fragments less than 0.08 inches [2 mm]). Larger **tephra** is heavier and **falls** closer to the **vent**. Larger blocks and **bombs** pose hazards to those close to the eruption such as at the 2014 Mount Ontake disaster in Japan discussed earlier.

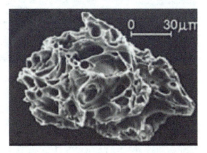

Figure 4.65: Micrograph of silica particle in volcanic ash. A cloud of these is capable of destroying an aircraft or automobile engine.

Hot **ash** poses an immediate danger to people, animals, plants, machines, roads, and buildings located close to the eruption. **Ash** is fine grained (< 2mm) and can travel airborne long distances away from the eruption site. Heavy accumulations of **ash** can cause buildings to collapse. In people, it may cause respiratory issues like silicosis. **Ash** is destructive to aircraft and automobile engines, which can disrupt transportation and shipping services. In 2010, the Eyjafjallajökull volcano in Iceland emitted a large **ash** cloud into the upper **atmosphere**, causing the largest air-travel disruption in northern Europe since World War II. No one was injured, but the service disruption was estimated to have cost the world economy billions of dollars.

Volcanic Gases

As **magma** rises to the surface the **confining** pressure decreases, and allows **dissolved** gases to escape into the **atmosphere**. Even **volcanoes** that are not actively erupting may emit hazardous gases, such as carbon dioxide (CO_2), sulfur dioxide (SO_2), hydrogen **sulfide** (H_2S), and hydrogen **halides** (HF, HCl, or HBr).

Carbon dioxide tends to sink and accumulate in depressions and basins. In **volcanic** areas known to emit carbon dioxide, low-lying areas may **trap** hazardous concentrations of this colorless and odorless gas. The Mammoth Mountain Ski Resort in California, is located within the Long Valley Caldera, is one such area of carbon dioxide-producing **volcanism**. In 2006, three ski patrol members died of suffocation caused by carbon dioxide after falling into a snow depression near a fumarole (info).

In rare cases, **volcanism** may create a sudden emission of gases without warning. Limnic eruptions (*limne* is Greek for lake), occur in crater lakes associated with active **volcanism**. The water in these lakes is supercharged with high concentrations of **dissolved** gases. If the water is physically jolted by a **landslide** or earthquake, it may **trigger** an immediate and **massive** release of gases out of **solution**. An analogous example would be what happens to vigorously shaken bottle of carbonated soda when the cap is opened. An infamous limnic eruption occurred in 1986 at Lake Nyos, Cameroon. Almost 2,000 people were killed by a **massive** release of carbon dioxide.

Lahars

Figure 4.66: Mud line shows the extent of lahars around Mount St. Helens.

Lahar is an Indonesian word and is used to describe a **volcanic** mudflow that forms from rapidly melting snow or **glaciers**. **Lahars** are slurries resembling wet concrete, and consist of water, **ash**, rock fragments, and other debris. These mudflows flow down the flanks of **volcanoes** or mountains covered with freshly-erupted **ash** and on steep slopes can reach speeds of up to 80 kph (50 mph).

Several major cities, including Tacoma, are located on prehistoric **lahar** flows that extend for many kilometers across the flood plains surrounding Mount Rainier in Washington (see map). A map of Mount Baker in Oregon shows a similar potential hazard for **lahar** flows (see map). A tragic scenario played out recently, in 1985, when a **lahar** from the Nevado del Ruiz **volcano** in Colombia buried the town of Armero and killed an estimated 23,000 people.

Monitoring

Geologists use various instruments to detect changes or indications that an eruption is imminent. The three videos show different types of **volcanic** monitoring used to predict eruptions 1) earthquake activity; 2) increases in gas emission; and 3) changes in land surface orientation and elevation.

Figure 4.67: Old lahars around Tacoma, Washington.

One video shows how monitoring earthquake frequency, especially special vibrational earthquakes called harmonic tremors, can detect **magma** movement and possible eruption. Another video shows how gas monitoring may be used to predict an eruption. A rapid increase of gas emission may indicate **magma** that is actively rising to surface and releasing **dissolved** gases out of **solution**, and that an eruption is imminent. The last video shows how a GPS unit and tiltmeter can detect land surface changes, indicating the **magma** is moving underneath it.

 One or more interactive elements has been excluded from this version of the text. You can view them online here:
https://www.youtube.com/watch?v=nlo-2JoNHrw

Video 4.8: Earthquake signals.

If you are using an offline version of this text, access this YouTube video via the QR code.

One or more interactive elements has been excluded from this version of the text. You can view them online here:
https://www.youtube.com/watch?v=owk4fWbw4qM

Video 4.9: Measuring gas emissions.

If you are using an offline version of this text, access this YouTube video via the QR code.

 One or more interactive elements has been excluded from this version of the text. You can view them online here:
https://www.youtube.com/watch?v=sNYQkxxd_0Q

Video 4.10: Using tiltmeters and GPS to monitor a volcano.

If you are using an offline version of this text, access this YouTube video via the QR code.

Take this quiz to check your comprehension of this section.

If you are using an offline version of this text, access the quiz for section 4.4 via the QR code.

 An interactive H5P element has been excluded from this version of the text. You can view it online here:
https://pressbooks.lib.vt.edu/introearthscience/?p=331#h5p-29

Summary

Igneous rock is divided into two major groups: **intrusive** rock that solidifies from underground **magma**, and **extrusive** rock formed from **lava** that erupts and cools on the surface. **Magma** is generated from **mantle** material at several **plate tectonics** situations by three types of melting: **decompression melting**, **flux melting**, or heat-induced melting. **Magma composition** is determined by differences in the melting temperatures of the **mineral** components (**Bowen's Reaction Series**). The processes affecting **magma composition** include **partial melting**, **magmatic differentiation**, **assimilation**, and **col-**

lision. **Volcanoes** come in a wide variety of shapes and sizes, and are classified by a multiple factors, including **magma composition**, and **plate tectonic** activity. Because **volcanism** presents serious hazards to human civilization, geologists carefully monitor **volcanic** activity to mitigate or avoid the dangers it presents.

Take this quiz to check your comprehension of this chapter.

If you are using an offline version of this text, access the quiz for chapter 4 via the QR code.

An interactive H5P element has been excluded from this version of the text. You can view it online here: https://pressbooks.lib.vt.edu/introearthscience/?p=331#h5p-30

URLs Listed Within This Chapter

USGS page of Mauna Loa: https://www.usgs.gov/volcanoes/mauna-loa

USGS page on Kilauea: https://www.usgs.gov/volcanoes/kilauea

Pyroclastic flows video: https://volcanoes.usgs.gov/vsc/movies/movie_101/PF_Animation.mp4

USGS page on volcano landslides that trigger waves and tsunamis: https://volcanoes.usgs.gov/Imgs/Jpg/Unzen/MayuyamaSlide_caption.html

Ski patrol's fatal fall into a volcanic fumarole: https://pubmed.ncbi.nlm.nih.gov/19364170/

Text References

1. Arndt, N.T., 1994, Chapter 1 Archean Komatiites, *in* K.C. Condie, editor, Developments in Precambrian Geology: Elsevier, p. 11–44.
2. Bateman, P.C., and Chappell, B.W., 1979, Crystallization, fractionation, and solidification of the Tuolumne Intrusive Series, Yosemite National Park, California: Geological Society of America Bulletin, v. 90, no. 5, p. 465–482., https://doi.org/10.1130/0016-7606(1979)90<465:CFASOT>2.0.CO;2
3. Boehler, R., 1996, Melting temperatures of the Earth's mantle and core: Earth's thermal structure: Annual Review of Earth and Planetary Sciences, v. 24, no. 1, p. 15–40., doi: 10.1146/annurev.earth.24.1.15.
4. Bowen, N.L., 1922, The Reaction Principle in Petrogenesis: J. Geol., v. 30, no. 3, p. 177–198.
5. Bowen, N.L., 1928, The evolution of the igneous rocks: Dover Publications, 334 p.
6. Carr, M.H., 1975, Geologic map of the Tharsis Quadrangle of Mars: IMAP.
7. Earle, S., 2015, Physical geology OER textbook: BC Campus OpenEd.
8. EarthScope, 2014, Mount Ontake Volcanic Eruption: Online, http://www.earthscope.org/science/geo-events/mount-ontake-volcanic-eruption, accessed July 2016.
9. Frankel, C., 2005, Worlds on Fire: Volcanoes on the Earth, the Moon, Mars, Venus and Io: Cambridge University Press, 396 p.
10. Glazner, A.F., Bartley, J.M., Coleman, D.S., Gray, W., and Taylor, R.Z., 2004, Are plutons assembled over millions of years by amalgamation from small magma chambers? GSA Today, v. 14, no. 4, p. 4., DOI: 10.1130/1052-5173(2004)014<0004:APAOMO>2.0.CO;2
11. Luongo, G., Perrotta, A., Scarpati, C., De Carolis, E., Patricelli, G., and Ciarallo, A., 2003, Impact of the AD 79 explosive

eruption on Pompeii, II. Causes of death of the inhabitants inferred by stratigraphic analysis and areal distribution of the human casualties: J. Volcanol. Geotherm. Res., v. 126, no. 3–4, p. 169–200.

12. Mueller, S., and Phillips, R.J., 1991, On the initiation of subduction: J. Geophys. Res. [Solid Earth], v. 96, no. B1, p. 651–665.

13. Peacock, M.A., 1931, Classification of Igneous Rock Series: The Journal of Geology, v. 39, no. 1, p. 54–67.

14. Perkins, S., 2011, 2010's Volcano-Induced Air Travel Shutdown Was Justified: Online, http://www.sciencemag.org/news/2011/04/2010s-volcano-induced-air-travel-shutdown-was-justified, accessed July 2016.

15. Peterson, D.W., and Tilling, R.I., 1980, Transition of basaltic lava from pahoehoe to aa, Kilauea Volcano, Hawaii: Field observations and key factors – ScienceDirect: J. Volcanol. Geotherm. Res., v. 7, no. 3–4, p. 271–293.

16. Petrini and Podladchikov, 2000, Lithospheric pressure–depth relationship in compressive regions of thickened crust: Journal of Metamorphic Geology, v. 18, no. 1, p. 67–77., doi: 10.1046/j.1525-1314.2000.00240.x.

17. Reid, J.B., Evans, O.C., and Fates, D.G., 1983, Magma mixing in granitic rocks of the central Sierra Nevada, California: Earth and Planetary Science Letters, v. 66, p. 243–261., doi: 10.1016/0012-821X(83)90139-5.

18. Rhodes, J.M., and Lockwood, J.P., 1995, Mauna Loa Revealed: Structure, Composition, History, and Hazards: Washington DC American Geophysical Union Geophysical Monograph Series, v. 92.

19. Scandone, R., Giacomelli, L., and Gasparini, P., 1993, Mount Vesuvius: 2000 years of volcanological observations: Journal of Volcanology and Geothermal Research, v. 58, p. 5–25.

20. Stovall, W.K., Wilkins, A.M., Mandeville, C.W., and Driedger, C.L., 2016, Fact Sheet.

21. Thorarinsson, S., 1969, The Lakagigar eruption of 1783: Bull. Volcanol., v. 33, no. 3, p. 910–929.

22. Tilling, R.I., 2008, The critical role of volcano monitoring in risk reduction: Adv. Geosci., v. 14, p. 3–11.

23. United States Geological Survey, 1999, Exploring the deep ocean floor: Online, http://pubs.usgs.gov/gip/dynamic/exploring.html, accessed July 2016.

24. United States Geological Survey, 2012, Black Rock Desert Volcanic Field: Online, http://volcanoes.usgs.gov/volcanoes/black_rock_desert/, accessed July 2016.

25. USGS, 2001, Dual volcanic tragedies in the Caribbean led to founding of HVO: Online, http://hvo.wr.usgs.gov/volcanowatch/archive/2001/01_05_03.html, accessed July 2016.

26. USGS, 2011, Volcanoes: Principal Types of Volcanoes: Online, http://pubs.usgs.gov/gip/volc/types.html, accessed July 2016.

27. USGS, 2012a, USGS: Volcano Hazards Program: Online, https://volcanoes.usgs.gov/vhp/hazards.html, accessed July 2016.

28. USGS, 2012b, Yellowstone Volcano Observatory: Online, https://volcanoes.usgs.gov/volcanoes/yellowstone/yellowstone_geo_hist_52.html, accessed July 2016.

29. USGS, 2016, Volcanoes General – What are the different types of volcanoes? Online, https://www2.usgs.gov/faq/categories/9819/2730, accessed March 2017.

30. USGS, 2017, The Volcanoes of Lewis and Clark – Mount St. Helens: Online, https://volcanoes.usgs.gov/observatories/cvo/Historical/LewisClark/Info/summary_mount_st_helens.shtml, accessed March 2017.

31. Wallace, P.J., 2005, Volatiles in subduction zone magmas: concentrations and fluxes based on melt inclusion and volcanic gas data: Journal of Volcanology and Geothermal Research, v. 140, no. 1–3, p. 217–240., doi: 10.1016/j.jvolgeores.2004.07.023.

32. Williams, H., 1942, The Geology of Crater Lake National Park, Oregon: With a Reconnaissance of the Cascade Range Southward to Mount Shasta: Carnegie institution.

Figure References

Figure 4.1: Lava flow in Hawai'i. Brocken Inaglory. 2007. CC BY-SA 3.0. https://commons.wikimedia.org/wiki/File:P%C4%81hoehoe_and_Aa_flows_at_Hawaii.jpg

Figure 4.2: Half Dome, an intrusive igneous batholith in Yosemite National Park. Jon Sullivan. 2004. Public domain. https://commons.wikimedia.org/wiki/File:Yosemite_20_bg_090404.jpg

Figure 4.3: Granite is a classic coarse-grained (phaneritic) intrusive igneous rock. James St. John. 2019. CC BY 2.0. https://commons.wikimedia.org/wiki/File:Granite_47_(49201189712).jpg

Figure 4.4: Basalt is a classic fine-grained extrusive igneous rock. James St. John. 2019. CC BY 2.0. https://commons.wikimedia.org/wiki/File:Basalt_3_(48674276863).jpg

Figure 4.5: Porphyritic texture. Jstuby. 2008. Public domain. https://commons.wikimedia.org/wiki/File:Olearyandesite.jpg

Figure 4.6: Pegmatitic texture. Jstuby. 2007. Public domain. https://commons.wikimedia.org/wiki/File:We-pegmatite.jpg

Figure 4.7: Scoria. Jonathan Zander (Digon3). 2008. CC BY-SA 3.0. https://commons.wikimedia.org/wiki/File:Scoria_Macro_Digon3.jpg

Figure 4.8: Pumice. deltalimatrieste. 2008. Public domain. https://commons.wikimedia.org/wiki/File:Pomice_di_veglia.jpg

Figure 4.9: Obsidian (volcanic glass). Note conchoidal fracture. Ji-Elle. 2011. CC BY-SA 3.0. https://commons.wikimedia.org/wiki/File:Lipari-Obsidienne_(5).jpg

Figure 4.10: Welded tuff. Wilson44691. 2010. Public domain. https://commons.wikimedia.org/wiki/File:HoleInThe-WallTuff.JPG

Figure 4.11: Mineral composition of common igneous rocks. Woudloper. 2009. Public domain. https://commons.wikimedia.org/wiki/File:Mineralogy_igneous_rocks_EN.svg

Figure 4.12: Igneous rock classification table with composition as vertical columns and texture as horizontal rows. Kindred Grey. 2022. Adapted from Belinda Madsen, An Introduction to Geology. OpenStax. Salt Lake Community College. CC BY-NC-SA 4.0.

Table 4.1: Aphanitic and phaneritic rock types with images. Quartz monzonite 36mw1037 by B.W. Hallett, V. F. Paskevich, L.J. Poppe, S.G. Brand, and D.S. Blackwood via USGS (Public domain, https://commons.wikimedia.org/wiki/File:Quartz_monzonite_36mw1037.jpg). PinkRhyolite by Michael C. Rygel, 2014 (CC BY-SA 3.0, https://commons.wikimedia.org/wiki/File:PinkRhyolite.tif). Diorite MA by Amcyrus2012, 2015 (CC BY 4.0, https://commons.wikimedia.org/wiki/File:Diorite_MA.JPG). Andesite by James St. John, 2014 (CC BY 2.0, https://flic.kr/p/oBkKSy). GabbroRockCreek1 by Mark A. Wilson, 2008 (Public domain, https://commons.wikimedia.org/wiki/File:GabbroRockCreek1.jpg). VesicularBasalt1 by Jstuby, 2008 (Public domain, https://commons.wikimedia.org/wiki/File:VesicularBasalt1.jpg).

Figure 4.13: Dike of olivine gabbro cuts across Baffin Island in the Canadian Arctic. Mike Beauregard. 2012. CC BY 2.0. https://en.wikipedia.org/wiki/File:Franklin_dike_on_northwestern_Baffin_Island.jpg

Figure 4.14: Igneous sill intruding between Paleozoic strata in Nova Scotia. Mikenorton. 2010. CC BY-SA 3.0. https://commons.wikimedia.org/wiki/File:Horton_Bluff_mid-Carboniferous_sill.JPG

Figure 4.15: Quartz monzonite in the Cretaceous of Montana, USA. James St. John. 2010. CC BY 2.0. https://commons.wikimedia.org/wiki/File:Butte_Quartz_Monzonite_(Late_Cretaceous,_76_Ma;_Rampart_Mountain,_northeast_of_Butte,_Montana,_USA)_1.jpg

Figure 4.16: Half Dome in Yosemite National Park, California, is a part of the Sierra Nevada batholith which is mostly made of granite. Jon Sullivan. 2004. Public domain. https://commons.wikimedia.org/wiki/File:Yosemite_20_bg_090404.jpg

Figure 4.17: The Henry Mountains in Utah are interpreted to be a laccolith, exposed by erosion of the overlying layers. Steven Mahoney. 2005. CC BY-SA 2.5. https://commons.wikimedia.org/wiki/File:Henry_Mountains,_Utah,_2005-06-01.jpg

Figure 4.18: Laccolith forms as a blister in between sedimentary strata. Erimus and Stannered. 2007. Public domain. https://commons.wikimedia.org/wiki/File:Laccolith.svg

Figure 4.19: Bowen's Reaction Series. Colivine. 2011. Public domain. https://commons.wikimedia.org/wiki/File:Bowen%27s_Reaction_Series.png

Figure 4.20: Olivine, the first mineral to crystallize in a melt. S kitahashi. 2006. CC BY-SA 3.0. https://commons.wikimedia.org/wiki/File:Peridot2.jpg

Figure 4.21: Norman L. Bowen. Unknown author. 1909. Public domain. https://commons.wikimedia.org/wiki/File:NormanL-Bowen_1909.jpg

Figure 4.22: Norman L. Bowen and his colleague working at the Carnegie Institution of Washington Geophysical Laboratory. Smithsonian Institution. 2010. Public domain. https://commons.wikimedia.org/wiki/File:(left_to_right)-_Norman_Levi_Bowen_(1887-1956)_and_Orville_Frank_Tuttle_(1916-1983)_(4730112454)_(cropped).jpg

Figure 4.23: Geothermal gradient. Bkilli1. 2013. CC BY-SA 3.0. https://commons.wikimedia.org/wiki/File:Temperature_schematic_of_inner_Earth.jpg

Figure 4.24: Pressure-temperature diagram showing temperature in degrees Celsius on the x-axis and depth below the surface in kilometers (km) on the y-axis. Kindred Grey. 2022. CC BY-SA 3.0. Adapted from Partial melting asthenosphere EN by Woudloper, 2010 (CC BY-SA 3.0, https://commons.wikimedia.org/wiki/File%3APartial_melting_asthenosphere_EN.svg).

Figure 4.25: Association of volcanoes with plate boundaries. Jose F. Vigil via USGS. 1997. Public domain. https://commons.wikimedia.org/wiki/File:Tectonic_plate_boundaries.png

Figure 4.26: Map of spreading ridges throughout the world. Eric Gaba. 2006. CC BY-SA 2.5. https://commons.wikimedia.org/wiki/File:Spreading_ridges_volcanoes_map-fr.svg

Figure 4.27: Pillow basalt on sea floor near Hawai'i. NOAA. 1988. Public domain. https://commons.wikimedia.org/wiki/File:Nur05018-Pillow_lavas_off_Hawaii.jpg

Figure 4.28: Black smoker hydrothermal vent with a colony of giant (6'+) tube worms. NOAA. 2006. Public domain. https://commons.wikimedia.org/wiki/File:Main_Endeavour_black_smoker.jpg

Figure 4.29: Distribution of hydrothermal vent fields. DeDuijn. 2016. CC BY-SA 4.0. https://commons.wikimedia.org/wiki/File:Distribution_of_hydrothermal_vent_fields.png

Figure 4.30: Distribution of volcanoes on the planet. USGS. 2007. Public domain. https://commons.wikimedia.org/wiki/File:Map_plate_tectonics_world.gif

Figure 4.31: Basaltic cinder cones of the Black Rock Desert near Beaver, Utah. Lee Siebert via Smithsonian Institution. 1996. Public domain. https://commons.wikimedia.org/wiki/File:Black_Rock_Desert_volcanic_field.jpg

Figure 4.32: Diagram showing a non-moving source of magma (mantle plume) and a moving overriding plate. Los688. 2008. Public domain. https://en.wikipedia.org/wiki/File:Hotspot(geology)-1.svg

Figure 4.33: The track of the Yellowstone hotspot, which shows the age of different eruptions in millions of years ago. Kelvin Case. 2013. CC BY 3.0. https://commons.wikimedia.org/wiki/File:HotspotsSRP_update2013.JPG

Figure 4.34: The Hawaiian–Emperor seamount and island chain. Ingo Wölbern. 2008. Public domain. https://commons.wikimedia.org/wiki/File:Hawaii-Emperor_engl.png

Figure 4.35: Mt. Shasta in Washington state with Shastina, its parasitic cone. Don Graham. 2013. CC BY-SA 2.0. https://commons.wikimedia.org/wiki/File:Mt._Shasta_and_Mt._Shastina,_CA_9-13_(26491330883).jpg

Figure 4.36: Oregon's Crater Lake was formed about 7700 years ago after the eruption of Mount Mazama. Zainubrazvi. 2006. CC BY-SA 3.0. https://en.wikipedia.org/wiki/File:Crater_lake_oregon.jpg

Figure 4.37: Kilauea in Hawai'i. Quinn Dombrowski. 2007. CC BY-SA 2.0. https://commons.wikimedia.org/wiki/File:Kilauea_Shield_Volcano_Hawaii_20071209A.jpg

Figure 4.38: Eruption of Kiluea in 2018 produced high viscosity lava shown here crossing a road. USGS. 2018. Public domain. https://commons.wikimedia.org/wiki/File:USGS_K%C4%ABlauea_multimediaFile-1955.jpg

Figure 4.39: Olympus Mons, an enormous shield volcano on Mars, the largest volcano in the solar system, standing about two and a half times higher than Everest is above sea level. NASA. 1978. Public domain. https://en.wikipedia.org/wiki/File:Olympus_Mons_alt.jpg

Figure 4.40: Ropey pahoehoe lava. Bbb. 2010. GNU Free Documentation License 1.2. https://de.wikivoyage.org/wiki/Datei:ReU_PtFournaise_Lavastr%C3%B6me.jpg

Figure 4.41: Blocky a'a lava. Librex. 2009. CC BY 2.0. https://commons.wikimedia.org/wiki/File:Lava_del_Volcan_Pacaya_2009-11-28.jpg

Figure 4.42: Volcanic fissure and flow, which could eventually form a lava tube. NPS. 2004. Public domain. https://commons.wikimedia.org/wiki/File:Volcano_q.jpg

Figure 4.43: Devils Tower in Wyoming has columnar jointing. Colin.faulkingham. 2005. Public domain. https://commons.wikimedia.org/wiki/File:Devils_Tower_CROP.jpg

Figure 4.44: Columnar jointing on Giant's Causeway in Ireland. Udri. 2014. CC BY-NC-SA 2.0. https://flic.kr/p/2j1mgwE

Figure 4.45: Mount Rainier towers over Tacoma, Washington. Lyn Topinka via USGS. 1984. Public domain. https://commons.wikimedia.org/wiki/File:Mount_Rainier_over_Tacoma.jpg

Figure 4.46: Mt. Fuji in Japan, a typical stratovolcano, symmetrical, increasing slope, visible crater at the top. Alpsdake. 2016. CC BY-SA 4.0. https://commons.wikimedia.org/wiki/File:Numazu_and_Mount_Fuji.jpg

Figure 4.47: Lava domes have started the rebuilding process at Mount St. Helens, Washington. Willie Scott via USGS. 2006. Public domain. https://commons.wikimedia.org/wiki/File:MSH06_aerial_crater_from_north_high_angle_09-12-06.jpg

Figure 4.48: Timeline of events at Mount Mazama. USGS and NPS. 2006. Public domain. https://commons.wikimedia.org/wiki/File:Mount_Mazama_eruption_timeline.PNG

Figure 4.49: Wizard Island sits in the caldera at Crater Lake. Don Graham. 2006. CC BY-SA 2.0. https://commons.wikimedia.org/wiki/File:Crater_Lake_National_Park,_OR_2006_(6539577313).jpg

Figure 4.50: Map of calderas and related rocks around Yellowstone. USGS. 1905. Public domain. https://www.usgs.gov/media/images/simplified-map-yellowstone-caldera

Figure 4.51: Several prominent ash beds found in North America, including three Yellowstone eruptions shaded pink (Mesa Falls, Huckleberry Ridge, and Lava Creek), the Bisho Tuff ash bed (brown dashed line), and the modern May 18th, 1980 ash fall (yellow). USGS. 2005. Public domain. https://commons.wikimedia.org/wiki/File:Yellowstone_volcano_-_ash_beds.svg

Figure 4.52: Sunset Crater, Arizona is a cinder cone. NPS. Unknown date. Public domain. https://commons.wikimedia.org/wiki/File:Sunset_Crater10.jpg

Figure 4.53: Soon after the birth of Parícutin in 1943. K. Segerstrom via USGS. 1943. Public domain. https://commons.wikimedia.org/wiki/File:Paricutin_30_612.jpg

Figure 4.54: Lava from Parícutin covered the local church and destroyed the town of San Juan, Mexico. Sparksmex. 2007. Public domain. https://commons.wikimedia.org/wiki/File:Paricutin2.jpg

Figure 4.55: World map of flood basalts. Williamborg. 2011. CC BY-SA 3.0. https://commons.wikimedia.org/wiki/File:Flood_Basalt_Map.jpg

Figure 4.56: Igneous rock types and related volcano types. Unknown author. Unknown date. "Personal use." https://www.seekpng.com/ipng/u2t4y3r500r5r5w7_table-of-igneous-rocks-and-related-volcano-types/

Figure 4.57: General diagram of volcanic hazards. USGS. 2008. Public domain. https://pubs.usgs.gov/fs/fs002-97/old%20html%20files/

Figure 4.58: Human remains from the 79 CE eruption of Vesuvius. Gary Todd. 2019. Public domain. https://commons.wikimedia.org/wiki/File:Pompeii_Ruins_Cast_of_Human_Victim_at_Villa_of_the_Mysteries_(48445486616).jpg

Figure 4.59: Mount St. Helens, the day before the May 18th, 1980 eruption (left) and 4 months after the major eruption (right). Mount St. Helens, one day before the devastating eruption by Harry Glicken, USGS/CVO, 1980 (Public domain, https://commons.wikimedia.org/wiki/File:Mount_St._Helens,_one_day_before_the_devastating_eruption.jpg). MSH80 st helens from johnston ridge 09-10-80 by Harry Glicken via USGS (Public domain, https://commons.wikimedia.org/wiki/File:MSH80_st_helens_from_johnston_ridge_09-10-80.jpg).

Figure 4.60: Image from the May 18, 1980, eruption of Mt. Saint Helens, Washington. Austin Post via USGS. 1980. Public domain. https://commons.wikimedia.org/wiki/File:MSH80_eruption_mount_st_helens_05-18-80-dramatic-edit.jpg

Figure 4.61: The material coming down from the eruption column is a pyroclastic flow. C.G. Newhall via USGS. 1984. Public domain. https://en.wikipedia.org/wiki/File:Pyroclastic_flows_at_Mayon_Volcano.jpg#:~:text=English%3A%20Pyroclastic%20flows%20at%20Mayon,50%20km%20toward%20the%20west.

Figure 4.62: The remains of St. Pierre. Angelo Heilprin. 1902. Public domain. https://commons.wikimedia.org/wiki/File:Pelee_1902_3.jpg

Figure 4.63: Sequence of events for Mount St. Helens, May 18th, 1980. Lyn Topinka via USGS. 1998. Public domain. https://commons.wikimedia.org/wiki/File:Msh_may18_sequence.gif

Figure 4.64: Aman sweeps ash from an eruption of Kelud, Indonesia. Crisco 1492. 2014. CC BY-SA 3.0. https://commons.wikimedia.org/wiki/File:Ash_in_Yogyakarta_during_the_2014_eruption_of_Kelud_01.jpg

Figure 4.65: Micrograph of silica particle in volcanic ash. USGS. 1980. Public domain. https://volcanoes.usgs.gov/volcanic_ash/components_ash.html

Figure 4.66: Mud line shows the extent of lahars around Mount St. Helens. USGS. 1980. Public domain. https://www.usgs.gov/media/images/lahars-resulting-may-18-1980-eruption-mount-st-helens

Figure 4.67: Old lahars around Tacoma, Washington. USGS. 1905. Public domain. https://www.usgs.gov/media/images/lahar-pathways-events-heading-mount-rainier-map-showing-t

5.

WEATHERING, EROSION, AND SEDIMENTARY ROCKS

Learning Objectives

By the end of this chapter, students will be able to:

- Describe how water is an integral part of all **sedimentary rock formation**.
- Explain how chemical and **mechanical weathering** turn **bedrock** into **sediment**.
- Differentiate the two main categories of sedimentary rocks: **clastic** rock formed from pieces of weathered **bedrock**; and chemical rock that precipitates out of **solution** by organic or inorganic means.
- Explain the importance of sedimentary structures and analysis of **depositional environments**, and how they provide insight into the Earth's history.

Sedimentary rock and the processes that create it, which include **weathering**, **erosion**, and **lithification**, are an integral part of understanding Earth Science. This is because the majority of the Earth's surface is made up of sedimentary rocks and their common predecessor, **sediments**. Even though sedimentary rocks can form in drastically different ways, their origin and creation have one thing in common, water.

5.1 The Unique Properties of Water

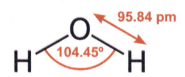

Figure 5.1: A model of a water molecule, showing the bonds between the hydrogen and oxygen.

Water plays a role in the formation of most **sedimentary rock**. It is one of the main agents involved in creating the **minerals** in **chemical sedimentary** rock. It also is a **weathering** and **erosion** agent, producing the grains that become **detrital sedimentary rock**. Several special properties make water an especially unique substance, and integral to the production of **sediments** and **sedimentary rock**.

The water molecule consists of two hydrogen atoms covalently **bonded** to one oxygen atom arranged in a specific and important geometry. The two hydrogen atoms are separated by an angle of about 105 degrees, and both are located to one side of the oxygen atom. This atomic arrangement, with the positively charged hydrogens on one side and negatively charged oxygen on the other side, gives the water molecule a property called **polarity**. Resembling a battery or a magnet, the molecule's positive-negative architecture leads to a whole suite of unique properties.

Polarity allows water molecules to stick to other substances. This is called **adhesion**. Water is also attracted to itself, a property called **cohesion**, which leads to water's most common form in the air, a droplet. **Cohesion** is responsible for creating surface **tension**, which various insects use to walk on water by distributing their weight across the surface.

Figure 5.2: Dew on a spider's web.

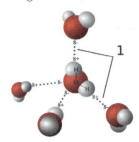

Figure 5.3: Hydrogen bonding between water molecules.

The fact that water is attracted to itself leads to another important property, one that is extremely rare in the natural world—the liquid form is denser than the solid form. The **polarity** of water creates a special type of weak **bonding** called **hydrogen bonds**. **Hydrogen bonds** allow the molecules in liquid water to sit close together. Water is densest at 4°C and is less dense above and below that **temperature**. As water solidifies into ice, the molecules must move apart in order to fit into the crystal lattice, causing water to expand and become less dense as it freezes. Because of this, ice floats and water at 4°C sinks, which keeps the oceans liquid and prevents them from freezing solid from the bottom up. This unique property of water keeps Earth, the water planet, habitable.

Even more critical for supporting life, water remains liquid over a very large range of temperatures, which is also a result of **cohesion**. Hydrogen **bonding** allows liquid water can absorb high amounts of energy before turning into vapor or gas. The wide range across which water remains a liquid, 0°C-100°C (32°F-212°F), is rarely exhibited in other substances. Without this high boiling point, liquid water as we know it would be constricted to narrow **temperature** zones on Earth, instead water is found from pole to pole. Further, water is the only substance that exists in all three phases, solid, liquid, and gas in Earth's surface environments.

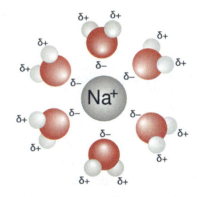

Figure 5.4: A sodium (Na) ion in solution.

Water is a **universal solvent**, meaning it dissolves more substances than any other commonly found, naturally occurring liquid. The water molecules use **polarity** and **hydrogen bonds** to pry ions away from the crystal lattice. Water is such a powerful solvent, it can **dissolve** even the strongest rocks and **minerals** given enough time.

 Take this quiz to check your comprehension of this section.

If you are using an offline version of this text, access the quiz for section 5.1 via the QR code.

An interactive H5P element has been excluded from this version of the text. You can view it online here:
https://pressbooks.lib.vt.edu/introearthscience/?p=430#h5p-31

5.2 Weathering and Erosion

Bedrock refers to the solid rock that makes up the Earth's outer **crust**. **Weathering** is a process that turns **bedrock** into smaller particles, called **sediment**. **Mechanical weathering** includes pressure expansion, **frost wedging**, **root wedging**, and salt expansion. Chemical **weathering** includes **carbonic acid** and **hydrolysis**, **dissolution**, and **oxidation**.

Erosion is a mechanical process, usually driven by water, wind, gravity, or ice, which transports **sediment** (and **soil**) from the place of **weathering**. Liquid water is the main agent of **erosion**. Gravity and **mass wasting** processes (see chapter 10) move rocks and **sediment** to new locations. Gravity and ice, in the form of **glaciers** (see chapter 14), move large rock fragments as well as fine **sediment**.

Erosion resistance is important in the creation of distinctive geological features. This is well-demonstrated in the cliffs of the Grand Canyon. The cliffs are made of rock left standing after less resistant materials have weathered and eroded away. Rocks with different levels of **erosion** resistance also create the unique-looking features called hoodoos in Bryce Canyon National Park and Goblin Valley State Park in Utah.

5.2.1 Mechanical Weathering

Mechanical weathering physically breaks **bedrock** into smaller pieces. The usual agents of **mechanical weathering** are pressure, **temperature**, freezing/thawing cycle of water, plant or animal activity, and salt evaporation.

Pressure Expansion

Bedrock buried deep within the Earth is under high pressure and **temperature**. When uplift and **erosion** brings **bedrock** to the surface, its **temperature** drops slowly, while its pressure drops immediately. The sudden pressure drop causes the rock to rapidly expand and crack; this is called pressure expansion. Sheeting or **exfoliation** is when the rock surface spalls off in layers. **Spheroidal weathering** is a type of **exfoliation** that produces rounded features and is caused when **chemical weathering** moves along **joints** in the **bedrock**.

Figure 5.5: The outer layer of this granite is fractured and eroding away, known as exfoliation.

Frost Wedging

Water seeps into cracks and fractures in rock.

When the water freezes, it expands about 9% in volume, which wedges apart the rock.

With repeated freeze/thaw cycles, rock breaks into pieces.

Figure 5.6: The process of frost wedging.

Frost wedging, also called **ice wedging**, uses the power of expanding ice to break apart rocks. Water works its way into various cracks, voids, and crevices. As the water freezes, it expands with great force, exploiting any weaknesses. When ice melts, the liquid water moves further into the widened spaces. Repeated cycles of freezing and melting eventually pry the rocks apart. The cycles can occur daily when fluctuations of **temperature** between day and night go from freezing to melting.

Root Wedging

Like **frost wedging**, **root wedging** happens when plant roots work themselves into cracks, prying the **bedrock** apart as they grow. Occasionally these roots may become fossilized. **Rhizolith** is the term for these roots preserved in the rock record. Tunneling organisms such as earthworms, termites, and ants are biological agents that induce **weathering** similar to **root wedging**.

Salt Expansion

Figure 5.7: The roots of this tree are demonstrating the destructive power of root wedging. Though this picture is a man-made rock (asphalt), it works on typical rock as well.

Salt expansion, which works similarly to **frost wedging**, occurs in areas of high evaporation or near-**marine** environments. Evaporation causes salts to **precipitate** out of **solution** and grow and expand into cracks in rock. Salt expansion is one of the causes of **tafoni**, a series of holes in a rock. Tafonis, cracks, and holes are weak points that become susceptible to increased **weathering**. Another phenomena that occurs when salt water evaporates can leave behind a square imprint preserved in a soft **sediment**, called a hopper crystal.

Figure 5.8: Tafoni from Salt Point, California.

5.2.2 Chemical Weathering

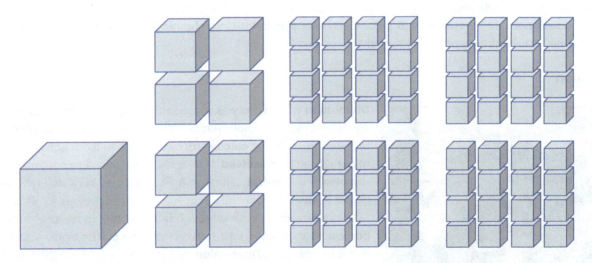

Figure 5.9: Each of these three groups of cubes has an equal volume. However, their surface areas are vastly different. On the left, the single cube has a length, width, and height of 4 units, giving it a surface area of 6(4×4)=96 and a volume of 4^3=64. The middle eight cubes have a length, width, and height of 2, meaning a surface area of 8(6(2×2))=8×24=192. They also have a volume of 8(2^3)=8×8=64. The 64 cubes on the right have a length, width, and height of 1, leading to a surface area of 64(6(1×1))=64×6=384. The volume remains unchanged, because 64(1^3)=64×1=64. The surface area to volume ratio (SA:V), which is related to the amount of material available for reactions, changes for each as well. On the left, it is 96/64=0.75 or 3:2. The center has a SA/V of 192/64=1.5, or 3:1. On the right, the SA:V is 384/64=6, or 6:1.

Chemical weathering is the dominate **weathering** process in warm, humid environments. It happens when water, oxygen, and other reactants chemically degrade the **mineral** components of **bedrock** and turn them into water-soluble ions which can then be transported by water. Higher temperatures accelerate **chemical weathering** rates.

Chemical and **mechanical weathering** work hand-in-hand via a fundamental concept called surface-area-to-volume ratio. **Chemical weathering** only occurs on rock surfaces because water and reactants cannot penetrate solid rock. **Mechanical weathering** penetrates **bedrock**, breaking large rocks into smaller pieces and creating new rock surfaces. This exposes more surface area to **chemical weathering**, enhancing its effects. In other words, higher surface-area-to-volume ratios produce higher rates of overall **weathering**.

Carbonic Acid and Hydrolysis

Carbonic acid (H_2CO_3) forms when carbon dioxide, the fifth-most abundant gas in the **atmosphere**, dissolves in water. This happens naturally in clouds, which is why **precipitation** is normally slightly acidic. **Carbonic acid** is an important agent in two **chemical weathering** reactions, **hydrolysis** and **dissolution**.

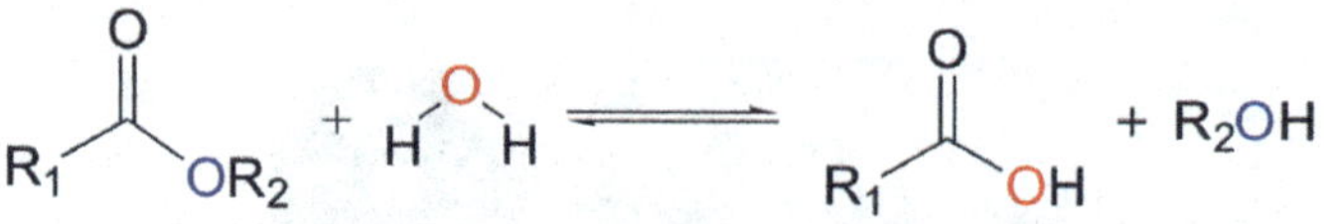

Figure 5.10: Generic hydrolysis diagram, where the bonds in mineral in question would represent the left side of the diagram.

Hydrolysis occurs via two types of reactions. In one reaction, water molecules ionize into positively charged H^{+1} and OH^{-1} ions and replace **mineral cations** in the crystal lattice. In another type of **hydrolysis**, **carbonic acid** molecules react directly with **minerals**, especially those containing silicon and aluminum (i.e. **Feldspars**), to form molecules of clay **minerals**.

Hydrolysis is the main process that breaks down **silicate** rock and creates clay **minerals**. The following is a **hydrolysis** reaction that occurs when silica-rich **feldspar** encounters **carbonic acid** to produce water-soluble clay and other ions:

feldspar + **carbonic acid** (in water) → clay + metal **cations** (Fe^{++}, Mg^{++}, Ca^{++}, Na^+, etc.) + bicarbonate **anions** (HCO_3^{-1}) + silica (SiO_2)

Clay **minerals** are platy **silicates** or phyllosilicates (see chapter 3) similar to micas, and are the main components of very fine-grained **sediment**. The **dissolved** substances may later **precipitate** into **chemical sedimentary** rocks like **evaporite** and **limestone**, as well as amorphous silica or **chert** nodules.

Dissolution

Figure 5.11: In this rock, a pyrite cube has dissolved (as seen with the negative "corner" impression in the rock), leaving behind small specks of gold.

Dissolution is a **hydrolysis** reaction that dissolves **minerals** in **bedrock** and leaves the ions in **solution**, usually in water. Some **evaporites** and **carbonates**, like salt and **calcite**, are more prone to this reaction; however, all **minerals** can be **dissolved**. Non-acidic water, having a neutral pH of 7, will **dissolve** any **mineral**, although it may happen very slowly. Water with higher levels of acid, naturally or man-made, dissolves rocks at a higher rate. Liquid water is normally slightly acidic due to the presence of **carbonic acid** and free H+ ions. Natural rainwater can be highly acidic, with pH levels as low as 2. **Dissolution** can be enhanced by a biological agent, such as when organisms like lichen and bacteria release organic acids onto the rocks they are attached to. Regions with high humidity (airborne moisture) and **precipitation** experience more **dissolution** due to greater contact time between rocks and water.

WEATHERING, EROSION, AND SEDIMENTARY ROCKS | 131

The **Goldich Dissolution Series** shows **chemical weathering** rates are associated to **crystallization** rankings in the **Bowen's Reaction Series** (see chapter 4). **Minerals** at the top of the Bowen series crystallize under high temperatures and pressures, and chemically **weather** at a faster rate than **minerals** ranked at the bottom. **Quartz**, a **felsic mineral** that crystallizes at 700°C, is very resistant to **chemical weathering**. High **crystallization**-point **mafic minerals**, such as **olivine** and **pyroxene** (1,250°C), **weather** relatively rapidly and more completely. **Olivine** and **pyroxene** are rarely found as end products of **weathering** because they tend to break down into elemental ions.

Figure 5.12: This mantle xenolith containing olivine (green) is chemically weathering by hydrolysis and oxidation into the pseudo-mineral iddingsite, which is a complex of water, clay, and iron oxides. The more altered side of the rock has been exposed to the environment longer.

Figure 5.13: Eroded karst topography in Minevre, France.

Dissolution is also noteworthy for the special geological features it creates. In places with abundant **carbonate bedrock**, **dissolution weathering** can produce a **karst** topography characterized by sinkholes or caves (see chapter 10).

Timpanogos Cave National Monument in Northern Utah is a well-known **dissolution** feature. The figure shows a cave **formation** created from **dissolution** followed by **precipitation**—groundwater **saturated** with **calcite** seeped into the cavern, where evaporation caused the **dissolved minerals** to **precipitate** out.

Figure 5.14: A formation called The Great Heart of Timpanogos in Timpanogos Cave National Monument.

Oxidation

Oxidation, the chemical reaction that causes rust in **metallic** iron, occurs geologically when iron atoms in a **mineral bond** with oxygen. Any **minerals** containing iron can be oxidized. The resultant iron **oxides** may permeate a rock if it is rich in iron **minerals**. **Oxides** may also form a coating that covers rocks and grains of **sediment**, or lines rock cavities and **fractures**. If the **oxides** are more susceptible to **weathering** than the original **bedrock**, they may create void spaces inside the rock mass or hollows on exposed surfaces.

Three commonly found **minerals** are produced by iron-**oxidation** reactions: red or grey hematite, brown goethite (pronounced "GUR-tite"), and yellow limonite. These iron **oxides** coat and bind **mineral** grains together into sedimentary rocks in a process called **cementation**, and often give these rocks a dominant color. They color the rock layers of the Colorado Plateau, as well as Zion, Arches, and Grand Canyon National Parks. These **oxides** can permeate a rock that is rich in iron-bearing **minerals** or can be a coating that forms in cavities or

Figure 5.15: Pyrite cubes are oxidized, becoming a new mineral goethite. In this case, goethite is a pseudomorph after pyrite, meaning it has taken the form of another mineral.

fractures. When the **minerals** replacing existing **minerals** in **bedrock** are resistant to **weathering**, iron concretions may occur in the rock. When **bedrock** is replaced by weaker **oxides**, this process commonly results in void spaces and weakness throughout the rock mass and often leaves hollows on exposed rock surfaces.

5.2.3 Erosion

Figure 5.16: A hoodoo near Moab, Utah.

Erosion is a mechanical process, usually driven by water, gravity, (see chapter 10), wind, or ice (see chapter 14) that removes **sediment** from the place of **weathering**. Liquid water is the main agent of **erosion**.

Erosion resistance is important in the creation of distinctive geological features. This is well demonstrated in the cliffs of the Grand Canyon. The cliffs are made of rock left standing after less resistant materials have weathered and eroded away. Rocks with different levels **erosion** resistant also create the unique-looking features called hoodoos in Bryce Canyon National Park and Goblin Valley State Park in Utah.

Figure 5.17: Grand Canyon from Mather Point.

5.2.4 Soil

Soil is a combination of air, water, **minerals**, and organic matter that forms at the transition between **biosphere** and **geosphere**. **Soil** is made when **weathering** breaks down **bedrock** and turns it into **sediment**. If **erosion** does not remove the **sediment** significantly, organisms can access the **mineral** content of the **sediments**. These organisms turn **minerals**, water, and atmospheric gases into organic substances that contribute to the **soil**.

Soil is an important **reservoir** for organic components necessary for plants, animals, and microorganisms to live. The organic component of **soil**, called **humus**, is a rich source of bioavailable nitrogen. Nitrogen is the most common **element** in the **atmosphere**, but it exists in a form most life forms are unable to use. Special bacteria found only in **soil** provide most nitrogen compounds that are usable, bioavailable, by life forms.

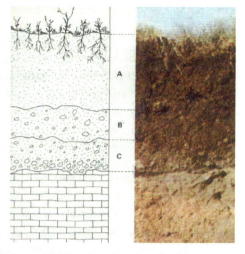

Figure 5.18: Sketch and picture of soil.

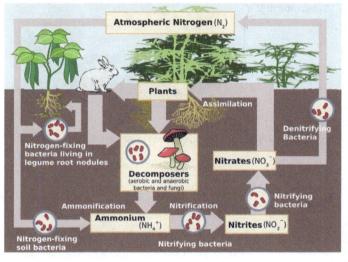

Figure 5.19: Schematic of the nitrogen cycle.

These nitrogen-fixing bacteria absorb nitrogen from the **atmosphere** and convert it into nitrogen compounds. These compounds are absorbed by plants and used to make DNA, amino acids, and enzymes. Animals obtain bioavailable nitrogen by eating plants, and this is the source of most of the nitrogen used by life. That nitrogen is an essential component of proteins and DNA. **Soils** range from poor to rich, depending on the amount of **humus** they contain. **Soil** productivity is determined by water and nutrient content. Freshly created **volcanic soils**, called andisols, and clay-rich **soils** that hold nutrients and water are examples of productive **soils**.

The nature of the **soil**, meaning its characteristics, is determined primarily by five components: 1) the mineralogy of the parent material; 2) topography, 3) **weathering**, 4) **climate**, and 5) the organisms that inhabit the **soil**. For example, **soil** tends to erode more rapidly on steep slopes so **soil** layers in these areas may be thinner than in flood plains, where it tends to accumulate. The quantity and chemistry of organic matter of **soil** affects how much and what varieties of life it can sustain. **Temperature** and **precipitation**, two major **weathering** agents, are dependent on **climate**. Fungi and bacteria contribute organic matter and the ability of **soil** to sustain life, interacting with plant roots to exchange nitrogen and other nutrients.

Figure 5.20: Agricultural terracing, as made by the Inca culture from the Andes, helps reduce erosion and promote soil formation, leading to better farming practices.

In well-formed **soils**, there is a discernable arrangement of distinct layers called **soil horizons**. These soil horizons can be seen in road cuts that expose the layers at the edge of the cut. Soil horizons make up the **soil profile**. Each **soil horizon** reflects **climate**, topography, and other soil-development factors, as well as its organic material and **mineral sediment composition**. The horizons are assigned names and letters. Differences in naming schemes depend on the area, **soil** type or research topic. The figure shows a simplified **soil profile** that uses commonly designated names and letters.

O Horizon: The top horizon is a thin layer of predominantly organic material, such as leaves, twigs, and other plant parts that are actively decaying into **humus**.

A Horizon: The next layer, called **topsoil**, consists of **humus** mixed with **mineral sediment**. As **precipitation** soaks down through this layer, it leaches out soluble chemicals. In wet climates with heavy **precipitation** this leaching out produces a separate layer called horizon E, the leaching or eluviation zone.

B Horizon: Also called **subsoil**, this layer consists of **sediment** mixed with **humus** removed from the upper layers. The **subsoil** is where **mineral sediment** is chemically weathered. The amount of organic material and degree of **weathering** decrease with depth. The upper **subsoil** zone, called **regolith**, is a porous mixture of **humus** and highly weathered **sediment**. In the lower zone, saprolite, scant organic material is mixed with largely unaltered **parent rock**.

C Horizon: This is **substratum** and is a zone of **mechanical weathering**. Here, **bedrock** fragments are physically broken but not chemically altered. This layer contains no organic material.

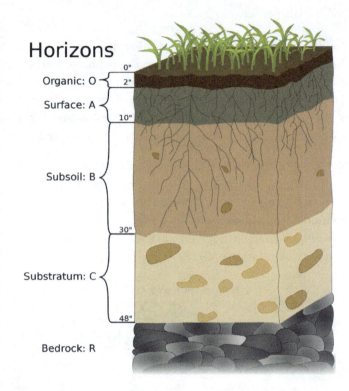

Figure 5.21: A simplified soil profile, showing labeled layers.

R Horizon: The final layer consists of unweathered, parent **bedrock** and fragments.

Figure 5.22: A sample of bauxite. Note the unweathered igneous rock in the center.

The United States governing body for agriculture, the USDA, uses a taxonomic classification to identify **soil** types, called **soil** orders. Xoxisols or laterite **soils** are nutrient-poor **soils** found in tropical regions. While poorly suited for growing crops, xosisols are home to most of the world's mineable aluminum **ore** (**bauxite**). Ardisol forms in dry climates and can develop layers of hardened **calcite**, called caliche. Andisols originate from **volcanic ash** deposits. Alfisols contain **silicate** clay **minerals**. These two **soil** orders are productive for farming due to their high content of **mineral** nutrients. In general, color can be an important factor in understanding **soil** conditions. Black **soils** tend to be anoxic, red oxygen-rich, and green oxygen-poor (i.e. reduced). This is true for many sedimentary rocks as well.

Not only is **soil** essential to **terrestrial** life in nature, but also human civilization via agriculture. Careless or uninformed human activity can seriously damage **soil**'s life-supporting properties. A prime example is the famous Dust Bowl disaster of the 1930s, which affected the midwestern United States. The damage occurred because of large-scale attempts develop prairieland in southern Kansas, Colorado, western Texas, and Oklahoma into farmland. Poor understanding of the region's geology, ecology, and **climate** led to farming practices that ruined the **soil profile**.

Figure 5.23: A dust storm approaches Stratford, Texas in 1935.

The prairie **soils** and **native** plants are well adapted to a relatively dry **climate**. With government encouragement, settlers moved in to homestead the region. They plowed vast areas of prairie into long, straight rows and planted grain. The plowing broke up the stable **soil profile** and destroyed the natural grasses and plants, which had long roots that anchored the **soil** layers. The grains they planted had shallower root systems and were plowed up every year, which made the **soil** prone to **erosion**. The plowed furrows were aligned in straight rows running downhill, which favored **erosion** and loss of **topsoil**.

The local **climate** does not produce sufficient **precipitation** to support non-**native** grain crops, so the farmers drilled wells and over-pumped water from the underground **aquifers**. The grain crops failed due to lack of water, leaving bare **soil** that was stripped from the ground by the prairie winds. Particles of midwestern prairie **soil** were deposited along the east coast and as far away as Europe. Huge dust storms called black blizzards made life unbearable, and the once-hopeful homesteaders left in droves. The setting for John Steinbeck's famous novel and John Ford's film, *The Grapes of Wrath*, is Oklahoma during this time. The lingering question is whether we have learned the lessons of the dust bowl, to avoid creating it again.

Take this quiz to check your comprehension of this section.

If you are using an offline version of this text, access the quiz for section 5.2 via the QR code.

 An interactive H5P element has been excluded from this version of the text. You can view it online here:
https://pressbooks.lib.vt.edu/introearthscience/?p=430#h5p-32

5.3 Sedimentary Rocks

Sedimentary rock is classified into two main categories: **clastic** and chemical. **Clastic** or **detrital** sedimentary rocks are made from pieces of **bedrock**, **sediment**, derived primarily by **mechanical weathering**. **Clastic** rocks may also include chemically weathered **sediment**. **Clastic** rocks are classified by grain shape, **grain size**, and **sorting**. Chemical sedimentary rocks are **precipitated** from water **saturated** with **dissolved minerals**. Chemical rocks are classified mainly by **composition** of **minerals** in the rock.

5.3.1 Lithification and Diagenesis

Lithification turns loose **sediment** grains, created by **weathering** and transported by **erosion**, into **clastic sedimentary rock** via three interconnected steps. **Deposition** happens when friction and gravity overcome the forces driving **sediment** transport, allowing **sediment** to accumulate. **Compaction** occurs when material continues to accumulate on top of the **sediment** layer, squeezing the grains together and driving out water. The mechanical **compaction** is aided by weak attractive forces between the smaller grains of **sediment**. **Groundwater** typically carries cementing agents into the **sediment**. These **minerals**, such as **calcite**, amorphous silica, or **oxides**, may have a different **composition** than the **sediment** grains. **Cementation** is the process of cementing **minerals** coating the **sediment** grains and gluing them together into a fused rock.

Figure 5.24: Geologic unconformity seen at Siccar Point on the east coast of Scotland.

Figure 5.25: Permineralization in petrified wood.

Diagenesis is an accompanying process to **lithification** and is a low-**temperature** form of rock **metamorphism** (see chapter 6). During **diagenesis**, **sediments** are chemically altered by heat and pressure. A classic example is aragonite ($CaCO_3$), a form of calcium **carbonate** that makes up most organic shells. When lithified aragonite undergoes **diagenesis**, the aragonite reverts to **calcite** ($CaCO_3$), which has the same chemical formula but a different crystalline structure. In **sedimentary rock** containing **calcite** and magnesium (Mg), **diagenesis** may **transform** the two **minerals** into dolomite ($CaMg(CO_3)_2$). **Diagenesis** may also reduce the **pore** space, or open volume, between **sedimentary rock** grains. The processes of **cementation**, **compaction**, and ultimately **lithification** occur within the realm of **diagenesis**, which includes the processes that turn organic material into **fossils**.

5.3.2 Detrital Sedimentary Rocks (Clastic)

Detrital or **clastic** sedimentary rocks consist of preexisting **sediment** pieces that comes from weathered bedrock. Most of this is mechanically weathered sediment, although some clasts may be pieces of chemical rocks. This creates some overlap between the two categories, since **clastic** sedimentary rocks may include chemical **sediments**. **Detrital** or **clastic** rocks are classified and named based on their **grain size**.

Grain Size

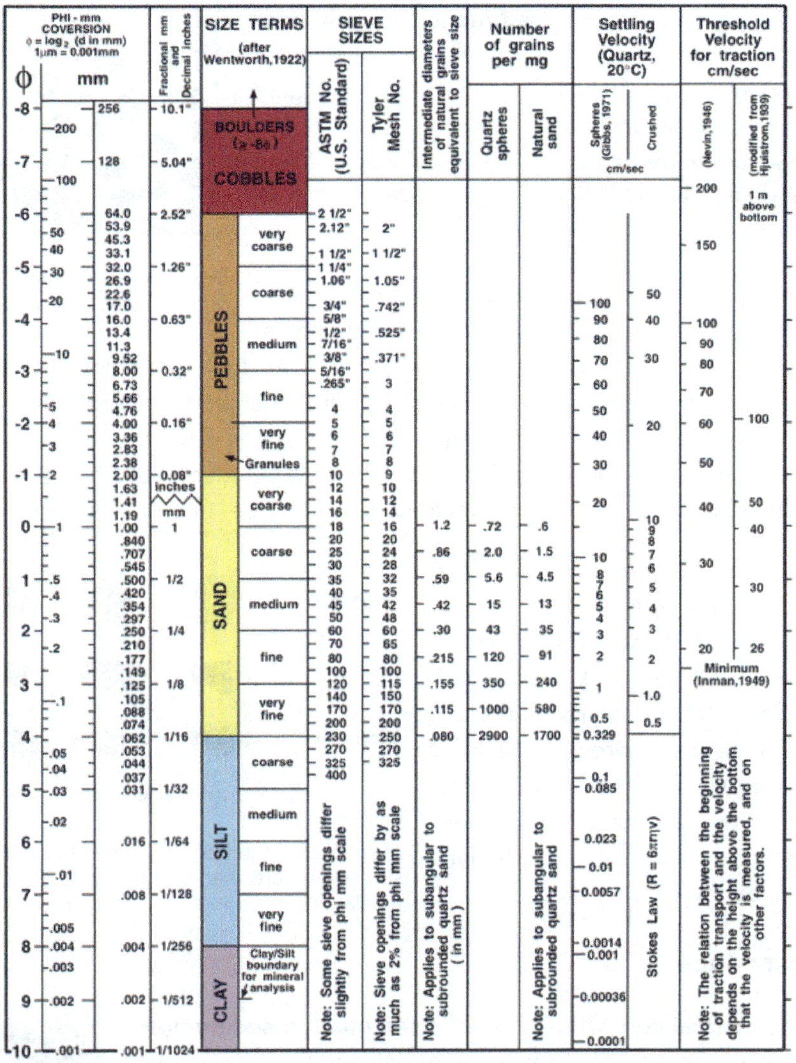

Figure 5.26: Size categories of sediments, known as the Wentworth scale.

Detrital rock is classified according to **sediment grain size**, which is **graded** from large to small on the Wentworth scale (see figure 5.26). **Grain size** is the average diameter of **sediment** fragments in **sediment** or rock. Grain sizes are delineated using a log base 2 scale. For example, the grain sizes in the pebble class are 2.52, 1.26, 0.63, 0.32, 0.16, and 0.08 inches, which correlate respectively to very coarse, coarse, medium, fine, and very fine granules. Large fragments, or clasts, include all grain sizes larger than 2 mm (5/64 in). These include, boulders, cobbles, granules, and gravel. Sand has a **grain size** between 2 mm and 0.0625 mm, about the lower limit of the naked eye's resolution. **Sediment** grains smaller than sand are called silt. Silt is unique; the grains can be felt with a finger or as grit between your teeth, but are too small to see with the naked eye.

Sorting and Rounding

Sorting describes the range of grain sizes within **sediment** or **sedimentary rock**. Geologists use the term "well sorted" to describe a narrow range of grain sizes, and "poorly sorted" for a wide range of grain sizes (see figure 5.27). It is important to note that **soil** engineers use similar terms with opposite definitions; well **graded sediment** consists of a variety of grain sizes, and poorly **graded sediment** has roughly the same grain sizes.

Figure 5.27: A well-sorted sediment (left) and a poorly-sorted sediment (right).

When reading the story told by rocks, geologists use **sorting** to interpret **erosion** or transport processes, as well as **deposition** energy. For example, wind-blown sands are typically extremely well sorted, while **glacial** deposits are typically poorly sorted. These characteristics help identify the type of **erosion** process that occurred. Coarse-grained **sediment** and poorly sorted rocks are usually found nearer to the source of **sediment**, while fine **sediments** are carried farther away. In a rapidly flowing mountain **stream** you would expect to see boulders and pebbles. In a lake fed by the **stream**, there should be sand and silt deposits. If you also find large boulders in the lake, this may indicate the involvement of another **sediment** transport process, such as **rockfall** caused by ice- or root-wedging.

Rounding is created when angular corners of rock fragments are removed from a piece of **sediment** due to abrasion during transport. Well-rounded **sediment** grains are defined as being free of all sharp edges. Very angular **sediment** retains the sharp corners. Most clast fragments start with some sharp edges due to the **bedrock**'s crystalline structure, and those points are worn down during transport. More rounded grains imply a longer **erosion** time or transport distance, or more energetic erosional process. **Mineral hardness** is also a factor in **rounding**.

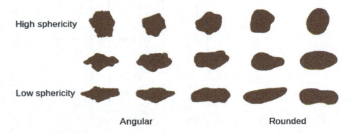

Figure 5.28: Degree of rounding in sediments. Sphericity refers to the spherical nature of an object, a completely different measurement unrelated to rounding.

Composition and Provenance

Composition describes the **mineral** components found in **sediment** or **sedimentary rock** and may be influenced by local geology, like **source rock** and hydrology. Other than clay, most **sediment** components are easily determined by visual inspection (see chapter 3). The most commonly found **sediment mineral** is **quartz** because of its low chemical reactivity and high **hardness**, making it resistant to **weathering**, and its ubiquitous occurrence in **continental bedrock**. Other commonly found **sediment** grains include **feldspar** and lithic fragments. Lithic fragments are pieces of fine-grained **bedrock**, and include **mud chips**, **volcanic** clasts, or pieces of **slate**.

Weathering of **volcanic rock** produces Hawaii's famous black (**basalt**) and green (**olivine**) sand beaches, which are rare elsewhere on Earth. This is because the local rock is **composed** almost entirely of **basalt** and provides an abundant source of dark colored clasts loaded with mafic minerals. According to the Goldich Dissolution Series, clasts high in **mafic minerals** are more easily destroyed compared to clasts **composed** of **felsic minerals** like **quartz**.

Figure 5.29: A sand grain made of basalt, known as a microlitic volcanic lithic fragment. Box is 0.25 mm. Top picture is plane-polarized light, bottom is cross-polarized light.

Figure 5.30: Hawiian beach composed of green olivine sand from weathering of nearby basaltic rock.

Geologists use **provenance** to discern the original source of **sediment** or **sedimentary rock**. **Provenance** is determined by analyzing **mineral composition** and types of **fossils** present, as well as textural features like **sorting** and **rounding**. **Provenance** is important for describing **tectonic** history, visualizing paleogeographic **formations**, unraveling an area's geologic history, or reconstructing past **supercontinents**.

In **quartz sandstone**, sometimes called **quartz arenite** (SiO_2), **provenance** may be determined using a rare, durable clast **mineral** called **zircon** ($ZrSiO_4$). **Zircon**, or zirconium **silicate**, contains traces of uranium, which can be used for age-dating the source **bedrock** that contributed **sediment** to the lithified **sandstone** rock (see chapter 7).

Classification of Clastic Rocks

Clastic rocks are classified according to the **grain size** of their **sediment**. Coarse-grained rocks contain clasts with a predominant **grain size** larger than sand. Typically, smaller **sediment** grains, collectively called **groundmass** or matrix, fill in much of the volume between the larger clasts, and hold the clasts together. **Conglomerates** are rocks containing coarse rounded clasts, and breccias contain angular clasts (see figure 5.31). Both **conglomerates** and breccias are usually poorly sorted.

Figure 5.31: Megabreccia in Titus Canyon, Death Valley National Park, California.

Figure 5.32: Enlarged image of frosted and rounded windblown sand grains.

Medium-grained rocks **composed** mainly of sand are called **sandstone**, or sometimes **arenite** if well sorted. **Sediment** grains in **sandstone** can having a wide variety of **mineral** compositions, roundness, and **sorting**. Some **sandstone** names indicate the rock's **mineral composition**. Quartz sandstone contains predominantly **quartz sediment** grains. **Arkose** is **sandstone** with significant amounts of **feldspar**, usually greater than 25%. **Sandstone** that contains **feldspar**, which weathers more quickly than **quartz**, is useful for analyzing the local geologic history. Greywacke is a term with conflicting definitions. **Greywacke** may refer to **sandstone** with a muddy matrix, or **sandstone** with many lithic fragments (small rock pieces).

Figure 5.33: The Rochester Shale, New York. Note the thin fissility in the layers.

Fine-grained rocks include **mudstone**, **shale**, **siltstone**, and **claystone**. **Mudstone** is a general term for rocks made of **sediment** grains smaller than sand (less than 2 mm). Rocks that are **fissile**, meaning they separate into thin sheets, are called **shale**. Rocks exclusively **composed** of silt or clay **sediment**, are called **siltstone** or **claystone**, respectively. These last two rock types are rarer than **mudstone** or **shale**.

Rock types found as a mixture between the main classifications, may be named using the less-common component as a descriptor. For example, a rock containing some silt but mostly rounded sand and gravel is called silty **conglomerate**. Sand-rich rock containing minor amounts of clay is called clayey **sandstone**.

Figure 5.34: Claystone laminations from Glacial Lake Missoula.

5.3.3 Chemical, Biochemical, and Organic

Chemical sedimentary rocks are formed by processes that do not directly involve **mechanical weathering** and **erosion**. **Chemical weathering** may contribute the **dissolved** materials in water that ultimately form these rocks. **Biochemical** and organic **sediments** are **clastic** in the sense that they are made from pieces of organic material that is deposited, buried, and lithified; however, they are usually classified as being chemically produced.

Inorganic **chemical sedimentary** rocks are made of **minerals precipitated** from ions **dissolved** in **solution**, and created without the aid of living organisms. Inorganic **chemical sedimentary** rocks form in environments where **ion** concentration, **dissolved** gasses, temperatures, or pressures are changing, which causes **minerals** to crystallize.

Biochemical sedimentary rocks are formed from shells and bodies of underwater organisms. The living organisms extract chemical components from the water and use them to build shells and other body parts. The components include aragonite, a **mineral** similar to and commonly replaced by **calcite**, and silica.

Organic sedimentary rocks come from organic material that has been deposited and lithified, usually underwater. The source materials are plant and animal remains that are transformed through burial and heat, and end up as **coal**, **oil**, and methane (**natural gas**).

Inorganic Chemical

Inorganic **chemical sedimentary** rocks are formed when **minerals precipitate** out of an aqueous **solution**, usually due to water evaporation. The **precipitate minerals** form various salts known as **evaporites**. For example, the Bonneville Salt Flats in Utah flood with winter rains and dry out every summer, leaving behind salts such as **gypsum** and **halite**. The **deposition** order of **evaporites** deposit is opposite to their solubility order, i.e. as water evaporates and increases the **mineral** concentration in **solution**, less soluble **minerals precipitate** out sooner than the highly soluble **minerals**. The **deposition** order and **saturation** percentages are depicted in the table, bearing in mind the process in nature may vary from laboratory derived values.

Figure 5.35: Salt-covered plain known as the Bonneville Salt Flats, Utah.

Mineral sequence	Percent seawater remaining after evaporation
Calcite	50%
Gypsum/anhydrite	20%
Halite	10%
Various potassium and magnesium salts	5%

Table 5.1: Deposition order and saturation percentages.

Figure 5.36: Ooids from Joulter's Cay, The Bahamas.

Figure 5.37: Limestone tufa towers along the shores of Mono Lake, California.

Calcium **carbonate**–**saturated** water precipitates porous masses of **calcite** called **tufa**. **Tufa** can form near degassing water and in saline lakes. Waterfalls downstream of springs often **precipitate tufa** as the turbulent water enhances degassing of carbon dioxide, which makes **calcite** less soluble and causes it to **precipitate**. Saline lakes concentrate calcium **carbonate** from a combination of wave action causing degassing, springs in the lakebed, and evaporation. In salty Mono Lake in California, **tufa** towers were exposed after water was diverted and lowered the lake levels.

Cave deposits like stalactites and stalagmites are another form of chemical **precipitation** of **calcite**, in a form called **travertine**. **Calcite** slowly precipitates from water to form the **travertine**, which often shows **banding**. This process is similar to the **mineral** growth on faucets in your home sink or shower that comes from hard (**mineral** rich) water. **Travertine** also forms at hot springs such as Mammoth Hot Spring in Yellowstone National Park.

Figure 5.39: Alternating bands of iron-rich and silica-rich mud, formed as oxygen combined with dissolved iron.

Figure 5.38: Travertine terraces of Mammoth Hot Springs, Yellowstone National Park, USA.

Banded iron formation deposits commonly formed early in Earth's history, but this type of **chemical sedimentary** rock is no longer being created. Oxygenation of the **atmosphere** and oceans caused free iron ions, which are water-soluble, to become oxidized and **precipitate** out of **solution**. The iron **oxide** was deposited, usually in **bands** alternating with layers of **chert**.

Chert, another commonly found **chemical sedimentary** rock, is usually produced from silica (SiO_2) **precipitated** from **groundwater**. Silica is highly insoluble on the surface of Earth, which is why **quartz** is so resistant to **chemical weathering**. Water deep underground is subjected to higher pressures and temperatures, which helps **dissolve** silica into an aqueous **solution**. As the **groundwater** rises toward or emerges at the surface the silica precipitates out, often as a cementing agent or into nodules. For example, the bases of the geysers in Yellowstone National Park are surrounded by silica deposits called geyserite or sinter. The silica is **dissolved** in water that is thermally heated by a relatively deep **magma** source. **Chert** can also form biochem-

Figure 5.40: A type of chert, flint, shown with a lighter weathered crust.

ically and is discussed in the **biochemical** subsection. **Chert** has many synonyms, some of which may have gem value such as jasper, flint, onyx, and agate, due to subtle differences in colors, striping, etc., but **chert** is the more general term used by geologists for the entire group.

Oolites are among the few **limestone** forms created by an inorganic chemical process, similar to what happens in **evaporite deposition**. When water is oversaturated with **calcite**, the **mineral** precipitates out around a nucleus, a sand grain or shell fragment, and forms little spheres called **ooids** (see figure 5.41). As evaporation continues, the **ooids** continue building concentric layers of **calcite** as they roll around in gentle currents.

Figure 5.41: Ooids forming an oolite.

Biochemical

Figure 5.42: Fossiliferous limestone (with brachiopods and bryozoans) from the Kope Formation of Ohio. Lower image is a section of the rock that has been etched with acid to emphasize the fossils.

Biochemical sedimentary rocks are not that different from **chemical sedimentary** rocks; they are also formed from ions **dissolved** in **solution**. However, **biochemical** sedimentary rocks rely on biological processes to extract the **dissolved** materials out of the water. Most macroscopic **marine** organisms use **dissolved minerals**, primarily aragonite (calcium **carbonate**), to build hard parts such as shells. When organisms die the hard parts settle as **sediment**, which become buried, compacted and cemented into rock.

This **biochemical** extraction and secretion is the main process for forming **limestone**, the most commonly occurring, non-**clastic sedimentary rock**. **Limestone** is mostly made of **calcite** ($CaCO_3$) and sometimes includes dolomite ($CaMgCO_3$), a close relative. Solid **calcite** reacts with hydrochloric acid by effervescing or fizzing. Dolomite only reacts to hydrochloric acid when ground into a powder, which can be done by scratching the rock surface (see chapter 3).

Limestone occurs in many forms, most of which originate from biological processes. Entire coral **reefs** and their ecosystems can be preserved in exquisite detail in **limestone** rock (see figure 5.42). **Fossiliferous limestone** contains many visible **fossils**. A type of **limestone** called **coquina** originates from beach sands made predominantly of shells that were then lithified. **Coquina** is **composed** of loosely-cemented shells and shell fragments. You can find beaches like this in modern tropical environments, such as the Bahamas. **Chalk** contains high concentrations of shells from a microorganism called a coccolithophore. **Micrite**, also known as microscopic **calcite** mud, is a very fine-grained **limestone** containing microfossils that can only be seen using a microscope.

Figure 5.43: Close-up on coquina.

Biogenetic **chert** forms on the deep **ocean floor**, created from **biochemical sediment** made of microscopic organic shells. This **sediment**, called ooze, may be calcareous (calcium **carbonate** based) or siliceous (silica-based) depending on the type of shells deposited. For example, the shells of radiolarians (zooplankton) and diatoms (phytoplankton) are made of silica, so they produce siliceous ooze.

Organic

Under the right conditions, intact pieces of organic material or material derived from organic sources, is preserved in the geologic record. Although not derived from **sediment**, this lithified organic material is associated with sedimentary **strata** and created by similar processes—burial, **compaction**, and **diagenesis**. C Deposits of these fuels develop in areas where organic material collects in large quantities. Lush swamplands can create conditions conducive to **coal formation**. Shallow-water, organic material-rich **marine sediment** can become highly productive **petroleum** and **natural gas** deposits. See chapter 16 for a more in-depth look at these **fossil**-derived energy sources.

Figure 5.44: Anthracite coal, the highest grade of coal.

Classification of Chemical Sedimentary Rocks

Figure 5.45: Gyprock, a rock made of the mineral gypsum. From the Castle formation of New Mexico.

In contrast to **detrital sediment**, chemical, **biochemical**, and organic sedimentary rocks are classified based on **mineral composition**. Most of these are monomineralic, **composed** of a single **mineral**, so the rock name is usually associated with the identifying **mineral**. **Chemical sedimentary** rocks consisting of **halite** are called rock salt. Rocks made of **Limestone (calcite)** is an exception, having elaborate subclassifications and even two competing classification methods: Folk Classification and Dunham Classification. The Folk Classification deals with rock grains and usually requires a specialized, petrographic microscope. The Dunham Classification is based on rock **texture**, which is visible to the naked eye or using a hand lens and is easier for field applications. Most **carbonate** geologists use the Dunham system.

Inorganic Clastic Sedimentary Rocks						
Texture	Grain size	Composition	Comments	Rock name	Map symbol	Picture
Clastic (fragmental)	Pebbles, cobbles, and/or boulders in a matrix of sand, silt and/or clay	Mostly quartz, feldspar, and clay minerals; may contain fragments of other rocks and minerals	Rounded fragments	Conglomerate		
			Angular fragments	Breccia		
	Sand (0.063 to 2 mm)		Fine to coarse in a variety of colors	Sandstone		
	Silt (0.039 to 0.063 mm)		Very fine grained, massive, usually dark	Siltstone		
	Clay (<0.0039 mm)		Compact, brittle, usually dark	Shale		
Chemically and/or Organically Formed Sedimentary Rocks						
Texture	Grain size	Composition	Comments	Rock name	Map symbol	Picture
Crystalline	Fine to coarse grains	Quartz	Chemical precipitates and evaporites	Chert		
		Halite		Rock salt		
		Gypsum		Rock gypsum		
		Dolomite		Dolostone*		
Crystalline or bioclastic	Microscopic to very coarse	Calcite	Biologic precipitates or cemented shell fragments	Limestone*		
Bioclastic	Clay (< 0.0039 mm)	Carbon	Black, compacted plant remains	Coal		
Bioclastic	Clay (< 0.0039 mm)	Clay and kerogen	Dark, may have oily smell or burn	Oil shale		

Other types of sandtone are arkose and graywacke. Varieties of limestone include chalk, coquina, micrite, travertine, oolite, tufa, and fossiliferous limestone.
* These react with dilute acid.

Figure 5.46: Sedimentary rock identification chart.

Take this quiz to check your comprehension of this section.

If you are using an offline version of this text, access the quiz for section 5.3 via the QR code.

 An interactive H5P element has been excluded from this version of the text. You can view it online here:
https://pressbooks.lib.vt.edu/introearthscience/?p=430#h5p-33

5.4 Sedimentary Structures

Sedimentary structures are visible textures or arrangements of **sediments** within a rock. Geologists use these structures to interpret the processes that made the rock and the environment in which it formed. They use **uniformitarianism** to usually compare sedimentary structures formed in modern environments to lithified counterparts in ancient rocks. Below is a summary discussion of common sedimentary structures that are useful for interpretations in the rock record.

5.4.1 Bedding Planes

Figure 5.47: Horizontal strata.

The most basic sedimentary structure is **bedding** planes, the planes that separate the layers or **strata** in sedimentary and some **volcanic** rocks. Visible in exposed outcroppings, each **bedding** plane indicates a change in **sediment deposition** conditions. This change may be subtle. For example, if a section of underlying **sediment** firms up, this may be enough to create a form a layer that is dissimilar from the overlying **sediment**. Each layer is called a **bed**, or stratum, the most basic unit of **stratigraphy**, the study of sedimentary layering.

As would be expected, **bed** thickness can indicate **sediment deposition** quantity and timing. Technically, a **bed** is a **bedding** plane thicker than 1 cm (0.4 in) and the smallest mappable unit. A layer thinner than 1 cm (0.4 in) is called a **lamina**. Varves are **bedding** planes created when **laminae** and **beds** are deposited in repetitive cycles, typically daily or seasonally. Varves are valuable geologic records of climatic histories, especially those found in lakes and **glacial** deposits.

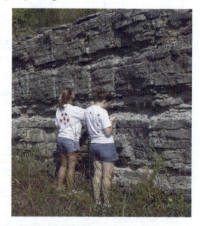

Figure 5.48: Students from the University of Wooster examine beds of Ordovician limestone in central Tennessee.

5.4.2 Graded Bedding

Figure 5.49: Image of the classic Bouma sequence. A=coarse- to fine-grained sandstone, possibly with an erosive base. B=laminated medium- to fine-grained sandstone. C=rippled fine-grained sandstone. D=laminated siltstone grading to mudstone.

Graded bedding refers to a sequence of increasingly coarse- or fine-grained **sediment** layers. **Graded bedding** often develops when **sediment deposition** occurs in an environment of decreasing energy. A **Bouma sequence** is **graded bedding** observed in **clastic** rock called **turbidite**. **Bouma sequence beds** are formed by **offshore sediment** gravity flows, which are underwater flows of **sediment**. These subsea density flows begin when **sediment** is stirred up by an energetic process and becomes a dense slurry of mixed grains. The **sediment** flow courses downward through submarine channels and canyons due to gravity acting on the density difference between the denser slurry and less dense surrounding seawater. As the flow reaches deeper ocean basins it slows down, loses energy, and deposits **sediment** in a **Bouma sequence** of coarse grains first, followed by increasingly finer grains (see figure 5.49).

5.4.3 Flow Regime and Bedforms

In fluid systems, such as moving water or wind, sand is the most easily transported and deposited **sediment** grain. Smaller particles like silt and clay are less movable by fluid systems because the tiny grains are chemically attracted to each other and stick to the underlying **sediment**. Under higher flow rates, the fine silt and clay **sediment** tends to stay in place and the larger sand grains get picked up and moved.

Bedforms are sedimentary structures created by fluid systems working on sandy **sediment**. **Grain size**, flow velocity, and **flow regime** or pattern interact to produce bedforms having unique, identifiable physical characteristics. Flow regimes are divided into upper and lower regimes, which are further divided into uppermost, upper, lower, and lowermost parts. The table below shows bedforms and their associated flow regimes. For example, the **dunes bedform** is created in the upper part of the lower **flow regime**.

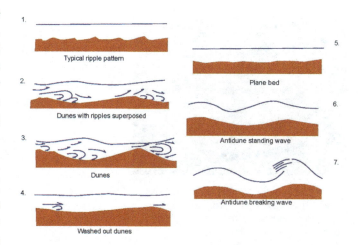

Figure 5.50: Bedforms from under increasing flow velocities.

Flow regime (part)	Bedform	Description
Lower (lowest)	Plane bed	Lower plane bed, flat laminations
Lower (lower)	Ripples	Small (with respect to flow) inclined layers dipping downflow
Lower (upper)	Dunes	Larger inclined cross beds, ±ripples, dipping downflow
Upper (lower)	Plane bed	Flat layers, can include lined-up grains (parting lineations)
Upper (upper)	Antidunes	Hard to preserve reverse dunes dipping shallowly upflow
Upper (uppermost)	Chutes/pools (rare)	Erosional, not really a bedform; rarely found preserved

Table 5.2: Bedforms and their associated flow regimes.

Plane Beds

Plane **beds** created in the lower **flow regime** are like **bedding** planes, on a smaller scale. The flat, parallel layers form as sandy **sediment** piles and move on top of layers below. Even non-flowing fluid systems, such as lakes, can produce **sediment** plane **beds**. Plane **beds** in the upper **flow regime** are created by fast-flowing fluids. They may look identical to lower-flow-regime **beds**; however, they typically show **parting lineations**, slight alignments of grains in rows and swaths, caused by high **sediment** transport rates that only occur in upper flow regimes.

Figure 5.51: Subtle lines across this sandstone (trending from the lower left to upper right) are parting lineations.

Ripples

Ripples are known by several names: **ripple** marks, **ripple** cross **beds**, or **ripple** cross **laminations**. The ridges or undulations in the **bed** are created as **sediment** grains pile up on top of the **plane bed**. With the exception of **dunes**, the scale of these **beds** is typically measured in centimeters. Occasionally, large flows like **glacial** lake outbursts, can produce **ripples** as tall as 20 m (66 ft).

Figure 5.52: Modern current ripple in sand from the Netherlands. The flow creates a steep side down current. In this image, the flow is from right to left.

First scientifically described by Hertha Ayrton, **ripple** shapes are determined by flow type and can be straight-crested, sinuous, or complex. Asymmetrical **ripples** form in a unidirectional flow. Symmetrical **ripples** are the result of an oscillating back-and-forth flow typical of intertidal swash zones. Climbing **ripples** are created from high sedimentation rates and appear as overlapping layers of **ripple** shapes (see figure 5.54).

Figure 5.53: A bidirectional flow creates this symmetrical wave ripple. From rocks in Nomgon, Mongolia. Note the crests of the ripples have been eroded away by subsequent flows in places.

Figure 5.54: Climbing ripple deposit from India.

Dunes

Dunes are very large and prominent versions of **ripples**, and typical examples of large **cross bedding**. **Cross bedding** happens when **ripples** or **dunes** pile atop one another, interrupting, and/or cutting into the underlying layers. Desert sand **dunes** are probably the first image conjured up by this category of **bedform**.

British geologist Agnold (1941) considered only Barchan and linear Seif **dunes** as the only true **dune** forms. Other workers have recognized transverse and **star dunes** as well as parabolic and **linear dunes** anchored by plants that are common in coastal areas as other types of **dunes**.

Figure 5.56: Modern sand dune in Morocco.

Figure 5.55: Lithified cross-bedded dunes from the high country of Zion National Park, Utah. The complexity of bedding planes results from the three-dimensional network of ancient dune flows.

Dunes are the most common sedimentary structure found within channelized flows of air or water. The biggest difference between **river dunes** and air-formed (desert) **dunes** is the depth of fluid **system**. Since the **atmosphere**'s depth is immense when compared to a **river** channel, desert **dunes** are much taller than those found in **rivers**. Some famous air-formed **dune** landscapes include the Sahara Desert, Death Valley, and the Gobi Desert.

As airflow moves **sediment** along, the grains accumulate on the **dune**'s windward surface (facing the wind). The angle of the windward side is typically shallower than the leeward (downwind) side, which has grains falling down over it. This difference in slopes can be seen in a **bed** cross-section and indicates the direction of the flow in the past. There are typically two styles of **dune beds**: the more common **trough** cross **beds** with curved windward surfaces, and rarer planar cross **beds** with flat windward surfaces.

In tidal locations with strong in-and-out flows, **dunes** can develop in opposite directions. This produces a feature called herringbone **cross bedding**.

Figure 5.57: Herringbone cross-bedding from the Mazomanie Formation, upper Cambrian of Minnesota.

Another **dune formation** variant occurs when very strong, hurricane-strength, winds agitate parts of the usually undisturbed seafloor. These **beds** are called **hummocky cross stratification** and have a 3D architecture of hills and valleys, with inclined and declined layering that matches the **dune** shapes.

Antidunes

Figure 5.59: Antidunes forming in Urdaibai, Spain.

Figure 5.58: Hummocky-cross stratification, seen as wavy lines throughout the middle of this rock face. Best example is just above the pencil in the center.

Antidunes are so named because they share similar characteristics with **dunes**, but are formed by a different, opposing process. While **dunes** form in lower flow regimes, **antidunes** come from fast-flowing upper flow regimes. In certain conditions of high flow rates, **sediment** accumulates upstream of a subtle **dip** instead of traveling downstream (see figure 5.59). **Antidunes** form in phase with the flow; in **rivers** they are marked by rapids in the current. **Antidunes** are rarely preserved in the rock record because the high flow rates needed to produce the **beds** also accelerate **erosion**.

5.4.4 Bioturbation

Bioturbation is the result of organisms burrowing through soft **sediment**, which disrupts the **bedding** layers. These tunnels are backfilled and eventually preserved when the **sediment** becomes rock. **Bioturbation** happens most commonly in shallow, **marine** environments, and can be used to indicate water depth.

Figure 5.60: Bioturbated dolomitic siltstone from Kentucky.

5.4.5 Mudcracks

Mudcracks occur in clay-rich **sediment** that is submerged underwater and later dries out. Water fills voids in the clay's crystalline structure, causing the **sediment** grains to swell. When this waterlogged **sediment** begins to dry out, the clay grains shrink. The **sediment** layer forms deep polygonal cracks with tapered openings toward the surface, which can be seen in profile. The cracks fill with new **sediment** and become visible veins running through the lithified rock. These dried-out clay **beds** are a major source of **mud chips**, small fragments of mud or **shale**, which commonly become **inclusions** in **sandstone** and **conglomerate**. What makes this sedimentary structure so important to geologists, is they only form in certain **depositional environments**—such as **tidal flats** that form underwater and are later exposed to air. Syneresis cracks are similar in

Figure 5.61: Lithified mudcracks from Maryland.

appearance to **mudcracks** but much rarer; they are formed when subaqueous (underwater) clay **sediment** shrinks.

5.4.6 Sole Marks

Figure 5.62: This flute cast shows a flow direction toward the upper right of the image, as seen by the bulge sticking down out of the layer above. The flute cast would have been scoured into a rock layer below that has been removed by erosion, leaving the sandy layer above to fill in the flute cast.

Sole marks are small features typically found in **river** deposits. They form at the base of a **bed**, the sole, and on top of the underlying **bed**. They can indicate several things about the **deposition** conditions, such as flow direction or **stratigraphic** up-direction (see Geopetal Structures section). **Flute casts** or **scour marks** are grooves carved out by the forces of fluid flow and **sediment** loads. The upstream part of the flow creates steep grooves and downstream the grooves are shallower. The grooves subsequently become filled by overlying **sediment**, creating a **cast** of the original hollow.

Formed similarly to **flute casts** but with a more regular and aligned shape, **groove casts** are produced by larger clasts or debris carried along in the water that scrape across the **sediment** layer. **Tool marks** come from objects like sticks carried in the fluid downstream or embossed into the **sediment** layer, leaving a depression that later fills with new **sediment**.

Figure 5.63: Groove casts at the base of a turbidite deposit in Italy.

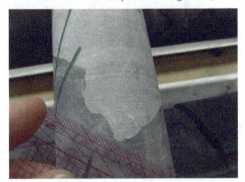

Figure 5.64: A drill core showing a load cast showing light-colored sand sticking down into dark mud.

Load **casts**, an example of soft-**sediment deformation**, are small indentations made by an overlying layer of coarse **sediment** grains or clasts intruding into a softer, finer-grained **sediment** layer.

5.4.7 Raindrop Impressions

Like their name implies, **raindrop impressions** are small pits or bumps found in soft **sediment**. While they are generally believed to be created by rainfall, they may be caused by other agents such as escaping gas bubbles.

Figure 5.65: Mississippian raindrop impressions over wave ripples from Nova Scotia.

5.4.8 Imbrication

Imbrication is a **stack** of large and usually flat clasts—cobbles, gravels, **mud chips**, etc.—that are aligned in the direction of fluid flow. The clasts may be stacked in rows, with their edges dipping down and flat surfaces aligned to face the flow (see figure 5.66). Or their flat surfaces may be parallel to the layer and long axes aligned with flow. Imbrications are useful for analyzing paleocurrents, or currents found in the geologic past, especially in **alluvial** deposits.

Figure 5.66: Cobbles in this conglomerate are positioned in a way that they are stacked on each other, which occurred as flow went from left to right.

5.4.9 Geopetal Structures

Geopetal structures, also called up-direction indicators, are used to identify which way was up when the **sedimentary rock** layers were originally formed. This is especially important in places where the rock layers have been deformed, tilted, or overturned. Well preserved **mudcracks**, **sole marks**, and **raindrop impressions** can be used to determine up direction. Other useful **geopetal** structures include:

- Vugs: Small voids in the rock that usually become filled during **diagenesis**. If the void is partially filled or filled in stages, it serves as a permanent record of a level bubble, frozen in time.
- **Cross bedding** – In places where **ripples** or **dunes** pile on top of one another, where one **cross bed** interrupts and/or cuts another below, this shows a cross-cutting relationship that indicates up direction.
- **Ripples**, **dunes**: Sometimes the **ripples** are preserved well enough to differentiate between the crests (top) and troughs (bottom).

Figure 5.67: This bivalve (clam) fossil was partially filled with tan sediment, partially empty. Later fluids filled in the fossil with white calcite minerals. The line between the sediment and the later calcite is paleo-horizontal.

- **Fossils**: Body **fossils** in life position, meaning the body parts are not scattered or broken, and **trace fossils** like footprints (see figure 5.68) can provide an up direction. Intact fossilized coral **reefs** are excellent up indicators because of their large size and easily distinguishable top and bottom. **Index fossils**, such as ammonites, can be used to age date **strata** and determine up direction based on relative rock ages.
- Vesicles – **Lava** flows eliminate gas upwards. An increase of vesicles toward the top of the flow indicates up.

Figure 5.68: Eubrontes trace fossil from Utah, showing the geopetal direction is into the image.

Take this quiz to check your comprehension of this section.

If you are using an offline version of this text, access the quiz for section 5.4 via the QR code.

> *An interactive H5P element has been excluded from this version of the text. You can view it online here:*
> *https://pressbooks.lib.vt.edu/introearthscience/?p=430#h5p-34*

5.5 Depositional Environments

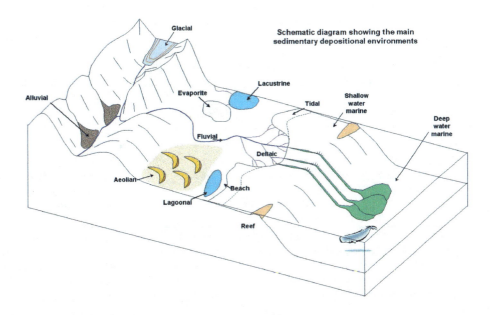

Figure 5.69: A representation of common depositional environments.

The ultimate goal of many **stratigraphy** studies is to understand the original **depositional environment**. Knowing where and how a particular **sedimentary rock** was formed can help geologists paint a picture of past environments—such as a mountain **glacier**, gentle **floodplain**, dry desert, or deep-sea **ocean floor**. The study of **depositional environments** is a complex endeavor; the table shows a simplified version of what to look for in the rock record.

Location	Sediment	Common rock types	Typical fossils	Sedimentary structures
Abyssal	Very fine muds and oozes, diatomaceous Earth	Chert	Diatoms	Few
Submarine fan	Graded Bouma sequences, alternating sand/mud	Clastic rocks	Rare	Channels, fan shape
Continental slope	Mud, possible sand, countourites	Shale, siltstone, limestone	Rare	Swaths
Lower shoreface	Laminated sand	Sandstone	Bioturbation	Hummocky cross beds
Upper shoreface	Planar sand	Sandstone	Bioturbation	Plane beds, cross beds
Littoral (beach)	Very well sorted sand	Sandstone	Bioturbation	Few
Tidal flat	Mud and sand with channels	Shale, mudstone, siltstone	Bioturbation	Mudcracks, symmetric ripples
Reef	Lime mud with coral	Limestone	Many, commonly coral	Few
Lagoon	Laminated mud	Shale	Many, bioturbation	Laminations
Delta	Channelized sand with mud, ±swamp	Clastic rocks	Many to few	Cross beds
Fluvial (river)	Sand and mud, can have larger sediments	Sandstone, conglomerate	Bone beds (rare)	Cross beds, channels, asymmetric ripples
Alluvial	Mud to boulders, poorly sorted	Clastic rocks	Rare	Channels, mud cracks
Lacustrine (lake)	Fine-grained laminations	Shale	Invertebrates, rare (deep) bone beds	Laminations
Paludal (swamp)	Plant material	Coal	Plant debris	Rare
Aeolian (dunes)	Very well sorted sand and silt	Sandstone	Rare	Cross beds (large)
Glacial	Mud to boulders, poorly sorted	Conglomerate (tillite)		Striations, drop stones

Table 5.3: Rock record and depositional environments.

5.5.1 Marine

Marine depositional environments are completely and constantly submerged in seawater. Their depositional characteristics are largely dependent on the depth of water with two notable exceptions, **submarine fans** and **turbidites**.

Abyssal

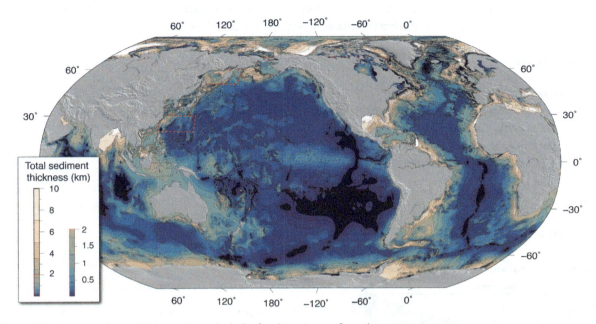

Figure 5.70: Marine sediment thickness. Note the lack of sediment away from the continents.

Abyssal sedimentary rocks form on the **abyssal plain**. The plain encompasses relatively flat **ocean floor** with some minor topographical features, called **abyssal** hills. These small seafloor mounts range 100 m to 20 km in diameter, and are possibly created by **extension**. Most **abyssal** plains do not experience significant fluid movement, so **sedimentary rock** formed there are very fine grained.

There are three categories of **abyssal sediment**. Calcareous oozes consist of **calcite**-rich plankton shells that have fallen to the **ocean floor**. An example of this type of **sediment** is **chalk**. Siliceous oozes are also made of plankton debris, but these organisms build their shells using silica or hydrated silica. In some cases such as with diatomaceous earth, **sediment** is deposited below the **calcite compensation depth**, a depth where **calcite** solubility increases. Any **calcite**-based shells are **dissolved**, leaving only silica-based shells. **Chert** is another common rock formed from these types of **sediment**. These two types of **abyssal sediment** are also classified as **biochemical** in origin (see biochemical section).

The third **sediment** type is pelagic clay. Very fine-grained clay particles, typically brown or red, descend through the water column very slowly. Pelagic clay **deposition** occurs in areas of remote open ocean, where there is little plankton accumulation.

Figure 5.71: Diatomaceous earth.

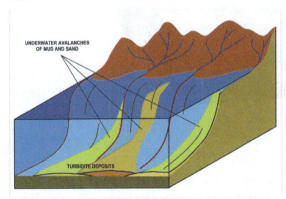

Figure 5.72: Turbidites inter-deposited within submarine fans.

Two notable exceptions to the fine-grained nature of **abyssal sediment** are **submarine fan** and **turbidite** deposits. **Submarine fans** occur **offshore** at the base of large **river** systems. They are initiated during times of low sea level, as strong **river** currents carve **submarine canyons** into the **continental shelf**. When sea levels rise, **sediment** accumulates on the shelf typically forming large, fan-shaped floodplains called deltas. Periodically, the **sediment** is disturbed creating dense slurries that flush down the underwater canyons in large gravity-induced events called **turbidites**. The **submarine fan** is formed by a network of **tur-**

bidites that deposit their **sediment** loads as the slope decreases, much like what happens above-water at **alluvial** fans and deltas. This sudden flushing transports coarser **sediment** to the **ocean floor** where they are otherwise uncommon. **Turbidites** are also the typical origin of **graded Bouma** sequences. (see chapter 5).

Continental Slope

Continental slope deposits are not common in the rock record. The most notable type of **continental slope** deposits are contourites. Contourites form on the slope between the **continental shelf** and deep **ocean floor**. Deep-water ocean currents deposit **sediment** into smooth drifts of various architectures, sometimes interwoven with **turbidites**.

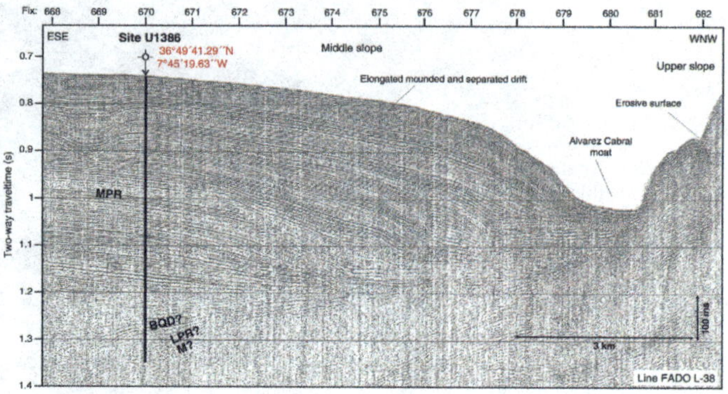

Lower Shoreface

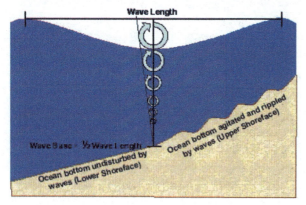

Figure 5.74: Diagram describing wavebase.

Figure 5.73: Contourite drift deposit imaged with seismic waves.

The lower **shoreface** lies below the normal depth of wave agitation, so the **sediment** is not subject to daily winnowing and **deposition**. These **sediment** layers are typically finely laminated, and may contain hummocky cross-stratification. Lower **shoreface beds** are affected by larger waves, such those as generated by hurricanes and other large storms.

Upper Shoreface

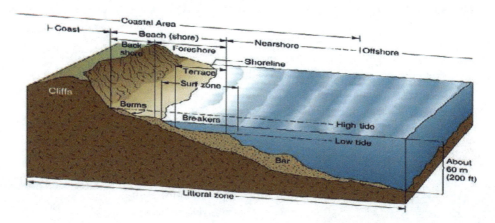

Figure 5.75: Diagram of zones of the shoreline.

The upper **shoreface** contains **sediments** within the zone of normal wave action, but still submerged below the beach environment. These **sediments** usually consist of very well sorted, fine sand. The main sedimentary structure is planar **bedding** consistent with the lower part of the upper **flow regime**, but it can also contain **cross bedding** created by **longshore currents**.

5.5.2 Transitional Coastline Environments

Transitional environments, more often called **shoreline** or **coastline** environments, are zones of complex interactions caused by ocean water hitting land. The **sediment** preservation potential is very high in these environments because **deposition** often occurs on the **continental shelf** and underwater. **Shoreline** environments are an important source of hydrocarbon deposits (**petroleum, natural gas**).

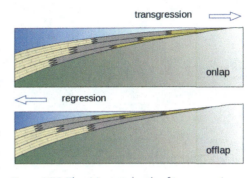

Figure 5.76: The rising sea levels of transgressions create onlapping sediments, regressions create offlapping.

The study of **shoreline depositional environments** is called **sequence stratigraphy**. **Sequence stratigraphy** examines depositional changes and 3D architectures associated with rising and falling sea levels, which is the main force at work in **shoreline** deposits. These sea-level fluctuations come from the daily tides, as well as **climate** changes and **plate tectonics**. A steady rise in sea level relative to the **shoreline** is called **transgression**. **Regression** is the opposite, a relative drop in sea level. Some common components of **shoreline** environments are **littoral** zones, **tidal flats**, **reefs**, **lagoons**, and deltas. For a more in-depth look at these environments, see chapter 12.

Littorals

The **littoral** zone, better known as the beach, consists of highly weathered, homogeneous, well-sorted sand grains made mostly of **quartz**. There are black sand and other types of sand beaches, but they tend to be unique exceptions rather than the rule. Because beach sands, past or present, are so highly evolved, the amount grain **weathering** can be discerned using the **minerals zircon**, tourmaline, and rutile. This tool is called the ZTR (**zircon**, tourmaline, rutile) index. The ZTR index is higher in more weathered beaches, because these relatively rare and **weather**-resistant **minerals** become **concentrated** in older beaches. In some beaches, the ZTR index is so high the sand can be harvested as an economically viable source of these **minerals**. The beach environment has no sedimentary structures, due to the constant bombardment of wave energy delivered by surf action. Beach **sediment** is moved around via multiple processes. Some beaches with high **sediment** supplies develop **dunes** nearby.

Figure 5.77: Lithified heavy mineral sand (dark layers) from a beach deposit in India.

Tidal Flats

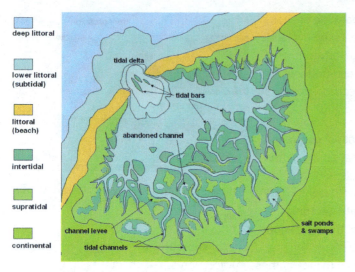

deep littoral

lower littoral (subtidal)

littoral (beach)

intertidal

supratidal

continental

tidal delta

tidal bars

abandoned channel

channel levee

tidal channels

salt ponds & swamps

Figure 5.78: General diagram of a tidal flat and associated features.

Tidal flats, or **mud flats**, are sedimentary environments that are regularly flooded and drained by ocean tides. **Tidal flats** have large areas of fine-grained **sediment** but may also contain coarser sands. **Tidal flat** deposits typically contain gradational **sediments** and may include multi-directional **ripple** marks. **Mudcracks** are also commonly seen due to the **sediment** being regularly exposed to air during low tides; the combination of **mudcracks** and **ripple** marks is distinctive to **tidal flats**.

Tidal water carries in **sediment**, sometimes focusing the flow through a narrow opening called a tidal inlet. Tidal channels, creek channels influenced by tides, can also **focus** tidally-induced flow. Areas of higher flow like inlets and tidal channels feature coarser grain sizes and larger **ripples**, which in some cases can develop into **dunes**.

Reefs

Reefs, which most people would immediately associate with tropical coral **reefs** found in the oceans, are not only made by living things. Natural buildups of sand or rock can also create **reefs**, similar to **barrier islands**. Geologically speaking, a **reef** is any topographically-elevated feature on the **continental shelf**, located oceanward of and separate from the beach. The term **reef** can also be applied to **terrestrial** (atop the **continental crust**) features. Capitol Reef National Park in Utah contains a topographic barrier, a **reef**, called the Waterpocket Fold.

Figure 5.79: Waterpocket fold, Capitol Reef National Park, Utah.

Most **reefs**, now and in the geologic past, originate from the biological processes of living organisms. The growth habits of coral **reefs** provide geologists important information about the past. The hard structures in coral **reefs** are built by soft-bodied **marine** organisms, which continually add new material and enlarge the **reef** over time. Under certain conditions, when the land beneath a **reef** is subsiding, the coral **reef** may grow around and through existing **sediment**, holding the **sediment** in place, and thus preserving the record of environmental and geological condition around it.

Figure 5.80: A modern coral reef.

Figure 5.81: The light blue reef is fringing the island of Vanatinai. As the island erodes away, only the reef will remain, forming a reef-bound seamount.

Sediment found in coral **reefs** is typically fine-grained, mostly **carbonate**, and tends to deposit between the intact coral skeletons. Water with high levels of silt or clay particles can inhibit the **reef** growth because coral organisms require sunlight to thrive; they host symbiotic algae called zooxanthellae that provide the coral with nourishment via photosynthesis. Inorganic **reef** structures have much more variable compositions. **Reefs** have a big impact on **sediment deposition** in **lagoon** environments since they are natural storm breaks, wave and storm buffers, which allows fine grains to settle and accumulate.

Reefs are found around shorelines and islands; coral **reefs** are particularly common in tropical locations. **Reefs** are also found around features known as **seamounts**, which is the base of an ocean island left standing underwater after the upper part is eroded away by waves. Examples include the Emperor Seamounts, formed millions of years ago over the Hawaiian **Hotspot**. **Reefs** live and grow along the upper edge of these flat-topped **seamounts**. If the **reef** builds up above sea level and completely encircles the top of the **seamount**, it is called a coral-ringed atoll. If the **reef** is submerged, due to **erosion**, **subsidence**, or sea level rise, the **seamount–reef** structure is called a guyot.

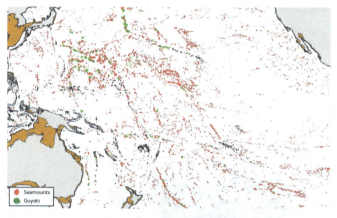

Figure 5.82: Seamounts and guyots in the North Pacific.

Lagoons

Lagoons are small bodies of seawater located inland from the **shore** or isolated by another geographic feature, such as a **reef** or **barrier island**. Because they are protected from the action of tides, currents, and waves, **lagoon** environments typically have very fine grained **sediments**. **Lagoons**, as well as **estuaries**, are ecosystems with high biological productivity. Rocks from these environments often includes **bioturbation** marks or **coal** deposits. Around **lagoons** where evaporation exceeds water inflow, salt flats, also known as sabkhas, and sand **dune** fields may develop at or above the high **tide** line.

Figure 5.83: Kara-Bogaz Gol lagoon, Turkmenistan.

Deltas

Figure 5.84: The Nile delta, in Egypt.

Figure 5.85: Birdfoot river-dominated delta of the Mississippi River.

Deltas form where **rivers** enter lakes or oceans and are of three basic shapes: **river**-dominated deltas, wave-dominated deltas, and **tide**-dominated deltas. The name **delta** comes from the Greek letter Δ (**delta**, uppercase), which resembles the triangular shape of the Nile **River delta**. The velocity of water flow is dependent on riverbed slope or **gradient**, which becomes shallower as the **river** descends from the mountains. At the point where a **river** enters an ocean or lake, its slope angle drops to zero degrees (0°). The flow velocity quickly drops as well, and **sediment** is deposited, from coarse clasts, to fine sand, and mud to form the **delta**. As one part of the **delta** becomes overwhelmed by **sediment**, the slow-moving flow gets diverted back and forth, over and over, and forms a spread out network of smaller distributary channels.

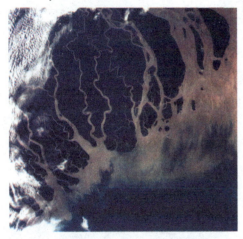

Figure 5.86: Tidal delta of the Ganges River.

Deltas are organized by the dominant process that controls their shape: **tide**-dominated, wave-dominated, or **river**-dominated. Wave-dominated deltas generally have smooth coastlines and beach-ridges on the land that represent previous shorelines. The Nile River delta is a wave-dominated type (see figure 5.84).

The Mississippi River delta is a **river**-dominated **delta**, shaped by levees along the **river** and its distributaries that confine the flow forming a shape called a birdfoot **delta**. Other times the tides or the waves can be a bigger factor, and can reshape the **delta** in various ways.

A **tide**-dominated **delta** is dominated by tidal currents. During flood stages when **rivers** have lots of water available, it develops distributaries that are separated by sand bars and sand ridges. The tidal **delta** of the Ganges River is the largest **delta** in the world.

5.5.3 Terrestrial

Terrestrial depositional environments are diverse. Water is a major factor in these environments, in liquid or frozen states, or even when it is lacking (arid conditions).

Fluvial

Fluvial (**river**) systems are formed by water flowing in channels over the land. They generally come in two main varieties: **meandering** or **braided**. In **meandering streams**, the flow carries **sediment** grains via a single channel that wanders back and forth across the **floodplain**. The **floodplain sediment** away from the channel is mostly fine grained material that only gets deposited during floods.

Figure 5.87: The Cauto River in Cuba. Note the sinuosity in the river, which is meandering.

Braided fluvial systems generally contain coarser **sediment** grains, and form a complicated series of intertwined channels that flow around gravel and sand bars (see chapter 11).

Figure 5.88: The braided Waimakariri river in New Zealand.

Alluvial

A distinctive characteristic of **alluvial** systems is the intermittent flow of water. **Alluvial** deposits are common in arid places with little **soil** development. Lithified **alluvial beds** are the primary **basin**-filling rock found throughout the **Basin and Range** region of the western United States. The most distinctive **alluvial** sedimentary deposit is the **alluvial** fan, a large cone of **sediment** formed by **streams** flowing out of dry mountain valleys into a wider and more open dry area. **Alluvial sediments** are typically poorly sorted and coarse grained, and often found near **playa** lakes or **aeolian** deposits (see chapter 13).

Figure 5.89: An alluvial fan spreads out into a broad alluvial plain. From Red Rock Canyon State Park, California.

Lacustrine

Lake systems and deposits, called **lacustrine**, form via processes somewhat similar to **marine** deposits, but on a much smaller scale. **Lacustrine** deposits are found in lakes in a wide variety of locations. Lake Baikal in southeast Siberia (Russia) is in a **tectonic basin**. Crater Lake (Oregon) sits in a **volcanic caldera**. The Great Lakes (northern United States) came from glacially carved and deposited **sediment**. Ancient Lake Bonneville (Utah) formed in a pluvial setting that during a **climate** that was relatively wetter and cooler than that of modern Utah. **Oxbow** lakes, named for their curved shape, originated in **fluvial** floodplains. **Lacustrine sediment** tends to be very fine grained and thinly laminated, with only minor contributions from wind-blown, current, and tidal deposits. When lakes dry out or evaporation outpaces **precipitation**, **playas** form. **Playa** deposits

Figure 5.90: Oregon's Crater Lake was formed about 7700 years ago after the eruption of Mount Mazama.

resemble those of normal lake deposits but contain more **evaporite minerals**. Certain **tidal flats** can have **playa**-type deposits as well.

Paludal

Paludal systems include bogs, marshes, swamps, or other wetlands, and usually contain lots of organic matter. **Paludal** systems typically develop in coastal environments, but are common occur in humid, low-lying, low-**latitude**, warm zones with large volumes of flowing water. A characteristic **paludal** deposit is a peat bog, a deposit rich in organic matter that can be converted into **coal** when lithified. **Paludal** environments may be associated with tidal, deltaic, **lacustrine**, and/or **fluvial deposition**.

Aeolian

Aeolian, sometimes spelled **eolian** or œolian, are deposits of windblown **sediments**. Since wind has a much lower carrying capacity than water, **aeolian** deposits typically consists of clast sizes from fine dust to sand. Fine silt and clay can cross very long distances, even entire oceans suspended in air.

With sufficient **sediment** influx, **aeolian** systems can potentially form large **dunes** in dry or wet conditions. The figure shows **dune** features and various types. British geologist Ralph A. Bagnold (1896-1990) considered only Barchan and linear Seif **dunes** as the only true **dune** forms. Other scientists recognize transverse, star, parabolic, and **linear dune** types. Parabolic and **linear dunes** grow from sand anchored by plants and are common in coastal areas.

Figure 5.91: Formation and types of dunes.

Compacted layers of wind-blown **sediment** is known as **loess**. **Loess** commonly starts as finely ground up rock flour created by **glaciers**. Such deposits cover thousands of square miles in the Midwestern United States. **Loess** may also form in desert regions (see chapter 13). Silt for the **Loess** Plateau in China came from the Gobi Desert in China and Mongolia.

Figure 5.92: Loess Plateau in China. The loess is so highly compacted that buildings and homes have been carved in it.

Glacial

Glacial sedimentation is very diverse, and generally consists of the most poorly-sorted **sediment** deposits found in nature. The main clast type is called **diamictite**, which literally means two sizes, referring to the unsorted mix of large and small rock fragments found in **glacial** deposits. Many **glacial tills**, glacially derived **diamictites**, include very finely-pulverized rock flour along with giant **erratic** boulders. The surfaces of larger clasts typically have striations from the rubbing, scraping, and polishing of surfaces by abrasion during the movement of **glacial** ice. **Glacial** systems are so large and produce so much **sediment**, they frequently create multiple, individualized **depositional environments**, such as **fluvial**, deltaic, **lacustrine**, pluvial, **alluvial**, and/or **aeolian** (see chapter 14).

Figure 5.93: Wide range of sediments near Athabaska Glacier, Jasper National Park, Alberta, Canada.

5.5.4 Facies

In addition to **mineral composition** and **lithification** process, geologists also classify **sedimentary rock** by its depositional characteristics, collectively called **facies** or lithofacies. Sedimentary **facies** consist of physical, chemical, and/or biological properties, including relative changes in these properties in adjacent **beds** of the same layer or geological age. Geologists analyze **sedimentary rock facies** to interpret the original **deposition** environment, as well as disruptive geological events that may have occurred after the rock layers were established.

It boggles the imagination to think of all the sedimentary **deposition** environments working next to each other, at the same time, in any particular region on Earth. The resulting **sediment beds** develop characteristics reflecting contemporaneous conditions at the time of **deposition**, which later may become preserved into the rock record. For example in the Grand Canyon, rock **strata** of the same geologic age includes many different **depositional environments**: beach sand, **tidal flat** silt, **offshore** mud, and farther **offshore limestone**. In other words, each sedimentary or **stratigraphic facies** presents recognizable characteristics that reflect specific, and different, **depositional environments** that were present at the same time.

Facies may also reflect depositional changes in the same location over time. During **periods** of rising sea level, called **marine transgression**, the **shoreline** moves inland as seawater covers what was originally dry land and creates new **offshore depositional environments**. When these **sediment beds** turn into **sedimentary rock**, the vertical **stratigraphy** sequence reveals beach lithofacies buried by **offshore** lithofacies.

Biological **facies** are remnants (**coal**, diatomaceous earth) or evidence (**fossils**) of living organisms. **Index fossils**, fossilized life forms specific to a particular environment and/or geologic time **period**, are an example of biological **facies**. The horizontal assemblage and vertical distribution of **fossils** are particularly useful for studying species evolution because **transgression**, **deposition**, burial, and **compaction** processes happen over a considerable geologic time range.

Fossil assemblages that show evolutionary changes greatly enhance our interpretation of Earth's ancient history by illustrating the **correlation** between **stratigraphic** sequence and geologic time scale. During the Middle **Cambrian period** (see chapter 7), regions around the Grand Canyon experienced **marine transgression** in a southeasterly direction (relative to current maps). This shift of the **shoreline** is reflected in the Tapeats Sandstone beach facies, Bright Angle Shale near-offshore mud **facies**, and Muav Limestone far-offshore facies. **Marine** organisms had plenty of time to evolve and adapt to their slowly changing environment; these changes are reflected in the biological **facies**, which show older life forms in the western regions of the canyon and younger life forms in the east.

Take this quiz to check your comprehension of this section.

If you are using an offline version of this text, access the quiz for section 5.5 via the QR code.

 An interactive H5P element has been excluded from this version of the text. You can view it online here:
https://pressbooks.lib.vt.edu/introearthscience/?p=430#h5p-35

Summary

Sedimentary rocks are grouped into two main categories: **clastic (detrital)** and chemical. **Clastic (detrital)** rocks are made of **mineral** clasts or **sediment** that lithifies into solid material. **Sediment** is produced by the mechanical or **chemical weathering** of **bedrock** and transported away from the source via **erosion**. **Sediment** that is deposited, buried, compacted, and sometimes cemented becomes **clastic** rock. **Clastic** rocks are classified by **grain size**; for example **sandstone** is made of sand-sized particles. **Chemical sedimentary** rocks comes from **minerals precipitated** out an aqueous **solution** and is classified according to **mineral composition**. The **chemical sedimentary** rock **limestone** is made of calcium **carbonate**. Sedimentary structures have textures and shapes that give insight on depositional histories. **Depositional environments** depend mainly on fluid transport systems and encompass a wide variety of underwater and above ground conditions. Geologists analyze depositional conditions, sedimentary structures, and rock records to interpret the paleogeographic history of a region.

Take this quiz to check your comprehension of this chapter.

If you are using an offline version of this text, access the quiz for chapter 5 via the QR code.

 An interactive H5P element has been excluded from this version of the text. You can view it online here:
https://pressbooks.lib.vt.edu/introearthscience/?p=430#h5p-36

Text References

1. Affolter, M.D., 2004, On the nature of volcanic lithic fragments: Definition source and evolution.
2. Ashley, G.M., 1990, Classification of large-scale subaqueous bedforms: a new look at an old problem-SEPM bedforms and bedding structures: J. Sediment. Res., v. 60, no. 1.
3. Ayrton, H., 1910, The origin and growth of ripple-mark: Proceedings of the Royal Society of London. Series A, Containing Papers of a Mathematical and Physical Character, v. 84, no. 571, p. 285–310.

4. Bagnold, R.A., 1941, The physics of blown sand and desert dunes: Methum, London, UK, p. 265.

5. Blatt, H., Middleton, G.V., and Murray, R., 1980, Origin of Sedimentary Rocks: Prentice-Hall, Inc., Englewood Cliffs, New Jersey, USA.

6. Bouma, A.H., Kuenen, P.H., and Shepard, F.P., 1962, Sedimentology of some flysch deposits: a graphic approach to facies interpretation: Elsevier Amsterdam.

7. Cant, D.J., 1982, Fluvial facies models and their application.

8. Dickinson, W.R., and Suczek, C.A., 1979, Plate tectonics and sandstone compositions: AAPG Bull., v. 63, no. 12, p. 2164–2182.

9. Dunham, R.J., 1962, Classification of carbonate rocks according to depositional textures.

10. Eisma, D., 1998, Intertidal deposits: River mouths, tidal flats, and coastal lagoons: CRC Marine Science, Taylor & Francis, CRC Marine Science.

11. Folk, R.L., 1974, Petrography of sedimentary rocks: Univ. Texas, Hemphill, Austin, Tex, v. 182.

12. Goldich, S.S., 1938, A study in rock-weathering: J. Geol., v. 46, no. 1, p. 17–58.

13. Hubert, J.F., 1962, A zircon-tourmaline-rutile maturity index and the interdependence of the composition of heavy mineral assemblages with the gross composition and texture of sandstones: J. Sediment. Res., v. 32, no. 3.

14. Johnson, C.L., Franseen, E.K., and Goldstein, R.H., 2005, The effects of sea level and palaeotopography on lithofacies distribution and geometries in heterozoan carbonates, south-eastern Spain: Sedimentology, v. 52, no. 3, p. 513–536., doi: 10.1111/j.1365-3091.2005.00708.x.

15. Karátson, D., Sztanó, O., and Telbisz, T., 2002, Preferred clast orientation in volcaniclastic mass-flow deposits: application of a new photo-statistical method: J. Sediment. Res., v. 72, no. 6, p. 823–835.

16. Klappa, C.F., 1980, Rhizoliths in terrestrial carbonates: classification, recognition, genesis and significance: Sedimentology, v. 27, no. 6, p. 613–629.

17. Longman, M.W., 1981, A process approach to recognizing facies of reef complexes.

18. Mckee, E.D., and Weir, G.W., 1953, Terminology for stratification and cross-stratification in sedimentary rocks: Geol. Soc. Am. Bull., v. 64, no. 4, p. 381–390.

19. Metz, R., 1981, Why not raindrop impressions? J. Sediment. Res., v. 51, no. 1.

20. Nichols, M.M., Biggs, R.B., and Davies, R.A.Jr., 1985, Estuaries, *in* Coastal Sedimentary Environments: Springer-Verlag: New York, p. 77–173.

21. Normark, W.R., 1978, Fan valleys, channels, and depositional lobes on modern submarine fans: characters for recognition of sandy turbidite environments: AAPG Bull., v. 62, no. 6, p. 912–931.

22. Pettijohn, F.J., and Potter, P.E., 2012, Atlas and glossary of primary sedimentary structures.

23. Plummer, P.S., and Gostin, V.A., 1981, Shrinkage cracks: desiccation or synaeresis? J. Sediment]. Res., v. 51, no. 4.

24. Reinson, G.E., 1984, Barrier-island and associated strand-plain systems, *in* Walker, R.G., editor, Facies Models: Geoscience Canada Reprint Series 1, p. 119–140.

25. Stanistreet, I.G., and McCarthy, T.S., 1993, The Okavango Fan and the classification of subaerial fan systems: Sediment. Geol., v. 85, no. 1, p. 115–133.

26. Stow, D.A.V., Faugères, J.-C., Viana, A., and Gonthier, E., 1998, Fossil contourites: a critical review: Sediment. Geol., v. 115, no. 1–4, p. 3–31.

27. Stow, D.A.V., and Piper, D.J.W., 1984, Deep-water fine-grained sediments: facies models: Geological Society, London, Special Publications, v. 15, no. 1, p. 611–646.

28. Udden, J.A., 1914, Mechanical composition of clastic sediments: Geol. Soc. Am. Bull., v. 25, no. 1, p. 655–744.

29. Wentworth, C.K., 1922, A scale of grade and class terms for clastic sediments: J. Geol., v. 30, no. 5, p. 377–392.

30. Yin, D., Peakall, J., Parsons, D., Chen, Z., Averill, H.M., Wignall, P., and Best, J., 2016, Bedform genesis in bedrock substrates: Insights into formative processes from a new experimental approach and the importance of suspension-dominated abrasion: Geomorphology, v. 255, p. 26–38.

Figure References

Figure 5.1: A model of a water molecule, showing the bonds between the hydrogen and oxygen. Dan Craggs. 2009. Public domain. https://commons.wikimedia.org/wiki/File:H2O_2D_labelled.svg

formation, leading to better farming practices. Unknown author. 2007. Public domain. https://www.wikiwand.com/en/Andean_civilizations#Media/File:Pisac006.jpg

Figure 5.21: A simplified soil profile, showing labeled layers. Wilsonbiggs. 2021. CC BY-SA 4.0. https://commons.wikimedia.org/wiki/File:Soil_Horizons.svg

Figure 5.22: A sample of bauxite. Werner Schellmann. 1965. CC BY-SA 2.5. https://commons.wikimedia.org/wiki/File:Bauxite_with_unweathered_rock_core._C_021.jpg

Figure 5.23: A dust storm approaches Stratford, Texas in 1935. George E. Marsh Album via NOAA. 1935. Public domain. https://commons.wikimedia.org/wiki/File:Dust_Storm_Texas_1935.jpg

Figure 5.24: Geologic unconformity seen at Siccar Point on the east coast of Scotland. dave souza. 2008. CC BY-SA 4.0. https://commons.wikimedia.org/wiki/File:Siccar_Point_red_capstone_closeup.jpg

Figure 5.25: Permineralization in petrified wood. Moondigger. 2005. CC BY-SA 2.5. https://commons.wikimedia.org/wiki/File:Petrified_forest_log_2_md.jpg

Figure 5.26: Size categories of sediments, known as the Wentworth scale. Jeffress Williams, Matthew A. Arsenault, Brian J. Buczkowski, Jane A. Reid, James G. Flocks, Mark A. Kulp, Shea Penland, and Chris J. Jenkins via USGS. 2011. Public domain. https://commons.wikimedia.org/wiki/File:Wentworth_scale.png

Figure 5.27: A well-sorted sediment (left) and a poorly-sorted sediment (right). Woudloper. 2009. CC BY-SA 3.0. https://commons.wikimedia.org/wiki/File:Sorting_in_sediment.svg

Figure 5.28: Degree of rounding in sediments. Woudloper. 2009. Public domain. https://commons.wikimedia.org/wiki/File:Rounding_%26_sphericity_EN.svg

Figure 5.29: A sand grain made of basalt, known as a microlitic volcanic lithic fragment. Matt Affolter (QFL247). 2009. CC BY-SA 3.0. https://commons.wikimedia.org/wiki/File:LvMS-Lvm.jpg

Figure 5.30: Hawiian beach composed of green olivine sand from weathering of nearby basaltic rock. Aren Elliott. 2018. CC BY-SA 4.0. https://commons.wikimedia.org/wiki/File:Panorama_of_Papakolea_green_sand_beach_in_Hawaii_2.jpg

Figure 5.31: Megabreccia in Titus Canyon, Death Valley National Park, California. NPS. Unknown date. Public domain. https://commons.wikimedia.org/wiki/File:Titus_Canyon_Narrows.jpg

Figure 5.32: Enlarged image of frosted and rounded windblown sand grains. Wilson44691. 2008. Public domain. https://commons.wikimedia.org/wiki/File:CoralPinkSandDunesSand.JPG

Figure 5.33: The Rochester Shale, New York. Wilson44691. 2015. Public domain. https://en.m.wikipedia.org/wiki/File:Rochester_Shale_Niagara_Gorge.jpg

Figure 5.34: Claystone laminations from Glacial Lake Missoula. Matt Affolter. 2010. CC BY-SA 3.0. https://commons.wikimedia.org/wiki/File:GLMsed.jpg

Figure 5.35: Salt-covered plain known as the Bonneville Salt Flats, Utah. Michael Pätzold. 2008. CC BY-SA 4.0. https://commons.wikimedia.org/wiki/File:Bonneville_salt_flats_pilot_peak.jpg

Figure 5.36: Ooids from Joulter's Cay, The Bahamas. Wilson44691. 2010. Public domain. https://commons.wikimedia.org/wiki/File:Ooids,_Joulter_Cays,_Bahamas.jpg

Figure 5.37: Limestone tufa towers along the shores of Mono Lake, California. Yukinobu Zengame. 2005. CC BY 2.0. https://commons.wikimedia.org/wiki/File:Limestone_towers_at_Mono_Lake,_California.jpg

Figure 5.74: Diagram describing wavebase. GregBenson. 2004. CC BY-SA 3.0. https://en.wikipedia.org/wiki/File:Wave-base.jpg

Figure 5.75: Diagram of zones of the shoreline. US government. 2012. Public domain. https://commons.wikimedia.org/wiki/File:Littoral_Zones.jpg

Figure 5.76: The rising sea levels of transgressions create onlapping sediments, regressions create offlapping. Woudloper. 2009. CC BY-SA 1.0. https://commons.wikimedia.org/wiki/File:Offlap_%26_onlap_EN.svg

Figure 5.77: Lithified heavy mineral sand (dark layers) from a beach deposit in India. Mark A. Wilson. 2008. Public domain. https://commons.wikimedia.org/wiki/File:HeavyMineralsBeachSand.jpg

Figure 5.78: General diagram of a tidal flat and associated features. Foxbat deinos. 2009. Public domain. https://commons.wikimedia.org/wiki/File:Tidal_flat_general_sketch.png

Figure 5.79: Waterpocket fold, Capitol Reef National Park, Utah. Bobak Ha'Eri. 2008. CC BY 3.0. https://commons.wikimedia.org/wiki/File:2008-0914-CapitolReef-WaterpocketFold1.jpg

Figure 5.80: A modern coral reef. Toby Hudson. 2010. CC BY-SA 3.0. https://commons.wikimedia.org/wiki/File:Coral_Outcrop_Flynn_Reef.jpg

Figure 5.81: The light blue reef is fringing the island of Vanatinai. NASA image by Jesse Allen and Rob Simmon, using data provided by the United States Geological Survey. 2002. Public domain. https://en.m.wikipedia.org/wiki/File:Vanatinai,_Louisiade_Archipelago.jpg

Figure 5.82: Seamounts and guyots in the North Pacific. PeterTHarris. 2015. CC BY-SA 3.0. https://commons.wikimedia.org/wiki/File:Distribution_of_seamounts_and_guyots_in_the_North_Pacific.pdf

Figure 5.83: Kara-Bogaz Gol lagoon, Turkmenistan. NASA. 1995. Public domain. https://commons.wikimedia.org/wiki/File:Kara-Bogaz_Gol_from_space,_September_1995.jpg

Figure 5.84: The Nile delta, in Egypt. Jacques Descloitres, MODIS Rapid Response Team, NASA/sh. 2003. Public domain. https://commons.wikimedia.org/wiki/File:Nile_River_and_delta_from_orbit.jpg

Figure 5.85: Birdfoot river-dominated delta of the Mississippi River. NASA. 2001. Public domain. https://commons.wikimedia.org/wiki/File:Mississippi_delta_from_space.jpg

Figure 5.86: Tidal delta of the Ganges River. NASA. 1994. Public domain. https://commons.wikimedia.org/wiki/File:Ganges_River_Delta,_Bangladesh,_India.jpg

Figure 5.87: The Cauto River in Cuba. Not home~commonswiki. 2007. Public domain. https://commons.wikimedia.org/wiki/File:Rio-cauto-cuba.JPG

Figure 5.88: The braided Waimakariri river in New Zealand. Greg O'Beirne. 2016. CC BY 2.5. https://www.wikiwand.com/simple/Braided_river#Media/File:Waimakariri01_gobeirne.jpg

Figure 5.89: An alluvial fan spreads out into a broad alluvial plain. Matt Affolter. 2010. CC BY-SA 3.0. https://commons.wikimedia.org/wiki/File:AlluvialPlain.JPG

Figure 5.90: Oregon's Crater Lake was formed about 7700 years ago after the eruption of Mount Mazama. Zainubrazvi. 2006. CC BY-SA 3.0. https://en.wikipedia.org/wiki/File:Crater_lake_oregon.jpg

Figure 5.91: Formation and types of dunes. NPS Natural Resources. 2018. Public domain. https://flic.kr/p/GAn1Dj

Figure 5.92: Loess Plateau in China. Till Niermann. 1987. CC BY-SA 3.0. https://en.wikipedia.org/wiki/File:Loess_landscape_china.jpg

Figure 5.93: Wide range of sediments near Athabaska Glacier, Jasper National Park, Alberta, Canada. Wing-Chi Poon. 2006. CC BY-SA 2.5. https://commons.wikimedia.org/wiki/File:Glacial_Transportation_and_Deposition.jpg

6.

METAMORPHIC ROCKS

Contributing Author: Dr. Peter Davis, Pacific Lutheran University

Learning Objectives

By the end of this chapter, students will be able to:

- Describe the **temperature** and pressure conditions of the **metamorphic** environment.
- Identify and describe the three principal **metamorphic** agents.
- Describe what **recrystallization** is and how it affects **mineral** crystals.
- Explain what **foliation** is and how it results from directed pressure and **recrystallization**.
- Explain the relationships among **slate**, **phyllite**, **schist**, and **gneiss** in terms of metamorphic grade.
- Define **index mineral**.
- Explain how **metamorphic facies** relate to **plate tectonic** processes.
- Describe what a contact **aureole** is and how **contact metamorphism** affects surrounding rock.
- Describe the role of **hydrothermal metamorphism** in forming **mineral** deposits and **ore** bodies.

Metamorphic rocks, *meta-* meaning change and *–morphos* meaning form, is one of the three rock categories in the **rock cycle** (see chapter 1). **Metamorphic rock** material has been changed by **temperature**, pressure, and/or fluids. The **rock cycle** shows that both **igneous** and sedimentary rocks can become **metamorphic** rocks. And **metamorphic** rocks themselves can be re-metamorphosed. Because **metamorphism** is caused by **plate tectonic** motion, **metamorphic rock** provides geologists with a history book of how past **tectonic** processes shaped our planet.

6.1 Metamorphic Processes

Metamorphism occurs when solid rock changes in **composition** and/or **texture** without the **mineral** crystals melting, which is how **igneous rock** is generated. **Metamorphic source rocks**, the rocks that experience the **metamorphism**, are called the **parent rock** or **protolith**, from *proto-* meaning first, and *lithos-* meaning rock. Most **metamorphic** processes take place deep underground, inside the earth's **crust**. During **metamorphism**, **protolith** chemistry is mildly

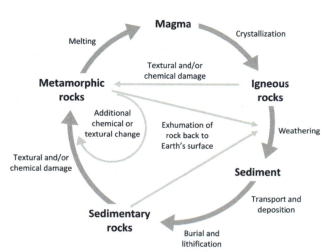

The rock cycle

Figure 6.1: Rock cycle showing the five materials (such as igneous rocks and sediment) and the processes by which one changes into another (such as weathering).

changed by increased **temperature** (heat), a type of pressure called **confining** pressure, and/or chemically reactive fluids. Rock **texture** is changed by heat, **confining** pressure, and a type of pressure called **directed stress**.

6.1.1 Temperature (Heat)

Temperature measures a substance's energy—an increase in **temperature** represents an increase in energy. **Temperature** changes affect the chemical equilibrium or **cation** balance in **minerals**. At high temperatures atoms may vibrate so vigorously they jump from one position to another within the crystal lattice, which remains intact. In other words, this atom swapping can happen while the rock is still solid.

The temperatures of **metamorphic rock** lies in between surficial processes (as in **sedimentary rock**) and **magma** in the **rock cycle**. Heat-driven **metamorphism** begins at temperatures as cold as 200°C, and can continue to occur at temperatures as high as 700°C-1,100°C. Higher temperatures would create **magma**, and thus, would no longer be a **metamorphic** process. **Temperature** increases with increasing depth in the Earth along a **geothermal gradient** (see chapter 4) and **metamorphic rock** records these depth-related **temperature** changes.

6.1.2 Pressure

Pressure is the force exerted over a unit area on a material. Like heat, pressure can affect the chemical equilibrium of **minerals** in a rock. The pressure that affects **metamorphic** rocks can be grouped into **confining** pressure and **directed stress**. **Stress** is a scientific term indicating a force. **Strain** is the result of this **stress**, including **metamorphic** changes within **minerals**.

Confining Pressure

Pressure exerted on rocks under the surface is due to the simple fact that rocks lie on top of one another. When pressure is exerted from rocks above, it is balanced from below and sides, and is called **confining** or **lithostatic** pressure. **Confining** pressure has equal pressure on all sides (see figure 6.2) and is responsible for causing chemical reactions to occur just like heat. These chemical reactions will cause new **minerals** to form.

Confining pressure is measured in bars and ranges from 1 bar at sea level to around 10,000 bars at the base of the **crust**. For **metamorphic** rocks, pressures range from a relatively low-pressure of 3,000 bars around 50,000 bars, which occurs around 15-35 kilometers below the surface.

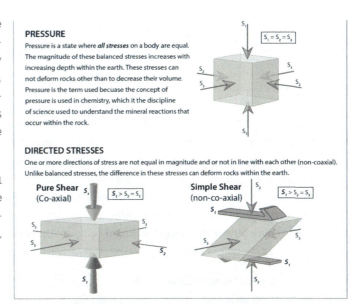

PRESSURE
Pressure is a state where *all stresses* on a body are equal. The magnitude of these balanced stresses increases with increasing depth within the earth. These stresses can not deform rocks other than to decrease their volume. Pressure is the term used becuase the concept of pressure is used in chemistry, which it the discipline of science used to understand the mineral reactions that occur within the rock.

DIRECTED STRESSES
One or more directions of stress are not equal in magnitude and or not in line with each other (non-coaxial). Unlike balanced stresses, the difference in these stresses can deform rocks within the earth.

Figure 6.2: Difference between pressure and stress and how they deform rocks. Pressure (or confining pressure) has equal stress (forces) in all directions and increases with depth under the Earth's surface. Under directed stress, some stress directions (forces) are stronger than others, and this can deform rocks.

Directed Stress

Directed stress, also called differential or **tectonic stress**, is an unequal balance of forces on a rock in one or more directions (see previous figure). Directed **stresses** are generated by the movement of lithospheric **plates**. **Stress** indicates a type of force acting on rock. **Strain** describes the resultant processes caused by **stress** and includes **metamorphic** changes in the **minerals**. In contrast to **confining** pressure, **directed stress** occurs at much lower pressures and does not generate chemical reactions that change **mineral composition** and atomic structure. Instead, **directed stress** modifies the **parent rock** at a mechanical level, changing the arrangement, size, and/or shape of the **mineral** crystals. These crystalline changes create identifying textures, which is shown in the figure below comparing the **phaneritic texture** of **igneous granite** with the **foliated texture** of **metamorphic gneiss**.

Figure 6.3: Pebbles (that used to be spherical or close to spherical) in quartzite deformed by directed stress.

Figure 6.4: An igneous rock granite (left) and foliated high-temperature and high-pressure metamorphic rock gneiss (right) illustrating a metamorphic texture.

Directed **stresses** produce rock textures in many ways. Crystals are rotated, changing their orientation in space. Crystals can get fractured, reducing their **grain size**. Conversely, they may grow larger as atoms migrate. Crystal shapes also become deformed. These mechanical changes occur via **recrystallization**, which is when **minerals dissolve** from an area of rock experiencing high **stress** and **precipitate** or regrow in a location having lower **stress**. For example, **recrystallization** increases **grain size** much like adjacent soap bubbles coalesce to form larger ones.

Recrystallization rearranges **mineral** crystals without fracturing the rock structure, deforming the rock like silly putty; these changes provide important clues to understanding the creation and movement of deep underground rock **faults**.

6.1.3 Fluids

A third **metamorphic** agent is chemically reactive fluids that are expelled by crystallizing **magma** and created by **metamorphic** reactions. These reactive fluids are made of mostly water (H_2O) and carbon dioxide (CO_2), and smaller amounts of potassium (K), sodium (Na), iron (Fe), magnesium (Mg), calcium (Ca), and aluminum (Al). These fluids react with **minerals** in the **protolith**, changing its chemical equilibrium and **mineral composition**, in a process similar to the reactions driven by heat and pressure. In addition to using **elements** found in the **protolith**, the chemical reaction may incorporate substances contributed by the fluids to create new **minerals**. In general, this style of **metamorphism**, in which fluids play an important role, is called **hydrothermal metamorphism** or **hydrothermal** alteration. Water actively participates in chemical reactions and allows extra mobility of the components in **hydrothermal** alteration.

Fluids-activated **metamorphism** is frequently involved in creating economically important **mineral** deposits that are located next to **igneous** intrusions or **magma** bodies. For example, the **mining** districts in the Cottonwood Canyons and **Mineral Basin** of northern Utah produce valuable **ores** such as argentite (silver **sulfide**), galena (lead **sulfide**), and chalcopyrite (copper iron **sulfide**), as well as the **native element** gold. These **mineral** deposits were created from the interaction between a granitic intrusion called the Little Cottonwood Stock and **country rock** consisting of mostly **limestone** and dolostone. Hot, circulating fluids expelled by the crystallizing **granite** reacted with and **dissolved** the surrounding **limestone** and dolostone, precipitating out new **minerals** created by the chemical reaction. **Hydrothermal** alternation of **mafic mantle** rock, such as **olivine** and **basalt**, creates the **metamorphic rock serpentinite**, a member of the serpentine subgroup of **minerals**. This **metamorphic** process happens at mid-ocean **spreading centers** where newly formed **oceanic crust** interacts with seawater.

Some **hydrothermal** alterations remove **elements** from the **parent rock** rather than deposit them. This happens when seawater circulates down through **fractures** in the fresh, still-hot **basalt**, reacting with and removing **mineral** ions from it. The **dissolved minerals** are usually ions that do not fit snugly in the **silicate** crystal structure, such as copper. The **mineral**-laden water emerges from the sea floor via **hydrothermal** vents called **black smokers**, named after the dark-colored precipitates produced when the hot **vent** water meets cold seawater (see chapter 4). Ancient **black smokers** were an important source of copper **ore** for the inhabitants of Cyprus (Cypriots) as early as 4,000 BCE, and later by the Romans.

Figure 6.5: Black smoker hydrothermal vent with a colony of giant (6'+) tube worms.

Take this quiz to check your comprehension of this section.

If you are using an offline version of this text, access the quiz for section 6.1 via the QR code.

 An interactive H5P element has been excluded from this version of the text. You can view it online here:
https://pressbooks.lib.vt.edu/introearthscience/?p=462#h5p-37

6.2 Metamorphic Textures

Metamorphic texture is the description of the shape and orientation of **mineral** grains in a **metamorphic rock**. Metamorphic rock textures are **foliated**, **non-foliated**, or lineated are described below.

Identify rock's foliation		Textural features	Mineral composition	Rock name	Parent rock
Foliated (layered texture)	Fine-grained or no visible grains	Flat, slaty cleavage is well developed. Dense, microscopic grains may exhibit slight sheen (or dull luster). Clanky sound when struck. Breaks into hard, flat sheets.	Fine, microscopic clay or mica	Slate	Shale
		Finely crystalline; micas hardly discernible, but impart a sheen or luster. Breaks along wavy surfaces.	Dark silicates and micas	Phyllite	Siltstone or shale
	Medium- to coarse-grained	Schistose texture. Foliation formed by alignment of visible crystals. Rock breaks along scaly foliation surfaces. Medium to fine-grained. Sparkling appearance.	Common minerals include chlorite, biotite, muscovite, garnet, and hornblende. Recognizable minerals used as part of rock name. Porphyroblasts common.	Mica schist / Garnet schist	Siltstone or shale
		Gneissic banding. Coarse-grained. Foliation present as minerals arranged into alternating light and dark layers giving the rock a banded texture in side view. Crystalline texture. No cleavage.	Light-colored quartz and feldspar; dark ferromagnesian minerals	Gneiss	Shale or granitic rocks
		Medium- to coarse-grained crystalline structure.	Crystals of amphibole (hornblende) in blade-like crystals	Amphibolite	Basalt, gabbro, or ultramafic igneous rocks
Non-foliated (no layered texture)	Fine-grained or no visible grains	Microcrystalline texture. Glassy black sheen. Conchoidal fracture. Low density.	Fine, tar-like, organic makeup	Anthracite coal	Coal
		Dense and dark-colored. Fine or microcrystalline texture. Very hard. Color can range from gray, gray-green to black.	Microscopic dark silicates	Hornfels	Many rock types
		Microcrystalline or no visible grains with smooth, wavy surfaces. May be dull or glossy. Usually shades of green.	Serpentine. May have fibrous asbestos visible.	Serpentinite	Ultramafic igneous rocks or peridotite
		Microcrystalline or no visible grains. Can be scratched with a fingernail. Shades of green, gray, brown, or white. Soapy feel.	Talc	Soapstone or talc schist	Ultramafic igneous rocks
	Fine- to coarse-grained	Crystalline. Hard (scratches glass). Breaks across grains. Sandy or sugary texture. Color variable; can be white, pink, buff, brown, red, purple.	Quartz grains fused together. Grains will not rub off like sandstone.	Quartzite	Quartz sandstone
		Finely crystalline (resembling a sugar cube) to medium or coarse texture. Color variable; white, pink, gray, among others. Fossils in some varieties.	Calcite or dolomite crystals tightly fused together. Calcite effervesces with HCl; dolomite effervesces only when powdered.	Marble	Limestone or dolostone
		Texture of conglomerate, but breaks across clasts as easily as around them. Pebbles may be stretched (lineated) or cut by rock cleavage.	Granules or pebbles are commonly granitic or jasper, chert, quartz, or quartzite.	Meta-conglomerate	Conglomerate

Table 6.1: Metamorphic rock identification table.

6.2.1 Foliation and Lineation

Foliation is a term used that describes **minerals** lined up in planes. Certain **minerals**, most notably the **mica** group, are mostly thin and planar by default. **Foliated** rocks typically appear as if the **minerals** are stacked like pages of a book, thus the use of the term 'folia', like a leaf. Other **minerals**, with hornblende being a good example, are longer in one direction, linear like a pencil or a needle, rather than a planar-shaped book. These linear objects can also be aligned within a rock. This is referred to as a **lineation**. Linear crystals, such as hornblende, tourmaline, or stretched **quartz** grains, can be arranged as part of a **foliation**, a **lineation**, or **foliation/lineation** together. If they lie on a plane with **mica**, but with no common or preferred direction, this is **foliation**. If the **minerals** line up and point in a common direction, but with no planar fabric, this is **lineation**. When **minerals** lie on a plane AND point in a common direction; this is both **foliation** and **lineation**.

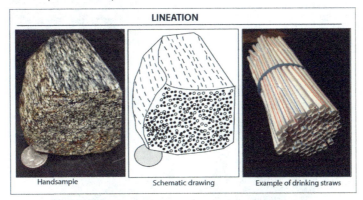

Figure 6.6: Example of lineation where minerals are aligned like a stack of straws or pencils.

Figure 6.7: An example of foliation WITH lineation.

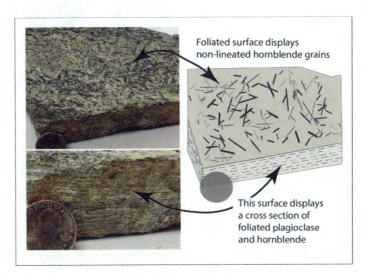

Figure 6.8: An example of foliation WITHOUT lineation.

Foliated metamorphic rocks are named based on the style of their foliations. Each rock name has a specific **texture** that defines and distinguishes it, with their descriptions listed below.

Slate is a fine-grained **metamorphic rock** that exhibits a **foliation** called **slaty cleavage** that is the flat orientation of the small platy crystals of **mica** and chlorite forming perpendicular to the direction of **stress**. The **minerals** in **slate** are too small to see with the unaided eye. The thin layers in **slate** may resemble sedimentary **bedding**, but they are a result of **directed stress** and may lie at angles to the original **strata**. In fact, original sedimentary layering may be partially or completely obscured by the **foliation**. Thin slabs of **slate** are often used as a building material for roofs and tiles.

Figure 6.9: Rock breaking along flat even planes.

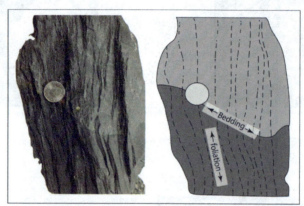

Figure 6.10: Foliation versus bedding. Foliation is caused by metamorphism. Bedding is a result of sedimentary processes. They do not have to align.

Phyllite is a **foliated metamorphic rock** in which platy **minerals** have grown larger and the surface of the **foliation** shows a sheen from light reflecting from the grains, perhaps even a wavy appearance, called crenulations. Similar to **phyllite** but with even larger grains is the **foliated metamorphic rock schist**, which has large platy grains visible as individual crystals. Common **minerals** are **muscovite**, **biotite**, and porphyroblasts of garnets. A porphyroblast is a large crystal of a particular **mineral** surrounded by small grains. **Schistosity** is a textural description of **foliation** created by the parallel alignment of platy visible grains. Some schists are named for their **minerals** such as **mica schist** (mostly micas), garnet **schist** (**mica schist** with garnets), and staurolite **schist** (**mica** schists with staurolite).

Figure 6.11: Phyllite with a small fold.

Figure 6.12: Schist.

Figure 6.13: Garnet staurolite muscovite schist.

Figure 6.14: Gneiss.

Gneissic banding is a **metamorphic foliation** in which visible **silicate minerals** separate into dark and light **bands** or lineations. These grains tend to be coarse and often folded. A rock with this **texture** is called **gneiss**. Since gneisses form at the highest temperatures and pressures, some **partial melting** may occur. This partially melted rock is a transition between **metamorphic** and **igneous** rocks called a **migmatite**.

Migmatites appear as dark and light **banded gneiss** that may be swirled or twisted some since some **minerals** started to melt. Thin accumulations of light colored rock layers can occur in a darker rock that are parallel to each other, or even cut across the **gneissic foliation**. The lighter colored layers are interpreted to be the result of the separation of a **felsic igneous** melt from the adjacent highly metamorphosed darker layers, or injection of a **felsic** melt from some distance away.

6.2.2 Non-foliated

Figure 6.15: Migmatite, a rock which was partially molten.

Figure 6.16: Marble.

Non-foliated textures do not have lineations, foliations, or other alignments of **mineral** grains. **Non-foliated metamorphic** rocks are typically **composed** of just one **mineral**, and therefore, usually show the effects of **metamorphism** with **recrystallization** in which crystals grow together, but with no preferred direction. The two most common examples of **non-foliated** rocks are **quartzite** and **marble**. Quartzite is a **metamorphic rock** from the **protolith sandstone**. In **quartzites**, the **quartz** grains from the original **sandstone** are enlarged and interlocked by **recrystallization**. A defining characteristic for distinguishing **quartzite** from **sandstone** is that when broken with a rock hammer, the **quartz** crystals break across the grains. In a **sandstone**, only a thin **mineral** cement holds the grains together, meaning that a broken piece of **sandstone** will leave the grains intact. Because most **sandstones** are rich in **quartz**, and **quartz** is a mechanically and chemically durable substance, **quartzite** is very hard and resistant to **weathering**.

Figure 6.17: Baraboo quartzite.

Marble is metamorphosed **limestone** (or dolostone) **composed** of **calcite** (or dolomite). **Recrystallization** typically generates larger interlocking crystals of **calcite** or dolomite. **Marble** and **quartzite** often look similar, but these **minerals** are considerably softer than **quartz**. Another way to distinguish **marble** from **quartzite** is with a drop of dilute hydrochloric acid. **Marble** will effervesce (fizz) if it is made of **calcite**.

A third **non-foliated** rock is **hornfels** identified by its dense, fine grained, hard, blocky or splintery **texture composed** of several **silicate minerals**. Crystals in **hornfels** grow smaller with **metamorphism**, and become so small that specialized study is required to identify them. These are common around **intrusive igneous** bodies and are hard to identify. The **protolith** of **hornfels** can be even harder to distinguish, which can be anything from **mudstone** to **basalt**.

Figure 6.18: Macro view of quartzite. Note the interconnectedness of the grains.

Figure 6.19: Unmetamorphosed, unconsolidated sand grains have space between the grains.

Take this quiz to check your comprehension of this section.

If you are using an offline version of this text, access the quiz for section 6.2 via the QR code.

 An interactive H5P element has been excluded from this version of the text. You can view it online here:
https://pressbooks.lib.vt.edu/introearthscience/?p=462#h5p-38

6.3 Metamorphic Grade

Metamorphic grade refers to the range of **metamorphic** change a rock undergoes, progressing from low (little **metamorphic** change) **grade** to high (significant metamorphic change) grade. Low-**grade metamorphism** begins at temperatures and pressures just above **sedimentary rock** conditions. The sequence **slate→phyllite→schist→gneiss** illustrates an increasing **metamorphic grade**.

Geologists use **index minerals** that form at certain temperatures and pressures to identify **metamorphic grade**. These **index minerals** also provide important clues to a rock's sedimentary **protolith** and the **metamorphic** conditions that created it. Chlorite, **muscovite**, **biotite**, garnet, and staurolite are **index minerals** representing a respective sequence of low-to-high **grade** rock. The figure shows a **phase diagram** of three **index minerals**—sillimanite, kyanite, and andalusite—with the same chemical formula (Al_2SiO_5) but having different crystal structures (**polymorphism**) created by different pressure and **temperature** conditions.

Figure 6.20: Garnet schist.

Complete this interactive activity to check your understanding.

If you are using an offline version of this text, access this interactive activity via the QR code.

 An interactive H5P element has been excluded from this version of the text. You can view it online here:
https://pressbooks.lib.vt.edu/introearthscience/?p=462#h5p-39

Some **metamorphic** rocks are named based on the highest **grade** of **index mineral** present. Chlorite **schist** includes the low-**grade index mineral** chlorite. **Muscovite schist** contains the slightly higher **grade muscovite**, indicating a greater degree of **metamorphism**. Garnet **schist** includes the high **grade index mineral** garnet, and indicating it has experienced much higher pressures and temperatures than chlorite.

Take this quiz to check your comprehension of this section.

If you are using an offline version of this text, access the quiz for section 6.3 via the QR code.

 An interactive H5P element has been excluded from this version of the text. You can view it online here:
https://pressbooks.lib.vt.edu/introearthscience/?p=462#h5p-40

6.4 Metamorphic Environments

As with **igneous** processes, **metamorphic** rocks form at different zones of pressure (depth) and **temperature** as shown on the pressure-**temperature** (P-T) diagram. The term **facies** is an **objective** description of a rock. In **metamorphic** rocks **facies** are groups of **minerals** called **mineral** assemblages. The names of **metamorphic facies** on the pressure-**temperature** diagram reflect **minerals** and **mineral** assemblages that are stable at these pressures and temperatures and provide information about the **metamorphic** processes that have affected the rocks. This is useful when interpreting the history of a **metamorphic rock**.

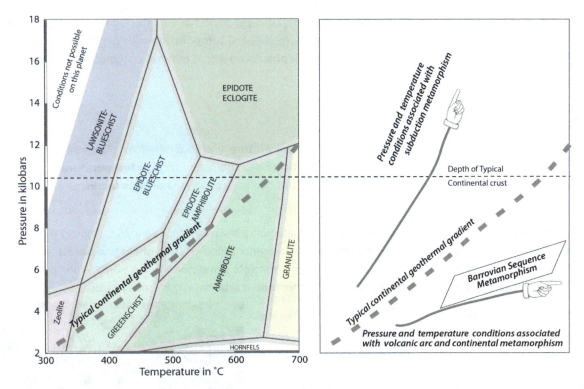

Figure 6.21: Pressure–temperature graphs of various metamorphic facies.

In the late 1800s, British geologist George Barrow mapped zones of **index minerals** in different **metamorphic** zones of an area that underwent **regional metamorphism**. Barrow outlined a progression of **index minerals**, named the Barrovian Sequence, that represents increasing **metamorphic grade**: chlorite (slates and phyllites) -> **biotite** (phyllites and schists) -> garnet (schists) -> staurolite (schists) -> kyanite (schists) -> sillimanite (schists and gneisses).

The first of the Barrovian sequence has a **mineral** group that is commonly found in the **metamorphic** greenschist **facies**. Greenschist rocks form under relatively low pressure and temperatures and represent the fringes of **regional metamorphism**. The "green" part of the name is derived from green **minerals** like chlorite, serpentine, and epidote, and the "**schist**" part is applied due to the presence of platy **minerals** such as **muscovite**.

Many different styles of **metamorphic facies** are recognized, tied to different geologic and **tectonic** processes. Recognizing these **facies** is the most direct way to interpret the **metamorphic** history of a rock. A simplified list of major **metamorphic facies** is given below.

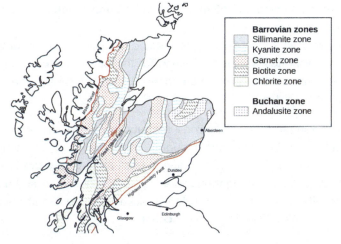

Figure 6.22: Barrovian sequence in Scotland.

6.4.1 Burial Metamorphism

Burial metamorphism occurs when rocks are deeply buried, at depths of more than 2000 meters (1.24 miles). **Burial metamorphism** commonly occurs in **sedimentary basins**, where rocks are buried deeply by overlying **sediments**. As an extension of **diagenesis**, a process that occurs during **lithification** (chapter 5), **burial metamorphism** can cause clay **minerals**, such as smectite, in **shales** to change to another clay **mineral** illite. Or it can cause **quartz sandstone** to metamorphose into the **quartzite** such the Big Cottonwood **Formation** in the Wasatch Range of Utah. This **formation** was deposited as ancient near-**shore** sands in the late **Proterozoic** (see chapter 7), deeply buried and metamorphosed to **quartzite**, folded, and later exposed at the surface in the Wasatch Range today. Increase of **temperature** with depth in combination with an increase of **confining** pressure produces low-**grade metamorphic** rocks with a **mineral** assemblages indicative of a zeolite **facies**.

6.4.2 Contact Metamorphism

Contact metamorphism occurs in rock exposed to high **temperature** and low pressure, as might happen when hot **magma** intrudes into or **lava** flows over pre-existing **protolith**. This combination of high **temperature** and low pressure produces numerous **metamorphic facies**. The lowest pressure conditions produce **hornfels facies**, while higher pressure creates greenschist, amphibolite, or granulite **facies**.

As with all **metamorphic rock**, the **parent rock texture** and chemistry are major factors in determining the final outcome of the **metamorphic** process, including what **index minerals** are present. Fine-grained **shale** and **basalt**, which happen to be chemically similar, characteristically recrystallize to produce **hornfels**. **Sandstone** (silica) surrounding an **igneous** intrusion becomes **quartzite** via **contact metamorphism**, and **limestone** (**carbonate**) becomes **marble**.

When **contact metamorphism** occurs deeper in the Earth, **metamorphism** can be seen as rings of **facies** around the intrusion, resulting in **aureoles**. These differences in **metamorphism** appear as distinct **bands** surrounding the intrusion, as can be seen around the Alta Stock in Little Cottonwood Canyon, Utah. The Alta Stock is a **granite** intrusion surrounded first by rings of the **index minerals amphibole** (tremolite) and **olivine** (forsterite), with a ring of talc (dolostone) located further away.

6.4.3 Regional Metamorphism

Figure 6.23: Contact metamorphism in outcrop.

Regional metamorphism occurs when **parent rock** is subjected to increased **temperature** and pressure over a large area, and is often located in mountain ranges created by converging **continental** crustal **plates**. This is the setting for the Barrovian sequence of rock **facies**, with the lowest **grade** of **metamorphism** occurring on the flanks of the mountains and highest **grade** near the **core** of the mountain range, closest to the **convergent** boundary.

An example of an old regional **metamorphic** environment is visible in the northern Appalachian Mountains while driving east from New York state through Vermont and into New Hampshire. Along this route the degree of **metamorphism** gradually increases from sedimentary **parent rock**, to low-**grade metamorphic rock**, then higher-**grade metamorphic rock**, and eventually the **igneous core**. The rock sequence is **sedimentary rock**, **slate**, **phyllite**, **schist**, **gneiss**, **migmatite**, and **granite**. In fact, New Hampshire is nicknamed the **Granite** State. The reverse sequence can be seen heading east, from eastern New Hampshire to the **coast**.

6.4.4 Subduction Zone Metamorphism

Subduction zone metamorphism is a type of **regional metamorphism** that occurs when a **slab** of **oceanic crust** is **subducted** under **continental crust** (see chapter 2). Because rock is a good insulator, the **temperature** of the descending **oceanic slab** increases slowly relative to the more rapidly increasing pressure, creating a **metamorphic** environment of high pressure and low **temperature**. Glaucophane, which has a distinctive blue color, is an **index mineral** found in **blueschist facies** (see **metamorphic facies** diagram). The California **Coast** Range near San Francisco has **blueschist–facies** rocks created by **subduction**-zone **metamorphism**, which include rocks made of **blueschist**, greenstone, and red **chert**. Greenstone, which is metamorphized **basalt**, gets its color from the **index mineral** chlorite.

Figure 6.24: Blueschist.

6.4.5 Fault Metamorphism

There are a range of **metamorphic** rocks made along **faults**. Near the surface, rocks are involved in repeated **brittle faulting** produce a material called *rock flour*, which is rock ground up to the particle size of flour used for food. At lower depths, **faulting** create cataclastites, chaotically-crushed mixes of rock material with little internal **texture**. At depths below **cataclasites**, where **strain** becomes **ductile**, **mylonites** are formed. **Mylonites** are **metamorphic** rocks created by dynamic **recrystallization** through directed **shear forces**, generally resulting in

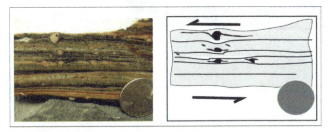

Figure 6.25: Mylonite.

a reduction of **grain size**. When larger, stronger crystals (like **feldspar**, **quartz**, garnet) embedded in a **metamorphic** matrix are **sheared** into an asymmetrical eye-shaped crystal, an **augen** is formed.

Figure 6.26: Examples of augens.

6.4.6 Shock Metamorphism

Shock (also known as impact) metamorphism is **metamorphism** resulting from **meteor** or other **bolide** impacts, or from a similar high-pressure shock event. **Shock metamorphism** is the result of very high pressures (and higher, but less extreme temperatures) delivered relatively rapidly. **Shock metamorphism** produces planar **deformation** features, tektites, shatter cones, and **quartz polymorphs**. **Shock metamorphism** produces planar **deformation** features (shock **laminae**), which are narrow planes of glassy material with distinct orientations found in **silicate mineral** grains. Shocked **quartz** has planar **deformation** features.

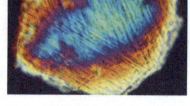

Figure 6.27: Shock lamellae in a quartz grain.

Figure 6.28: Shatter cone.

Shatter cones are cone-shaped pieces of rock created by dynamic branching **fractures** caused by impacts. While not strictly a **metamorphic** structure, they are common around **shock metamorphism**. Their diameter can range from microscopic to several meters. Fine-grained rocks with shatter cones show a distinctive horsetail pattern.

Shock metamorphism can also produce **index minerals**, though they are typically only found via microscopic analysis. The **quartz polymorphs** coesite and stishovite are indicative of **impact metamorphism**. As discussed in chapter 3, **polymorphs** are **minerals** with the same **composition** but different crystal structures. Intense pressure (> 10 GPa) and moderate to high temperatures (700-1200 °C) are required to form these **minerals**.

Shock metamorphism can also produce glass. Tektites are gravel-size glass grains ejected during an impact event. They resemble **volcanic** glass but, unlike **volcanic** glass, tektites contain no water or **phenocrysts**, and have a different bulk and isotopic chemistry. Tektites contain partially melted **inclusions** of shocked **mineral** grains. Although all are melt glasses, tektites are also chemically distinct from trinitite, which is produced from thermonuclear detonations, and fulgurites, which are produced by lightning strikes. All geologic glasses not derived from **volcanoes** can be called with the general term pseudotachylytes, a name which can also be applied to glasses created by **faulting**. The term pseudo in this context means 'false' or 'in the appearance of', a **volcanic rock** called tachylite because the material observed looks like a **volcanic rock**, but is produced by significant **shear** heating.

Figure 6.29: Tektites.

Take this quiz to check your comprehension of this section.

If you are using an offline version of this text, access the quiz for section 6.4 via the QR code.

 An interactive H5P element has been excluded from this version of the text. You can view it online here:
https://pressbooks.lib.vt.edu/introearthscience/?p=462#h5p-41

One or more interactive elements has been excluded from this version of the text. You can view them online here:
https://www.youtube.com/watch?v=Ncr-46YX-N0

Video 6.1: Identifying metamorphic rocks.

If you are using an offline version of this text, access this YouTube video via the QR code.

One or more interactive elements has been excluded from this version of the text. You can view them online here:
https://www.youtube.com/watch?v=HUydPhIaQQU

Video 6.2: Identifying metamorphic rocks.

If you are using an offline version of this text, access this YouTube video via the QR code.

Summary

Metamorphism is the process that changes existing rocks (called **protoliths**) into new rocks with new **minerals** and new textures. Increases in **temperature** and pressure are the main causes of **metamorphism**, with fluids adding important mobilization of materials. The primary way **metamorphic** rocks are identified is with **texture**. **Foliated** textures come from platy **minerals** forming planes in a rock, while **non-foliated metamorphic** rocks have no internal fabric. **Grade** describes the amount of **metamorphism** in a rock, and **facies** are a set of **minerals** that can help guide an observer to an interpretation of the **metamorphic** history of a rock. Different **tectonic** or geologic environments cause **metamorphism**, including collisions, **subduction**, **faulting**, and even impacts from space.

Take this quiz to check your comprehension of this chapter.

If you are using an offline version of this text, access the quiz for chapter 6 via the QR code.

 An interactive H5P element has been excluded from this version of the text. You can view it online here: https://pressbooks.lib.vt.edu/introearthscience/?p=462#h5p-42

Text References

1. Bucher, K., and Grapes, R., 2011, Petrogenesis of metamorphic rocks: Springer, 341 p.
2. Jeong, I.-K., Heffner, R.H., Graf, M.J., and Billinge, S.J.L., 2003, Lattice dynamics and correlated atomic motion from the atomic pair distribution function: Phys. Rev. B Condens. Matter, v. 67, no. 10, p. 104301.
3. Marshak, S., 2009, Essentials of Geology, 3rd or 4th Edition.
4. Proctor, B.P., McAleer, R., Kunk, M.J., and Wintsch, R.P., 2013, Post-Taconic tilting and Acadian structural overprint of the classic Barrovian metamorphic gradient in Dutchess County, New York: Am. J. Sci., v. 313, no. 7, p. 649–682.
5. Timeline of Art History, 2007, Reference Reviews, v. 21, no. 8, p. 45–45.

Figure References

Figure 6.1: Rock cycle showing the five materials (such as igneous rocks and sediment) and the processes by which one changes into another (such as weathering). Kindred Grey. 2022. CC BY 4.0.

Figure 6.2: Difference between pressure and stress and how they deform rocks. Peter Davis. 2017. CC BY-NC-SA 4.0. https://slcc.pressbooks.pub/introgeology/chapter/6-metamorphic-rocks/

Figure 6.3: Pebbles (that used to be spherical or close to spherical) in quartzite deformed by directed stress. Peter Davis. 2017. CC BY-NC-SA 4.0. https://slcc.pressbooks.pub/introgeology/chapter/6-metamorphic-rocks/

Figure 6.4: An igneous rock granite (left) and foliated high-temperature and high-pressure metamorphic rock gneiss (right) illustrating a metamorphic texture. Peter Davis. 2017. CC BY-NC-SA 4.0. https://slcc.pressbooks.pub/introgeology/chapter/6-metamorphic-rocks/

Figure 6.5: Black smoker hydrothermal vent with a colony of giant (6'+) tube worms. NOAA. 2006. Public domain. https://commons.wikimedia.org/wiki/File:Main_Endeavour_black_smoker.jpg

Table 6.1: Metamorphic rock identification table. Kindred Grey. 2022. Table data from Belinda Madsen's graphic in CH 6 of An Introduction to Geology. OpenStax. Salt Lake Community College. CC BY-NC-SA 4.0.

Figure 6.6: Example of lineation where minerals are aligned like a stack of straws or pencils. Peter Davis. 2017. CC BY-NC-SA 4.0. https://slcc.pressbooks.pub/introgeology/chapter/6-metamorphic-rocks/

Figure 6.7: An example of foliation WITH lineation. Peter Davis. 2017. CC BY-NC-SA 4.0. https://slcc.pressbooks.pub/introgeology/chapter/6-metamorphic-rocks/

Figure 6.8: An example of foliation WITHOUT lineation. Peter Davis. 2017. CC BY-NC-SA 4.0. https://slcc.pressbooks.pub/introgeology/chapter/6-metamorphic-rocks/

Figure 6.9: Rock breaking along flat even planes. Uta Baumfelder. 2010. Public domain. https://commons.wikimedia.org/wiki/File:Ehemaliger_Schiefertagebau_am_Brand.JPG

Figure 6.10: Foliation versus bedding. Peter Davis. 2017. CC BY-NC-SA 4.0. https://slcc.pressbooks.pub/introgeology/chapter/6-metamorphic-rocks/

Figure 6.11: Phyllite with a small fold. Peter Davis. 2017. CC BY-NC-SA 4.0. https://slcc.pressbooks.pub/introgeology/chapter/6-metamorphic-rocks/

Figure 6.12: Schist. Michael C. Rygel. 2012. CC BY-SA 3.0. https://commons.wikimedia.org/wiki/File:Schist_detail.jpg

Figure 6.13: Garnet staurolite muscovite schist. Peter Davis. 2017. CC BY-NC-SA 4.0. https://slcc.pressbooks.pub/introgeology/chapter/6-metamorphic-rocks/

Figure 6.14: Gneiss. Siim Sepp. 2005. CC BY-SA 3.0. https://commons.wikimedia.org/wiki/File:Gneiss.jpg

Figure 6.15: Migmatite, a rock which was partially molten. Peter Davis. 2017. CC BY-NC-SA 4.0. https://slcc.pressbooks.pub/introgeology/chapter/6-metamorphic-rocks/

Figure 6.16: Marble. Peter Davis. 2017. CC BY-NC-SA 4.0. https://slcc.pressbooks.pub/introgeology/chapter/6-metamorphic-rocks/

Figure 6.17: Baraboo quartzite. Peter Davis. 2017. CC BY-NC-SA 4.0. https://slcc.pressbooks.pub/introgeology/chapter/6-metamorphic-rocks/

Figure 6.18: Macro view of quartzite. Manishwiki15. 2012. CC BY-SA 3.0. https://commons.wikimedia.org/wiki/File:Sample_of_Quartzite.JPG

Figure 6.19: Unmetamorphosed, unconsolidated sand grains have space between the grains. Wilson44691. 2008. Public domain. https://commons.wikimedia.org/wiki/File:CoralPinkSandDunesSand.JPG

Figure 6.20: Garnet schist. Graeme Churchard (GOC53). 2005. CC BY 2.0. https://commons.wikimedia.org/wiki/File:Garnet_Mica_Schist_Syros_Greece.jpg

Figure 6.21: Pressure–temperature graphs of various metamorphic facies. Peter Davis. 2017. CC BY-NC-SA 4.0. https://slcc.pressbooks.pub/introgeology/chapter/6-metamorphic-rocks/

Figure 6.22: Barrovian sequence in Scotland. Woudloper. 2009. CC BY-SA 3.0. https://commons.wikimedia.org/wiki/File:Scotland_metamorphic_zones_EN.svg

Figure 6.23: Contact metamorphism in outcrop. Random Tree. 2012. Public domain. https://commons.wikimedia.org/wiki/File:Metamorphic_Aureole_in_the_Henry_Mountains.JPG

Figure 6.24: Blueschist. Peter Davis. 2017. CC BY-NC-SA 4.0. https://slcc.pressbooks.pub/introgeology/chapter/6-metamorphic-rocks/

Figure 6.25: Mylonite. Peter Davis. 2017. CC BY-NC-SA 4.0. https://slcc.pressbooks.pub/introgeology/chapter/6-metamorphic-rocks/

Figure 6.26: Examples of augens. Peter Davis. 2017. CC BY-NC-SA 4.0. https://slcc.pressbooks.pub/introgeology/chapter/6-metamorphic-rocks/

Figure 6.27: Shock lamellae in a quartz grain. Glen A. Izett. 2000. Public domain. https://commons.wikimedia.org/wiki/File:820qtz.jpg

Figure 6.28: Shatter cone. JMGastonguay. 2014. CC BY-SA 4.0. https://commons.wikimedia.org/wiki/File:Shatter-ConeCharlevoix1.jpg

Figure 6.29: Tektites. Brocken Inaglory. 2007. CC BY-SA 3.0. https://en.wikipedia.org/wiki/File:Two_tektites.JPG

7.

GEOLOGIC TIME

Learning Objectives

By the end of this chapter, students should be able to:

- Explain the difference between relative time and numeric time.
- Describe the five principles of **stratigraphy**.
- Apply **relative dating** principles to a block diagram and interpret the sequence of geologic events.
- Define an **isotope**, and explain **alpha decay**, **beta decay**, and **electron capture** as mechanisms of **radioactive** decay.
- Describe how radioisotopic dating is accomplished and list the four key **isotopes** used.
- Explain how carbon-14 forms in the **atmosphere** and how it is used in dating recent events.
- Explain how scientists know the numeric age of the Earth and other events in Earth history.
- Explain how sedimentary sequences can be dated using radioisotopes and other techniques.
- Define a **fossil** and describe types of **fossils** preservation.
- Outline how natural selection takes place as a mechanism of evolution.
- Describe **stratigraphic correlation**.
- List the **eons**, **eras**, and **periods** of the geologic time scale and explain the purpose behind the divisions.
- Explain the relationship between time units and corresponding rock units—**chronostratigraphy** versus **lithostratigraphy**.

The geologic time scale and basic outline of Earth's history were worked out long before we had any scientific means of assigning numerical age units, like years, to events of Earth history. Working out Earth's history depended on realizing some key principles of relative time. Nicolas Steno (1638-1686) introduced basic principles of **stratigraphy**, the study of layered rocks, in 1669. William Smith (1769-1839), working with the **strata** of English **coal mines**, noticed that **strata** and their sequence were consistent throughout the region. Eventually he produced the first national geologic map of Britain, becoming known as "the Father of English Geology." Nineteenth-century scientists developed a relative time scale using Steno's principles, with names derived from the characteristics of the rocks in those areas. The figure of this geologic time scale shows the names of the units and subunits. Using this time scale, geologists can place all events of Earth history in order without ever knowing their numerical ages. The specific events within Earth history are discussed in chapter 8.

Figure 7.1: Nicolas Steno, c. 1670.

7.1 Relative Dating

EON	ERA	PERIOD	MILLIONS OF YEARS AGO
Phanerozoic	Cenozoic	Quaternary	--- 1.6 ---
		Tertiary	--- 66 ---
	Mesozoic	Cretaceous	---138 ---
		Jurassic	--- 205 ---
		Triassic	--- 240 ---
	Paleozoic	Permian	--- 290 ---
		Pennsylvanian	---330 ---
		Mississippian	--- 360 ---
		Devonian	---410 ---
		Silurian	--- 435 ---
		Ordovician	--- 500 ---
		Cambrian	--- 570 ---
Proterozoic	Late Proterozoic Middle Proterozoic Early Proterozoic		--- 2500---
Archean	Late Archean Middle Archean Early Archean		---3800?---
Pre-Archean			

Figure 7.2: Geologic time scale.

Relative dating is the process of determining if one rock or geologic event is older or younger than another, without knowing their specific ages—i.e., how many years ago the object was formed. The principles of relative time are simple, even obvious now, but were not generally accepted by scholars until the scientific revolution of the 17th and 18th centuries. James Hutton (see chapter 1) realized geologic processes are slow and his ideas on **uniformitarianism** (i.e., "the present is the key to the past") provided a basis for interpreting rocks of the Earth using scientific principles.

7.1.1 Relative Dating Principles

Stratigraphy is the study of layered sedimentary rocks. This section discusses principles of relative time used in all of geology, but are especially useful in **stratigraphy**.

Principle of Superposition: In an otherwise undisturbed sequence of sedimentary **strata**, or rock layers, the layers on the bottom are the oldest and layers above them are younger.

Principle of Original Horizontality: Layers of rocks deposited from above, such as **sediments** and **lava** flows, are originally laid down horizontally. The exception to this principle is at the margins of basins, where the **strata** can slope slightly downward into the **basin**.

Figure 7.3: Lower strata are older than those lying on top of them.

Figure 7.4: Lateral continuity.

Principle of Lateral Continuity: Within the depositional **basin**, **strata** are continuous in all directions until they thin out at the edge of that **basin**. Of course, all **strata** eventually end, either by hitting a geographic barrier, such as a ridge, or when the depositional process extends too far from its source, either a **sediment** source or a **volcano**. **Strata** that are cut by a canyon later remain continuous on either side of the canyon.

Principle of Cross-Cutting Relationships: **Deformation** events like **folds**, **faults** and **igneous** intrusions that cut across rocks are younger than the rocks they cut across.

Principle of Inclusions: When one rock **formation** contains pieces or **inclusions** of another rock, the included rock is older than the **host rock**.

Principle of Fossil Succession: Evolution has produced a succession of unique **fossils** that correlate to the units of the geologic time scale. Assemblages of **fossils** contained in **strata** are unique to the time they lived, and can be used to correlate rocks of the same age across a wide geographic distribution. Assemblages of **fossils** refers to groups of several unique **fossils** occurring together.

Figure 7.5: Dark dike cutting across older rocks, the lighter of which is younger than the grey rock.

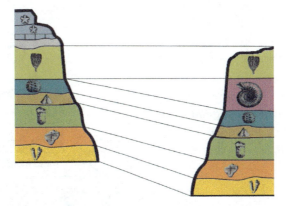

Figure 7.6: Fossil succession showing correlation among strata.

7.1.2 Grand Canyon Example

Figure 7.7: The Grand Canyon of Arizona.

The Grand Canyon of Arizona illustrates the **stratigraphic** principles. The photo shows layers of rock on top of one another in order, from the oldest at the bottom to the youngest at the top, based on the **principle of superposition**. The predominant white layer just below the canyon rim is the Coconino Sandstone. This layer is laterally continuous, even though the intervening canyon separates its outcrops. The rock layers exhibit the **principle of lateral continuity**, as they are found on both sides of the Grand Canyon which has been carved by the Colorado River.

The diagram called "Grand Canyon's Three Sets of Rocks" shows a cross-section of the rocks exposed on the walls of the Grand Canyon, illustrating the **principle of cross-cutting relationships**, **superposition**, and **original horizontality**. In the lowest parts of the Grand Canyon are the oldest sedimentary **formations**, with **igneous** and **metamorphic** rocks at the bottom. The **principle of cross-cutting relationships** shows the sequence of these events. The **metamorphic schist** (#16) is the oldest rock **formation** and the cross-cutting **granite** intrusion (#17) is younger. As seen in the figure, the other layers on the walls of the Grand Canyon are numbered in reverse order with #15 being the oldest and #1 the youngest. This illustrates the **principle of superposition**. The Grand Canyon region lies in Colorado Plateau, which is characterized by horizontal or nearly horizontal **strata**, which follows the **principle of original horizontality**. These rock **strata** have been barely disturbed from their original **deposition**, except by a broad regional uplift.

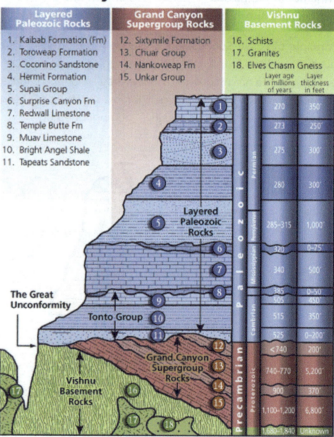

Figure 7.8: The rocks of the Grand Canyon.

The photo of the Grand Canyon here show **strata** that were originally deposited in a flat layer on top of older **igneous** and **metamorphic** "**basement**" rocks, per the **original horizontality** principle. Because the **formation** of the **basement** rocks and the **deposition** of the overlying **strata** is not continuous but broken by events of **metamorphism**, intrusion, and **erosion**, the contact between the **strata** and the older **basement** is termed an **unconformity**. An **unconformity** represents a **period** during which **deposition** did not occur or **erosion** removed rock that had been deposited, so there are no rocks that represent events of Earth history during that span of time at that place. **Unconformities** appear in cross sections and **stratigraphic** columns as wavy lines between **formations**. **Unconformities** are discussed in the next section.

Figure 7.9: The red, layered rocks of the Grand Canyon Supergroup overlying the dark-colored rocks of the Vishnu schist represents a type of unconformity called a nonconformity.

7.1.3 Unconformities

Figure 7.10: All three of these formations have a disconformity at the two contacts between them. The pinching Temple Butte is the easiest to see the erosion, but even between the Muav and Redwall there is an unconformity.

There are three types of **unconformities**, **nonconformity**, **disconformity**, and **angular unconformity**. A **nonconformity** occurs when **sedimentary rock** is deposited on top of **igneous** and **metamorphic** rocks as is the case with the contact between the **strata** and **basement** rocks at the bottom of the Grand Canyon.

The **strata** in the Grand Canyon represent alternating **marine transgressions** and **regressions** where sea level rose and fell over millions of years. When the sea level was high **marine strata** formed. When sea-level fell, the land was exposed to **erosion** creating an **unconformity**. In the Grand Canyon cross-section, this **erosion** is shown as heavy wavy lines between the various numbered **strata**. This is a type of **unconformity** called a **disconformity**, where either non-**deposition** or **erosion** took place. In other words, layers of rock that could have been present, are absent. The time that could have been represented by such layers is instead represented by the **disconformity**. Disconformities are **unconformities** that occur between parallel layers of **strata** indicating either a **period** of no **deposition** or **erosion**.

The **Phanerozoic strata** in most of the Grand Canyon are horizontal. However, near the bottom horizontal **strata** overlie tilted **strata**. This is known as the Great **Unconformity** and is an example of an **angular unconformity**. The lower **strata** were tilted by **tectonic** processes that disturbed their **original horizontality** and caused the **strata** to be eroded. Later, horizontal **strata** were deposited on top of the tilted **strata** creating the **angular unconformity**.

Here are three graphical illustrations of the three types of **unconformity**.

Figure 7.11: In the lower part of the picture is an angular unconformity in the Grand Canyon known as the Great Unconformity. Notice flat lying strata over dipping strata.

Disconformity, where is a break or stratigraphic absence between **strata** in an otherwise parallel sequence of **strata**.

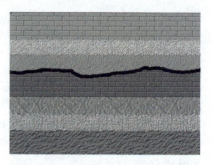

Figure 7.12: Disconformity.

Nonconformity, where sedimentary **strata** are deposited on crystalline (**igneous** or **metamorphic**) rocks.

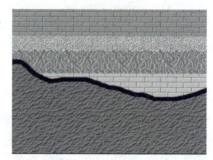

Figure 7.13: Nonconformity (the lower rocks are igneous or metamorphic).

Angular unconformity, where sedimentary **strata** are deposited on a terrain developed on sedimentary **strata** that have been deformed by tilting, folding, and/or **faulting**, so that they are no longer horizontal.

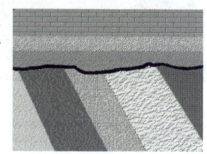

Figure 7.14: Angular unconformity.

7.1.3 Applying Relative Dating Principles

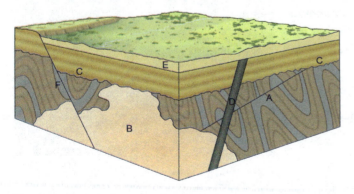

Figure 7.15: Block diagram to apply relative dating principles. The wavy rock is an old metamorphic gneiss, A and F are faults, B is an igneous granite, D is a basaltic dike, and C and E are sedimentary strata.

In the block diagram, the sequence of geological events can be determined by using the relative-dating principles and known properties of **igneous**, sedimentary, **metamorphic rock** (see chapter 4, chapter 5, and chapter 6). The sequence begins with the folded **metamorphic gneiss** on the bottom. Next, the gneiss is cut and displaced by the **fault** labeled A. Both the **gneiss** and **fault** A are cut by the **igneous** granitic intrusion called **batholith** B; its irregular outline suggests it is an **igneous** granitic intrusion emplaced as **magma** into the **gneiss**. Since **batholith** B cuts both the **gneiss** and **fault** A, **batholith** B is younger than the other two rock **formations**. Next, the **gneiss**, **fault** A, and **batholith** B were eroded forming a **nonconformity** as shown with the wavy line. This **unconformity** was actually an ancient landscape surface on which **sedimentary rock** C was subsequently deposited perhaps by a **marine transgression**. Next, **igneous** basaltic **dike** D cut through all rocks except **sedimentary rock** E. This shows that there is a **disconformity** between sedimentary rocks C and E. The top of **dike** D is level with the top of layer C, which establishes that **erosion** flattened the landscape prior to the **deposition** of layer E, creating a **disconformity** between rocks D and E. **Fault** F cuts across all of the older rocks B, C and E, producing a **fault scarp**, which is the low ridge on the upper-left side of the diagram. The final events affecting this area are current **erosion** processes working on the land surface, **rounding** off the edge of the **fault scarp**, and producing the modern landscape at the top of the diagram.

Take this quiz to check your comprehension of this section.

If you are using an offline version of this text, access the quiz for section 7.1 via the QR code.

 An interactive H5P element has been excluded from this version of the text. You can view it online here:
https://pressbooks.lib.vt.edu/introearthscience/?p=517#h5p-43

7.2 Absolute Dating

Figure 7.16: Canada's Nuvvuagittuq Greenstone Belt may have the oldest rocks and oldest evidence life on Earth, according to recent studies.

Relative time allows scientists to tell the story of Earth events, but does not provide specific numeric ages, and thus, the rate at which geologic processes operate. Based on Hutton's **principle of uniformitarianism** (see chapter 1), early geologists surmised geological processes work slowly and the Earth is very old. **Relative dating** principles was how scientists interpreted Earth history until the end of the 19th Century. Because science advances as technology advances, the discovery of **radioactivity** in the late 1800s provided scientists with a new scientific tool called radioisotopic dating. Using this new technology, they could assign specific time units, in this case years, to **mineral** grains within a rock. These numerical values are not dependent on comparisons with other rocks such as with **relative dating**, so this dating method is called **absolute dating**. There are several types of **absolute dating** discussed in this section but radioisotopic dating is the most common and therefore is the **focus** on this section.

7.2.1 Radioactive Decay

All **elements** on the Periodic Table of Elements (see chapter 3) contain **isotopes**. An **isotope** is an atom of an **element** with a different number of neutrons. For example, hydrogen (H) always has 1 proton in its nucleus (the atomic number), but the number of neutrons can vary among the **isotopes** (0, 1, 2). Recall that the number of neutrons added to the atomic number gives the atomic mass. When hydrogen has 1 proton and 0 neutrons it is sometimes called protium (^{1}H), when hydrogen has 1 proton and 1 neutron it is called deuterium (^{2}H), and when hydrogen has 1 proton and 2 neutrons it is called tritium (^{3}H).

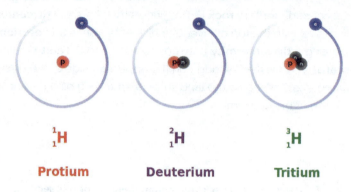

Figure 7.17: Three isotopes of hydrogen.

Many **elements** have both stable and unstable **isotopes**. For the hydrogen example, ^{1}H and ^{2}H are stable, but ^{3}H is unstable. Unstable **isotopes**, called **radioactive isotopes**, spontaneously decay over time releasing subatomic particles or energy in a process called **radioactive** decay. When this occurs, an unstable **isotope** becomes a more stable **isotope** of another **element**. For example, carbon-14 (^{14}C) decays to nitrogen-14 (^{14}N).

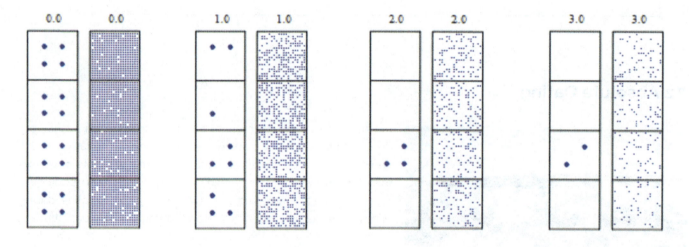

Figure 7.18: Simulation of half-life. On each left column, simulations with only a few atoms. On the right, simulations with many atoms.

The **radioactive** decay of any individual atom is a completely unpredictable and random event. However, some rock specimens have an enormous number of **radioactive isotopes**, perhaps trillions of atoms, and this large group of **radioactive isotopes** does have a predictable pattern of **radioactive** decay. The **radioactive** decay of *half* of the **radioactive isotopes** in this group takes a specific amount of time. The time it takes for half of the atoms in a substance to decay is called the **half-life**. In other words, the **half-life** of an **isotope** is the amount of time it takes for half of a group of unstable **isotopes** to decay to a stable **isotope**. The **half-life** is constant and measurable for a given **radioactive isotope**, so it can be used to calculate the age of a rock. For example, the **half-life** uranium-238 (^{238}U) is 4.5 billion years and the **half-life** of ^{14}C is 5,730 years.

The principles behind this dating method require two key assumptions. First, the **mineral** grains containing the **isotope** formed at the same time as the rock, such as **minerals** in an **igneous rock** that crystallized from **magma**. Second, the **mineral** crystals remain a closed **system**, meaning they are not subsequently altered by **elements** moving in or out of them.

These requirements place some constraints on the kinds of rock suitable for dating, with **igneous rock** being the best. **Metamorphic** rocks are crystalline, but the processes of **metamorphism** may reset the clock and derived ages may represent a smear of different **metamorphic** events rather than the age of original **crystallization**. **Detrital** sedimentary rocks contain clasts from separate **parent rocks** from unknown locations and derived ages are thus meaningless. However, sedimentary rocks with **precipitated minerals**, such as **evaporites**, may contain **elements** suitable for radioisotopic dating. **Igneous pyroclastic** layers and **lava** flows within a sedimentary sequence can be used to date the sequence. Cross-cutting **igneous** rocks and **sills**

Granite   **Gneiss**

Figure 7.19: Granite (left) and gneiss (right). Dating a mineral within the granite would give the crystallization age of the rock, while dating the gneiss might reflect the timing of metamorphism.

can be used to bracket the ages of affected, older sedimentary rocks. The resistant **mineral zircon**, found as clasts in many ancient sedimentary rocks, has been successfully used for establishing very old dates, including the age of Earth's oldest known rocks. Knowing that **zircon minerals** in metamorphosed **sediments** came from older rocks that are no longer available for study, scientists can date **zircon** to establish the age of the pre-**metamorphic source rocks**.

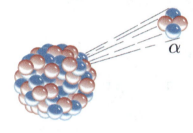

α

There are several ways **radioactive** atoms decay. We will consider three of them here—**alpha decay**, **beta decay**, and **electron capture**. **Alpha decay** is when an alpha particle, which consists of two protons and two neutrons, is emitted from the nucleus of an atom. This also happens to be the nucleus of a helium atom; helium gas may get trapped in the crystal lattice of a **mineral** in which **alpha decay** has taken place. When an atom loses two protons from its nucleus, lowering its atomic number, it is transformed into an **element** that is two atomic numbers lower on the Periodic Table of the Elements.

Figure 7.20: An alpha decay: Two protons and two neutrons leave the nucleus.

Group→	1	2	3		4	5	6	7	8	9	10	11	12	13	14	15	16	17	18
↓Period																			
1	1 H																		2 He
2	3 Li	4 Be												5 B	6 C	7 N	8 O	9 F	10 Ne
3	11 Na	12 Mg												13 Al	14 Si	15 P	16 S	17 Cl	18 Ar
4	19 K	20 Ca	21 Sc		22 Ti	23 V	24 Cr	25 Mn	26 Fe	27 Co	28 Ni	29 Cu	30 Zn	31 Ga	32 Ge	33 As	34 Se	35 Br	36 Kr
5	37 Rb	38 Sr	39 Y		40 Zr	41 Nb	42 Mo	43 Tc	44 Ru	45 Rh	46 Pd	47 Ag	48 Cd	49 In	50 Sn	51 Sb	52 Te	53 I	54 Xe
6	55 Cs	56 Ba	57 La	*	72 Hf	73 Ta	74 W	75 Re	76 Os	77 Ir	78 Pt	79 Au	80 Hg	81 Tl	82 Pb	83 Bi	84 Po	85 At	86 Rn
7	87 Fr	88 Ra	89 Ac	**	104 Rf	105 Db	106 Sg	107 Bh	108 Hs	109 Mt	110 Ds	111 Rg	112 Cn	113 Nh	114 Fl	115 Mc	116 Lv	117 Ts	118 Og

		58 Ce	59 Pr	60 Nd	61 Pm	62 Sm	63 Eu	64 Gd	65 Tb	66 Dy	67 Ho	68 Er	69 Tm	70 Yb	71 Lu
*		58 Ce	59 Pr	60 Nd	61 Pm	62 Sm	63 Eu	64 Gd	65 Tb	66 Dy	67 Ho	68 Er	69 Tm	70 Yb	71 Lu
**		90 Th	91 Pa	92 U	93 Np	94 Pu	95 Am	96 Cm	97 Bk	98 Cf	99 Es	100 Fm	101 Md	102 No	103 Lr

Figure 7.21: Periodic table of the elements.

The loss of four particles, in this case two neutrons and two protons, also lowers the mass of the atom by four. For example **alpha decay** takes place in the unstable **isotope** ^{238}U, which has an atomic number of 92 (92 protons) and mass number of 238 (total of all protons and neutrons). When ^{238}U spontaneously emits an alpha particle, it becomes thorium-234 (^{234}Th) The **radioactive** decay product of an **element** is called its **daughter isotope** and the original **element** is called the **parent isotope**. In this case, ^{238}U is the **parent isotope** and ^{234}Th is the **daughter isotope**. The **half-life** of ^{238}U is 4.5 billion years, i.e., the time it takes for half of the **parent isotope** atoms to decay into the **daughter isotope**. This isotope of uranium, ^{238}U, can be used for **absolute dating** the oldest materials found on Earth, and even **meteorites** and materials from the earliest events in our **solar system**.

Beta Decay

Beta decay is when a neutron in its nucleus splits into an electron and a proton. The electron is emitted from the nucleus as a beta ray. The new proton increases the **element**'s atomic number by one, forming a new **element** with the same atomic mass as the **parent isotope**. For example, ^{234}Th is unstable and undergoes **beta decay** to form protactinium-234 (^{234}Pa), which also undergoes **beta decay** to form uranium-234 (^{234}U). Notice these are all **isotopes** of different **elements** but they have the same atomic mass of 234. The decay process of **radioactive elements** like uranium keeps producing **radioactive parents** and **daughters** until a stable, or non-radioactive, daughter is formed. Such a series is called a **decay chain**. The **decay chain** of the **radioactive parent isotope** ^{238}U progresses through a series of alpha (red arrows on the adjacent figure) and beta decays (blue arrows), until it forms the stable **daughter isotope**, lead-206 (^{206}Pb).

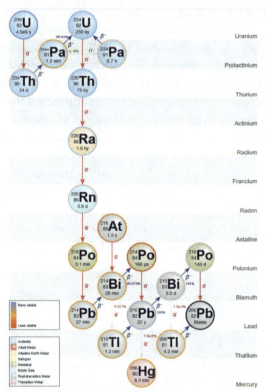

Figure 7.22: Decay chain of U-238 to stable Pb-206 through a series of alpha and beta decays.

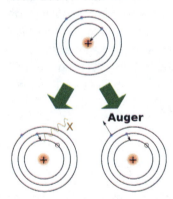

Figure 7.23: The two paths of electron capture.

Electron capture is when a proton in the nucleus captures an electron from one of the electron shells and becomes a neutron. This produces one of two different effects: 1) an electron jumps in to fill the missing spot of the departed electron and emits an X-ray, or 2) in what is called the Auger process, another electron is released and changes the atom into an **ion**. The atomic number is reduced by one and mass number remains the same. An example of an **element** that decays by **electron capture** is potassium-40 (^{40}K). **Radioactive** ^{40}K makes up a tiny percentage (0.012%) of naturally occurring potassium, most of which not **radioactive**. ^{40}K decays to argon-40 (^{40}Ar) with a **half-life** of 1.25 billion years, so it is very useful for dating geological events. Below is a table of some of the more commonly-used **radioactive** dating **isotopes** and their half-lives.

Elements	Parent symbol	Daughter symbol	Half-life
Uranium-238/Lead-206	^{238}U	^{206}Pb	4.5 billion years
Uranium-235/Lead-207	^{235}U	^{207}Pb	704 million years
Potassium-40/Argon-40	^{40}K	^{40}Ar	1.25 billion years
Rubidium-87/Strontium-87	^{87}Rb	^{87}Sr	48.8 billion years
Carbon-14/Nitrogen-14	^{14}C	^{14}N	5,730 years

Table 7.1: Some common isotopes used for radioisotopic dating.

7.2.2 Radioisotopic Dating

For a given a sample of rock, how is the dating procedure carried out? The parent and **daughter isotopes** are separated out of the **mineral** using chemical extraction. In the case of uranium, ^{238}U and ^{235}U **isotopes** are separated out together, as are the ^{206}Pb and ^{207}Pb with an instrument called a **mass spectrometer**.

Figure 7.24: Mass spectrometer instrument.

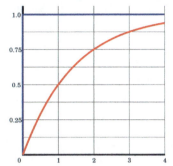

Figure 7.25: Graph of the amount of half life versus the amount of daughter isotope.

Here is a simple example of age calculation using the daughter-to-parent ratio of **isotopes**. When the **mineral** initially forms, it consists of 0% daughter and 100% **parent isotope**, so the daughter-to-parent ratio (D/P) is 0. After one **half-life**, half the parent has decayed so there is 50% daughter and 50% parent, a 50/50 ratio, with D/P = 1. After two half-lives, there is 75% daughter and 25% parent (75/25 ratio) and D/P = 3. This can be further calculated for a series of half-lives as shown in the table. The table does not show more than 10 half-lives because after about 10 half-lives, the amount of remaining parent is so small it becomes too difficult to accurately measure via chemical analysis. Modern applications of this method have achieved remarkable accuracies of plus or minus two million years in 2.5 billion years (that's ±0.055%). Applying the uranium/lead technique in any given sample analysis provides two separate clocks running at the same time, ^{238}U and ^{235}U. The existence of these two clocks in the same sample gives a cross-check between the two. Many geological samples contain multiple parent/daughter pairs, so cross-checking the clocks confirms that radioisotopic dating is highly reliable.

Half lives (#)	Parent present (%)	Daughter present (%)	Daughter/parent ratio	Parent/daughter ratio
Start the clock	100	0	0	Infinite
1	50	50	1	1
2	25	75	3	0.33
3	12.5	87.5	7	0.143
4	6.25	93.75	15	0.0667
5	3.125	96.875	31	0.0325
10	0.098	99.9	1023	0.00098

Table 7.2: Ratio of parent to daughter in terms of half-life.

Another radioisotopic dating method involves carbon and is useful for dating archaeologically important samples containing organic substances like wood or bone. Radiocarbon dating, also called carbon dating, uses the unstable **isotope** carbon-14 (^{14}C) and the stable **isotope** carbon-12 (^{12}C). Carbon-14 is constantly being created in the **atmosphere** by the interaction of cosmic particles with atmospheric nitrogen-14 (^{14}N). Cosmic particles such as neutrons **strike** the nitrogen nucleus, kicking out a proton but leaving the neutron in the nucleus. The **collision** reduces the atomic number by one, changing it from seven to six, changing the nitrogen into carbon with the same mass number of 14. The ^{14}C quickly **bonds** with oxygen (O) in the **atmosphere** to form carbon dioxide ($^{14}CO_2$) that mixes with other atmospheric carbon dioxide ($^{12}CO_2$) and this mix of gases is incorporated

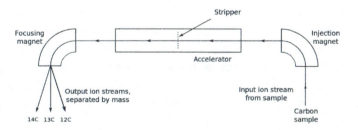

Figure 7.26: Schematic of carbon going through a mass spectrometer.

into living matter. While an organism is alive, the ratio of $^{14}C/^{12}C$ in its body doesn't really change since CO_2 is constantly exchanged with the **atmosphere**. However, when it dies, the radiocarbon clock starts ticking as the ^{14}C decays back to ^{14}N by **beta decay**, which has a **half-life** of 5,730 years. The radiocarbon dating technique is thus useful for 57,300 years or so, about 10 half-lives back.

Radiocarbon dating relies on daughter-to-parent ratios derived from a known quantity of parent ^{14}C. Early applications of carbon dating assumed the production and concentration of ^{14}C in the **atmosphere** remained fairly constant for the last 50,000 years. However, it is now known that the amount of parent ^{14}C levels in the **atmosphere** has varied. Comparisons of carbon ages with tree-ring data and other data for known events have allowed reliable calibration of the radiocarbon dating method. Taking into account carbon-14 baseline levels must be calibrated against other reliable dating methods, carbon dating has been shown to be a reliable method for dating archaeological specimens and very recent geologic events.

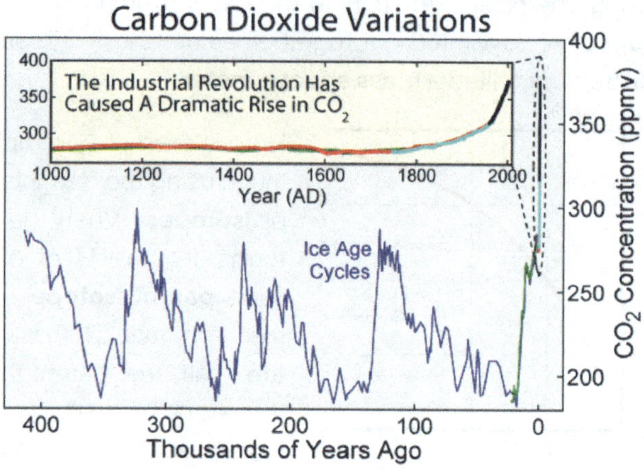

Figure 7.27: Carbon dioxide concentrations over the last 400,000 years.

7.2.3 Age of the Earth

Figure 7.28: Artist's impression of the Earth in the Hadean.

The work of Hutton and other scientists gained attention after the Renaissance (see chapter 1), spurring exploration into the idea of an ancient Earth. In the late 19th century William Thompson, a.k.a. Lord Kelvin, applied his knowledge of physics to develop the assumption that the Earth started as a hot molten sphere. He estimated the Earth is 98 million years old, but because of uncertainties in his calculations stated the age as a range of between 20 and 400 million years. This animation illustrates how Kelvin calculated this range and why his numbers were so far off, which has to do with unequal heat transfer within the Earth. It has also been pointed out that Kelvin failed to consider pliability and **convection** in the Earth's **mantle** as a heat transfer mechanism. Kelvin's estimate for Earth's age was considered plausible but not without challenge, and the discovery of **radioactivity** provided a more accurate method for determining ancient ages.

In the 1950's, Clair Patterson (1922–1995) thought he could determine the age of the Earth using **radioactive isotopes** from **meteorites**, which he considered to be early **solar system** remnants that were present at the time Earth was forming. Patterson analyzed **meteorite** samples for uranium and lead using a **mass spectrometer**. He used the uranium/lead dating technique in determining the age of the Earth to be 4.55 billion years, give or take about 70 million (± 1.5%). The current estimate for the age of the Earth is 4.54 billion years, give or take 50 million (± 1.1%). It is remarkable that Patterson, who was still a graduate student at the University of Chicago, came up with a result that has been little altered in over 60 years, even as technology has improved dating methods.

7.2.4 Dating Geological Events

Radioactive isotopes of **elements** that are common in **mineral** crystals are useful for radioisotopic dating. The uranium/lead method, with its two cross-checking clocks, is most often used with crystals of the **mineral zircon** ($ZrSiO_4$) where uranium can substitute for zirconium in the crystal lattice. **Zircon** is resistant to **weathering** which makes it useful for dating geological events in ancient rocks. During **metamorphic** events, **zircon** crystals may form multiple crystal layers, with each layer recording the isotopic age of an event, thus tracing the progress of the several **metamorphic** events.

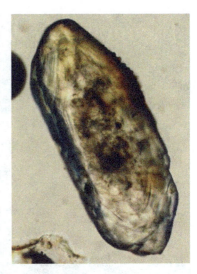

Figure 7.29: Photomicrograph of zircon crystal.

Geologists have used **zircon** grains to do some amazing studies that illustrate how scientific conclusions can change with technological advancements. **Zircon** crystals from Western Australia that formed when the crust first differentiated from the **mantle** 4.4 billion years ago have been determined to be the oldest known rocks. The **zircon** grains were incorporated into metasedimentary host rocks, sedimentary rocks showing signs of having undergone partial metamorphism. The host rocks were not very old but the embedded zircon grains were created 4.4 billion years ago, and survived the subsequent processes of **weathering**, **erosion**, **deposition**, and **metamorphism**. From other properties of the **zircon** crystals, researchers concluded that not only were continental rocks exposed above sea level, but also that conditions on the early Earth were cool enough for liquid water to exist on the surface. The presence of liquid water allowed the processes of weathering and erosion to take place. Researchers at UCLA studied 4.1 billion-year-old **zircon** crystals and found carbon in the **zircon** crystals that may be biogenic in origin, meaning that life may have existed on Earth much earlier than previously thought.

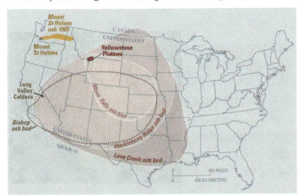

Figure 7.30: Several prominent ash beds found in North America, including three Yellowstone eruptions shaded pink (Mesa Falls, Huckleberry Ridge, and Lava Creek), the Bisho Tuff ash bed (brown dashed line), and the modern May 18th, 1980 ash fall (yellow).

Igneous rocks best suited for radioisotopic dating because their primary **minerals** provide dates of **crystallization** from **magma**. **Metamorphic** processes tend to reset the clocks and smear the **igneous rock**'s original date. **Detrital** sedimentary rocks are less useful because they are made of **minerals** derived from multiple parent sources with potentially many dates. However, scientists can use **igneous** events to date sedimentary sequences. For example, if sedimentary **strata** are between a **lava** flow and **volcanic ash bed** with radioisotopic dates of 54 million years and 50 million years, then geologists know the sedimentary **strata** and its **fossils** formed between 54 and 50 million years ago. Another example would be a 65 million year old **volcanic dike** that cut across sedimentary **strata**. This provides an upper limit age on the sedimentary **strata**, so this **strata** would be older than 65 million years. Potassium is common in **evaporite sediments** and has been used for potassium/argon dating. Primary sedimentary **minerals** containing **radioactive isotopes** like ^{40}K, has provided dates for important geologic events.

7.2.5 Other Absolute Dating Techniques

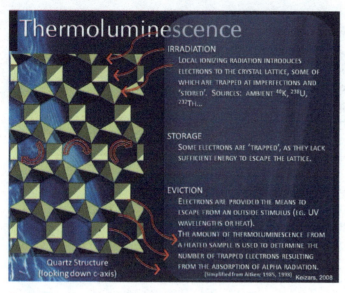

Figure 7.31: Thermoluminescence, a type of luminescence dating.

Luminescence (aka Thermoluminescence): Radioisotopic dating is not the only way scientists determine numeric ages. Luminescence dating measures the time elapsed since some **silicate minerals**, such as coarse-**sediments** of **silicate minerals**, were last exposed to light or heat at the surface of Earth. All buried **sediments** are exposed to radiation from normal background radiation from the decay process described above. Some of these electrons get trapped in the crystal lattice of **silicate minerals** like **quartz**. When exposed at the surface, ultraviolet radiation and heat from the Sun releases these electrons, but when the **minerals** are buried just a few inches below the surface, the electrons get trapped again. Samples of coarse **sediments** collected just a few feet below the surface are analyzed by stimulating them with light in a lab. This stimulation releases the trapped electrons as a photon of light which is called luminescence. The amount luminescence released indicates how long the **sediment** has been buried. Luminescence dating is only useful for dating young **sediments** that are less than 1 million years old. In Utah, luminescence dating is used to determine when coarse-grained **sediment** layers were buried near a **fault**. This is one technique used to determine the **recurrence** interval of large earthquakes on **faults** like the Wasatch Fault that primarily cut coarse-grained material and lack buried organic **soils** for radiocarbon dating.

Fission Track: Fission track dating relies on damage to the crystal lattice produced when unstable ^{238}U decays to the **daughter product** ^{234}Th and releases an alpha particle. These two decay products move in opposite directions from each other through the crystal lattice leaving a visible track of damage. This is common in uranium-bearing **mineral** grains such as apatite. The tracks are large and can be visually counted under an optical microscope. The number of tracks correspond to the age of the grains. Fission track dating works from about 100,000 to 2 billion (1×10^5 to 2×10^9) years ago. Fission track dating has also been used as a second clock to confirm dates obtained by other methods.

Figure 7.32: Apatite from Mexico.

Take this quiz to check your comprehension of this section.

If you are using an offline version of this text, access the quiz for section 7.2 via the QR code.

  An interactive H5P element has been excluded from this version of the text. You can view it online here:
https://pressbooks.lib.vt.edu/introearthscience/?p=517#h5p-44

7.3 Fossils and Evolution

Figure 7.33: Archaeopteryx lithographica, specimen displayed at the Museum für Naturkunde in Berlin.

Fossils are any evidence of past life preserved in rocks. They may be actual remains of body parts (rare), impressions of soft body parts, **casts** and **molds** of body parts (more common), body parts replaced by **mineral** (common) or evidence of animal behavior such as footprints and burrows. The body parts of living organisms range from the hard bones and shells of animals, soft cellulose of plants, soft bodies of jellyfish, down to single cells of bacteria and algae. Which body parts can be preserved? The vast majority of life today consists soft-bodied and/or single celled organisms, and will not likely be preserved in the geologic record except under unusual conditions. The best environment for preservation is the ocean, yet **marine** processes can **dissolve** hard parts and scavenging can reduce or eliminate remains. Thus, even under ideal conditions in the ocean, the likelihood of preservation is quite limited. For **terrestrial** life, the possibility of remains being buried and preserved is even more limited. In other words, the **fossil** record is incomplete and records only a small percentage of life that existed. Although incomplete, **fossil** records are used for **stratigraphic correlation**, using the **Principle of Faunal Succession**, and provide a method used for establishing the age of a **formation** on the Geologic Time Scale.

7.3.1 Types of Preservation

Remnants or impressions of hard parts, such as a **marine** clam shell or dinosaur bone, are the most common types of **fossils**. The original material has almost always been replaced with new **minerals** that preserve much of the shape of the original shell, bone, or cell. The common types of **fossil** preservation are **actual preservation**, **permineralization**, **molds** and **casts**, **carbonization**, and **trace fossils**.

Actual preservation is a rare form of fossilization where the original materials or hard parts of the organism are preserved. Preservation of soft-tissue is very rare since these organic materials easily disappear because of bacterial decay. Examples of **actual preservation** are unaltered biological materials like insects in amber or original **minerals** like mother-of-pearl on the interior of a shell. Another example is mammoth skin and hair preserved in post-glacial deposits in the Arctic regions. Rare mummification has left fragments of soft tissue, skin, and sometimes even blood vessels of dinosaurs, from which proteins have been isolated and evidence for DNA fragments have been discovered.

Figure 7.34: The trilobites had a hard exoskeleton, and is an early arthropod, the same group that includes modern insects, crustaceans, and arachnids.

Figure 7.35: Mosquito preserved in amber.

Permineralization occurs when an organism is buried, and then **elements** in **groundwater** completely impregnate all spaces within the body, even cells. Soft body structures can be preserved in great detail, but stronger materials like bone and teeth are the most likely to be preserved. Petrified wood is an example of detailed cellulose structures in the wood being preserved. The University of California Berkeley website has more information on **permineralization**.

Figure 7.36: Permineralization in petrified wood.

Molds and **casts** form when the original material of the organism dissolves and leaves a cavity in the surrounding rock. The shape of this cavity is an **external mold**. If the mold is subsequently filled with **sediments** or a **mineral precipitate**, the organism's external shape is preserved as a **cast**. Sometimes internal cavities of organisms, such internal **casts** of clams, snails, and even skulls are preserved as internal **casts** showing details of soft structures. If the chemistry is right, and burial is rapid, **mineral** nodules form around soft structures preserving the three-dimensional detail. This is called **authigenic mineralization**.

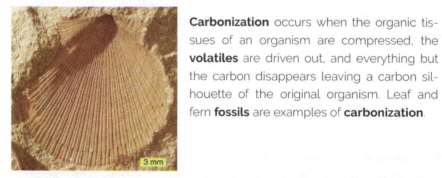

Figure 7.37: External mold of a clam.

Carbonization occurs when the organic tissues of an organism are compressed, the **volatiles** are driven out, and everything but the carbon disappears leaving a carbon silhouette of the original organism. Leaf and fern **fossils** are examples of **carbonization**.

Figure 7.38: Carbonized leaf.

Figure 7.39: Dinosaur tracks as a record of its passing.

Trace fossils are indirect evidence left behind by an organism, such as burrows and footprints, as it lived its life. **Ichnology** is specifically the study of prehistoric animal tracks. Dinosaur tracks testify to their presence and movement over an area, and even provide information about their size, gait, speed, and behavior. Burrows dug by tunneling organisms tell of their presence and mode of life. Other **trace fossils** include fossilized feces called coprolites and stomach stones called gastroliths that provide information about diet and habitat.

Figure 7.40: Fossil animal droppings (coprolite).

7.3.2 Evolution

Evolution has created a variety of ancient **fossils** that are important to **stratigraphic correlation**. (see chapter 7 and chapter 5) This section is a brief discussion of the process of evolution. The British naturalist Charles Darwin (1809-1882) recognized that life forms evolve into progeny life forms. He proposed natural selection—which operated on organisms living under environmental conditions that posed challenges to survival—was the mechanism driving the process of evolution forward.

The basic classification unit of life is the species: a population of organisms that exhibit shared characteristics and are capable of reproducing fertile offspring. For a species to survive, each individual within a particular population is faced with challenges posed by the environment and must survive them long enough to reproduce. Within the natural variations present in the population, there may be individuals possessing characteristics that give them some advantage in facing the environmental challenges. These individuals are more likely to reproduce and pass these favored characteristics on to successive generations. If sufficient individuals in a population fail to surmount the challenges of the environment and the population cannot produce

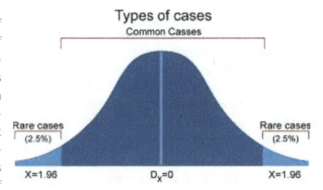

Figure 7.41: Variation within a population.

enough viable offspring, the species becomes **extinct**. The average lifespan of a species in the **fossil** record is around a million years. That life still exists on Earth shows the role and importance of evolution as a natural process in meeting the continual challenges posed by our dynamic Earth. If the inheritance of certain distinctive characteristics is sufficiently favored over time, populations may become genetically isolated from one another, eventually resulting in the evolution of separate species. This genetic isolation may also be caused by a geographic barrier, such as an island surrounded by ocean. This **theory** of evolution by natural selection was elaborated by Darwin in his book *On the Origin of Species* (see chapter 1). Since Darwin's original ideas, technology has provided many tools and mechanisms to study how evolution and speciation take place and this arsenal of tools is growing. Evolution is well beyond the **hypothesis** stage and is a well-established **theory** of modern science.

Variation within populations occurs by the natural mixing of genes through sexual reproduction or from naturally occurring mutations. Some of this genetic variation can introduce advantageous characteristics that increase the individual's chances of survival. While some species in the **fossil** record show little morphological change over time, others show gradual or punctuated changes, within which **intermediate** forms can be seen.

Take this quiz to check your comprehension of this section.

If you are using an offline version of this text, access the quiz for section 7.3 via the QR code.

 An interactive H5P element has been excluded from this version of the text. You can view it online here:
https://pressbooks.lib.vt.edu/introearthscience/?p=517#h5p-45

7.4 Correlation

Correlation is the process of establishing which sedimentary **strata** are of the same age but geographically separate. **Correlation** can be determined by using magnetic polarity reversals (chapter 2), rock types, unique rock sequences, or **index fossils**. There are four main types of **correlation**: **stratigraphic**, lithostratigraphic, chronostratigraphic, and biostratigraphic.

7.4.1 Stratigraphic Correlation

Stratigraphic correlation is the process of establishing which sedimentary **strata** are the same age at distant geographical areas by means of their **stratigraphic** relationship. Geologists construct geologic histories of areas by mapping and making **stratigraphic** columns-a detailed description of the **strata** from bottom to top. An example of **stratigraphic** relationships and **correlation** between Canyonlands National Park and Zion National Park in Utah.

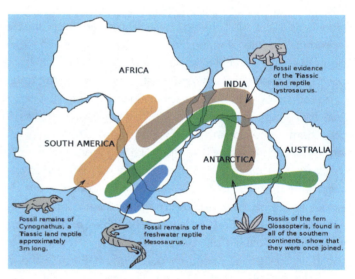

Figure 7.42: Image showing fossils that connect the continents of Gondwana (the southern continents of Pangea).

At Canyonlands, the Navajo Sandstone overlies the Kayenta Formation which overlies the cliff-forming Wingate Formation. In Zion, the Navajo Sandstone overlies the Kayenta Formation which overlies the cliff-forming Moenave Formation. Based on the **stratigraphic** relationship, the Wingate and Moenave Formations correlate. These two **formations** have unique names because their **composition** and outcrop pattern is slightly different. Other **strata** in the Colorado Plateau and their sequence can be recognized and correlated over thousands of square miles.

The Grand Staircase

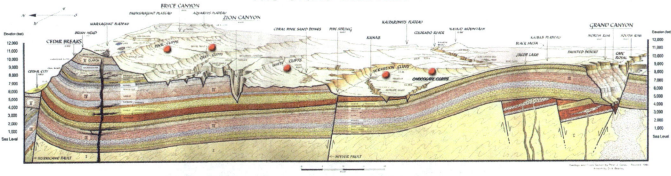

Figure 7.43: Correlation of strata along the Grand Staircase from the Grand Canyon to Zion Canyon, Bryce Canyon, and Cedar Breaks.

7.4.2 Lithostratigraphic Correlation

Figure 7.44: View of Navajo Sandstone from Angel's Landing in Zion National Park.

Lithostratigraphic correlation establishes a similar age of **strata** based on the lithology that is the **composition** and physical properties of that **strata**. *Lithos* is Greek for stone and -logy comes from the Greek word for doctrine or science. **Lithostratigraphic correlation** can be used to correlate whole **formations** long distances or can be used to correlate smaller **strata** within formations to trace their extent and regional **depositional environments**.

For example, the Navajo Sandstone, which makes up the prominent walls of Zion National Park, is the same Navajo Sandstone in Canyonlands because the lithology of the two are identical even though they are hundreds of miles apart. Extensions of the same Navajo Sandstone formation are found miles away in other parts of southern Utah, including Capitol Reef and Arches National Parks. Further, this same **formation** is the called the Aztec Sandstone in Nevada and Nugget Sandstone near Salt Lake City because they are lithologically distinct enough to warrant new names.

Figure 7.45: Stevens Arch in the Navajo Sandstone at Coyote Gulch some 125 miles away from Zion National Park.

7.4.3 Chronostratigraphic Correlation

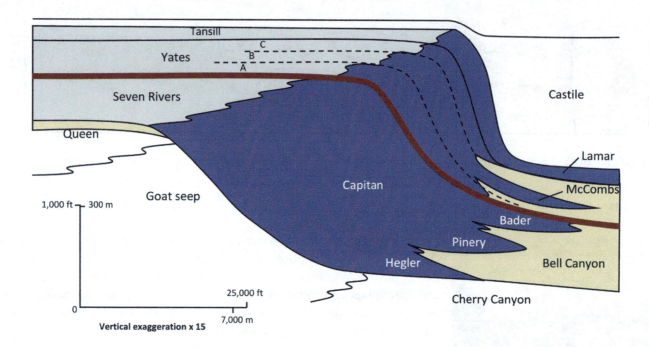

Figure 7.46: Cross-section of the Permian El Capitan Reef at Guadalupe National Monument, Texas. The red line shows a chronostratigraphic time line that represents a snapshot in time in which the shallow marine lagoon/back reef area (light blue), main Capitan reef (dark blue), and deep marine silstones (yellow) were all being deposited at the same time.

Chronostratigraphic correlation matches rocks of the same age, even though they are made of different lithologies. Different lithologies of sedimentary rocks can form at the same time at different geographic locations because **depositional environments** vary geographically. For example, at any one time in a **marine** setting there could be this sequence of **depositional environments** from beach to deep **marine**: beach, near **shore** area, shallow **marine lagoon**, **reef**, slope, and deep **marine**. Each **depositional environment** will have a unique **sedimentary rock formation**. On the figure of the Permian Capitan Reef at Guadalupe National Monument in West Texas, the red line shows a chronostratigraphic time line that represents a snapshot in time. Shallow-water **marine lagoon**/back **reef** area is light blue, the main Capitan **reef** is dark blue, and deep-water **marine siltstone** is yellow. All three of these unique lithologies were forming at the same time in **Permian** along this red timeline.

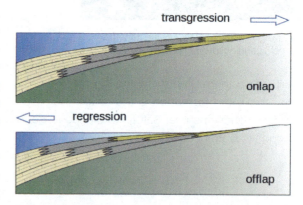

Figure 7.47: The rising sea levels of transgressions create onlapping sediments, regressions create offlapping. Ocean water is shown in blue so the time line is on the surface below the water. At the same time sandstone (buff color), limestone (gray), and shale (mustard color) are all forming at different depths of water.

7.4.4 Biostratigraphic Correlation

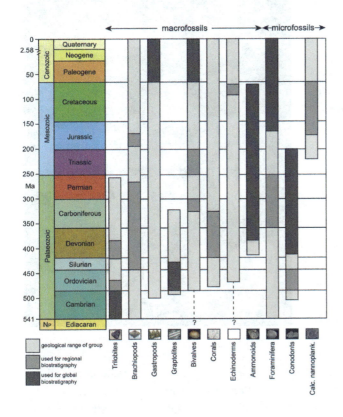

Figure 7.48: Index fossils used for biostratigraphic correlation.

Biostratigraphic correlation uses **index fossils** to determine **strata** ages. **Index fossils** represent assemblages or groups of organisms that were uniquely present during specific intervals of geologic time. Assemblages is referring a group of **fossils**. **Fossils** allow geologists to assign a **formation** to an absolute date range, such as the **Jurassic Period** (199 to 145 million years ago), rather than a relative time scale. In fact, most of the geologic time ranges are mapped to **fossil** assemblages. The most useful **index fossils** come from lifeforms that were geographically widespread and had a species lifespan that was limited to a narrow time interval. In other words, **index fossils** can be found in many places around the world, but only during a narrow time frame. Some of the best **fossils** for **biostratigraphic correlation** are microfossils, most of which came from single-celled organisms.

As with microscopic organisms today, they were widely distributed across many environments throughout the world. Some of these microscopic organisms had hard parts, such as exoskeletons or outer shells, making them better candidates for preservation.

Figure 7.49: Foraminifera, microscopic creatures with hard shells.

Foraminifera, single celled organisms with calcareous shells, are an example of an especially useful **index fossil** for the **Cretaceous Period** and **Cenozoic Era**.

Conodonts are another example of microfossils useful for **biostratigraphic correlation** of the **Cambrian** through **Triassic Periods**. Conodonts are tooth-like phosphatic structures of an eel-like multi-celled organism that had no other preservable hard parts. The conodont-bearing creatures lived in shallow **marine** environments all over the world. Upon death, the phosphatic hard parts were scattered into the rest of the **marine sediments**. These distinctive tooth-like structures are easily collected and separated from **limestone** in the laboratory.

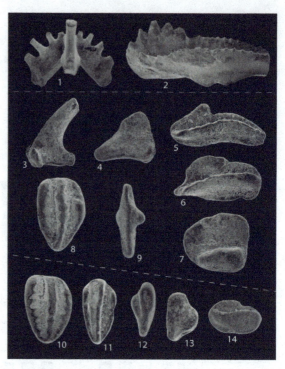

Figure 7.50: Conodonts.

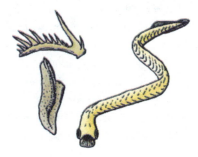

Figure 7.51: Artist reconstruction of the conodont animal.

Because the conodont creatures were so widely abundant, rapidly evolving, and readily preserved in **sediments**, their **fossils** are especially useful for correlating **strata**, even though knowledge of the actual animal possessing them is sparse. Scientists in the 1960s carried out a fundamental **biostratigraphic correlation** that tied **Triassic** conodont zonation into ammonoids, which are **extinct** ancient cousins of the pearly nautilus. Up to that point ammonoids were the only standard for **Triassic correlation**, so cross-referencing micro- and macro-**index fossils** enhanced the reliability of **biostratigraphic correlation** for either type. That conodont study went on to establish the use of conodonts to internationally correlate **Triassic strata** located in Europe, Western North America, and the Arctic Islands of Canada.

7.4.5 Geologic Time Scale

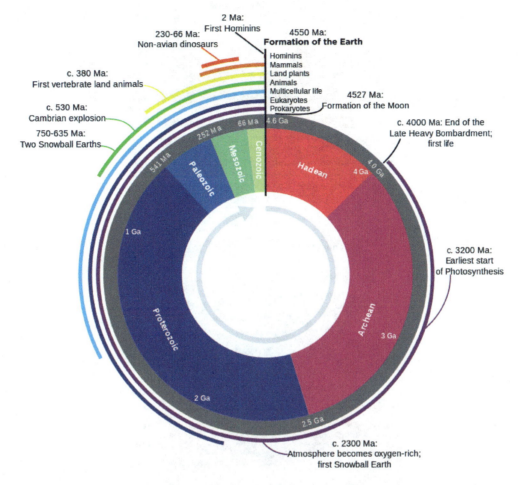

Figure 7.52: Geologic time on Earth, represented circularly, to show the individual time divisions and important events. Ga=billion years ago, Ma=million years ago.

Geologic time has been subdivided into a series of divisions by geologists. **Eon** is the largest division of time, followed by **era**, **period**, **epoch**, and age. The partitions of the geologic time scale is the same everywhere on Earth; however, rocks may or may not be present at a given location depending on the geologic activity going on during a particular **period** of time. Thus, we have the concept of time vs. rock, in which time is an unbroken continuum but rocks may be missing and/or unavailable for study. The figure of the geologic time scale, represents time flowing continuously from the beginning of the Earth, with the time units presented in an unbroken sequence. But that does not mean there are rocks available for study for all of these time units.

EON	ERA	PERIOD		EPOCH		Ma
Phanerozoic	Cenozoic	Quaternary		Holocene		0.011
				Pleistocene	Late	0.8
					Early	2.4
		Tertiary	Neogene	Pliocene	Late	3.6
					Early	5.3
				Miocene	Late	11.2
					Middle	16.4
					Early	23.0
			Paleogene	Oligocene	Late	28.5
					Early	34.0
				Eocene	Late	41.3
					Middle	49.0
					Early	55.8
				Paleocene	Late	61.0
					Early	65.5
	Mesozoic	Cretaceous		Late		99.6
				Early		145
		Jurassic		Late		161
				Middle		176
				Early		200
		Triassic		Late		228
				Middle		245
				Early		251
	Paleozoic	Permian		Late		260
				Middle		271
				Early		299
		Pennsylvanian		Late		306
				Middle		311
				Early		318
		Mississippian		Late		326
				Middle		345
				Early		359
		Devonian		Late		385
				Middle		397
				Early		416
		Silurian		Late		419
				Early		423
		Ordovician		Late		428
				Middle		444
				Early		488
		Cambrian		Late		501
				Middle		513
				Early		542
Precambrian	Proterozoic	Late		Neoproterozoic (Z)		1000
		Middle		Mesoproterozoic (Y)		1600
		Early		Paleoproterozoic (X)		2500
	Archean	Late				3200
		Early				4000
	Hadean					

Figure 7.53: Geologic time scale with ages shown.

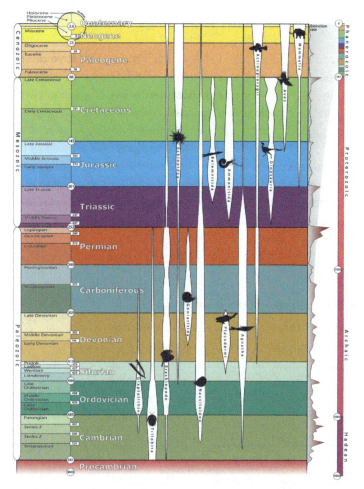

Figure 7.54: Names from the geologic time scale applied to taxonomical diversity of some major animal taxa.

The geologic time scale was developed during the 19th century using the principles of **stratigraphy**. The relative order of the time units was determined before geologist had the tools to assign numerical ages to **periods** and events. **Biostratigraphic correlation** using **fossils** to assign **era** and **period** names to sedimentary rocks on a worldwide scale. With the expansion of science and technology, some geologists think the influence of humanity on natural processes has become so great they are suggesting a new geologic time **period**, known as the **Anthropocene**.

Take this quiz to check your comprehension of this section.

If you are using an offline version of this text, access the quiz for section 7.4 via the QR code.

 An interactive H5P element has been excluded from this version of the text. You can view it online here: https://pressbooks.lib.vt.edu/introearthscience/?p=517#h5p-46

Summary

Events in Earth history can be placed in sequence using the five principles of **relative dating**. The geologic time scale was completely worked out in the 19th Century using these principles without knowing any actual numeric ages for the events. The discovery of **radioactivity** in the late 1800s enabled **absolute dating**, the assignment of numerical ages to events in the Earth's history, using decay of unstable **radioactive isotopes**. Accurately interpreting radioisotopic dating data depends on the type of rock tested and accurate assumptions about **isotope** baseline values. With a combination of relative and **absolute dating**, the history of geological events, age of Earth, and a geologic time scale have been determined with considerable accuracy. **Stratigraphic correlation** is additional tool used for understanding how **depositional environments** change geographically. Geologic time is vast, providing plenty of time for the evolution of various lifeforms, and some of these have become preserved as **fossils** that can be used for **biostratigraphic correlation**. The geologic time scale is continuous, although the rock record may be broken because rocks representing certain time **periods** may be missing.

Take this quiz to check your comprehension of this chapter.

If you are using an offline version of this text, access the quiz for chapter 7 via the QR code.

An interactive H5P element has been excluded from this version of the text. You can view it online here:
https://pressbooks.lib.vt.edu/introearthscience/?p=517#h5p-47

URLs Linked Within This Chapter

Animation 1: Why is it Hot Underground? [Video: 1:49] https://www.youtube.com/watch?v=mOSpRzW2i_4

The University of California Berkeley website: https://ucmp.berkeley.edu/paleo/fossils/permin.html

Text References

1. Allison, P.A., and Briggs, D.E.G., 1993, Exceptional fossil record: Distribution of soft-tissue preservation through the Phanerozoic: Geology, v. 21, no. 6, p. 527–530.
2. Bell, E.A., Boehnke, P., Harrison, T.M., and Mao, W.L., 2015, Potentially biogenic carbon preserved in a 4.1 billion-year-old zircon: Proc. Natl. Acad. Sci. U. S. A., v. 112, no. 47, p. 14518–14521.
3. Brent Dalrymple, G., 1994, The Age of the Earth: Stanford University Press.
4. Burleigh, R., 1981, W. F. Libby and the development of radiocarbon dating: Antiquity, v. 55, no. 214, p. 96–98.
5. Christopher B. DuRoss, Stephen F. Personius, Anthony J. Crone, Susan S. Olig, and William R. Lund, 2011, Integration of Paleoseismic Data from Multiple Sites to Develop an Objective Earthquake Chronology: Application to the Weber Segment of the Wasatch Fault Zone, Utah: Bulletin of the Seismological Society of America, v. 101, no. 6, p. 2765–2781., doi: 0.1785/0120110102.
6. Dass, C., 2007, Basics of mass spectrometry, *in* Fundamentals of Contemporary Mass Spectrometry: John Wiley & Sons, Inc., p. 1–14.
7. Elston, D.P., Billingsley, G.H., and Young, R.A., 1989, Geology of Grand Canyon, Northern Arizona (with Colorado River Guides): Lees Ferry to Pierce Ferry, Arizona: Amer Geophysical Union.
8. Erickson, J., Coates, D.R., and Erickson, H.P., 2014, An introduction to fossils and minerals: seeking clues to the Earth's

past: Facts on File science library, Facts On File, Incorporated, Facts on File science library.

9. Geyh, M.A., and Schleicher, H., 1990, Absolute Age Determination: Physical and Chemical Dating Methods and Their Application, 503 pp: Spring-er-Verlag, New York.

10. Ireland, T., 1999, New tools for isotopic analysis: Science, v. 286, no. 5448, p. 2289–2290.

11. Jackson, P.W., and of London, G.S., 2007, Four Centuries of Geological Travel: The Search for Knowledge on Foot, Bicycle, Sledge and Camel: Geological Society special publication, Geological Society, Geological Society special publication.

12. Jaffey, A.H., Flynn, K.F., Glendenin, L.E., Bentley, W.C., and others, 1971, Precision measurement of half-lives and specific activities of U 235 and U 238: Phys. Rev. C Nucl. Phys.

13. Léost, I., Féraud, G., Blanc-Valleron, M.M., and Rouchy, J.M., 2001, First absolute dating of Miocene Langbeinite evaporites by 40Ar/39Ar laser step-heating:[K2Mg2 (SO4) 3] Stebnyk Mine (Carpathian Foredeep Basin): Geophys. Res. Lett., v. 28, no. 23, p. 4347–4350.

14. Mosher, L.C., 1968, Triassic conodonts from Western North America and Europe and Their Correlation: J. Paleontol., v. 42, no. 4, p. 895–946.

15. Oberthür, T., Davis, D.W., Blenkinsop, T.G., and Höhndorf, A., 2002, Precise U–Pb mineral ages, Rb–Sr and Sm–Nd systematics for the Great Dyke, Zimbabwe—constraints on late Archean events in the Zimbabwe craton and Limpopo belt: Precambrian Res., v. 113, no. 3–4, p. 293–305.

16. Patterson, C., 1956, Age of meteorites and the earth: Geochim. Cosmochim. Acta, v. 10, no. 4, p. 230–237.

17. Schweitzer, M.H., Wittmeyer, J.L., Horner, J.R., and Toporski, J.K., 2005, Soft-tissue vessels and cellular preservation in Tyrannosaurus rex: Science, v. 307, no. 5717, p. 1952–1955.

18. Valley, J.W., Peck, W.H., King, E.M., and Wilde, S.A., 2002, A cool early Earth: Geology, v. 30, no. 4, p. 351–354.

19. Whewell, W., 1837, History of the Inductive Sciences: From the Earliest to the Present Times: J.W. Parker, 492 p.

20. Wilde, S.A., Valley, J.W., Peck, W.H., and Graham, C.M., 2001, Evidence from detrital zircons for the existence of continental crust and oceans on the Earth 4.4 Gyr ago: Nature, v. 409, no. 6817, p. 175–178.

21. Winchester, S., 2009, The Map That Changed the World: William Smith and the Birth of Modern Geology: Harper-Collins.

Figure References

Figure 7.1: Nicolas Steno, c. 1670. Justus Sustermans. ca. 1666 and 1677; uploaded in 2012. Public domain. https://commons.wikimedia.org/wiki/File:Portrait_of_Nicolas_Stenonus.jpg

Figure 7.2: Geologic time scale. USGS. 1997. Public domain. https://pubs.usgs.gov/gip/fossils/numeric.html

Figure 7.3: Lower strata are older than those lying on top of them. Wilson44691. 2009. Public domain. https://commons.wikimedia.org/wiki/File:IsfjordenSuperposition.jpg

Figure 7.4: Lateral continuity. Roger Bolsius. 2013. CC BY-SA 3.0. https://en.wikipedia.org/wiki/File:Grand_Canyon_Panorama_2013.jpg

Figure 7.5: Dark dike cutting across older rocks, the lighter of which is younger than the grey rock. Thomas Eliasson of Geological Survey of Sweden. 2008. CC BY 2.0. https://commons.wikimedia.org/wiki/File:Multiple_Igneous_Intrusion_Phases_Kosterhavet_Sweden.jpg

Figure 7.6: Fossil succession showing correlation among strata. 2010 דק. CC BY-SA 3.0. https://commons.wikimedia.org/wiki/File:Faunal_sucession.jpg

Figure 7.7: The Grand Canyon of Arizona. Jean-Christophe BENOIST. 2012. CC BY 3.0. https://commons.wikimedia.org/wiki/File:Grand_Canyon_-_Hopi_Point.JPG

Figure 7.8: The rocks of the Grand Canyon. NPS. 2018. Public domain. https://www.nps.gov/articles/age-of-rocks-in-grand-canyon.htm

Figure 7.9: The red, layered rocks of the Grand Canyon Supergroup overlying the dark-colored rocks of the Vishnu schist represents a type of unconformity called a nonconformity. Simeon87. 2012. CC BY-SA 3.0. https://commons.wikimedia.org/wiki/File:Grand_Canyon_with_Snow_4.JPG

Figure 7.10: All three of these formations have a disconformity at the two contacts between them. NPS. 2009. Public domain. https://commons.wikimedia.org/wiki/File:Redwall,_Temple_Butte_and_Muav_formations_in_Grand_Canyon.jpg

Figure 7.11: In the lower part of the picture is an angular unconformity in the Grand Canyon known as the Great Unconformity. Doug Dolde. 2008. Public domain. https://commons.wikimedia.org/wiki/File:View_from_Lipan_Point.jpg

Figure 7.12: Disconformity. 2008 .דק. CC BY-SA 3.0. https://commons.wikimedia.org/wiki/File:Disconformity.jpg

Figure 7.13: Nonconformity (the lower rocks are igneous or metamorphic). 2008 .דק. CC BY-SA 3.0. https://commons.wikimedia.org/wiki/File:Nonconformity.jpg

Figure 7.14: Angular unconformity. 2008 .דק. CC BY-SA 3.0. https://commons.wikimedia.org/wiki/File:Angular_unconformity.jpg

Figure 7.15: Block diagram to apply relative dating principles. Woudloper. 2009. CC BY-SA 1.0. https://commons.wikimedia.org/wiki/File:Cross-cutting_relations.svg

Figure 7.16: Canada's Nuvvuagittuq Greenstone Belt may have the oldest rocks and oldest evidence life on Earth, according to recent studies. NASA. 2008. Public domain. https://commons.wikimedia.org/wiki/File:Nuvvuagittuq_belt_rocks.jpg

Figure 7.17: Three isotopes of hydrogen. Dirk Hünniger. 2016. CC BY-SA 3.0. https://commons.wikimedia.org/wiki/File:Hydrogen_Deuterium_Tritium_Nuclei_Schmatic-en.svg

Figure 7.18: Simulation of half-life. Sbyrnes321. 2010. Public domain. https://commons.wikimedia.org/wiki/File:Halflife-sim.gif

Figure 7.19: Granite (left) and gneiss (right). Fjæregranitt3 by Friman, 2007 (CC BY-SA 3.0, https://commons.wikimedia.org/wiki/File:Fj%C3%A6regranitt3.JPG). Gneiss by Siim Sepp, 2005 (CC BY-SA 3.0, https://commons.wikimedia.org/wiki/File:Gneiss.jpg).

Figure 7.20: An alpha decay: Two protons and two neutrons leave the nucleus. Inductiveload. 2007. Public domain. https://commons.wikimedia.org/wiki/File:Alpha_Decay.svg

Figure 7.21: Periodic table of the elements. Sandbh. 2017. CC BY-SA 4.0. https://en.wikipedia.org/wiki/File:Periodic_Table_Chart_with_less_active_and_active_nonmetals.png

Figure 7.22: Decay chain of U-238 to stable Pb-206 through a series of alpha and beta decays. ThaLibster. 2017. CC BY-SA 4.0. https://en.wikipedia.org/wiki/File:Decay_Chain_of_Uranium-238.svg

Figure 7.23: The two paths of electron capture. Pamputt. 2015. CC BY-SA 4.0. https://en.wikipedia.org/wiki/File:Atomic_rearrangement_following_an_electron_capture.svg

Figure 7.24: Mass spectrometer instrument. Archives CAMECA. 2006. CC BY-SA 3.0. https://commons.wikimedia.org/wiki/File:IMS3F_pbmf.JPG

Figure 7.25: Graph of the amount of half life versus the amount of daughter isotope. Krishnavedala. 2015. Public domain. https://en.wikipedia.org/wiki/File:Half_times.svg

Figure 7.26: Schematic of carbon going through a mass spectrometer. Miko Christie. 2013. CC BY-SA 3.0. https://commons.wikimedia.org/wiki/File:Accelerator_mass_spectrometer_schematic_for_radiocarbon.svg

Figure 7.27: Carbon dioxide concentrations over the last 400,000 years. Robert A. Rohde. 2011. CC BY-SA 3.0. https://commons.wikimedia.org/wiki/File:Carbon_Dioxide_400kyr.svg

Figure 7.28: Artist's impression of the Earth in the Hadean. Tim Bertelink. 2016. CC BY-SA 4.0. https://commons.wikimedia.org/wiki/File:Hadean.png

Figure 7.29: Photomicrograph of zircon crystal. Denniss. 2006. CC BY-SA 3.0. https://commons.wikimedia.org/wiki/File:Zircon_microscope.jpg

Figure 7.30: Several prominent ash beds found in North America, including three Yellowstone eruptions shaded pink (Mesa Falls, Huckleberry Ridge, and Lava Creek), the Bisho Tuff ash bed (brown dashed line), and the modern May 18th, 1980 ash fall (yellow). USGS. 2005. Public domain. https://commons.wikimedia.org/wiki/File:Yellowstone_volcano_-_ash_beds.svg

Figure 7.31: Thermoluminescence, a type of luminescence dating. Zkeizars. 2008. CC BY-SA 4.0. https://commons.wikimedia.org/wiki/File:Keizars_TLexplained2.jpg

Figure 7.32: Apatite from Mexico. Robert M. Lavinsky. Before March 2010. CC BY-SA 3.0. https://commons.wikimedia.org/wiki/File:Apatite-(CaF)-280343.jpg

Figure 7.33: Archaeopteryx lithographica, specimen displayed at the Museum für Naturkunde in Berlin. H. Raab. 2009. CC BY-SA 3.0. https://commons.wikimedia.org/wiki/File:Archaeopteryx_lithographica_(Berlin_specimen).jpg

Figure 7.34: The trilobites had a hard exoskeleton, and is an early arthropod, the same group that includes modern insects, crustaceans, and arachnids. Wilson44691. 2010. Public domain. https://commons.wikimedia.org/wiki/File:ElrathiakingiUtahWheelerCambrian.jpg

Figure 7.35: Mosquito preserved in amber. Didier Desouens. 2010. CC BY-SA 4.0. https://commons.wikimedia.org/wiki/File:Ambre_Dominique_Moustique.jpg

Figure 7.36: Permineralization in petrified wood. Moondigger. 2005. CC BY-SA 2.5. https://commons.wikimedia.org/wiki/File:Petrified_forest_log_2_md.jpg

Figure 7.37: External mold of a clam. Wilson44691. 2007. Public domain. https://commons.wikimedia.org/wiki/File:Aviculopecten_subcardiformis01.JPG

Figure 7.38: Carbonized leaf. Wilson44691. 2008. Public domain. https://commons.wikimedia.org/wiki/File:ViburnumFossil.jpg

Figure 7.39: Dinosaur tracks as a record of its passing. Ballista. 2006. CC BY-SA 3.0. https://commons.wikimedia.org/wiki/File:Cheirotherium_prints_possibly_Ticinosuchus.JPG

Figure 7.40: Fossil animal droppings (coprolite). USGS. 2008. Public domain. https://commons.wikimedia.org/wiki/File:Coprolite.jpg

Figure 7.41: Variation within a population. Inglesenargentina. 2006. Public domain. https://en.wikipedia.org/wiki/File:Bell-shaped-curve.JPG

Figure 7.42: Image showing fossils that connect the continents of Gondwana (the southern continents of Pangea). Osvaldocangaspadilla. 2010. Public domain. https://commons.wikimedia.org/wiki/File:Snider-Pellegrini_Wegener_fossil_map.svg

Figure 7.43: Correlation of strata along the Grand Staircase from the Grand Canyon to Zion Canyon, Bryce Canyon, and Cedar Breaks. NPS. 2005. Public domain. https://commons.wikimedia.org/wiki/File:Grand_Staircase-big.jpg

Figure 7.44: View of Navajo Sandstone from Angel's Landing in Zion National Park. Diliff. 2004. CC BY-SA 3.0. https://en.wikipedia.org/wiki/File:Zion_angels_landing_view.jpg

Figure 7.45: Stevens Arch in the Navajo Sandstone at Coyote Gulch some 125 miles away from Zion National Park. G. Thomas. 2007. Public domain. https://en.wikipedia.org/wiki/File:StevensArchUT.jpg

Figure 7.46: Cross-section of the Permian El Capitan Reef at Guadalupe National Monument, Texas. Kindred Grey. 2022. Adapted under fair use from Garber, R.A., Grover, G.A., & Harris, P.M. (1989). Geology of the Capitan Shelf Margin – Subsurface Data from the Northern Delaware Basin (DOI:10.2110/cor.89.13.0003).

Figure 7.47: The rising sea levels of transgressions create onlapping sediments, regressions create offlapping. Woudloper. 2009. CC BY-SA 1.0. https://commons.wikimedia.org/wiki/File:Offlap_%26_onlap_EN.svg

Figure 7.48: Index fossils used for biostratigraphic correlation. David Bond. 2016. CC BY-SA 4.0. https://commons.wikimedia.org/wiki/File:Biostratigraphic_index_fossils_01.svg

Figure 7.49: Foraminifera, microscopic creatures with hard shells. Hans Hillewaert. 2011. CC BY-SA 4.0. https://en.wikipedia.org/wiki/File:Quinqueloculina_seminula.jpg

Figure 7.50: Conodonts. USGS. 2007. Public domain. https://commons.wikimedia.org/wiki/File:Conodonts.jpg

Figure 7.51: Artist reconstruction of the conodont animal. Philippe Janvier. 1997. CC BY 3.0. https://commons.wikimedia.org/wiki/File:Euconodonta.gif

Figure 7.52: Geologic time on Earth, represented circularly, to show the individual time divisions and important events. Woudloper; adapted by Hardwigg. 2010. Public domain. https://commons.wikimedia.org/wiki/File:Geologic_Clock_with_events_and_periods.svg

Figure 7.53: Geologic time scale with ages shown. USGS. 2009. Public domain. https://commons.wikimedia.org/wiki/File:Geologic_time_scale.jpg

Figure 7.54: Names from the geologic time scale applied to taxonomical diversity of some major animal taxa. Frederik Lerouge. 2015. CC BY-SA 4.0. https://commons.wikimedia.org/wiki/File:Geologic_Time_Scale.png

8.

EARTH HISTORY

By the end of this chapter, students should be able to:

- Describe the turbulent beginning of Earth during the **Hadean** and **Archean Eons**.
- Identify the transition to modern **atmosphere**, **plate tectonics**, and evolution that occurred in the **Proterozoic Eon**.
- Describe the **Paleozoic** evolution and **extinction** of invertebrates with hard parts, fish, amphibians, reptiles, tetrapods, and land plants; and **tectonics** and sedimentation associated with the **supercontinent Pangea**.
- Describe the **Mesozoic** evolution and **extinction** of birds, dinosaurs, and mammmals; and **tectonics** and sedimentation associated with the breakup of **Pangea**.
- Describe the **Cenozoic** evolution of mammals and birds, paleoclimate, and **tectonics** that shaped the modern world.

Entire courses and careers have been based on the wide-ranging topics covering Earth's history. Throughout the long history of Earth, change has been the norm. Looking back in time, an untrained eye would see many unfamiliar life forms and terrains. The main topics studied in Earth history are paleogeography, paleontology, and paleoecology and paleoclimatology—respectively, past landscapes, past organisms, past ecosystems, and past environments. This chapter will cover briefly the origin of the universe and the 4.6 billion year history of Earth. This Earth history will **focus** on the major physical and biological events in each **Eons** and **Era**.

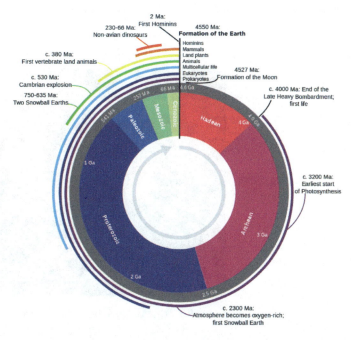

Figure 8.1: Geologic time on Earth, represented circularly, to show the individual time divisions and important events. Ga=billion years ago, Ma=million years ago.

8.1 Hadean Eon

Geoscientists use the geological time scale to assign relative age names to events and rocks, separating major events in Earth's history based on significant changes as recorded in rocks and **fossils**. This section summarizes the most notable events of each major time interval. For a breakdown on how these time intervals are chosen and organized, see chapter 7.

The **Hadean Eon**, named after the Greek god and ruler of the underworld Hades, is the oldest **eon** and dates from 4.5–4.0 billion years ago.

This time represents Earth's earliest history, during which the planet was characterized by a partially molten surface, **volcanism**, and asteroid impacts. Several mechanisms made the newly forming Earth incredibly hot: gravitational **compression**, **radioactive** decay, and asteroid impacts. Most of this initial heat still exists inside the Earth. The **Hadean** was originally defined as the birth of the planet occurring 4.0 billion years ago and preceding the existence of many rocks and life forms. However, geologists have dated **minerals** at 4.4 billion years, with evidence that liquid water was present. There is possibly even evidence of life existing over 4.0 billion years ago. However, the most reliable record for early life, the microfossil record, starts at 3.5 billion years ago.

EON	ERA	PERIOD	EPOCH		Ma
Phanerozoic	Cenozoic	Quaternary	Holocene		0.011
			Pleistocene	Late	0.8
				Early	2.4
		Tertiary (Neogene)	Pliocene	Late	3.6
				Early	5.3
			Miocene	Late	11.2
				Middle	16.4
				Early	23.0
		Tertiary (Paleogene)	Oligocene	Late	28.5
				Early	34.0
			Eocene	Late	41.3
				Middle	49.0
				Early	55.8
			Paleocene	Late	61.0
				Early	65.5
	Mesozoic	Cretaceous	Late		99.6
			Early		145
		Jurassic	Late		161
			Middle		176
			Early		200
		Triassic	Late		228
			Middle		245
			Early		251
	Paleozoic	Permian	Late		260
			Middle		271
			Early		299
		Pennsylvanian	Late		306
			Middle		311
			Early		318
		Mississippian	Late		326
			Middle		345
			Early		359
		Devonian	Late		385
			Middle		397
			Early		416
		Silurian	Late		419
			Early		423
		Ordovician	Late		428
			Middle		444
			Early		488
		Cambrian	Late		501
			Middle		513
			Early		542
Precambrian	Proterozoic	Late	Neoproterozoic (Z)		1000
		Middle	Mesoproterozoic (Y)		1600
		Early	Paleoproterozoic (X)		2500
	Archean	Late			3200
		Early			4000

Figure 8.2: Geological time scale with ages shown.

Figure 8.3: Artist's impression of the Earth in the Hadean.

8.1.1 Origin of Earth's Crust

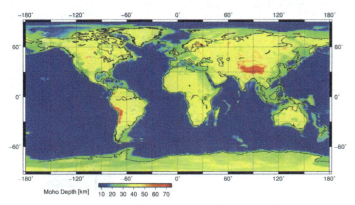

Figure 8.4: The global map of the depth of the moho.

As Earth cooled from its molten state, **minerals** started to crystallize and settle resulting in a separation of **minerals** based on density and the creation of the **crust**, **mantle**, and **core**. The earliest Earth was chiefly molten material and would have been rounded by gravitational forces so it resembled a ball of **lava** floating in space. As the outer part of the Earth slowly cooled, the high melting-point **minerals** (see Bowen's Reaction Series in chapter 4) formed solid slabs of early **crust**. These slabs were probably unstable and easily reabsorbed into the liquid **magma** until the Earth cooled enough to allow numerous larger fragments to form a thin primitive **crust**. Scientists generally assume this **crust** was **oceanic** and **mafic** in **composition**, and littered with impacts, much like the Moon's current **crust**. There is still some debate over when **plate tectonics** started, which would have led to the **formation** of **continental** and **felsic crust**. Regardless of this, as Earth cooled and solidified, less dense **felsic minerals** floated to the surface of the Earth to form the **crust**, while the denser **mafic** and **ultramafic** materials sank to form the **mantle** and the highest-density iron and nickel sank into the **core**. This differentiated the Earth from a homogenous planet into a heterogeneous one with layers of **felsic crust**, **mafic crust**, **ultramafic mantle**, and iron and nickel **core**.

8.1.2 Origin of the Moon

Figure 8.5: Dark side of the Moon.

Several unique features of Earth's Moon have prompted scientists to develop the current **hypothesis** about its **formation**. The Earth and Moon are tidally locked, meaning that as the Moon orbits, one side always faces the Earth and the opposite side is not visible to us. Also and most importantly, the chemical compositions of the Earth and Moon show nearly identical **isotope** ratios and volatile content. Apollo missions returned from the Moon with rocks that allowed scientists to conduct very precise comparisons between Moon and Earth rocks. Other bodies in the **solar system** and **meteorites** do not share the same degree of similarity and show much higher variability. If the Moon and Earth formed together, this would explain why they are so chemically similar.

Many ideas have been proposed for the origin of the Moon: The Moon could have been captured from another part of the **solar system** and formed in place together with the Earth, or the Moon could have been ripped out of the early Earth. None of proposed explanations can account for all the evidence. The currently prevailing **hypothesis** is the giant-impact **hypothesis**. It proposes a body about half of Earth's size must have shared at least parts of Earth's orbit and collided with it, resulting in a violent mixing and scattering of material from both objects. Both bodies would be composed of a combination of materials, with more of the lower density splatter coalescing into the Moon. This may explain why the Earth has a higher density and thicker **core** than the Moon.

Figure 8.6: Artist's concept of the giant impact from a Mars-sized object that could have formed the moon.

One or more interactive elements has been excluded from this version of the text. You can view them online here: https://www.youtube.com/watch?v=UIKmSQqp8wY

Video 8.1: Evolution of the Moon.

If you are using an offline version of this text, access this YouTube video via the QR code.

8.1.3 Origin of Earth's Water

Explanations for the origin of Earth's water include **volcanic** outgassing, comets, and **meteorites**. The **volcanic** outgassing **hypothesis** for the origin of Earth's water is that it originated from inside the planet, and emerged via **tectonic** processes as vapor associated with **volcanic** eruptions. Since all **volcanic** eruptions contain some water vapor, at times more than 1% of the volume, these alone could have created Earth's surface water. Another likely source of water was from space. Comets are a mixture of dust and ice, with some or most of that ice being frozen water. Seemingly dry meteors can contain small but measurable amounts of water, usually trapped in their **mineral** structures. During heavy bombardment **periods** later in Earth's history, its cooled surface was pummeled by comets and **meteorites**, which could be why so much water exists above ground. There isn't a definitive answer for what process is the source of ocean water. Earth's water isotopically matches water found in **meteorites** much better than that of comets. However, it is hard to know if Earth processes could have changed the water's isotopic signature over the last 4-plus billion years. It is possible that all three sources contributed to the origin of Earth's water.

Figure 8.7: Water vapor leaves comet 67P/Churyumov–Gerasimenko.

Take this quiz to check your comprehension of this section.

If you are using an offline version of this text, access the quiz for section 8.1 via the QR code.

 An interactive H5P element has been excluded from this version of the text. You can view it online here: https://pressbooks.lib.vt.edu/introearthscience/?p=602#h5p-50

8.2 Archean Eon

The **Archean Eon**, which lasted from 4.0–2.5 billion years ago, is named after the Greek word for beginning. This **eon** represents the beginning of the rock record. Although there is current evidence that rocks and **minerals** existed during the **Hadean Eon**, the **Archean** has a much more robust rock and **fossil** record.

8.2.1 Late Heavy Bombardment

Figure 8.8: Artist's impression of the Archean.

Objects were chaotically flying around at the start of the **solar system**, building the planets and moons. There is evidence that after the planets formed, about 4.1–3.8 billion years ago, a second large spike of asteroid and comet impacted the Earth and Moon in an event called **late heavy bombardment**. **Meteorites** and comets in stable or semi-stable orbits became unstable and started impacting objects throughout the **solar system**. In addition, this event is called the lunar cataclysm because most of the Moons craters are from this event. During **late heavy bombardment**, the Earth, Moon, and all planets in the **solar system** were pummeled by material from the asteroid and Kuiper belts. Evidence of this bombardment was found within samples collected from the Moon.

Figure 8.9: 2015 image from NASA's New Horizons probe of Pluto. The lack of impacts found on the Tombaugh Regio (the heart-shaped plain, lower right) has been inferred as being younger than the Late Heavy Bombardment and the surrounding surface due to its lack of impacts.

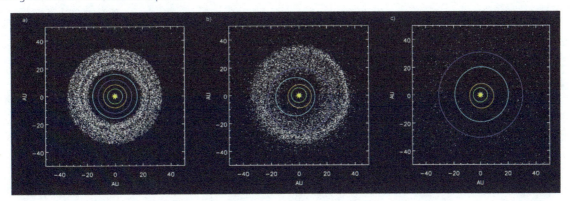

Figure 8.10: Simulation of before, during, and after the late heavy bombardment.

It is universally accepted that the **solar system** experienced extensive asteroid and comet bombardment at its start; however, some other process must have caused the second increase in impacts hundreds of millions of years later. A leading **theory** blames gravitational **resonance** between Jupiter and Saturn for disturbing orbits within the asteroid and Kuiper belts based on a similar process observed in the Eta Corvi star **system**.

8.2.2 Origin of the Continents

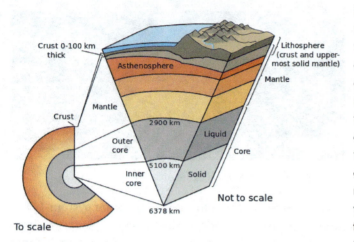

Figure 8.11: The layers of the Earth. Physical layers include lithosphere and asthenosphere; chemical layers are crust, mantle, and core.

In order for **plate tectonics** to work as it does currently, it necessarily must have continents. However, the easiest way to create **continental** material is via **assimilation** and differentiation of existing continents (see chapter 4). This chicken-and-egg quandary over how continents were made in the first place is not easily answered because of the great age of **continental** material and how much evidence has been lost during **tectonics** and **erosion**. While the timing and specific processes are still debated, **volcanic** action must have brought the first **continental** material to the Earth's surface during the **Hadean**, 4.4 billion years ago. This model does not solve the problem of **continent formation**, since **magmatic differentiation** seems to need thicker **crust**. Nevertheless, the continents formed by some incremental process during the early history of Earth. The best idea is that density differences allowed lighter **felsic** materials to float upward and heavier **ultramafic** materials and **metallic** iron to sink. These density differences led to the layering of the Earth, the layers that are now detected by **seismic** studies. Early protocontinents accumulated **felsic** materials as developing **plate–tectonic** processes brought lighter material from the **mantle** to the surface.

The first solid evidence of modern **plate tectonics** is found at the end of the **Archean**, indicating at least some **continental lithosphere** must have been in place. This evidence does not necessarily mark the starting point of **plate tectonics**; remnants of earlier **tectonic** activity could have been erased by the **rock cycle**.

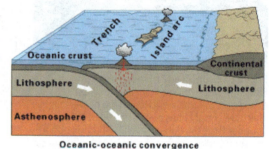

Oceanic-oceanic convergence

Figure 8.12: Subduction of an oceanic plate beneath another oceanic plate, forming a trench and an island arc. Several island arcs might combine and eventually evolve into a continent.

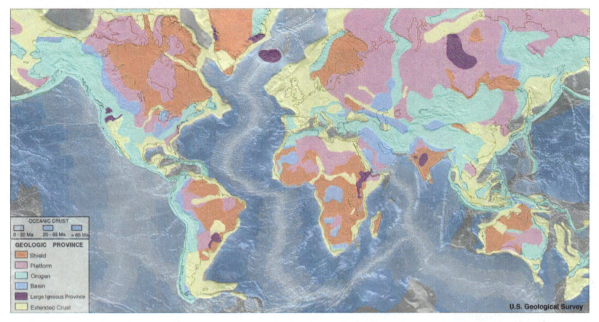

Figure 8.13: Geologic provinces with the Shield (orange) and Platform (pink) comprising the Craton, the stable interior of continents.

The stable interiors of the current continents are called **cratons** and were mostly formed in the **Archean Eon**. A **craton** has two main parts: the **shield**, which is crystalline **basement** rock near the surface, and the **platform** made of sedimentary rocks covering the **shield**. Most **cratons** have remained relatively unchanged with most **tectonic** activity having occurred around **cratons** instead of within them. Whether they were created by **plate tectonics** or another process, **Archean** continents gave rise to the **Proterozoic** continents that now dominate our planet.

The general guideline as to what constitutes a **continent** and differentiates **oceanic** from **continental crust** is under some debate. At passive margins, **continental crust** grades into **oceanic crust** at passive margins, making a distinction difficult. Even island-arc and hot-spot material can seem more closely related to **continental crust** than **oceanic**. Continents usually have a **craton** in the middle with **felsic igneous** rocks. There is evidence that submerged masses like Zealandia, that includes present-day New Zealand, would be considered a **continent**. **Continental crust** that does not contain a **craton** is called a **continental** fragment, such as the island of Madagascar off the east coast of Africa.

Figure 8.14: The continent of Zealandia.

8.2.3 First Life on Earth

Figure 8.15: Fossils of microbial mats from Sweden.

Life most likely started during the late **Hadean** or early **Archean Eons**. The earliest evidence of life are chemical signatures, microscopic filaments, and microbial mats. Carbon found in 4.1 billion year old **zircon** grains have a chemical signature suggesting an organic origin. Other evidence of early life are 3.8–4.3 billion-year-old microscopic filaments from a **hydrothermal vent** deposit in Quebec, Canada. While the chemical and microscopic filaments evidence is not as robust as **fossils**, there is significant **fossil** evidence for life at 3.5 billion years ago. These first well-preserved **fossils** are photosynthetic microbial mats, called **stromatolites**, found in Australia.

Although the origin of life on Earth is unknown, **hypotheses** include a chemical origin in the early **atmosphere** and ocean, deep-sea **hydrothermal** vents, and delivery to Earth by comets or other objects. One **hypothesis** is that life arose from the chemical environment of the Earth's early **atmosphere** and oceans, which was very different than today. The oxygen-free **atmosphere** produced a reducing environment with abundant methane, carbon dioxide, sulfur, and nitrogen compounds. This is what the **atmosphere** is like on other bodies in the **solar system**. In the famous Miller-Urey **experiment**, researchers simulated early Earth's **atmosphere** and lightning within a sealed vessel. After igniting sparks within the vessel, they discovered the **formation** of amino acids, the fundamental building blocks of proteins. In 1977, when scientists discovered an isolated ecosystem around **hydrothermal** vents on a deep-sea **mid-ocean ridge** (see chapter 4), it opened the door for another explanation of the origin of life. The **hydrothermal** vents have a unique ecosystem of critters with **chemosynthesis** as the foundation of the food chain instead of photosynthesis. The ecosystem is deriving its

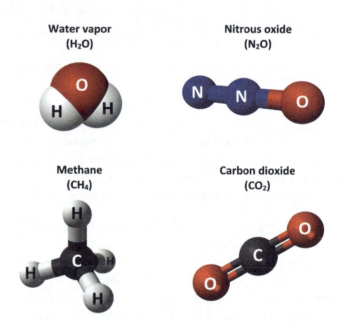

Figure 8.16: Greenhouse gases were more common in Earth's early atmosphere.

energy from hot chemical-rich waters pouring out of underground towers. This suggests that life could have started on the deep **ocean floor** and derived energy from the heat from the Earth's interior via **chemosynthesis**. Scientists have since expanded the search for life to more unconventional places, like Jupiter's icy moon Europa.

> *One or more interactive elements has been excluded from this version of the text. You can view them online here:*
> *https://www.youtube.com/watch?v=qyt6llcOJLo*

Video 8.2: Animation of the original Miller-Urey 1959 experiment that simulated the early atmosphere and created amino acids from simple elements and compounds.

If you are using an offline version of this text, access this YouTube video via the QR code.

Another possibility is that life or its building blocks came to Earth from space, carried aboard comets or other objects. Amino acids, for example, have been found within comets and **meteorites**. This intriguing possibility also implies a high likelihood of life existing elsewhere in the cosmos.

Take this quiz to check your comprehension of this section.

If you are using an offline version of this text, access the quiz for section 8.2 via the QR code.

> *An interactive H5P element has been excluded from this version of the text. You can view it online here:*
> *https://pressbooks.lib.vt.edu/introearthscience/?p=602#h5p-51*

8.3 Proterozoic Eon

Figure 8.17: Diagram showing the main products and reactants in photosynthesis. The one product that is not shown is sugar, which is the chemical energy that goes into constructing the plant, and the energy that is stored in the plant which is used later by the plant or by animals that consume the plant.

The **Proterozoic Eon**, meaning "earlier life," comes after the **Archean Eon** and ranges from 2.5 billion to 541 million years old. During this time, most of the central parts of the continents had formed and **plate tectonic** processes had started. Photosynthesis by microbial organisms, such as single-celled cyanobacteria, had been slowly adding oxygen to the oceans. As cyanobacteria evolved into multicellular organisms, they completely transformed the oceans and later the **atmosphere** by adding **massive** amounts of free oxygen gas (O_2) and initiated what is called the **Great Oxygenation Event** (GOE). This drastic environmental change decimated the anaerobic bacteria, which could not survive in the presence of free oxygen. On the other hand, aerobic organisms could thrive in ways they could not earlier.

An oxygenated world also changed the chemistry of the planet in significant ways. For example, iron remained in **solution** in the non-oxygenated environment of the earlier **Archean Eon**. In chemistry, this is known as a reducing environment. Once the environment was oxygenated, iron combined with free oxygen to form solid precipitates of iron **oxide**, such as the **mineral** hematite or magnetite. These precipitates accumulated into large **mineral** deposits with red **chert** known as **banded**-iron **formations**, which are dated at about 2 billion years.

The **formation** of iron **oxide minerals** and red **chert** (see figure 8.18) in the oceans lasted a long time and prevented oxygen levels from increasing significantly, since **precipitation** took the oxygen out of the water and deposited it into the rock **strata**. As oxygen continued to be produced and **mineral precipitation** leveled off, **dissolved** oxygen gas eventually **saturated** the oceans and started bubbling out into the **atmosphere**. Oxygenation of the **atmosphere** is the single biggest event that distinguishes the **Archean** and **Proterozoic** environments. In addition to changing **mineral** and ocean chemistry, the GOE is also tabbed as triggering Earth's first **glaciation** event around 2.1 billion years ago, the Huron Glaciation. Free oxygen reacted with methane in the

Figure 8.18: Alternating bands of iron-rich and silica-rich mud, formed as oxygen combined with dissolved iron.

atmosphere to produce carbon dioxide. Carbon dioxide and methane are called greenhouse gases because they **trap** heat within the Earth's **atmosphere**, like the insulated glass of a greenhouse. Methane is a more effective insulator than carbon dioxide, so as the proportion of carbon dioxide in the **atmosphere** increased, the **greenhouse effect** decreased, and the planet cooled.

8.3.1 Rodinia

By the **Proterozoic Eon**, lithospheric plates had formed and were moving according to **plate tectonic** forces that were similar to current times. As the moving **plates** collided, the ocean basins closed to form a **supercontinent** called **Rodinia**. The **supercontinent** formed about 1 billion years ago and broke up about 750 to 600 million years ago, at the end of the **Proterozoic**. One of the resulting fragments was a **continental** mass called **Laurentia** that would later become North America. Geologists have reconstructed **Rodinia** by matching and aligning ancient mountain chains, assembling the pieces like a jigsaw puzzle, and using paleomagnetics to orient to magnetic north.

The disagreements over these complex reconstructions is exemplified by geologists proposing at least six different models for the breakup of **Rodinia** to create Australia, Antarctica, parts of China, the Tarim **craton** north of the Himalaya, Siberia, or the Kalahari **craton** of eastern Africa. This breakup created lots of shallow-water, biologically favorable environments that fostered the evolutionary break-throughs marking the start of the next **eon**, the **Phanerozoic**.

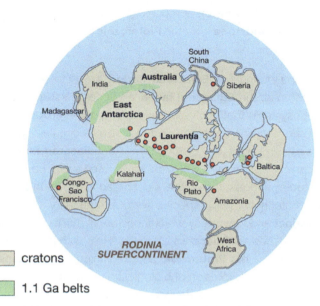

Figure 8.19: One possible reconstruction of Rodinia 1.1 billion years ago.

8.3.2 Life Evolves

Early life in the **Archean** and earlier is poorly documented in the **fossil** record. Based on chemical evidence and evolutionary **theory**, scientists propose this life would have been single-celled photosynthetic organisms, such as the cyanobacteria that created **stromatolites**. Cyanobacteria produced free oxygen in the **atmosphere** through photosynthesis. Cyanobacteria, archaea, and bacteria are **prokaryotes**—primitive organisms made of single cells that lack cell nuclei and other organelles.

Figure 8.20: Modern cyanobacteria (as stromatolites) in Shark Bay, Australia.

A large evolutionary step occurred during the **Proterozoic Eon** with the appearance of **eukaryotes** around 2.1 to 1.6 billion years ago. **Eukaryotic** cells are more complex, having nuclei and organelles. The nuclear DNA is capable of more complex replication and regulation than that of **prokaryotic** cells. The organelles include mitochondria for producing energy and chloroplasts for photosynthesis. The **eukaryote** branch in the tree of life gave rise to fungi, plants, and animals.

Figure 8.21: Fossil stromatolites in Saratoga Springs, New York.

Another important event in Earth's biological history occurred about 1.2 billion years ago when **eukaryotes** invented sexual reproduction. Sharing genetic material from two reproducing individuals, male and female, greatly increased genetic variability in their offspring. This genetic mixing accelerated evolutionary change, contributing to more complexity among individual organisms and within ecosystems (see chapter 7).

Proterozoic land surfaces were barren of plants and animals and geologic processes actively shaped the environment differently because land surfaces were not protected by leafy and woody vegetation. For example, rain and **rivers** would have caused **erosion** at much higher rates on land surfaces devoid of plants. This resulted in thick accumulations of pure **quartz sandstone** from the **Proterozoic Eon** such as the extensive **quartzite formations** in the **core** of the Uinta Mountains in Utah.

Fauna during the **Ediacaran Period**, 635.5 to 541 million years ago are known as the **Ediacaran fauna**, and offer a first glimpse at the diversity of ecosystems that evolved near the end of the **Proterozoic**. These soft-bodied organisms were among the first multicellular life forms and probably were similar to jellyfish or worm-like. **Ediacaran fauna** did not have hard parts like shells and were not well preserved in the rock records. However, studies suggest they were widespread in the Earth's oceans. Scientists still debate how many species were evolutionary dead-ends that became **extinct** and how many were ancestors of modern groupings. The transition of soft-bodied **Ediacaran** life to life forms with hard body parts occurred at the end of the **Proterozoic** and beginning of the **Phanerozoic Eons**. This evolutionary explosion of biological diversity made a dramatic difference in scientists' ability to understand the history of life on Earth.

Figure 8.22: Dickinsonia, a typical Ediacaran fossil.

Take this quiz to check your comprehension of this section.

If you are using an offline version of this text, access the quiz for section 8.3 via the QR code.

 An interactive H5P element has been excluded from this version of the text. You can view it online here: https://pressbooks.lib.vt.edu/introearthscience/?p=602#h5p-52

8.4 Phanerozoic Eon: Paleozoic Era

Figure 8.23: The trilobites had a hard exoskeleton, and is an early arthropod, the same group that includes modern insects, crustaceans, and arachnids.

The **Phanerozoic Eon** is the most recent, 541 million years ago to today, and means "visible life" because the **Phanerozoic** rock record is marked by an abundance of **fossils**. **Phanerozoic** organisms had hard body parts like claws, scales, shells, and bones that were more easily preserved as **fossils**. Rocks from the older **Precambrian** time are less commonly found and rarely include **fossils** because these organisms had soft body parts. **Phanerozoic** rocks are younger, more common, and contain the majority of extant **fossils**. The study of rocks from this **eon** yields much greater detail. The **Phanerozoic** is subdivided into three **eras**, from oldest to youngest they are **Paleozoic** ("ancient life"), **Mesozoic** ("middle life"), and **Cenozoic** ("recent life") and the remaining three chapter headings are on these three important **eras**.

Life in the early **Paleozoic Era** was dominated by **marine** organisms but by the middle of the **era** plants and animals evolved to live and reproduce on land. Fish evolved jaws and fins evolved into jointed limbs. The development of lungs allowed animals to emerge from the sea and become the first air-breathing tetrapods (four-legged animals) such as amphibians. From amphibians evolved reptiles with the amniotic egg. From reptiles evolved an early ancestor to birds and mammals and their scales became feathers and fur. Near the end of the **Paleozoic Era**, the **Carboniferous Period** had some

Figure 8.24: Trilobites, by Heinrich Harder, 1916.

of the most extensive forests in Earth's history. Their fossilized remains became the **coal** that powered the industrial revolution

8.4.1 Paleozoic Tectonics and Paleogeography

Figure 8.25: Laurentia, which makes up the North American craton.

During the **Paleozoic Era**, sea-levels rose and fell four times. With each sea-level rise, the majority of North America was covered by a shallow tropical ocean. Evidence of these submersions are the abundant **marine** sedimentary rocks such as **limestone** with **fossils** corals and **ooids**. Extensive sea-level **falls** are documented by widespread **unconformities**. Today, the midcontinent has extensive **marine** sedimentary rocks from the **Paleozoic** and western North America has thick layers of **marine limestone** on block faulted mountain ranges such as Mt. Timpanogos near Provo, Utah.

The assembly of **supercontinent Pangea**, sometimes spelled **Pangaea**, was completed by the late **Paleozoic Era**. The name **Pangea** was originally coined by Alfred Wegener and means "all land." **Pangea** is the when all of the major continents were grouped together as one by a series of **tectonic** events including **subduction** island-**arc** accretion, and **continental** collisions, and ocean-**basin** closures. In North America, these **tectonic** events occurred on the east **coast** and are known as the Taconic, Acadian, Caledonian, and Alleghanian orogenies. The Appalachian Mountains are the erosional remnants of these mountain building events in North America. Surrounding **Pangea** was a global ocean **basin** known as the Panthalassa. Continued **plate** movement extended the ocean into **Pangea**, forming a large bay called the Tethys Sea that eventually divided the land mass into two smaller **supercontinents**, Laurasia and Gondwana. Laurasia consisted of **Laurentia** and Eurasia, and Gondwana consisted of the remaining continents of South America, Africa, India, Australia, and Antarctica.

Figure 8.26: A reconstruction of Pangaea, showing approximate positions of modern continents.

 One or more interactive elements has been excluded from this version of the text. You can view them online here: https://www.youtube.com/watch?v=ovT90wYrVk4

Video 8.3: Animation of plate movement the last 3.3 billion years. Pangea occurs at the 4:40 mark.

If you are using an offline version of this text, access this YouTube video via the QR code.

While the east coast of North America was tectonically active during the **Paleozoic Era**, the west coast remained mostly inactive as a **passive margin** during the early **Paleozoic**. The western edge of North American **continent** was near the present-day Nevada-Utah border and was an expansive shallow **continental shelf** near the paleoequator. However, by the **Devonian Period**, the Antler **orogeny** started on the west coast and lasted until the Pennsylvanian **Period**. The Antler **orogeny** was a **volcanic island arc** that was accreted onto western North America with the **subduction** direction away from North America. This created a mountain range on the west coast of North American called the Antler highlands and was the first part of building the land in the west that would eventually make most of California, Oregon, and Washington states. By the late **Paleozoic**, the Sonoma **orogeny** began on the west coast and was another **collision** of an **island arc**. The Sonoma **orogeny** marks the change in **subduction** direction to be toward North America with a **volcanic arc** along the entire west coast of North America by late **Paleozoic** to early **Mesozoic Eras**.

By the end of the **Paleozoic Era**, the east coast of North America had a very high mountain range due to **continental collision** and the creation of **Pangea**. The west coast of North America had smaller and isolated **volcanic** highlands associated with **island arc** accretion. During the **Mesozoic Era**, the size of the mountains on either side of North America would flip, with the west coast being a more tectonically active **plate boundary** and the east coast changing into a **passive margin** after the breakup of **Pangea**.

8.4.2 Paleozoic Evolution

The beginning of the **Paleozoic Era** is marked by the first appearance of hard body parts like shells, spikes, teeth, and scales; and the appearance in the rock record of most animal phyla known today. That is, most basic animal body plans appeared in the rock record during the **Cambrian Period**. This sudden appearance of biological diversity is called the **Cambrian Explosion**. Scientists debate whether this sudden appearance is more from a rapid evolutionary diversification as a result of a warmer **climate** following the late **Proterozoic glacial** environments, better preservation and fossilization of hard parts, or artifacts of a more complete and recent rock record. For example, fauna may have been diverse during the **Ediacaran Period**, setting the state for the **Cambrian Explosion**, but they lacked hard body parts and would have left few **fossils** behind. Regardless, during the **Cambrian Period** 541–485 million years ago marked the appearance of most animal phyla.

Figure 8.27: Anomalocaris reconstruction by the MUSE science museum in Italy.

Figure 8.28: Original plate from Walcott's 1912 description of Opabinia, with labels: fp = frontal appendage, e = eye, ths = thoracic somites, i = intestine, ab = abdominal segment.

One of the best **fossil** sites for the **Cambrian Explosion** was discovered in 1909 by Charles Walcott (1850–1927) in the Burgess **Shale** in western Canada. The Burgess **Shale** is a **Lagerstätte**, a site of exceptional **fossil** preservation that includes impressions of soft body parts. This discovery allowed scientists to study **Cambrian** animals in immense detail because soft body parts are not normally preserved and fossilized. Other **Lagerstätte** sites of similar age in China and Utah have allowed scientist to form a detailed picture of **Cambrian** biodiversity. The biggest mystery surrounds animals that do not fit existing lineages and are unique to that time. This includes many famous fossilized creatures: the first compound-eyed trilobites; *Wiwaxia*, a creature covered in spiny **plates**; *Hallucigenia*, a walking worm with spikes; *Opabinia*, a five-eyed arthropod with a grappling claw; and *Anomalocaris*, the alpha predator of its time, complete with grasping appendages and circular **mouth** with sharp **plates**. Most notably appearing during the **Cambrian** is an important ancestor to humans. A segmented worm called *Pikaia* is thought to be the earliest ancestor of the **Chordata** phylum that includes **vertebrates**, animals with backbones.

By the end of the **Cambrian**, mollusks, brachiopods, nautiloids, gastropods, graptolites, echinoderms, and trilobites covered the sea floor. Although most animal phyla appeared by the **Cambrian**, the biodiversity at the family, genus, and species level was low until the **Ordovician Period**. During the Great Ordovician Biodiversification Event, **vertebrates** and invertebrates (animals without backbone) became more diverse and complex at family, genus, and

Figure 8.29: A modern coral reef.

species level. The cause of the rapid speciation event is still debated but some likely causes are a combination of warm temperatures, expansive **continental** shelves near the equator, and more **volcanism** along the **mid-ocean ridges**. Some have shown evidence that an asteroid breakup event and consequent heavy **meteorite** impacts correlate with this diversification event. The additional **volcanism** added nutrients to ocean water helping support a robust ecosystem. Many life forms and ecosystems that would be recognizable in current times appeared at this time. Mollusks, corals, and arthropods in particular multiplied to dominate the oceans.

Figure 8.30: Guadalupe National Park is made of a giant fossil reef.

One important evolutionary advancement during the **Ordovician Period** was reef-building organisms, mostly colonial coral. Corals took advantage of the ocean chemistry, using **calcite** to build large structures that resembled modern **reefs** like the Great Barrier Reef off the coast of Australia. These reefs housed thriving ecosystems of organisms that swam around, hid in, and crawled over them. Reefs are important to paleontologists because of their preservation potential, **massive** size, and in-place ecosystems. Few other **fossils** offer more diversity and complexity than **reef** assemblages.

According to evidence from **glacial** deposits, a small **ice age** caused sea-levels to drop and led to a major **mass extinction** by the end of the **Ordovician**. This is the earliest of five **mass extinction** events documented in the **fossil** record. During this **mass extinction**, an unusually large number of species abruptly disappear in the **fossil** record (see video below).

 One or more interactive elements has been excluded from this version of the text. You can view them online here: https://www.youtube.com/watch?v=a09mOAKXvJs

Video 8.4: 3-minute video describing mass extinctions and how they are defined.

If you are using an offline version of this text, access this YouTube video via the QR code.

Life bounced back during the **Silurian period**. The **period**'s major evolutionary event was the development of jaws from the forward pair of gill arches in bony fishes and sharks. Hinged jaws allowed fish to exploit new food sources and ecological niches. This **period** also included the start of armored fishes, known as the placoderms. In addition to fish and jaws, **Silurian** rocks provide the first evidence of **terrestrial** or land-dwelling plants and animals. The first vascular plant, *Cooksonia*, had woody tissues, **pores** for gas exchange, and veins for water and food transport. Insects, spiders, scorpions, and crustaceans began to inhabit moist, freshwater **terrestrial** environments.

Figure 8.31: The placoderm Bothriolepis panderi from the Devonian of Russia.

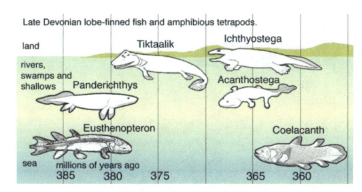

Figure 8.32: Several different types of fish and amphibians that led to walking on land.

The **Devonian Period** is called the Age of Fishes due to the rise in plated, jawed, and lobe-finned fishes. The lobe-finned fishes, which were related to the modern lungfish and coelacanth, are important for their eventual evolution into tetrapods, four-limbed **vertebrate** animals that can walk on land. The first lobe-finned land-walking fish, named *Tiktaalik*, appeared about 385 million years ago and serves as a transition **fossil** between fish and early tetrapods. Though Tiktaalik was clearly a fish, it had some tetrapod structures as well. Several **fossils** from the **Devonian** are more tetrapod like than fish like but these weren't fully **terrestrial**. The first fully **terrestrial** tetrapod arrived in the Mississippian (early **Carboniferous**) **period**.

By the Mississippian (early **Carboniferous**) **period**, tetrapods had evolved into two main groups, amphibians and amniotes, from a common tetrapod ancestor. The amphibians were able to breathe air and live on land but still needed water to nurture their soft eggs. The first reptile (an amniote) could live and reproduce entirely on land with hard-shelled eggs that wouldn't dry out.

Land plants had also evolved into the first trees and forests. Toward the end of the **Devonian**, another **mass extinction** event occurred. This **extinction**, while severe, is the least temporally defined, with wide variations in the timing of the event or events. **Reef** building organisms were the hardest hit, leading to dramatic changes in **marine** ecosystems.

The next time **period**, called the **Carboniferous** (North American geologists have subdivided this into the Mississippian and Pennsylvanian **periods**), saw the highest levels of oxygen ever known, with forests (e.g., ferns, club mosses) and swamps dominating the landscape. This helped cause the largest arthropods ever, like the millipede *Arthropleura*, at 2.5 meters (6.4 feet) long! It also saw the rise of a new group of animals, the reptiles. The evolutionary advantage that reptiles have over amphibians is the amniote egg (egg with a protective shell), which allows them to rely on non-aquatic environments for reproduction. This widened the **terrestrial** reach of reptiles compared to amphibians.

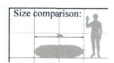

Figure 8.33: A reconstruction of the giant arthropod (insects and their relatives) Arthropleura.

This booming life, especially plant life, created cooling temperatures as carbon dioxide was removed from the **atmosphere**. By the middle **Carboniferous**, these cooler temperatures led to an **ice age** (called the Karoo **Glaciation**) and less-productive forests. The reptiles fared much better than the amphibians, leading to their diversification. This **glacial** event lasted into the early **Permian**.

Figure 8.34: Reconstruction of Dimetrodon.

By the **Permian**, with **Pangea** assembled, the **supercontinent** led to a dryer **climate**, and even more diversification and domination by the reptiles. The groups that developed in this warm **climate** eventually radiated into dinosaurs. Another group, known as the synapsids, eventually evolved into mammals. Synapsids, including the famous sail-backed *Dimetrodon* are commonly confused with dinosaurs. Pelycosaurs (of the Pennsylvanian to early **Permian** like *Dimetrodon*) are the first group of synapsids that exhibit the beginnings of mammalian characteristics such as well-differentiated dentition: incisors, highly developed canines in lower and upper jaws and cheek teeth, premolars and molars. Starting in the late **Permian**, a second group of synapsids, called the therapsids (or mammal-like reptiles) evolve, and become the ancestors to mammals.

Permian Mass Extinction

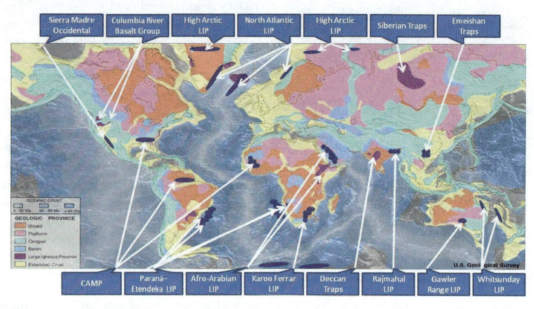

Figure 8.35: World map of flood basalts. Note the largest is the Siberian Traps.

The end of the **Paleozoic era** is marked by the largest **mass extinction** in earth history. The **Paleozoic era** had two smaller **mass extinctions**, but these were not as large as the **Permian Mass Extinction**, also known as the **Permian-Triassic Extinction Event**. It is estimated that up to 96% of **marine** species and 70% of land-dwelling (**terrestrial**) **vertebrates** went extinct. Many famous organisms, like sea scorpions and trilobites, were never seen again in the **fossil** record. What caused such a widespread **extinction** event? The exact cause is still debated, though the leading idea relates to extensive **volcanism** associated with the Siberian **Traps**, which are one of the largest deposits of **flood basalts** known on Earth, dating to the time of the **extinction** event. The eruption size is estimated at over 3 million cubic kilometers that is approximately 4,000,000 times larger than the famous 1980 Mt. St. Helens eruption in Washington. The unusually large **volcanic** eruption would have contributed a large amount of toxic gases, aerosols, and greenhouse gasses into the **atmosphere**. Further, some evidence suggests that the **volcanism** burned vast **coal** deposits releasing methane (a greenhouse gas) into the **atmosphere**. As discussed in chapter 15, greenhouse gases cause the **climate** to warm. This extensive addition of greenhouse gases from the Siberian **Traps** may have caused a runaway **greenhouse effect** that rapidly changed the **climate**, acidified the oceans, disrupted food chains, disrupted carbon cycling, and caused the largest **mass extinction**.

Take this quiz to check your comprehension of this section.

If you are using an offline version of this text, access the quiz for section 8.4 via the QR code.

 An interactive H5P element has been excluded from this version of the text. You can view it online here:
https://pressbooks.lib.vt.edu/introearthscience/?p=602#h5p-53

8.5 Phanerozoic Eon: Mesozoic Era

Following the **Permian Mass Extinction**, the **Mesozoic** ("middle life") was from 252 million years ago to 66 million years ago. As **Pangea** started to break apart, mammals, birds, and flowering plants developed. The **Mesozoic** is probably best known as the age of reptiles, most notably, the dinosaurs.

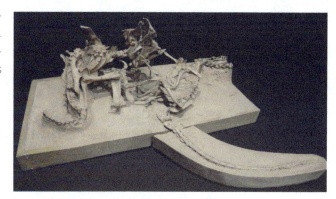

Figure 8.36: Perhaps the greatest fossil ever found, a velociraptor attacked a protoceratops, and both were fossilized mid sequence.

8.5.1 Mesozoic Tectonics and Paleogeography

Figure 8.37: Pangea breaking up.

Pangea started breaking up (in a region that would become eastern Canada and United States) around 210 million years ago in the Late **Triassic**. Clear evidence for this includes the age of the **sediments** in the Newark Supergroup **rift** basins and the Palisades **sill** of the eastern part of North America and the age of the Atlantic **ocean floor**. Due to sea-floor spreading, the oldest rocks on the Atlantic's floor are along the **coast** of northern Africa and the east coast of North America, while the youngest are along the **mid-ocean ridge**.

This age pattern shows how the Atlantic Ocean opened as the young Mid-Atlantic Ridge began to create the seafloor. This means the Atlantic ocean started opening and was first formed here. The southern Atlantic opened next, with South America separating from central and southern Africa. Last (happening after the **Mesozoic** ended) was the northernmost Atlantic, with Greenland and Scandinavia parting ways. The breaking points of each **rifted plate** margin eventually turned into the passive **plate** boundaries of the east coast of the Americas today.

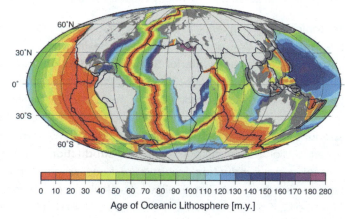

Figure 8.38: Age of oceanic lithosphere, in millions of years. Notice the differences in the Atlantic Ocean along the coasts of the continents.

One or more interactive elements has been excluded from this version of the text. You can view them online here: https://www.youtube.com/watch?v=6o1HawAOTEI

Video 8.5: Video of Pangea breaking apart and plates moving to their present locations.

If you are using an offline version of this text, access this YouTube video via the QR code.

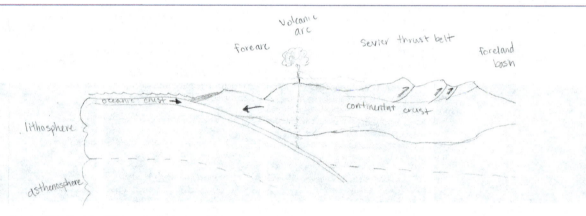

Figure 8.39: Sketch of the major features of the Sevier Orogeny.

In western North America, an active **plate** margin had started with **subduction**, controlling most of the **tectonics** of that region in the **Mesozoic**. Another possible island-**arc collision** created the Sonoman **Orogeny** in Nevada during the latest **Paleozoic** to the **Triassic**. In the **Jurassic**, another island-**arc collision** caused the Nevadan **Orogeny**, a large Andean-style **volcanic arc** and thrust belt. The Sevier Orogeny followed in the Cretaceous, which was mainly a volcanic arc to the west and a thin-skinned fold and thrust belt to the east, meaning stacks of shallow **faults** and **folds** built up the topography. Many of the structures in the Rocky Mountains today date from this **orogeny**.

Tectonics had an influence in one more important geographic feature in North America: the **Cretaceous** Western Interior Foreland **Basin**, which flooded during high sea levels forming the **Cretaceous Interior Seaway**. **Subduction** from the west was the Farallon Plate, an **oceanic plate** connected to the Pacific Plate (seen today as remnants such as the Juan de Fuca Plate, off the coast of the Pacific Northwest). **Subduction** was shallow at this time because a very young, hot and less dense portion of the Farallon **plate** was **subducted**. This shallow **subduction** caused a downwarping in the central part of North America. High sea levels due to shallow **subduction**, and increasing rates of seafloor spreading and **subduction**, high temperatures, and melted ice also contributed to the high sea levels. These factors allowed a shallow epicontinental seaway that extended from the Gulf of Mexico to the Arctic Ocean to divide North America into two separate land masses, Laramidia to the west and Appalachia to the east, for 25 million years. Many of the **coal** deposits in Utah and Wyoming formed from swamps along the shores of this seaway. By the end of the **Cretaceous**, cooling temperatures caused the seaway to regress.

Figure 8.40: The Cretaceous Interior Seaway in the mid-Cretaceous.

8.5.2 Mesozoic Evolution

Figure 8.41: A Mesozoic scene from the late Jurassic.

The **Mesozoic era** is dominated by reptiles, and more specifically, the dinosaurs. The **Triassic** saw devastated ecosystems that took over 30 million years to fully re-emerge after the **Permian Mass Extinction**. The first appearance of many modern groups of animals that would later flourish occurred at this time. This includes frogs (amphibians), turtles (reptiles), **marine** ichthyosaurs and plesiosaurs (**marine** reptiles), mammals, and the archosaurs. The archosaurs ("ruling reptiles") include ancestral groups that went **extinct** at the end of the **Triassic**, as well as the flying pterosaurs, crocodilians, and the dinosaurs. Archosaurs, like the placental mammals after them, occupied all major environments: **terrestrial** (dinosaurs), in the air (pterosaurs), aquatic (crocodilians) and even fully **marine** habitats (**marine** crocodiles). The pterosaurs, the first **vertebrate** group to take flight, like the dinosaurs and mammals, start small in the **Triassic**.

At the end of the **Triassic**, another **mass extinction** event occurred, the fourth major **mass extinction** in the geologic record. This was perhaps caused by the Central Atlantic Magmatic Province **flood basalt**. The end-**Triassic extinction** made certain lineages go extinct and helped spur the evolution of survivors like mammals, pterosaurs (flying reptiles), ichthyosaurs/plesiosaurs/mosasaurs (**marine** reptiles), and dinosaurs.

Figure 8.42: A drawing of the early plesiosaur Agustasaurus from the Triassic of Nevada.

Figure 8.43: Reconstruction of the small (<5") Megazostrodon, one of the first animals considered to be a true mammal.

Mammals, as previously mentioned, got their start from a reptilian synapsid ancestor possibly in the late **Paleozoic**. Mammals stayed small, in mainly nocturnal niches, with insects being their largest prey. The development of warm-blooded circulation and fur may have been a response to this lifestyle.

In the **Jurassic**, species that were previously common, flourished due to a warmer and more tropical **climate**. The dinosaurs were relatively small animals in the **Triassic period** of the **Mesozoic**, but became truly **massive** in the **Jurassic**. Dinosaurs are split into two groups based on their hip structure, i.e. orientation of the pubis and ischium bones in relationship to each other. This is referred to as the "reptile hipped" saurischians and the "bird hipped" ornithischians. This has recently been brought into question by a new idea for dinosaur lineage.

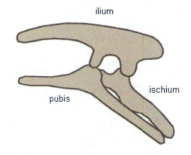

Figure 8.44: Closed structure of a ornithischian hip, which is similar to a birds.

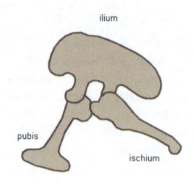

Figure 8.45: Open structure of a saurischian hip, which is similar to a lizards.

Most of the dinosaurs of the **Triassic** were saurischians, but all of them were bipedal. The major adaptive advantage dinosaurs had was changes in the hip and ankle bones, tucking the legs under the body for improved locomotion as opposed to the semi-erect gait of crocodiles or the sprawling posture of reptiles. In the **Jurassic**, limbs (or a lack thereof) were also important to another group of reptiles, leading to the evolution of *Eophis*, the oldest snake.

There is a paucity of dinosaur **fossils** from the Early and Middle **Jurassic**, but by the Late **Jurassic** they were dominating the planet. The saurischians diversified into the giant herbivorous (plant-eating) long-necked sauropods weighing up to 100 tons and bipedal carnivorous theropods, with the possible exception of the *Therizinosaurs*. All of the ornithischians (e.g *Stegosaurus, Iguanodon, Triceratops, Ankylosaurus, Pachycephhlosaurus*) were herbivorous with a strong tendency to have a "turtle-like" beak at the tips of their mouths.

Figure 8.47: Archaeopteryx lithographica, specimen displayed at the Museum für Naturkunde in Berlin.

Figure 8.46: Therizinosaurs, like Beipiaosaurus (shown in this restoration), are known for their enormous hand claws.

The pterosaurs grew and diversified in the **Jurassic**, and another notable arial organism developed and thrived in the **Jurassic**: birds. When *Archeopteryx* was found in the Solnhofen **Lagerstätte** of Germany, a seeming dinosaur-bird hybrid, it started the conversation on the origin of birds. The idea that birds evolved from dinosaurs occurred very early in the history of research into evolution, only a few years after Darwin's *On the Origin of Species*. This study used a remarkable **fossil** of *Archeopteryx* from a transitional animal between dinosaurs and birds. Small meat-eating theropod dinosaurs were likely the branch that became birds due to their similar features. A significant debate still exists over how and when powered flight evolved. Some have stated a running-start model, while others have favored a tree-leaping gliding model or even a semi-combination: flapping to aid in climbing.

The **Cretaceous** saw a further diversification, specialization, and domination of the dinosaurs and other fauna. One of the biggest changes on land was the transition to angiosperm-dominated flora. Angiosperms, which are plants with flowers and seeds, had originated in the **Cretaceous**, switching many plains to grasslands by the end of the **Mesozoic**. By the end of the **period**, they had replaced gymnosperms (evergreen trees) and ferns as the dominant plant in the world's forests. Haplodiploid eusocial insects (bees and ants) are descendants from **Jurassic** wasp-like ancestors that co-evolved with the flowering plants during this time **period**. The breakup of **Pangea** not only shaped our modern world's geography, but biodiversity at the time as well. Throughout the **Mesozoic**, animals on the isolated, now separated island continents (formerly parts of **Pangea**), took strange evolutionary turns. This includes giant titanosaurian sauropods (*Argentinosaurus*) and theropods (*Giganotosaurus*) from South America.

Figure 8.48: Reconstructed skeleton of Argentinosaurus, from Naturmuseum Senckenberg in Germany.

K-T Extinction

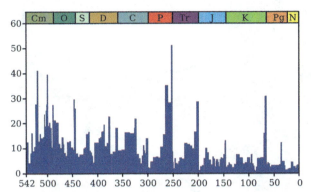

Figure 8.49: Graph of the rate of extinctions. Note the large spike at the end of the Cretaceous (labeled as K).

Similar to the end of the **Paleozoic era**, the **Mesozoic Era** ended with the K-Pg **Mass Extinction** (previously known as the **K-T Extinction**) 66 million years ago. This **extinction** event was likely caused by a large **bolide** (an extraterrestrial impactor such as an asteroid, **meteoroid**, or comet) that collided with earth. Ninety percent of plankton species, 75% of plant species, and all the dinosaurs went **extinct** at this time.

Figure 8.50: Artist's depiction of an impact event.

One of the strongest pieces of evidence comes from the **element** iridium. Quite rare on Earth, and more common in **meteorites**, it has been found all over the world in higher concentrations at a particular layer of rock that formed at the time of the K-T boundary. Soon other scientists started to find evidence to back up the claim. Melted rock spheres, a special type of "shocked" **quartz** called stishovite, that only is found at impact sites, was found in many places around the world. The huge impact created a strong thermal pulse that could be responsible for global forest fires, strong acid rains, a corresponding abundance of ferns, the first colonizing plants after a forest fire, enough debris thrown into the air to significantly cool temperatures afterward, and a 2-km high **tsunami** inferred from deposits found from Texas to Alabama.

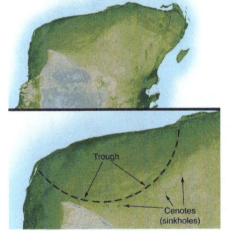

Figure 8.51: The land expression of the Chicxulub crater. The other side of the crater is within the Gulf of México.

Still, with all this evidence, one large piece remained missing: the crater where the **bolide** impacted. It was not until 1991 that the crater was confirmed using **petroleum** company geophysical data. Even though it is the third largest confirmed crater on Earth at roughly 180 km wide, the **Chicxulub Crater** was hard to find due to being partially underwater and partially obscured by the dense forest canopy of the Yucatan Peninsula. Coring of the center of the impact called the peak ring contained **granite**, indicating the impact was so powerful that it lifted **basement sediment** from the **crust** several miles toward the surface. In 2010, an international team of scientists reviewed 20 years of research and blamed the impact for the **extinction**.

With all of this information, it seems like the case would be closed. However, there are other events at this time which could have partially aided the demise of so many organisms. For example, sea levels are known to be slowly decreasing at the time of the K-T event, which is tied to **marine** extinctions, though any study on gradual vs. sudden changes in the **fossil** record is flawed due to the incomplete nature of the **fossil** record. Another big event at this time was the Deccan Traps **flood basalt volcanism** in India. At over 1.3 million cubic kilometers of material, it was certainly a large source of material hazardous to ecosystems at the time, and it has been suggested as at least partially responsible for the **extinction**. Some have found the impact and eruptions too much of a coincidence, and have even linked the two together.

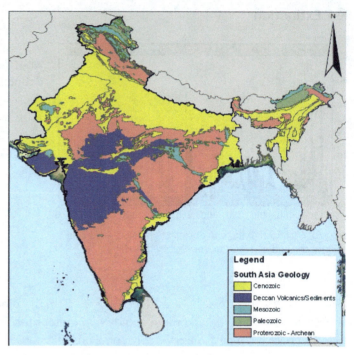

Figure 8.52: Geology of India, showing purple as Deccan Traps-related rocks.

Take this quiz to check your comprehension of this section.

If you are using an offline version of this text, access the quiz for section 8.5 via the QR code.

An interactive H5P element has been excluded from this version of the text. You can view it online here:
https://pressbooks.lib.vt.edu/introearthscience/?p=602#h5p-54

8.6 Phanerozoic Eon: Cenozoic Era

The **Cenozoic**, meaning "new life," is known as the age of mammals because it is in this **era** that mammals came to be a dominant and large life form, including human ancestors. Birds, as well, flourished in the open niches left by the dinosaur's demise. Most of the **Cenozoic** has been relatively warm, with the main exception being the **ice age** that started about 2.558 million years ago and (despite recent warming) continues today. **Tectonic** shifts in the west caused **volcanism**, but eventually changed the long-standing **subduction** zone into a **transform** boundary.

Figure 8.53: Paraceratherium, seen in this reconstruction, was a massive (15-20 ton, 15 foot tall) ancestor of rhinos.

8.6.1 Cenozoic Tectonics and Paleogeography

 One or more interactive elements has been excluded from this version of the text. You can view them online here: https://www.youtube.com/watch?v=IDTBY5WDELg

Video 8.6: Animation of the last 38 million years of movement in western North America. Note, that after the ridge is subducted, convergent turns to transform (with divergent inland).

If you are using an offline version of this text, access this YouTube video via the QR code.

In the **Cenozoic**, the **plates** of the Earth moved into more familiar places, with the biggest change being the closing of the Tethys Sea with **collisions** such as the Alps, Zagros, and Himalaya, a **collision** that started about 57 million years ago, and continues today. Maybe the most significant **tectonic** feature that occurred in the **Cenozoic** of North America was the conversion of the west **coast** of California from a **convergent** boundary **subduction** zone to a **transform** boundary. **Subduction** off the **coast** of the western United States, which had occurred throughout the **Mesozoic**, had continued in the **Cenozoic**. After the Sevier **Orogeny** in the late **Mesozoic**, a subsequent **orogeny**

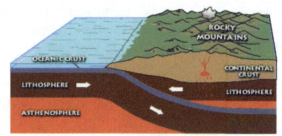

Figure 8.54: Shallow subduction during the Laramide Orogeny.

called the Laramide **Orogeny**, occurred in the early **Cenozoic**. The Laramide was **thick-skinned**, different than the Sevier **Orogeny**. It involved deeper crustal rocks, and produced bulges that would become mountain ranges like the Rockies, Black Hills, Wind River Range, Uinta Mountains, and the San Rafael Swell. Instead of descending directly into the **mantle**, the **subducting plate** shallowed out and moved eastward beneath the **continental plate** affecting the overlying **continent** hundreds of miles east of the **continental** margin and building high mountains. This occurred because the **subducting plate** was so young and near the **spreading center** and the density of the **plate** was therefore low and **subduction** was hindered.

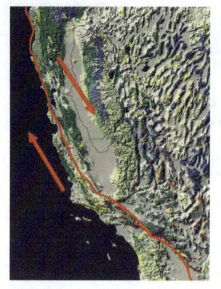

Figure 8.55: Map of the San Andreas fault, showing relative motion.

As the **mid-ocean ridge** itself started to subduct, the relative motion had changed. **Subduction** caused a relative convergence between the **subducting** Farallon **plate** and the North American **plate**. On the other side of the **mid-ocean ridge** from the Farallon **plate** was the Pacific **plate**, which was moving away from the North American **plate**. Thus, as the **subduction** zone consumed the **mid-ocean ridge**, the relative movement became **transform** instead of **convergent**, which went on to become the San Andreas Fault System. As the San Andreas grew, it caused east-west directed **extensional** forces to spread over the western United States, creating the **Basin and Range** province. The **transform fault** switched position over the last 18 million years, twisting the mountains around Los Angeles, and new **faults** in the southeastern California deserts may become a future San Andreas-style **fault**. During this switch from **subduction** to **transform**, the nearly horizontal Farallon **slab** began to sink into the **mantle**. This caused magmatism as the **subducting slab** sank, allowing **asthenosphere** material to rise around it. This event is called the Oligocene ignimbrite flare-up, which was one of the most significant **periods** of **volcanism** ever, including the largest single confirmed eruption, the 5000 cubic kilometer Fish Canyon Tuff.

8.6.2 Cenozoic Evolution

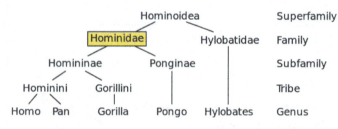

Figure 8.56: Family tree of Hominids (Hominadae).

There are five groups of early mammals in the **fossil** record, based primarily on **fossil** teeth, the hardest bone in **vertebrate** skeletons. For the purpose of this text, the most important group are the Eupantotheres, that diverge into the two main groups of mammals, the marsupials (like *Sinodelphys*) and placentals or eutherians (like *Eomaia*) in the **Cretaceous** and then diversified in the **Cenozoic**. The marsupials dominated on the isolated island continents of South America and Australia, and many went **extinct** in South America with the introduction of placental mammals. Some well-known mammal groups have been highly studied with interesting evolutionary stories in the **Cenozoic**. For example, horses started small with four toes, ended up larger and having just one toe. Cetaceans (**marine** mammals like whales and dolphins) started on land from small bear-like (mesonychids) creatures in the early **Cenozoic** and gradually took to water. However, no study of evolution has been more studied than human evolution. Hominids, the name for human-like primates, started in eastern Africa several million years ago.

The first critical event in this story is an environmental change from jungle to more of a savanna, probably caused by changes in Indian Ocean circulation. While bipedalism is known to have evolved before this shift, it is generally believed that our bipedal ancestors (like *Australopithecus*) had an advantage by covering ground more easily in a more open environment compared to their non-bipedal evolutionary cousins. There is also a growing body of evidence, including the famous "Lucy" **fossil** of an Australopithecine, that our early ancestors lived in trees. Arboreal animals usually demand a high intelligence to navigate through a three-dimensional world. It is from this lineage that humans evolved, using endurance running as a means to acquire more resources and possibly even hunt. This can explain many uniquely human features, from our long legs, strong achilles, lack of lower gut protection, and our wide range of running efficiencies.

Now that the hands are freed up, the next big step is a large brain. There have been arguments from a switch to more meat eating, cooking with fire, tool use, and even the construct of society itself to explain this increase in brain size. Regardless of how, it was this increased cognitive power that allowed humans to reign as their ancestors moved out of Africa and explored the world, ultimately entering the Americas through land bridges like the Bering Land Bridge. The details of this worldwide migration and the different branches of the hominid evolutionary tree are very complex, and best reserved for its own course.

Figure 8.57: Lucy skeleton, showing real fossil (brown) and reconstructed skeleton (white).

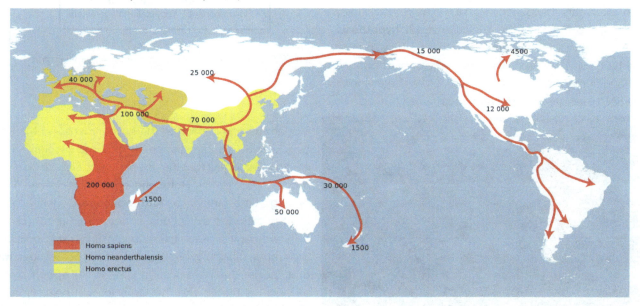

Figure 8.58: The hypothesized movement of the homo genus. Years are marked as to the best guess of the timing of movement.

Anthropocene and Extinction

Humans have had an influence on the Earth, its ecosystems and **climate**. Yet, human activity can not explain all of the changes that have occurred in the recent past. The start of the **Quaternary period**, the last and current **period** of the **Cenozoic**, is marked by the start of our current **ice age** 2.58 million years ago. During this time **period**, **ice sheets** advanced and retreated, most likely due to **Milankovitch cycles** (see chapter 15). Also at this time, various cold-adapted megafauna emerged (like giant sloths, saber-tooth cats, and woolly mammoths), and most of them went **extinct** as the Earth warmed from the most recent **glacial** maximum. A long-standing debate is over the cause of these and other extinctions. Is **climate** warming to blame, or were they caused by humans? Certainly, we know of recent human extinctions of animals like the dodo or passenger pigeon. Can we connect modern extinctions to extinctions in the recent past? If so, there are several ideas as to how this happened. Possibly the most widely accepted and oldest is the hunting/overkill **hypothesis**. The idea behind this **hypothesis** is that humans hunted large herbivores for food, then carnivores could not find food, and human arrival times in locations has been shown to be tied to increased **extinction** rates in many cases.

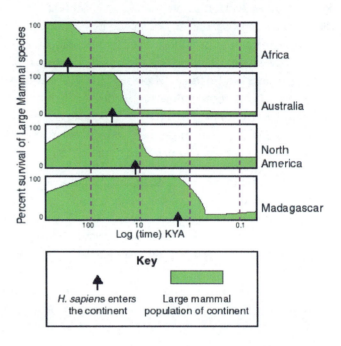

Figure 8.59: Graph showing abundance of large mammals and the introduction of humans.

Figure 8.60: Bingham Canyon Mine, Utah. This open pit mine is the largest man-made removal of rock in the world.

Modern human impact on the environment and the Earth as a whole is unquestioned. In fact, many scientists are starting to suggest that the rise of human civilization ended and/or replaced the **Holocene epoch** and defines a new geologic time interval: the **Anthropocene**. Evidence for this change includes extinctions, increased tritium (hydrogen with two neutrons) due to nuclear testing, rising pollutants like carbon dioxide, more than 200 never-before seen **mineral** species that have occurred only in this **epoch**, materials such as plastic and metals which will be long lasting "**fossils**" in the geologic record, and large amounts of earthen material moved. The biggest scientific debate with this topic is the starting point. Some say that humans' invention of agriculture would be recognized in geologic strata and that should be the starting point, around 12,000 years ago. Others link the start of the industrial revolution and the subsequent addition of vast amounts of carbon dioxide in the **atmosphere**. Either way, the idea is that alien geologists visiting Earth in the distant future would easily recognize the impact of humans on the Earth as the beginning of a new geologic **period**.

Take this quiz to check your comprehension of this section.

If you are using an offline version of this text, access the quiz for section 8.6 via the QR code.

An interactive H5P element has been excluded from this version of the text. You can view it online here:
https://pressbooks.lib.vt.edu/introearthscience/?p=602#h5p-55

Summary

The changes that have occurred since the inception of Earth are vast and significant. From the oxygenation of the **atmosphere**, the progression of life forms, the assembly and deconstruction of several **supercontinents**, to the **extinction** of more life forms than exist today, having a general understanding of these changes can put present change into a more rounded perspective.

Take this quiz to check your comprehension of this chapter.

If you are using an offline version of this text, access the quiz for chapter 8 via the QR code.

An interactive H5P element has been excluded from this version of the text. You can view it online here:
https://pressbooks.lib.vt.edu/introearthscience/?p=602#h5p-56

URLs Linked Within This Chapter

Charles Walcott (1850–1927): http://www.nasonline.org/member-directory/deceased-members/20000936.html

Text References

1. Alvarez, L.W., Alvarez, W., Asaro, F., and Michel, H.V., 1980, Extraterrestrial cause for the cretaceous-tertiary extinction: Science, v. 208, no. 4448, p. 1095–1108.
2. Beerling, D., 2008, The emerald planet: how plants changed Earth's history: OUP Oxford.
3. Boyce, J.W., Liu, Y., Rossman, G.R., Guan, Y., Eiler, J.M., Stolper, E.M., and Taylor, L.A., 2010, Lunar apatite with terrestrial volatile abundances: Nature, v. 466, no. 7305, p. 466–469.
4 Brueckner, H K., and Snyder, W.S., 1985, Structure of the Havallah sequence, Golconda allochthon, Nevada: Evidence for prolonged evolution in an accretionary prism: Geol. Soc. Am. Bull., v. 96, no. 9, p. 1113–1130.
5. Brusatte, S.L., Benton, M.J., Ruta, M., and Lloyd, G.T., 2008, The first 50 Myr of dinosaur evolution: macroevolutionary

pattern and morphological disparity: Biol. Lett., v. 4, no. 6, p. 733–736.

6. Canup, R.M., and Asphaug, E., 2001, Origin of the Moon in a giant impact near the end of the Earth's formation: Nature, v. 412, no. 6848, p. 708–712.

7. Clack, J.A., 2009, The Fish–Tetrapod Transition: New Fossils and Interpretations: Evolution: Education and Outreach, v. 2, no. 2, p. 213–223., doi: 10/cz257q.

8. Cohen, K.M., Finney, S.C., Gibbard, P.L., and Fan, J.-X., 2013, The ICS International Chronostratigraphic Chart: Episodes, v. 36, no. 3, p. 199–204.

9. Colbert, E.H., and Morales, M.A., 1991, History of the Backboned Animals Through Time: New York: Wiley.

10. De Laubenfels, M.W., 1956, Dinosaur extinction: one more hypothesis: J. Paleontol.

11. Gomes, R., Levison, H.F., Tsiganis, K., and Morbidelli, A., 2005, Origin of the cataclysmic Late Heavy Bombardment period of the terrestrial planets: Nature, v. 435, no. 7041, p. 466–469.

12. Hatcher, R.D., Jr, Thomas, W.A., and Viele, G.W., 1989, The Appalachian-Ouachita Orogen in the United States: Geological Society of America.

13. Hosono, N., Karato, S., Makino, J., and Saitoh, T.R., 2019, Terrestrial magme ocean origin of the Moon: Nature Geoscience, p. 1., doi: 10.1038/s41561-019-0354-2.

14. Hsiao, E., 2004, Possibility of life on Europa.

15. Hubble, E., 1929, A relation between distance and radial velocity among extra-galactic nebulae: Proc. Natl. Acad. Sci. U. S. A., v. 15, no. 3, p. 168–173.

16. Ingersoll, R.V., 1982, Triple-junction instability as cause for late Cenozoic extension and fragmentation of the western United States: Geology, v. 10, no. 12, p. 621–624.

17. Johnson, C.M., 1991, Large-scale crust formation and lithosphere modification beneath Middle to Late Cenozoic calderas and volcanic fields, western North America: J. Geophys. Res. [Solid Earth], v. 96, no. B8, p. 13485–13507.

18. Kass, M.S., 1999, Prognathodon stadtmani:(Mosasauridae) a new species from the Mancos Shale (lower Campanian) of western Colorado: Vertebrate Paleontology in Utah, Utah Geological.

19. Livaccari, R.F., 1991, Role of crustal thickening and extensional collapse in the tectonic evolution of the Sevier-Laramide orogeny, western United States: Geology, v. 19, no. 11, p. 1104–1107.

20. McMenamin, M.A., and Schulte McMenamin, D.L., 1990, The Emergence of Animals: The Cambrian Breakthrough: Columbia University Press.

21. Mitrovica, J.X., Beaumont, C., and Jarvis, G.T., 1989, Tilting of continental interiors by the dynamical effects of subduction: Tectonics.

22. Rücklin, M., Donoghue, P.C.J., Johanson, Z., Trinajstic, K., Marone, F., and Stampanoni, M., 2012, Development of teeth and jaws in the earliest jawed vertebrates: Nature, v. 491, no. 7426, p. 748–751.

23. Sahney, S., and Benton, M.J., 2008, Recovery from the most profound mass extinction of all time: Proc. Biol. Sci., v. 275, no. 1636, p. 759–765.

24. Salaris, M., and Cassisi, S., 2005, Evolution of stars and stellar populations: John Wiley & Sons.

25. Schoch, R.R., 2012, Amphibian Evolution: The life of Early Land Vertebrates: Wiley-Blackwell.

26. Sharp, B.J., 1958, MINERALIZATION IN THE INTRUSIVE ROCKS IN LITTLE COTTONWOOD CANYON, UTAH: GSA Bulletin, v. 69, no. 11, p. 1415–1430., doi: 10.1130/0016-7606(1958)69[1415:MITIRI]2.0.CO;2.

27. Wiechert, U., Halliday, A.N., Lee, D.C., Snyder, G.A., Taylor, L.A., and Rumble, D., 2001, Oxygen isotopes and the moon-forming giant impact: Science, v. 294, no. 5541, p. 345–348.

28. Wilde, S.A., Valley, J.W., Peck, W.H., and Graham, C.M., 2001, Evidence from detrital zircons for the existence of continental crust and oceans on the Earth 4.4 Gyr ago: Nature, v. 409, no. 6817, p. 175–178.

29. Wood, R.A., 2019, The rise of Animals.: Scientific American, v. 320, no. 6, p. 24–31.

Figure References

Figure 8.1: Geologic time on Earth, represented circularly, to show the individual time divisions and important events. Woudloper; adapted by Hardwigg. 2010. Public domain. https://commons.wikimedia.org/wiki/File:Geologic_Clock_with_events_and_periods.svg

Figure 8.2: Geological time scale with ages shown. USGS. 2009. Public domain. https://commons.wikimedia.org/wiki/File:Geologic_time_scale.jpg

Figure 8.3: Artist's impression of the Earth in the Hadean. Tim Bertelink. 2016. CC BY-SA 4.0. https://commons.wikimedia.org/wiki/File:Hadean.png

Figure 8.4: The global map of the depth of the moho. AllenMcC. 2013. CC BY-SA 3.0. https://commons.wikimedia.org/wiki/File:Mohomap.png

Figure 8.5: Dark side of the Moon. Apollo 16 astronauts via NASA. 1972. Public domain. https://en.wikipedia.org/wiki/File:Back_side_of_the_Moon_AS16-3021.jpg

Figure 8.6: Artist's concept of the giant impact from a Mars-sized object that could have formed the moon. NASA/JPL-Caltech. 2017. Public domain. https://www.nasa.gov/multimedia/imagegallery/image_feature_1454.html

Figure 8.7: Water vapor leaves comet 67P/Churyumov–Gerasimenko. ESA/Rosetta/NAVCAM. 2015. CC BY-SA 3.0 IGO. https://commons.wikimedia.org/wiki/File:Comet_on_7_July_2015_NavCam.jpg

Figure 8.8: Artist's impression of the Archean. Tim Bertelink. 2017. CC BY-SA 4.0. https://commons.wikimedia.org/wiki/File:Archean.png

Figure 8.9: 2015 image from NASA's New Horizons probe of Pluto. NASA / Johns Hopkins University Applied Physics Laboratory / Southwest Research Institute. 2015. Public domain. https://commons.wikimedia.org/wiki/File:Nh-pluto-in-true-color_2x_JPEG-edit-frame.jpg

Figure 8.10: Simulation of before, during, and after the late heavy bombardment. Kesäperuna. 2019. CC BY-SA 3.0. https://commons.wikimedia.org/wiki/File:Lhborbits.png

Figure 8.11: The layers of the Earth. Drlauraguertin. 2015. CC BY-SA 3.0. https://wiki.seg.org/wiki/File:Earthlayers.png

Figure 8.12: Subduction of an oceanic plate beneath another oceanic plate, forming a trench and an island arc. USGS. 1999. Public domain. https://commons.wikimedia.org/wiki/File:Oceanic-continental_convergence_Fig21oceancont.gif

Figure 8.13: Geologic provinces with the Shield (orange) and Platform (pink) comprising the Craton, the stable interior of continents. USGS. 2005. Public domain. https://commons.wikimedia.org/wiki/File:World_geologic_provinces.jpg

Figure 8.14: The continent of Zealandia. NOAA. 2006. Public domain. https://commons.wikimedia.org/wiki/File:Zealandia_topography.jpg

Figure 8.15: Fossils of microbial mats from Sweden. Smith609. 2008. CC BY-SA 3.0. https://commons.wikimedia.org/wiki/File:Runzelmarken.jpg

Figure 8.16: Greenhouse gases were more common in Earth's early atmosphere. Kindred Grey. 2022. CC BY 4.0. Water molecule 3D by Dbc334, 2006 (Public domain, https://commons.wikimedia.org/wiki/File:Water_molecule_3D.svg). Nitrous-oxide-dimensions-3D-balls by Ben Mills, 2007 (Public domain, https://commons.wikimedia.org/wiki/File:Nitrous-oxide-dimensions-3D-balls.png). Methane-CRC-MW-3D-balls by Ben Mills, 2009 (Public domain, https://en.m.wikipedia.org/wiki/File:Methane-CRC-MW-3D-balls.png). Carbon dioxide 3D ball by Jynto, 2011 (Public domain, https://commons.wikimedia.org/wiki/File:Carbon_dioxide_3D_ball.png).

Figure 8.17: Diagram showing the main products and reactants in photosynthesis. At09kg. 2011. CC BY-SA 3.0. https://commons.wikimedia.org/wiki/File:Photosynthesis.gif

Figure 8.18: Alternating bands of iron rich and silica rich mud, formed as oxygen combined with dissolved iron. Wilson44691. 2008. Public domain. https://commons.wikimedia.org/wiki/File:MichiganBIF.jpg

Figure 8.19: One possible reconstruction of Rodinia 1.1 billion years ago. John Goodge. 2011. Public domain. https://commons.wikimedia.org/wiki/File:Rodinia_reconstruction.jpg

Figure 8.20: Modern cyanobacteria (as stromatolites) in Shark Bay, Australia. Paul Harrison. 2005. CC BY-SA 3.0. https://en.wikipedia.org/wiki/File:Stromatolites_in_Sharkbay.jpg

Figure 8.21: Fossil stromatolites in Saratoga Springs, New York. Rygel, M.C. 2005. CC BY-SA 3.0. https://commons.wikimedia.org/wiki/File:Stromatolites_hoyt_mcr1.JPG

Figure 8.22: Dickinsonia, a typical Ediacaran fossil. Verisimilus. 2007. CC BY-SA 3.0. https://commons.wikimedia.org/wiki/File:DickinsoniaCostata.jpg

Figure 8.23: The trilobites had a hard exoskeleton, and is an early arthropod, the same group that includes modern insects, crustaceans, and arachnids. Wilson44691. 2010. Public domain. https://commons.wikimedia.org/wiki/File:ElrathiakingiUtahWheelerCambrian.jpg

Figure 8.24: Trilobites, by Heinrich Harder, 1916. Heinrich Harder. 1916. Public domain. https://commons.wikimedia.org/wiki/File:Trilobite_Heinrich_Harder.jpg

Figure 8.25: Laurentia, which makes up the North American craton. USGS. 2005. Public domain. https://commons.wikimedia.org/wiki/File:North_america_craton_nps.gif

Figure 8.26: A reconstruction of Pangaea, showing approximate positions of modern continents. Kieff. 2009. CC BY-SA 3.0. https://commons.wikimedia.org/wiki/File:Pangaea_continents.svg

Figure 8.27: Anomalocaris reconstruction by the MUSE science museum in Italy. Matteo De Stefano/MUSE. 2016. CC BY-SA 3.0. https://commons.wikimedia.org/wiki/File:Anomalocaris_canadensis_-_reconstruction_-_MUSE.jpg

Figure 8.28: Original plate from Walcott's 1912 description of Opabinia, with labels: fp = frontal appendage, e = eye, ths = thoracic somites, i = intestine, ab = abdominal segment. Charles Doolittle Walcott. 1912. Public domain. https://commons.wikimedia.org/wiki/File:Opabinia_regalis_-_Walcott_Cambrian_Geology_and_Paleontology_II_plate_28_.jpg

Figure 8.29: A modern coral reef. Toby Hudson. 2010. CC BY-SA 3.0. https://commons.wikimedia.org/wiki/File:Coral_Outcrop_Flynn_Reef.jpg

Figure 8.30: Guadalupe National Park is made of a giant fossil reef. Zereshk. 2007. CC BY-SA 3.0. https://commons.wikimedia.org/wiki/File:Guadalupe_Nima2.JPG

Figure 8.31: The placoderm Bothriolepis panderi from the Devonian of Russia. Haplochromis. 2007. CC BY-SA 3.0. https://commons.wikimedia.org/wiki/File:Bothriolepis_panderi.jpg

Figure 8.32: Several different types of fish and amphibians that led to walking on land. Dave Souza; adapted by Pixelsquid. 2020. CC BY-SA 3.0. https://commons.wikimedia.org/wiki/File:Fishapods.svg

Figure 8.33: A reconstruction of the giant arthropod (insects and their relatives) Arthropleura. Tim Bertelink. 2016. CC BY-SA 4.0. https://commons.wikimedia.org/wiki/File:Arthropleura.png

Figure 8.34: Reconstruction of Dimetrodon. Max Bellomio. 2019. CC BY-SA 4.0. https://commons.wikimedia.org/wiki/File:Dimetrodon_grandis_3D_Model_Reconstruction.png

Figure 8.35: World map of flood basalts. Williamborg. 2011. CC BY-SA 3.0. https://commons.wikimedia.org/wiki/File:Flood_Basalt_Map.jpg

Figure 8.36: Perhaps the greatest fossil ever found, a velociraptor attacked a protoceratops, and both were fossilized mid sequence. Yuya Tamai. 2014. CC BY 2.0. https://commons.wikimedia.org/wiki/File:Fighting_dinosaurs_(1).jpg

Figure 8.37: Animation showing Pangea breaking up. USGS. 2005. Public domain. https://commons.wikimedia.org/wiki/File:Pangea_animation_03.gif

Figure 8.38: Age of oceanic lithosphere, in millions of years. Muller, R.D., M. Sdrolias, C. Gaina, and W.R. Roest (2008) Age, spreading rates and spreading symmetry of the world's ocean crust, Geochem. Geophys. Geosyst., 9, Q04006, doi:10.1029/2007GC001743. CC BY-SA 3.0. https://commons.wikimedia.org/wiki/File:Age_of_oceanic_lithosphere.jpg

Figure 8.39: Sketch of the major features of the Sevier Orogeny. Pinkcorundum. 2011. Public domain. https://www.wikiwand.com/en/Sevier_orogeny#Media/File:Sevierorogeny.jpg

Figure 8.40: The Cretaceous Interior Seaway in the mid-Cretaceous. By William A. Cobban and Kevin C. McKinney, USGS. 2004. Public domain. https://commons.wikimedia.org/wiki/File:Cretaceous_seaway.png

Figure 8.41: A Mesozoic scene from the late Jurassic. Gerhard Boeggemann. 2006. CC BY-SA 2.5. https://commons.wikimedia.org/wiki/File:Europasaurus_holgeri_Scene_2.jpg

Figure 8.42: A drawing of the early plesiosaur Agustasaurus from the Triassic of Nevada. Nobu Tamura. 2008. CC BY 3.0. https://commons.wikimedia.org/wiki/File:Augustasaurus_BW.jpg

Figure 8.43: Reconstruction of the small (<5") Megazostrodon, one of the first animals considered to be a true mammal. Theklan. 2017. CC BY-SA 4.0. https://commons.wikimedia.org/wiki/File:Megazostrodon_sp._Natural_History_Museum_-_London.jpg

Figure 8.44: Closed structure of a ornithischian hip, which is similar to a birds. Fred the Oyster. 2014. CC BY-SA 4.0. https://commons.wikimedia.org/wiki/File:Ornithischia_pelvis_structure.svg

Figure 8.45: Open structure of a saurischian hip, which is similar to a lizards. Fred the Oyster. 2014. CC BY-SA 4.0. https://commons.wikimedia.org/wiki/File:Saurischia_pelvis_structure.svg

Figure 8.46: Therizinosaurs, like Beipiaosaurus (shown in this restoration), are known for their enormous hand claws. Matt Martyniuk. 2009. CC BY-SA 3.0. https://commons.wikimedia.org/wiki/File:Beipiao1mmartyniuk.png

Figure 8.47: Archaeopteryx lithographica, specimen displayed at the Museum für Naturkunde in Berlin. H. Raab. 2009. CC BY-SA 3.0. https://commons.wikimedia.org/wiki/File:Archaeopteryx_lithographica_(Berlin_specimen).jpg

Figure 8.48: Reconstructed skeleton of Argentinosaurus, from Naturmuseum Senckenberg in Germany. Fva K. 2010. CC BY-NC-ND 3.0. https://commons.wikimedia.org/wiki/File:Argentinosaurus_DSC_2943.jpg

Figure 8.49: Graph of the rate of extinctions. Smith609. 2008. CC BY-SA 3.0. https://commons.wikimedia.org/wiki/File:Extinction_intensity.svg

Figure 8.50: Artist's depiction of an impact event. Made by Fredrik. Cloud texture from public domain NASA image. 2004. Public domain. https://commons.wikimedia.org/wiki/File:Impact_event.jpg

Figure 8.51: The land expression of the Chicxulub crater. NASA/JPL-Caltech. 2000. Public domain. https://commons.wikimedia.org/wiki/File:Yucatan_chix_crater.jpg

Figure 8.52: Geology of India, showing purple as Deccan Traps-related rocks. CamArchGrad. 2007. Public domain. https://en.wikipedia.org/wiki/File:India_Geology_Zones.jpg

Figure 8.53: Paraceratherium, seen in this reconstruction, was a massive (15-20 ton, 15 foot tall) ancestor of rhinos. Tim Bertelink. 2016. CC BY-SA 4.0. https://commons.wikimedia.org/wiki/File:Indricotherium.png

Figure 8.54: Shallow subduction during the Laramide Orogeny. Melanie Moreno, USGS. 2006. Public domain. https://commons.wikimedia.org/wiki/File:Shallow_subduction_Laramide_orogeny.png

Figure 8.55: Map of the San Andreas fault, showing relative motion. Kate Barton, David Howell, and Joe Vigil via USGS. 2006. Public domain. https://commons.wikimedia.org/wiki/File:Sanandreas.jpg

Figure 8.56: Family tree of Hominids (Hominadae). Fred the Oyster. 2014. CC BY-SA 4.0. https://commons.wikimedia.org/wiki/File:Hominidae_chart.svg

Figure 8.57: Lucy skeleton, showing real fossil (brown) and reconstructed skeleton (white). Andrew. 2007. CC BY-SA 2.0. https://commons.wikimedia.org/wiki/File:Lucy_Skeleton.jpg

Figure 8.58: The hypothesized movement of the homo genus. NordNordWest. 2014. Public domain. https://commons.wikimedia.org/wiki/File:Spreading_homo_sapiens_la.svg

Figure 8.59: Graph showing abundance of large mammals and the introduction of humans. ElinWhitneySmith. 2006. Public domain. https://commons.wikimedia.org/wiki/File:Extinctions_Africa_Austrailia_NAmerica_Madagascar.gif

Figure 8.60: Bingham Canyon Mine, Utah. Doc Searls. 2016. CC BY 2.0. https://commons.wikimedia.org/wiki/File:Bingham_Canyon_mine_2016.jpg

9

CRUSTAL DEFORMATION AND EARTHQUAKES

By the end of this chapter, students should be able to:

- Differentiate between **stress** and **strain**.
- Identify the three major types of **stress**.
- Differentiate between **brittle**, **ductile**, and **elastic deformation**.
- Describe the geological map symbol used for **strike** and **dip** of **strata**.
- Name and describe different **fold** types.
- Differentiate the three major **fault** types and describe their associated movements.
- Explain how **elastic rebound** relates to earthquakes.
- Describe different **seismic wave** types and how they are measured.
- Explain how humans can induce seismicity.
- Describe how **seismographs** work to record earthquake waves.
- From **seismograph** records, locate the epicenter of an earthquake.
- Explain the difference between earthquake **magnitude** and intensity.
- List earthquake factors that determine ground shaking and destruction.
- Identify secondary earthquake hazards.
- Describe notable historical earthquakes.

Crustal **deformation** occurs when applied forces exceed the internal strength of rocks, physically changing their shapes. Those forces are called **stress**, and the physical changes they create are called **strain**. Forces involved in **tectonic** processes as well as gravity and **igneous pluton** emplacement produce **strains** in rocks that include **folds**, **fractures**, and **faults**. When rock experiences large amounts of **shear stress** and breaks with rapid, **brittle deformation**, energy is released in the form of **seismic** waves, commonly known as an earthquake.

9.1 Stress and Strain

Stress is the force exerted per unit area and **strain** is the physical change that results in response to that force. When applied **stress** is greater than the internal strength of rock, **strain** results in the form of **deformation** of the rock caused by the **stress**. **Strain** in rocks can be represented as a change in rock volume and/or rock shape, as well as fracturing the rock. There are three types of **stress**: **tensional**, **compressional**, and **shear**. **Tensional stress** involves forces pulling in opposite directions, which results in **strain** that stretches and thins rock. **Compressional stress** involves forces pushing together, and **compressional strain** shows up as rock folding and thickening. **Shear stress** involves transverse forces; the **strain** shows up as opposing blocks or regions of material moving past each other.

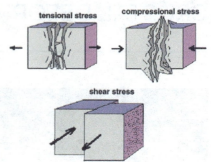

Figure 9.1: Types of stress. Clockwise from top left: tensional stress, compressional stress, and shear stress, and some examples of resulting strain.

Type of stress	Associated plate boundary type (see chapter 2)	Resulting strain	Associated fault and offset types
Tensional	Divergent	Stretching and thinning	Normal
Compressional	Convergent	Shortening and thickening	Reverse
Shear	Transform	Tearing	Strike-slip

Table 9.1: Types of stress and resulting strain.

Take this quiz to check your comprehension of this section.

If you are using an offline version of this text, access the quiz for section 9.1 via the QR code.

 An interactive H5P element has been excluded from this version of the text. You can view it online here:
https://pressbooks.lib.vt.edu/introearthscience/?p=669#h5p-57

9.2 Deformation

When rocks are **stressed**, the resulting **strain** can be elastic, **ductile**, or **brittle**. This change is generally called **deformation**. **Elastic deformation** is **strain** that is reversible after a **stress** is released. For example, when you stretch a rubber **band**, it elastically returns to its original shape after you release it. **Ductile deformation** occurs when enough **stress** is applied to a material that the changes in its shape are permanent, and the material is no longer able to revert to its original shape. For example, if you bend a metal bar too far, it can be permanently bent out of shape. The point at which **elastic deformation** is surpassed and **strain** becomes permanent is called the **yield point**. In the figure, **yield point** is where the line transitions from **elastic deformation** to **ductile deformation** (the end of the dashed line). **Brittle deformation** is another critical point of no return, when rock integrity fails and the rock **fractures** under increasing **stress**.

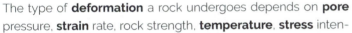

Figure 9.2: Different materials deform differently when stress is applied. Material "A" has relatively little deformation when undergoing large amounts of stress, before undergoing plastic deformation, and finally brittlely failing. Material "B" only elastically deforms before brittlely failing. Material "C" undergoes significant plastic deformation before finally failing brittlely.

The type of **deformation** a rock undergoes depends on **pore** pressure, **strain** rate, rock strength, **temperature**, **stress** intensity, time, and **confining** pressure. **Pore** pressure is exerted on the rock by fluids in the open spaces or **pores** embedded within rock or **sediment**. **Strain** rate measures how quickly a material is deformed. For example, applying **stress** slowly makes it is easier to bend a piece of wood without breaking it. Rock strength measures how easily a rock deforms under **stress**. **Shale** has low strength and **granite** has high strength. Removing heat, or decreasing the **temperature**, makes materials more rigid and susceptible to **brittle deformation**. On the other hand, heating materials make them more **ductile** and less **brittle**. Heated glass can be bent and stretched.

Factor	Strain response
Increase temperature	More ductile
Increase strain rate	More brittle
Increase rock strength	More brittle

Table 9.2: Relationship between factors operating on rock and the resulting strains.

Take this quiz to check your comprehension of this section.

If you are using an offline version of this text, access the quiz for section 9.2 via the QR code.

 An interactive H5P element has been excluded from this version of the text. You can view it online here:
https://pressbooks.lib.vt.edu/introearthscience/?p=669#h5p-58

9.3 Geological Maps

Geologic maps are two dimensional (2D) representations of geologic **formations** and structures at the Earth's surface, including **formations**, **faults**, **folds**, inclined **strata**, and rock types. **Formations** are recognizable rock units. Geologists use geologic maps to represent where geologic **formations**, **faults**, **folds**, and inclined rock units are. Geologic **formations** are recognizable, mappable rock units. Each **formation** on the map is indicated by a color and a label. For examples of geologic maps, see the Utah Geological Survey (UGS) geologic map viewer.

Formation labels include symbols that follow a specific protocol. The first one or more letters are uppercase and represent the geologic time **period** of the **formation**. More than one uppercase letter indicates the **formation** is associated with multiple time **periods**. The following lowercase letters represent the **formation** name, abbreviated rock description, or both.

9.3.1 Cross Sections

Cross sections are subsurface interpretations made from surface and subsurface measurements. Maps display geology in the horizontal plane, while cross sections show subsurface geology in the vertical plane. For more information on cross sections, check out the AAPG wiki.

9.3.2 Strike and Dip

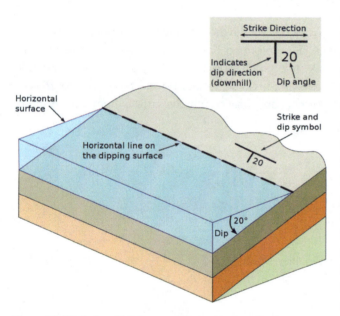

Figure 9.4: Attitude symbol on geologic map (with compass directions for reference) showing strike of N30°E and dip of 45° to the SE.

Geologists use a special symbol called **strike** and **dip** to represent inclined **beds**. **Strike** and **dip** map symbols look like the capital letter *T*, with a short trunk and extra-wide top line. The short trunk represents the **dip** and the top line represents the **strike**. **Dip** is the angle that a **bed** plunges into the Earth from the horizontal. A number next to the symbol represents **dip** angle. One way to visualize the **strike** is to think about a line made by standing water on the inclined layer. That line is horizontal and lies on a compass direction that has some angle with respect to true north or south (see figure 9.3). The **strike** angle is that angle measured by a special compass. E.g., N 30° E (read north 30 degrees east) means the horizontal line points northeast at an angle of

Figure 9.3: "Strike" and "dip" are words used to describe the orientation of rock layers with respect to North/South and horizontal.

30° from true north. The **strike** and **dip** symbol is drawn on the map at the **strike** angle with respect to true north on the map. The **dip** of the inclined layer represents the angle down to the layer from horizontal, in the figure 45° SE (read dipping 45 degrees to the SE). The direction of **dip** would be the direction a ball would roll if set on the layer and released. A horizontal rock **bed** has a **dip** of 0° and a vertical **bed** has a **dip** of 90°. **Strike** and **dip** considered together are called rock attitude.

This video illustrates geologic structures and associated map symbols.

One or more interactive elements has been excluded from this version of the text. You can view them online here:
https://www.youtube.com/watch?v=UzZFMWH-lSQ

Video 9.1: Folds, dip, and strike.

If you are using an offline version of this text, access this YouTube video via the QR code.

Take this quiz to check your comprehension of this section.

If you are using an offline version of this text, access the quiz for section 9.3 via the QR code.

 An interactive H5P element has been excluded from this version of the text. You can view it online here:
https://pressbooks.lib.vt.edu/introearthscience/?p=669#h5p-59

9.4 Folds

Geologic **folds** are layers of rock that are curved or bent by **ductile deformation**. **Folds** are most commonly formed by **compressional** forces at depth, where hotter temperatures and higher **confining** pressures allow **ductile deformation** to occur.

Folds are described by the orientation of their axes, **axial planes**, and limbs. The plane that splits the **fold** into two halves is known as the **axial plane**. The **fold axis** is the line along which the bending occurs and is where the **axial plane** intersects the folded **strata**. The **hinge line** follows the line of greatest bend in a **fold**. The two sides of the **fold** are the **fold** limbs.

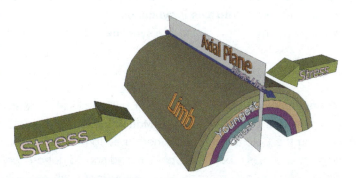

Figure 9.5: Model of anticline. Oldest beds are in the center and youngest on the outside. The axial plane intersects the center angle of bend. The hinge line follows the line of greatest bend, where the axial plane intersects the outside of the fold.

Symmetrical **folds** have a vertical **axial plane** and limbs have equal but opposite dips. Asymmetrical **folds** have dipping, non-vertical **axial planes**, where the limbs **dip** at different angles. Overturned **folds** have steeply dipping **axial planes** and the limbs **dip** in the same direction but usually at different **dip** angles. Recumbent **folds** have horizontal or nearly horizontal **axial planes**. When the **axis** of the **fold** plunges into the ground, the **fold** is called a plunging **fold**. **Folds** are classified into five categories: **anticline**, **syncline**, **monocline**, **dome**, and **basin**.

9.4.1 Anticline

Anticlines are arch-like, or A-shaped, **folds** that are convex-upward in shape. They have downward curving limbs and **beds** that **dip** down and away from the central **fold axis**. In **anticlines**, the oldest rock **strata** are in the center of the **fold**, along the **axis**, and the younger **beds** are on the outside. Since geologic maps show the intersection of surface topography with underlying geologic structures, an **anticline** on a geologic map can be identified by both the attitude of the **strata** forming the **fold** and the older age of the rocks inside the **fold**. An antiform has the same shape as an **anticline**, but the relative ages of the **beds** in the **fold** cannot be determined. **Oil** geologists are interested in **anticlines** because they can form **oil traps**, where **oil** migrates up along the limbs of the **fold** and accumulates in the high point along the **fold axis**.

Figure 9.6: An Anticline near Bcharre, Lebanon.

9.4.2 Syncline

https://sketchfab.com/models/3f0259ea2c6b4807a32fe3c950d13324/embed

3D Model 9.1: Synclinal fold.

If you are using an offline version of this text, access this interactive 3D model via the QR code, or visit https://sketchfab.com/3d-models/synclinal-fold-macigno-formation-3f0259ea2c6b4807a32fe3c950d13324.

Synclines are **trough**-like, or U shaped, **folds** that are concave-upward in shape. They have **beds** that **dip** down and in toward the central **fold axis**. In **synclines**, older rock is on the outside of the **fold** and the youngest rock is inside of the **fold axis**. A synform has the shape of a **syncline** but like an antiform, does not have distinguishable age zones.

9.4.3 Monocline

Monoclines are step-like **folds**, in which flat rocks are upwarped or downwarped, then continue flat. **Monoclines** are relatively common on the Colorado Plateau where they form "**reefs**," which are ridges that act as topographic barriers and should not be confused with ocean **reefs** (see chapter 5). Capitol **Reef** is an example of a **monocline** in Utah. **Monoclines** can be caused by bending of shallower sedimentary **strata** as **faults** grow below them. These **faults** are commonly called "blind **faults**" because they end before reaching the surface and can be either normal or **reverse faults**.

Figure 9.7: Monocline at Colorado National Monument.

9.4.4 Dome

A **dome** is a symmetrical to semi-symmetrical upwarping of rock **beds**. **Domes** have a shape like an inverted bowl, similar to an architectural **dome** on a building. Examples of **domes** in Utah include the San Rafael Swell, Harrisburg Junction **Dome**, and Henry Mountains. **Domes** are formed from compressional forces, underlying **igneous** intrusions (see chapter 4), by salt diapirs, or even impacts, like upheaval **dome** in Canyonlands National Park.

9.4.5 Basin

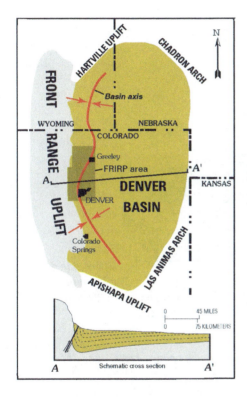

Figure 9.9: The Denver Basin is an active sedimentary basin at the eastern extent of the Rocky Mountains. As sediment accumulates, the basin subsides, creating a basin-shape of beds that are all dipping towards the center of the basin.

A **basin** is the inverse of a **dome**, a bowl-shaped depression in a rock **bed**. The Uinta Basin in Utah is an example of a **basin**. Some structural basins are also **sedimentary basins** that collect large quantities of sediment over time. Sedimentary basins can form as a result of folding but are much more commonly produced in mountain building, forming between mountain blocks or via **faulting**. Regardless of the cause, as the **basin** sinks or subsides, it can accumulate more **sediment** because the weight of the **sediment** causes more **subsidence** in a positive-feedback loop. There are active **sedimentary basins** all over the world. An example of a rapidly subsiding **basin** in Utah is the Oquirrh Basin, dated to the Pennsylvanian-**Permian** age, which has accumulated over 9,144 m (30,000 ft) of **fossiliferous sandstones**, **shales**, and **limestones**. These **strata** can be seen in the Wasatch Mountains along the east side of Utah Valley, especially on Mt. Timpanogos and in Provo Canyon.

Figure 9.8: This prominent circular feature in the Sahara desert of Mauritania has attracted attention since the earliest space missions because it forms a conspicuous bull's-eye in the otherwise rather featureless expanse of the desert. Initially interpreted as a meteorite impact structure because of its high degree of circularity, it is now thought to be merely a symmetrical uplift (circular anticline) that has been laid bare by erosion.

Take this quiz to check your comprehension of this section.

If you are using an offline version of this text, access the quiz for section 9.4 via the QR code.

 An interactive H5P element has been excluded from this version of the text. You can view it online here:
https://pressbooks.lib.vt.edu/introearthscience/?p=669#h5p-60

9.5 Faults

Faults are the places in the **crust** where **brittle deformation** occurs as two blocks of rocks move relative to one another. Normal and **reverse faults** display vertical, also known as **dip-slip**, motion. **Dip-slip** motion consists of relative up-and-down movement along a dipping **fault** between two blocks, the **hanging wall** and **footwall**. In a **dip-slip system**, the **footwall** is below the **fault** plane and the **hanging wall** is above the **fault** plane. A good way to remember this is to imagine a **mine** tunnel running along a **fault**; the **hanging wall** would be where a miner would hang a lantern and the **footwall** would be at the miner's feet.

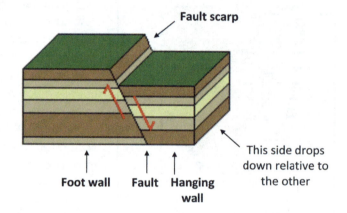

Figure 9.10: Common terms used for normal faults. Normal faults form when the hanging wall move down relative to the footwall.

Faulting as a term refers to rupture of rocks. Such ruptures occur at **plate** boundaries but can also occur in **plate** interiors as well. **Faults** slip along the **fault** plane. The **fault scarp** is the **offset** of the surface produced where the **fault** breaks through the surface. **Slickensides** are polished, often grooved surfaces along the **fault** plane created by friction during the movement.

A **joint** or **fracture** is a plane of **brittle deformation** in rock created by movement that is not **offset** or **sheared**. **Joints** can result from many processes, such as cooling, depressurizing, or folding. **Joint** systems may be regional affecting many square miles.

9.5.1 Normal Faults

Normal **faults** move by a vertical motion where the hanging-wall moves downward relative to the **footwall** along the **dip** of the **fault**. Normal **faults** are created by **tensional** forces in the crust. Normal faults and tensional forces commonly occur at **divergent plate** boundaries, where the **crust** is being stretched by **tensional stresses** (see chapter 2). Examples of normal **faults** in Utah are the Wasatch Fault, the Hurricane Fault, and other **faults** bounding the valleys in the **Basin and Range** province.

Figure 9.11: Example of a normal fault in an outcrop of the Pennsylvanian Honaker Trail Formation near Moab, Utah.

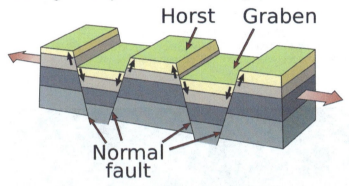

Figure 9.12: Faulting that occurs in the crust under tensional stress.

Grabens, **horsts**, and **half-grabens** are blocks of **crust** or rock bounded by normal **faults** (see chapter 2). **Grabens** drop down relative to adjacent blocks and create valleys. **Horsts** rise up relative to adjacent down-dropped blocks and become areas of higher topography. Where occurring together, **horsts** and **grabens** create a symmetrical pattern of valleys surrounded by normal **faults** on both sides and mountains. **Half-grabens** are a one-sided version of a **horst** and **graben**, where blocks are tilted by a **normal fault** on one side, creating an asymmetrical valley-mountain arrangement. The mountain-valleys of the **Basin and Range** Province of Western Utah and Nevada consist of a series of full and **half-grabens** from the Salt Lake Valley to the Sierra Nevada Mountains.

Normal **faults** do not continue clear into the **mantle**. In the **Basin and Range** Province, the **dip** of a **normal fault** tends to decrease with depth, i.e., the **fault** angle becomes shallower and more horizontal as it goes deeper. Such decreasing dips happen when large amounts of **extension** occur along very low-angle normal **faults**, known as detachment **faults**. The normal **faults** of the **Basin and Range**, produced by **tension** in the **crust**, appear to become detachment **faults** at greater depths.

9.5.2 Reverse Faults

In **reverse faults**, **compressional** forces cause the **hanging wall** to move up relative to the **footwall**. A **thrust fault** is a **reverse fault** where the **fault** plane has a low **dip** angle of less than 45°. Thrust **faults** carry older rocks on top of younger rocks and can even cause repetition of rock units in the **stratigraphic** record.

Convergent plate boundaries with **subduction** zones create a special type of "reverse" **fault** called a **megathrust fault** where denser **oceanic crust** drives down beneath less dense overlying **crust**. **Megathrust faults** cause the largest **magnitude** earthquakes yet measured and commonly cause **massive** destruction and **tsunamis**.

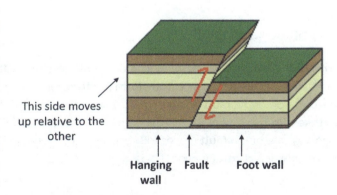

This side moves up relative to the other

Hanging wall   Fault   Foot wall

Figure 9.13: Simplified block diagram of a reverse fault.

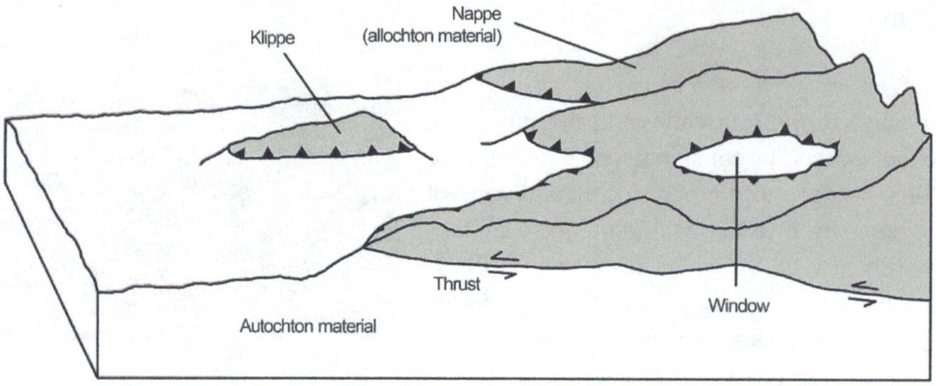

Figure 9.14: Terminology of thrust faults (low-angle reverse faults). A klippe is the remnant of the hangingwall (aka nappe), where the surrounding material has been eroded away. A window is where part of the hangingwall has been eroded away to expose the footwall (autochton). Note the symbol shows flags on the overlying thrust plate.

Figure 9.15: Thrust fault in the North Qilian Mountains (Qilian Shan). The blueish rock is a thick fault gouge of basement, the redish stuff is above the fault plane. Everything thrust over the brown quaternary conglomerates (right part of the picture). The fault plane dips 65 degrees to the South.

9.5.3 Strike–Slip Faults

Strike-slip faults have side-to-side motion. **Strike-slip faults** are most commonly associated with **transform plate** boundaries and are prevalent in **transform fracture** zones along **mid-ocean ridges**. In pure **strike-slip** motion, **fault** blocks on either side of the **fault** do not move up or down relative to each other, rather move laterally, side to side. The direction of **strike-slip** movement is determined by an observer standing on a block on one side of the **fault**. If the block on the opposing side of the **fault** moves left relative to the observer's block, this is called **sinistral** motion. If the opposing block moves right, it is **dextral** motion.

One or more interactive elements has been excluded from this version of the text. You can view them online here:
https://pressbooks.lib.vt.edu/introearthscience/?p=669#video-669-1

Video 9.2: Video showing motion in a strike-slip fault.

If you are using an offline version of this text, access this YouTube video via the QR code.

Bends along **strike-slip faults** create areas of **compression** or **tension** between the sliding blocks (see chapter 2). **Tensional stresses** create transtensional features with normal **faults** and basins, such as the Salton Sea in California. **Compressional stresses** create **transpressional** features with **reverse faults** and cause small-scale mountain building, such as the San Gabriel Mountains in California. The **faults** that splay off **transpression** or **transtension** features are known as **flower structures**.

An example of a **dextral**, **right-lateral strike-slip fault** is the San Andreas Fault, which denotes a **transform** boundary between the North American and Pacific **plates**. An example of a **sinistral**, **left-lateral strike-slip fault** is the Dead Sea **fault** in Jordan and Israel.

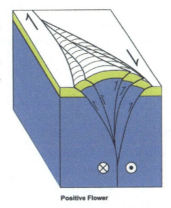
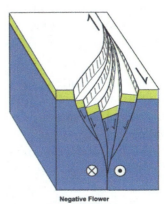

Positive Flower   Negative Flower

Figure 9.16: Flower structures created by strike-slip faults. Depending on the relative movement in relation to the bend in the fault, flower structures can create basins or mountains.

 One or more interactive elements has been excluded from this version of the text. You can view them online here:
https://www.youtube.com/watch?v=qlk7IfYMufs

Video 9.3: Video showing how faults are classified.

If you are using an offline version of this text, access this YouTube video via the QR code.

Take this quiz to check your comprehension of this section.

If you are using an offline version of this text, access the quiz for section 9.5 via the QR code.

 An interactive H5P element has been excluded from this version of the text. You can view it online here:
https://pressbooks.lib.vt.edu/introearthscience/?p=669#h5p-61

9.6 Earthquake Essentials

Earthquakes are felt at the surface of the Earth when energy is released by blocks of rock sliding past each other, i.e. **faulting** has occurred. **Seismic energy** thus released travels through the Earth in the form of **seismic** waves. Most earthquakes occur along active **plate** boundaries. **Intraplate** earthquakes (not along **plate** boundaries) occur and are still poorly understood. The USGS Earthquakes Hazards Program has a real time map showing the most recent earthquakes.

9.6.1 How Earthquakes Happen

The release of **seismic energy** is explained by the **elastic rebound theory**. When rock is **strained** to the point that it undergoes **brittle deformation**, The place where the initial offsetting rupture takes place between the **fault** blocks is called the **focus**. This **offset** propagates along the **fault**, which is known as the **fault** plane.

The **fault** blocks of persistent **faults** like the Wasatch **Fault** (Utah), that show recurring movements, are locked together by friction. Over hundreds to thousands of years, **stress** builds up along the **fault** until it overcomes frictional resistance, rupturing the rock and initiating **fault** movement. The deformed unbroken rocks snap back toward their original shape in a process called **elastic rebound.**

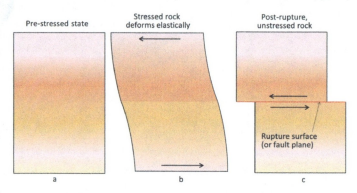

Figure 9.17: Process of elastic rebound: a) Undeformed state, b) accumulation of elastic strain, and c) brittle failure and release of elastic strain.

Think of bending a stick until it breaks; stored energy is released, and the broken pieces return to near their original orientation.

Bending, the **ductile deformation** of the rocks near a **fault**, reflects a build-up of **stress**. In earthquake-prone areas like California, **strain** gauges are used to measure this bending and help seismologists, scientists who study earthquakes, understand more about predicting them. In locations where the **fault** is not locked, **seismic stress** causes continuous, gradual displacement between the **fault** blocks called **fault creep**. **Fault creep** occurs along some parts of the San Andreas Fault (California).

After an initial earthquake, continuous application of **stress** in the **crust** causes elastic energy to begin to build again during a **period** of inactivity along the **fault**. The accumulating elastic **strain** may be periodically released to produce small earthquakes on or near the main **fault** called **foreshocks**. **Foreshocks** can occur hours or days before a large earthquake, or may not occur at all. The main release of energy during the major earthquake is known as the **mainshock**. **Aftershocks** may follow the **mainshock** to adjust new **strain** produced during the **fault** movement and generally decrease over time.

9.6.2 Focus and Epicenter

The earthquake **focus**, also called **hypocenter**, is the initial point of rupture and displacement of the rock moves from the **hypocenter** along the **fault** surface. The earthquake **focus** or **hypocenter** is the point along the **fault** plane from which initial **seismic** waves spread outward and is always at some depth below the ground surface. From the **focus**, rock displacement propagates up, down, and laterally along the **fault** plane. This displacement produces shock waves called **seismic** waves. The larger the displacement between the opposing **fault** blocks and the further the displacement propagates along the **fault** surface, the more **seismic energy** is released and the greater the amount and time of shaking is produced. The **epicenter** is the location on the Earth's surface vertically above the **focus**. This is the location that most news reports give because it is the center of the area where people are affected.

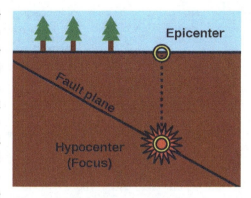

Figure 9.18: The hypocenter is the point along the fault plane in the subsurface from which seismic energy emanates. The epicenter is the point on land surface vertically above the hypocenter.

9.6.3 Seismic Waves

To understand earthquakes and how earthquake energy moves through the Earth, consider the basic properties of waves. Waves describe how energy moves through a medium, such as rock or unconsolidated **sediments** in the case of earthquakes. Wave **amplitude** indicates the **magnitude** or height of earthquake motion. **Wavelength** is the distance between two successive peaks of a wave. Wave frequency is the number of repetitions of the motion over a **period** of time, cycles per time unit. **Period**, which is the amount of time for a wave to travel one **wavelength**, is the inverse of frequency. When multiple waves combine, they can interfere with each other (see figure 9.19). When waves combine in sync, they produce constructive interference, where the influence of one wave adds to and magnifies the other. If waves are out of sync, they produce destructive interference, which diminishes the amplitudes of both waves. If two combined waves have the same **amplitude** and frequency but are one-half **wavelength** out of sync, the resulting destructive interference can eliminate each wave. These processes of wave **amplitude**, frequency, **period**, and constructive and destructive interference determine the **magnitude** and intensity of earthquakes.

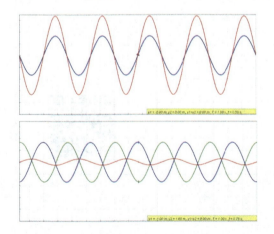

Figure 9.19: Example of constructive and destructive interference; note red line representing the results of interference.

Seismic waves are the physical expression of energy released by the **elastic rebound** of rock within displaced **fault** blocks and are felt as an earthquake. **Seismic** waves occur as **body waves** and **surface waves**. **Body waves** pass underground through the Earth's interior body and are the first **seismic** waves to propagate out from the **focus**. Body waves include primary (P) waves and secondary (S) waves. **P waves** are the fastest **body waves** and move through rock via **compression**, very much like sound waves move through air. Rock particles move forward and back during passage of the **P waves**, enabling them to travel through solids, liquids, plasma, and gases. **S waves** travel slower, following **P waves**, and propagate as **shear** waves that move rock particles from side to side. Because they are restricted to lateral movement, **S waves** can only travel through solids but not liquids, plasma, or gases.

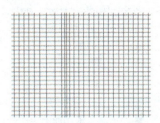

Figure 9.20: P-waves are compressional.

During an earthquake, **body waves** pass through the Earth and into the **mantle** as a sub-spherical wave front. Considering a point on a wave front, the path followed by a specific point on the spreading wave front is called a **seismic** ray and a **seismic** ray reaches a specific **seismograph** located at one of thousands of **seismic** monitoring stations scattered over the Earth. Density increases with depth in the Earth, and since **seismic** velocity increases with density, a process called **refraction** causes earthquake rays to curve away from the vertical and bend back toward the surface, passing through different bodies of rock along the way.

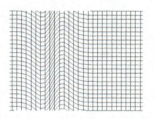

Figure 9.21: S waves are shear.

Surface waves are produced when **body waves** from the **focus strike** the Earth's surface. Surface waves travel along the Earth's surface, radiating outward from the **epicenter**. **Surface waves** take the form of rolling waves called Raleigh Waves and side to side waves called Love Waves (watch videos for wave propagation animations). **Surface waves** are produced primarily as the more energetic **S waves strike** the surface from below with some **surface wave** energy contributed by **P waves** (videos courtesy blog.Wolfram.com). **Surface waves** travel more slowly than **body waves** and because of their complex horizontal and vertical movement, **surface waves** are responsible for most of the damage caused by an earthquake. **Love waves** produce predominantly horizontal ground shaking and, ironically from their name, are the most destructive. **Rayleigh waves** produce an elliptical motion with longitudinal dilation and **compression**, like ocean waves. However, Raleigh waves cause rock particles to move in a direction opposite to that of water particles in ocean waves.

The Earth has been described as ringing like a bell after an earthquake with earthquake energy reverberating inside it. Like other waves, **seismic** waves refract (bend) and bounce (reflect) when passing through rocks of differing densities. **S waves**, which cannot move through liquid, are blocked by the Earth's liquid **outer core**, creating an S wave shadow zone on the side of the planet opposite to the earthquake **focus**. **P waves**, on the other hand, pass through the **core**, but are refracted into the **core** by the difference of density at the **core–mantle** boundary. This has the effect of creating a cone shaped **P wave** shadow zone on parts of the other side of the Earth from the **focus**.

 One or more interactive elements has been excluded from this version of the text. You can view them online here:
https://vimeo.com/153300296?embedded=true&source=vimeo_logo&owner=48249274

Video 9.4: Body and surface waves of 2011 Tohoku earthquake.

If you are using an offline version of this text, access this video via the QR code.

9.6.4 Induced Seismicity

Earthquakes known as **induced seismicity** occur near **natural gas** extraction sites because of human activity. Injection of waste fluids in the ground, commonly a byproduct of an extraction process for **natural gas** known as **fracking**, can increase the outward pressure that liquid in the **pores** of a rock exerts, known as **pore** pressure. The increase in **pore** pressure decreases the frictional forces that keep rocks from sliding past each other, essentially lubricating **fault** planes. This effect is causing earthquakes to occur near injection sites, in a human induced activity known as **induced seismicity**. The significant increase in drilling activity in the central United States has spurred the requirement for the disposal of significant amounts of waste drilling fluid, resulting in a measurable change in the cumulative number of earthquakes experienced in the region.

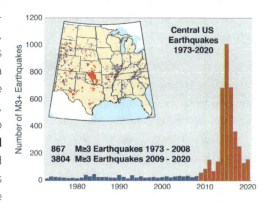

Figure 9.22: Frequency of earthquakes in the central United States. Note the sharp increase in the number of earthquakes from 2010 to 2020.

Take this quiz to check your comprehension of this section.

If you are using an offline version of this text, access the quiz for section 9.6 via the QR code.

An interactive H5P element has been excluded from this version of the text. You can view it online here:
https://pressbooks.lib.vt.edu/introearthscience/?p=669#h5p-62

9.7 Measuring Earthquakes

9.7.1 Seismographs

One or more interactive elements has been excluded from this version of the text. You can view them online here:
https://www.youtube.com/watch?v=83GOKn7kWXM

Video 9.5: Animation of a horizontal seismograph.

If you are using an offline version of this text, access this YouTube video via the QR code.

People feel approximately 1 million earthquakes a year, usually when they are close to the source and the earthquake registers at least **moment magnitude** 2.5. Major earthquakes of **moment magnitude** 7.0 and higher are extremely rare. The U. S. Geological Survey (USGS) Earthquakes Hazards Program real-time map shows the location and **magnitude** of recent earthquakes around the world.

To accurately study **seismic** waves, geologists use **seismographs** that can measure even the slightest ground vibrations. Early 20th-century seismograms use a weighted pen (pendulum) suspended by a long **spring** above a recording device fixed solidly to the ground. The recording device is a rotating drum mounted with a continuous strip of paper. During an earthquake, the suspended pen stays motionless and records ground movement on the paper strip. The resulting graph a seismogram. Digital versions use magnets, wire coils, electrical sensors, and digital signals instead of mechanical pens, springs, drums, and paper. A **seismograph** array is multiple **seismographs** configured to measure vibrations in three directions: north-south (x axis), east-west (y axis), and up-down (z axis).

 One or more interactive elements has been excluded from this version of the text. You can view them online here:
https://www.youtube.com/watch?v=DX5VXGmdnAg

Video 9.6: Animation of a vertical seismograph.

If you are using an offline version of this text, access this YouTube video via the QR code.

To pinpoint the location of an earthquake **epicenter**, seismologists use the differences in arrival times of the P, S, and **surface waves**. After an earthquake, **P waves** will appear first on a seismogram, followed by **S waves**, and finally **surface waves**, which have the largest **amplitude**. It is important to note that **surface waves** lose energy quickly, so they are not measurable at great distances from the **epicenter**. These time differences determine the distance but not the direction of the epicenter. By using wave arrival times recorded on **seismographs** at multiple stations, seismologists can apply triangulation to pin point the location of the **epicenter** of an earthquake. At least three seismograph stations are needed for triangulation. The distance from each station to the **epicenter** is plotted as the radius of a circle. The epicenter is demarked where the circles intersect. This method also works in 3D, using multi-**axis** seismographs and sphere radii to calculate the underground depth of the **focus**.

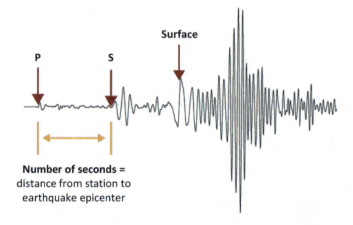

Figure 9.23: A seismogram showing the arrivals of the P, S, and surface waves.

One or more interactive elements has been excluded from this version of the text. You can view them online here:
https://www.youtube.com/watch?v=oBS7BKqHRhs&t=1s

Video 9.7: This video shows the method of triangulation to locate the epicenter of an earthquake.

If you are using an offline version of this text, access this YouTube video via the QR code.

9.7.2 Seismograph Network

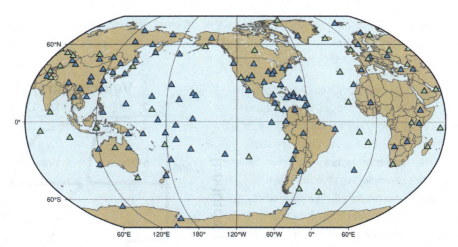

Figure 9.24: Global network of seismic stations. Note that this map does not show all of the world's seismic stations, just one of the networks of stations scientists use to measure seismic activity.

The International Registry of Seismograph Stations lists more than 20,000 **seismographs** on the planet. By comparing data from multiple **seismographs**, scientists can map the properties of the inside of the Earth, detect detonations of large explosive devices, and predict **tsunamis**. The Global Seismic Network, a worldwide set of linked **seismographs** that electronically distribute real-time data, includes more than 150 stations that meet specific design and precision standards. The USArray is a network of hundreds of permanent and transportable **seismographs** in the United States that are used to map the subsurface activity of earthquakes (see video).

Along with monitoring for earthquakes and related hazards, the Global **Seismograph** Network helps detect nuclear weapons testing, which is monitored by the Comprehensive Nuclear Test Ban Treaty Organization. Most recently, **seismographs** have been used to determine nuclear weapons testing by North Korea.

One or more interactive elements has been excluded from this version of the text. You can view them online here:
https://www.youtube.com/watch?v=5xc-rNOISQE

Video 9.8: Nepal earthquake (M7.9) ground motion visualization.

If you are using an offline version of this text, access this YouTube video via the QR code.

9.7.3 Seismic Tomography

Very much like a CT (Computed Tomography) scan uses X-rays at different angles to image the inside of a body, **seismic tomography** uses **seismic** rays from thousands of earthquakes that occur each year, passing at all angles through masses of rock, to generate images of internal Earth structures.

Using the assumption that the earth consists of homogenous layers, geologists developed a model of expected properties of earth materials at every depth within the earth called the PREM (Preliminary Reference Earth Model). These properties include **seismic wave** transmission velocity, which is dependent on rock density and elasticity. In the **mantle**, **temperature** differences affect rock density. Cooler rocks have a higher density and therefore transmit **seismic** waves faster. Warmer rocks have a lower density and transmit earthquake waves slower. When the arrival times of **seismic** rays at individual seismic stations are compared to arrival times predicted by PREM, differences are called **seismic anomalies** and can be measured for bodies of rock within the earth from seismic rays passing through them at stations of the **seismic** network. Because

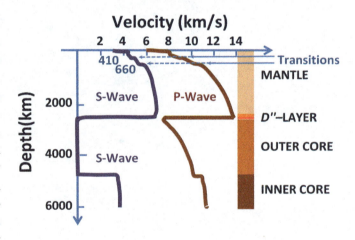

Figure 9.25: Speed of seismic waves with depth in the earth. Two thousand kilometers is 1240 miles.

seismic rays travel at all angles from lots of earthquakes and arrive at lots of stations of the **seismic** network, like CT scans of the body, variations in the properties of the rock bodies allow 3D images to be constructed of the rock bodies through which the rays passed. Seismologists are thus able to construct 3D images of the interior of the Earth..

For example, seismologists have mapped the Farallon Plate, a **tectonic plate** that **subducted** beneath North America during the last several million years, and the Yellowstone **magma chamber**, which is a product of the Yellowstone **hot spot** under the North American **continent**. Peculiarities of the Farallon **Plate subduction** are thought to be responsible for many features of western North America including the Rocky Mountains (see chapter 8).

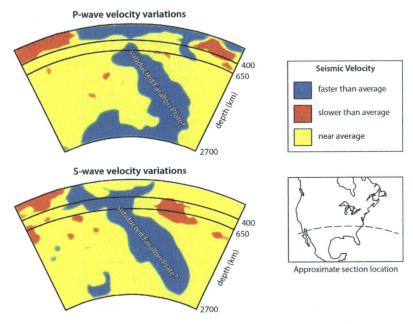

Figure 9.26: Simplified and interpreted P- and S-wave velocity variations in the mantle across southern North America showing the subducted Farallon Plate.

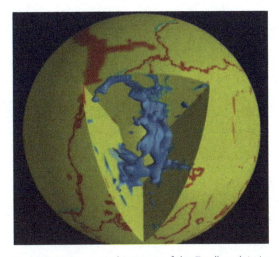

Figure 9.27: Tomographic image of the Farallon plate in the mantle below North America.

9.7.4 Earthquake Magnitude and Intensity

Richter Scale

Magnitude is the measure of the energy released by an earthquake. The **Richter scale** (M_L), the first and most well-known **magnitude** scale, was developed by Charles F. Richter (1900-1985) at the California Institute of Technology. This was the **magnitude** scale used historically by early seismologists. Used by early seismologists, **Richter magnitude** (M_L) is determined from the maximum **amplitude** of the pen tracing on the seismogram recording. Adjustments for **epicenter** distance from the **seismograph** are made using the arrival-time differences of S and **P waves**.

The **Richter Scale** is logarithmic, based on powers of 10. This means an increase of one Richter unit represents a 10-fold increase in **seismic**-wave **amplitude** or in other words, a **magnitude** 6 earthquake shakes the ground 10 times more than a magnitude 5. However, the *actual energy released* for each **magnitude** unit is 32 times greater, which means a **magnitude** 6 earthquake releases 32 times more energy than a **magnitude** 5.

The **Richter Scale** was developed for earthquakes in Southern California, using local **seismographs**. It has limited applications for larger distances and very large earthquakes. Therefore, most agencies no longer use the **Richter Scale**. **Moment magnitude** (M_W), which is measured using **seismic** arrays and generates values comparable to the **Richter Scale**, is more accurate for measuring earthquakes across the Earth, including large earthquakes, although they require more time to calculate. News media often report Richter magnitudes right after an earthquake occurs even though scientific calculations now use moment magnitudes.

Moment Magnitude Scale

The **moment magnitude** scale depicts the absolute size of earthquakes, comparing information from multiple locations and using a measurement of actual energy released calculated from cross-sectional area of rupture, amount of slippage, and the rigidity of the rocks. Because each earthquake occurs in a unique geologic setting and the rupture area is often hard to measure, estimates of **moment magnitude** can take days or even months to calculate.

Like the **Richter Scale**, the **moment magnitude** scale is logarithmic. **Magnitude** values of the two scales are approximately equal, except for very large earthquakes. Both scales are used for reporting earthquake **magnitude**. The **Richter Scale** provides a quick **magnitude** estimate immediately following the quake and thus, is usually reported in news accounts. **Moment magnitude** calculations take much longer but are more accurate and thus, more useful for scientific analysis.

One or more interactive elements has been excluded from this version of the text. You can view them online here: https://www.youtube.com/watch?v=HL3KGK5eqaw

Video 9.9: Moment magnitude explained.

If you are using an offline version of this text, access this YouTube video via the QR code.

Modified Mercalli Intensity Scale

The **Modified Mercalli Intensity Scale** (MMI) is a **qualitative** rating of ground-shaking intensity based on observable structural damage and people's perceptions. This scale uses a *I* (Roman numeral one) rating for the lowest intensity and *X* (ten) for the highest (see table) and can vary depending on **epicenter** location and population density, such as urban versus rural settings. Historically, scientists used the MMI Scale to categorize earthquakes before they developed **quantitative** measurements of **magnitude**. Intensity maps show locations of the most severe damage, based on residential questionnaires, local news articles, and on-site assessment reports.

Intensity	Shaking	Description/damage
I	Not felt	Not felt except by a very few under especially favorable conditions.
II	Weak	Felt only by a few persons at rest, especially on upper floors of buildings.
III	Weak	Felt quite noticeably by persons indoors, especially on upper floors of buildings. Many people do not recognize it as an earthquake. Standing motor cars may rock slightly. Vibrations similar to the passing of a truck. Duration estimated.
IV	Light	Felt indoors by many, outdoors by few during the day. At night, some awakened. Dishes, windows, doors disturbed; walls make cracking sound. Sensation like heavy truck striking building. Standing motor cars rocked noticeably.
V	Moderate	Felt by nearly everyone; many awakened. Some dishes, windows broken. Unstable objects overturned. Pendulum clocks may stop.
VI	Strong	Felt by all, many frightened. Some heavy furniture moved; a few instances of fallen plaster. Damage slight.
VII	Very strong	Damage negligible in buildings of good design and construction; slight to moderate in well-built ordinary structures; considerable damage in poorly built or badly designed structures; some chimneys broken.
VIII	Severe	Damage slight in specially designed structures; considerable damage in ordinary substantial buildings with partial collapse. Damage great in poorly built structures. Fall of chimneys, factory stacks, columns, monuments, walls. Heavy furniture overturned.
IX	Violent	Damage considerable in specially designed structures; well-designed frame structures thrown out of plumb. Damage great in substantial buildings, with partial collapse. Buildings shifted off foundations.
X	Extreme	Some well-built wooden structures destroyed; most masonry and frame structures destroyed with foundations. Rails bent.

Table 9.3: Abridged Mercalli Scale from USGS General Interest Publication 1989-288-913.

Shake Maps

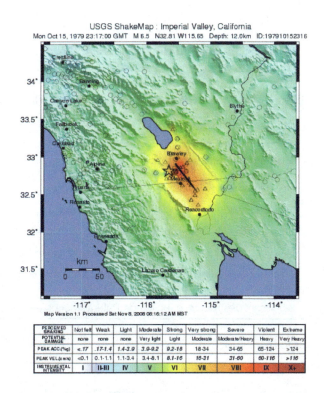

Figure 9.28: Example of a shake map.

Shake maps, written ShakeMaps by the USGS, use high-quality, computer-interpolated data from **seismograph** networks to show areas of intense shaking. Shake maps are useful in the crucial minutes after an earthquake, as they show emergency personnel where the greatest damage likely occurred and help them locate possibly damaged gas lines and other utility facilities.

Take this quiz to check your comprehension of this section.

If you are using an offline version of this text, access the quiz for section 9.7 via the QR code.

 An interactive H5P element has been excluded from this version of the text. You can view it online here:
https://pressbooks.lib.vt.edu/introearthscience/?p=669#h5p-63

9.8 Earthquake Risk

9.8.1 Factors That Determine Shaking

Earthquake **magnitude** is an absolute value that measures pure energy release. Intensity however, i.e. how much the ground shakes, is a determined by several factors.

Earthquake magnitude—In general, the larger the **magnitude**, the stronger the shaking and the longer the shaking will last.

This table is taken from from the USGS and shows scales of **magnitude** and **Mercalli** Intensity, and descriptions of shaking and resulting damage.

Magnitude	Modified Mercalli Intensity	Shaking/damage description
1.0-3.0	I	Only felt by a very few.
3.0-3.9	II-III	Noticeable indoors, especially on upper floors.
4.0-4.9	IV-V	Most to all feel it. Dishes, doors, cars shake and possibly break.
5.0-5.9	VI-VII	Everyone feels it. Some items knocked over or broken. Building damage possible.
6.0-6.9	VII-IX	Frightening amounts of shaking. Significant damage especially with poorly constructed buildings.
≥ 7.0	≥ VIII	Significant destruction of buildings. Potential for objects to be thrown in air from shaking.

Table 9.4: Mercalli Intensity as it relates to magnitude.

Location and direction—Shaking is more severe closer to the **epicenter**. The severity of shaking is influenced by the location of the observer relative to **epicenter**, direction of rupture propagation, and path of greatest rupture.

Local geologic conditions—Seismic waves are affected by the nature of the ground materials through which they pass. Different materials respond differently to an earthquake. Think of shaking a block of Jello versus a meatloaf, one will jiggle much more when hit by waves of the same **amplitude**. The ground's response to shaking depends on the degree of substrate consolidation. Solid sedimentary, **igneous**, or **metamorphic bedrock** shakes less than unconsolidated **sediments**.

One or more interactive elements has been excluded from this version of the text. You can view them online here:
https://www.youtube.com/watch?v=536xSZ_XkSs

Video 9.10: This video shows how different substrates behave in response to different seismic waves and their potential for destruction.

If you are using an offline version of this text, access this YouTube video via the QR code.

Seismic waves move fastest through consolidated **bedrock**, slower through unconsolidated **sediments**, and slowest through unconsolidated **sediments** with a high water content. Seismic energy is transmitted by **wave velocity** and **amplitude**. When seismic waves slow down, energy is transferred to the **amplitude**, increasing the motion of **surface waves**, which in turn amplifies ground shaking.

Focus depth—Deeper earthquakes cause less surface shaking because much of their energy, transmitted as **body waves**, is lost before reaching the surface. Recall that **surface waves** are generated by P and **S waves** impacting the Earth's surface.

9.8.2 Factors that Determine Destruction

Just as certain conditions will impact intensity of ground-shaking, several factors affect how much destruction is caused.

Building materials—The flexibility of a building material determines its resistance to earthquake damage. Unreinforced masonry (URM) is the material most devastated by ground shaking. Wood framing fastened with nails bends and flexes during **seismic wave** passage and is more likely to survive intact. Steel also has the ability to deform elastically before **brittle** failure. The Fix the Bricks campaign in Salt Lake City, Utah has good information on URMs and earthquake safety.

Intensity and duration—Greater shaking and duration of shaking causes more destruction than lower and shorter shaking.

Figure 9.29: Example of devastation on unreinforced masonry by seismic motion.

Resonance—**Resonance** occurs when **seismic wave** frequency matches a building's natural shaking frequency and increases the shaking happened in the 1985 Mexico City Earthquake, where buildings of heights between 6 and 15 stories were especially vulnerable to earthquake damage. Skyscrapers designed with earthquake resilience have dampers and base isolation features to reduce **resonance**.

Resonance is influenced by the properties of the building materials. Changes in the structural integrity of a building can alter **resonance**. Conversely, changes in measured **resonance** can indicate a potentially altered structural integrity.

These two videos discuss why buildings fall during earthquakes and a modern procedure to reduce potential earthquake destruction for larger buildings.

 One or more interactive elements has been excluded from this version of the text. You can view them online here:
https://www.youtube.com/watch?v=H4VQuI_SmCg

Video 9.11: Why do buildings fall in earthquakes?

If you are using an offline version of this text, access this YouTube video via the QR code.

 One or more interactive elements has been excluded from this version of the text. You can view them online here:
https://www.youtube.com/watch?v=DP7fB1I7UwE

Video 9.12: Base isolators.

If you are using an offline version of this text, access this YouTube video via the QR code.

9.8.3 Earthquake Recurrence

A long hiatus in activity on along a **fault** segment with a history of recurring earthquakes is known as a **seismic** gap. The lack of activity may indicate the **fault** segment is locked, which may produce a buildup of **strain** and higher probability of an earthquake recurring. Geologists dig earthquake trenches across **faults** to estimate the frequency of past earthquake occurrences. Trenches are effective for faults with relatively long **recurrence** intervals, roughly 100s to 10,000s of years between significant earthquakes. Trenches are less useful in areas with more frequent earthquakes because they usually have more recorded data.

9.8.4 Earthquake Distribution

This video shows the distribution of significant earthquakes on the Earth during the years 2010 through 2012. Like **volcanoes**, earthquakes tend to aggregate around active boundaries of **tectonic plates**. The exception is **intraplate** earthquakes, which are comparatively rare.

Figure 9.30: Fault trench near Teton Fault. Trenches allow geologists to see a cross section of a fault and to use dating techniques to determine how frequently earthquakes occur.

 One or more interactive elements has been excluded from this version of the text. You can view them online here: https://www.youtube.com/watch?v=Wc6vtj4yYcY

Video 9.13: 2010-2012 Earthquake visualization map.

If you are using an offline version of this text, access this YouTube video via the QR code.

Subduction zones—Subduction zones, found at **convergent plate** boundaries, are where the largest and deepest earthquakes, called **megathrust** earthquakes, occur. Examples of **subduction**-zone earthquake areas include the Sumatran Islands, Aleutian Islands, west coast of South America, and Cascadia **Subduction** Zone off the coast of Washington and Oregon. See chapter 2 for more information about **subduction** zones.

Collision zones—Collisions between converging **continental plates** create broad earthquake zones that may generate deep, large earthquakes from the remnants of past **subduction** events or other deep-crustal processes. The Himalayan Mountains (northern border of the Indian subcontinent) and Alps (southern Europe and Asia) are active regions of **colli-sion**-zone earthquakes. See chapter 2 for more information about **collision** zones.

Transform boundaries—Strike-slip faults created at **transform** boundaries produce moderate-to-large earthquakes, usually having a maximum **moment magnitude** of about 8. The San Andreas Fault (California) is an example of a **trans-form**-boundary earthquake zone. Haiti's Enriquillo Plantain Garden fault system, which caused the 2010 earthquake near Port-au-Prince (see below), and Septentrional Fault, which destroyed Cap-Haïtien in 1842 and shook Cuba in 2020, are also

transform faults. Other examples are the Alpine Fault (New Zealand) and Anatolian Faults (Turkey). See chapter 2 for more information about **transform** boundaries.

Divergent boundaries—Continental rifts and **mid-ocean ridges** found at **divergent** boundaries generally produce moderate earthquakes. Examples of active earthquake zones include the East African Rift System (southwestern Asia through eastern Africa), Iceland, and **Basin and Range** province (Nevada, Utah, California, Arizona, and northwestern Mexico). See chapter 2 for more information about **divergent** boundaries.

Intraplate earthquakes—Intraplate earthquakes are not found near **tectonic plate** boundaries, but generally occur in areas of weakened **crust** or **concentrated tectonic stress**. The New Madrid **seismic** zone, which covers Missouri, Illinois, Tennessee, Arkansas, and Indiana, is thought to represent the failed Reelfoot **rift**. The failed **rift** zone weakened the **crust**, making it more responsive to **tectonic plate** movement and interaction. Geologists theorize the infrequently occurring earthquakes are produced by low **strain** rates.

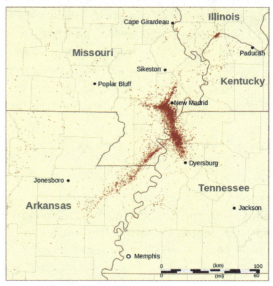

Figure 9.31: High density of earthquakes in the New Madrid seismic zone.

9.8.5 Secondary Hazards Caused by Earthquakes

Most earthquake damage is caused by ground shaking and fault block displacement. In addition, there are secondary hazards that endanger structures and people, in some cases after the shaking stops.

Figure 9.32: Buildings toppled from liquefaction during a 7.5 magnitude earthquake in Japan.

Buildings toppled from liquefaction during a 7.5 magnitude earthquake in Japan.

Liquefaction—Liquefaction occurs when water-**saturated**, unconsolidated **sediments**, usually silt or sand, become fluid-like from shaking. The shaking breaks the **cohesion** between grains of **sediment**, creating a slurry of particles suspended in water. Buildings settle or tilt in the liquified **sediment**, which looks very much like quicksand in the movies. **Liquefaction** also creates sand **volcanoes**, cone-shaped features created when liquefied sand is squirted through an overlying and usually finer-grained layer.

One or more interactive elements has been excluded from this version of the text. You can view them online here:
https://www.youtube.com/watch?v=b_aIm5oi5eA

Video 9.14: This video demonstrates how liquefaction takes place.

If you are using an offline version of this text, access this YouTube video via the QR code.

This video shows **liquefaction** occurring during the 2011 earthquake in Japan.

 One or more interactive elements has been excluded from this version of the text. You can view them online here:
https://www.youtube.com/watch?v=rn3oAvmZY8k

Video 9.15: Liquefaction during the 2011 earthquake in Japan.

If you are using an offline version of this text, access this YouTube video via the QR code.

Figure 9.33: As the ocean depth becomes shallower, the wave slows down and pile up on top of itself, making large, high-amplitude waves.

Tsunamis—Among the most devastating natural disasters are **tsunamis**, earthquake-induced ocean waves. When the sea floor is **offset** by **fault** movement or an underwater **landslide**, the ground displacement lifts a volume of ocean water and generates the **tsunami** wave. Ocean wave behavior, which includes **tsunamis**, is covered in chapter 12. **Tsunami** waves are fast-moving with low **amplitude** in deep ocean water but grow significantly in **amplitude** in the shallower waters approaching **shore**. When a **tsunami** is about to **strike** land, the drawback of the **trough** preceding the **wave crest** causes the water to recede dramatically from **shore**. Tragically, curious people wander out and follow the disappearing water, only to be overcome by an oncoming wall of water that can be upwards of a 30 m (100 ft) high. Early warning systems help mitigate the loss of life caused by **tsunamis**.

Landslides—Shaking can **trigger landslides** (see chapter 10). In 1992 a **moment magnitude** 5.9 earthquake in St. George, Utah, caused a **landslide** that destroyed several structures in the Balanced Rock Hills subdivision in Springville, Utah.

Seiches—Seiches are waves generated in lakes by earthquakes. The shaking may cause water to slosh back-and-forth or sometimes change the lake depth. Seiches in Hebgen Lake during a 1959 earthquake caused major destruction to nearby structures and roads.

This video shows a seich generated in a swimming pool by an earthquake in Nepal in 2015.

Figure 9.34: Schoolhouse in Thistle, Utah destroyed by a landslide.

 One or more interactive elements has been excluded from this version of the text. You can view them online here: https://www.youtube.com/watch?v=27GMnYEWL0M

Video 9.16: A seich generated in a swimming pool by an earthquake in Nepal in 2015.

If you are using an offline version of this text, access this YouTube video via the QR code.

Land elevation changes—**Elastic rebound** and displacement along the **fault** plane can cause significant land elevation changes, such as **subsidence** or upheaval. The 1964 Alaska earthquake produced significant land elevation changes, with the differences in height between the **hanging wall** and **footwall** ranging from one to several meters (3–30 ft). The Wasatch Mountains in Utah represent an accumulation of **fault scarps** created a few meters at a time, over a few million years.

Take this quiz to check your comprehension of this section.

If you are using an offline version of this text, access the quiz for section 9.8 via the QR code.

 An interactive H5P element has been excluded from this version of the text. You can view it online here: https://pressbooks.lib.vt.edu/introearthscience/?p=669#h5p-64

9.9 Case Studies

Video explaining the **seismic** activity and hazards of the Intermountain Seismic Belt and the Wasatch Fault, a large **intraplate** area of **seismic** activity.

 One or more interactive elements has been excluded from this version of the text. You can view them online here:
https://www.youtube.com/watch?v=DByPiCkznE0

Video 9.17: Activities of the Intermountain Seismic Belt and the Wasatch Fault.

If you are using an offline version of this text, access this YouTube video via the QR code.

9.9.1 North American Earthquakes

Basin and Range earthquakes: Earthquakes in the **Basin and Range** Province, from the Wasatch Fault (Utah) to the Sierra Nevada (California), occur primarily in normal **faults** created by **tensional** forces. The Wasatch Fault, which defines the eastern extent of the **Basin and Range** province, has been studied as an earthquake hazard for more than 100 years.

New Madrid earthquakes (1811-1812): Historical accounts of earthquakes in the New Madrid **seismic** zone date as far back as 1699 and earthquakes continue to be reported in modern times. A sequence of large (M_w >7) occurred from December 1811 to February 1812 in the New Madrid area of Missouri. The earthquakes damaged houses in St. Louis, affected the **stream** course of the Mississippi River, and leveled the town of New Madrid. These earthquakes were the result of **intraplate seismic** activity.

Charleston (1886): The 1886 earthquake in Charleston South Carolina was a **moment magnitude** 7.0, with a **Mercalli** intensity of X, caused significant ground motion, and killed at least 60 people. This **intraplate** earthquake was likely associated with ancient **faults** created during the breakup of **Pangea**. The earthquake caused significant **liquefaction**. Scientists estimate the **recurrence** of destructive earthquakes in this area with an interval of approximately 1500 to 1800 years.

Great San Francisco earthquake and eire (1906): On April 18, 1906, a large earthquake, with an estimated **moment magnitude** of 7.8 and MMI of X, occurred along the San Andreas Fault near San Francisco California. There were multiple **aftershocks** followed by devastating fires, resulting in about 80% of the city being destroyed. Geologists G.K. Gilbert and Richard L. Humphrey, working independently, arrived the day following the earthquake and took measurements and photographs.

Figure 9.35: Remains of San Francisco after the 1906 earthquake and fire.

Alaska (1964): The 1964 Alaska earthquake, **moment magnitude** 9.2, was one of the most powerful earthquakes ever recorded. The earthquake originated in a **megathrust fault** along the Aleutian **subduction** zone. The earthquake caused large areas of land **subsidence** and uplift, as well as significant **mass wasting**.

Video from the USGS about the 1964 Alaska earthquake.

One or more interactive elements has been excluded from this version of the text. You can view them online here:
https://www.youtube.com/watch?v=lE2j10xyOgI

Video 9.18: The 1964 Alaska earthquake.

If you are using an offline version of this text, access this YouTube video via the QR code.

Loma Prieta (1989): The Loma Prieta, California, earthquake was created by movement along the San Andreas Fault. The **moment magnitude** 6.9 earthquake was followed by a **magnitude** 5.2 **aftershock**. It caused 63 deaths, buckled portions of the several freeways, and collapsed part of the San Francisco-Oakland Bay Bridge.

This video shows how shaking propagated across the Bay Area during the 1989 Loma Prieta earthquake.

One or more interactive elements has been excluded from this version of the text. You can view them online here:
https://www.youtube.com/watch?v=VuQSs7QHL28

Video 9.19: How shaking propagated across the Bay Area during the 1989 Loma Prieta earthquake.

If you are using an offline version of this text, access this YouTube video via the QR code.

This video shows destruction caused by the 1989 Loma Prieta earthquake.

One or more interactive elements has been excluded from this version of the text. You can view them online here:
https://www.youtube.com/watch?v=L6jYgqLyIPw

Video 9.20: Destruction caused by the 1989 Loma Prieta earthquake.

If you are using an offline version of this text, access this YouTube video via the QR code.

9.9.2 Global Earthquakes

Many of history's largest earthquakes occurred in **megathrust** zones, such as the Cascadia Subduction Zone (Washington and Oregon coasts) and Mt. Rainier (Washington).

Shaanxi, China (1556): On January 23, 1556 an earthquake of an approximate **moment magnitude** 8 hit central China, killing approximately 830,000 people in what is considered the most deadly earthquake in history. The high death toll was attributed to the collapse of cave dwellings (*yaodong*) built in **loess** deposits, which are large banks of windblown, compacted **sediment** (see chapter 5). Earthquakes in this are region are believed to have a **recurrence** interval of 1000 years.

Lisbon, Portugal (1755): On November 1, 1755 an earthquake with an estimated **moment magnitude** range of 8–9 struck Lisbon, Portugal, killing between 10,000 to 17,400 people. The earthquake was followed by a **tsunami**, which brought the total death toll to between 30,000-70,000 people.

Valdivia, Chile (1960): The May 22, 1960 earthquake was the most powerful earthquake ever measured, with a **moment magnitude** 9.4–9.6 and lasting an estimated 10 minutes. It triggered **tsunamis** that destroyed houses across the Pacific Ocean in Japan and Hawaii and caused vents to erupt on the Puyehue-Cordón Caulle (Chile).

Video describing the **tsunami** produced by the 1960 Chili earthquake.

One or more interactive elements has been excluded from this version of the text. You can view them online here:
https://www.youtube.com/watch?v=RHYbprZAIWo

Video 9.21: Tsunami produced by the 1960 Chili earthquake.

If you are using an offline version of this text, access this YouTube video via the QR code.

Tangshan, China (1976): Just before 4 a.m. (Beijing time) on July 28, 1976 a **moment magnitude** 7.8 earthquake struck Tangshan (Hebei Province), China, and killed more than 240,000 people. The high death-toll is attributed to people still being asleep or at home and most buildings being made of unreinforced masonry.

Sumatra, Indonesia (2004): On December 26, 2004, slippage of the Sunda **megathrust fault** generated a **moment magnitude** 9.0–9.3 earthquake off the **coast** of Sumatra, Indonesia. This **megathrust** fault is created by the Australia **plate subducting** below the Sunda plate in the Indian Ocean. The resultant **tsunamis** created **massive** waves as tall as 24 m (79 ft) when they reached the **shore** and killed more than an estimated 200,000 people along the Indian Ocean **coastline**.

Haiti (2010): The **moment magnitude** 7 earthquake that occurred on January 12, 2010, was followed by many **aftershocks** of **magnitude** 4.5 or higher. More than 200,000 people are estimated to have died as result of the earthquake. The widespread infrastructure damage and crowded conditions contributed to a cholera outbreak, which is estimated to have caused thousands more deaths.

Tōhoku, Japan (2011): Because most Japanese buildings are designed to tolerate earthquakes, the **moment magnitude** 9.0 earthquake on March 11, 2011, was not as destructive as the **tsunami** it created. The **tsunami** caused more than 15,000 deaths and tens of billions of dollars in damage, including the destructive meltdown of the Fukushima nuclear power plant.

Take this quiz to check your comprehension of this section.

If you are using an offline version of this text, access the quiz for section 9.9 via the QR code.

 An interactive H5P element has been excluded from this version of the text. You can view it online here:
https://pressbooks.lib.vt.edu/introearthscience/?p=669#h5p-65

Summary

Geologic **stress**, applied force, comes in three types: **tension**, **shear**, and **compression**. **Strain** is produced by **stress** and produces three types of **deformation**: elastic, **ductile**, and **brittle**. Geological maps are two-dimensional representations of surface **formations** which are the surface expression of three-dimensional geologic structures in the subsurface. The map symbol called **strike** and **dip** or rock attitude indicates the orientation of rock **strata** with reference to north-south and horizontal. Folded rock layers are categorized by the orientation of their limbs, **fold** axes and **axial planes**. **Faults** result when **stress** forces exceed rock integrity and friction, leading to **brittle deformation** and breakage. The three major **fault** types are described by the movement of their fault blocks: normal, **strike-slip**, and reverse.

Earthquakes, or **seismic** activity, are caused by sudden **brittle deformation** accompanied by **elastic rebound**. The release of energy from an earthquake **focus** is generated as **seismic** waves. P and **S waves** travel through the Earth's interior. When they **strike** the outer **crust**, they create **surface waves**. Human activities, such as mining and nuclear detonations, can also cause **seismic** activity. **Seismographs** measure the energy released by an earthquake using a logarithmic scale of **magnitude** units; the **Moment Magnitude** Scale has replaced the original **Richter Scale**. Earthquake intensity is the perceived effects of ground shaking and physical damage. The location of earthquake foci is determined from triangulation readings from multiple **seismographs**.

Earthquake rays passing through rocks of the Earth's interior and measured at the **seismographs** of the worldwide **Seismic** Network allow 3-D imaging of buried rock masses as **seismic** tomographs.

Earthquakes are associated with **plate tectonics**. They usually occur around the active **plate** boundaries, including zones of **subduction**, **collision**, and **transform** and **divergent** boundaries. Areas of **intraplate** earthquakes also occur. The damage caused by earthquakes depends on a number of factors, including **magnitude**, location and direction, local conditions, building materials, intensity and duration, and **resonance**. In addition to damage directly caused by ground shaking, secondary earthquake hazards include **liquefaction**, **tsunamis**, **landslides**, seiches, and elevation changes.

Take this quiz to check your comprehension of this chapter.

If you are using an offline version of this text, access the quiz for chapter 9 via the QR code.

An interactive H5P element has been excluded from this version of the text. You can view it online here:
https://pressbooks.lib.vt.edu/introearthscience/?p=669#h5p-66

URLs Linked Within This Chapter

UGS geologic map viewer: https://geology.utah.gov/apps/intgeomap/

AAPG wiki: https://wiki.aapg.org/Cross_section

USGS Earthquakes Hazards Program: https://earthquake.usgs.gov/earthquakes/map/?extent=27.60567,-132.97852&extent=51.91717,-97.25098&range=week&magnitude=all&listOnlyShown=true&timeZone=utc&settings=true

Raleigh waves: Propagation of Seismic Waves: Rayleigh waves. [Video: 0:15] https://www.youtube.com/watch?v=6yXgfY-HAS7c

Love waves: Propagation of Seismic Waves: Love waves. [Video: 0:15] https://www.youtube.com/watch?v=t7wJu0Kts7w

Blog.Wolfram.com: https://blog.wolfram.com/

International Registry of Seismograph Stations: http://www.isc.ac.uk/registries/

Global Seismic Network: https://www.usgs.gov/programs/earthquake-hazards/gsn-global-seismographic-network

USArray: http://www.usarray.org/

Comprehensive Nuclear Test Ban Treaty Organization: https://www.ctbto.org/

Fix the Bricks: https://www.slc.gov/em/fix-the-bricks

Text References

1. Christenson, G.E., 1995, The September 2, 1992 ML 5.8 St. George earthquake, Washington County, Utah: Utah Geological Survey Circular 88, 48 p.

2. Coleman, J.L., and Cahan, S.M., 2012, Preliminary catalog of the sedimentary basins of the United States: U.S. Geological Survey Open-File Report 1111, 27 p.

3. Earle, S., 2015, Physical geology OER textbook: BC Campus OpenEd.

4. Feldman, J., 2012, When the Mississippi Ran Backwards: Empire, Intrigue, Murder, and the New Madrid Earthquakes of 1811 and 1812: Free Press, 320 p.

5. Fuller, M.L., 1912, The New Madrid earthquake: Central United States Earthquake Consortium Bulletin 494, 129 p.

6. Gilbert, G.K., and Dutton, C.E., 1877, Report on the geology of the Henry Mountains: Washington, U.S. Government Printing Office, 160 p.

7. Hildenbrand, T.G., and Hendricks, J.D., 1995, Geophysical setting of the Reelfoot rift and relations between rift structures and the New Madrid seismic zone: U.S. Geological Survey Professional Paper 1538-E, 36 p.

8. Means, W.D., 1976, Stress and Strain – Basic Concepts of Continuum Mechanics: Berlin, Springe, 273 p.

9. Ressetar, R. (Ed.), 2013, The San Rafael Swell and Henry Mountains Basin: geologic centerpiece of Utah: Utah Geological Association, Utah Geological Association, 250 p.

10. Satake, K., and Atwater, B.F., 2007, Long-Term Perspectives on Giant Earthquakes and Tsunamis at Subduction Zones: Annual Review of Earth and Planetary Sciences, v. 35, no. 1, p. 349–374., doi: 10.1146/annurev.earth.35.031306.140302.

11. Talwani, P., and Cox, J., 1985, Paleoseismic evidence for recurrence of Earthquakes near Charleston, South Carolina: Science, v. 229, no. 4711, p. 379–381.

Figure References

Figure 9.1: Types of stress. Michael Kimberly, North Carolina State University via USGS. 2021. Public domain. https://www.usgs.gov/media/images/stresstypesgif

Figure 9.2: Different materials deform differently when stress is applied. Steven Earle. 2019. CC BY. Figure 12.1.1 from https://opentextbc.ca/physicalgeology2ed/chapter/12-1-stress-and-strain/

Figure 9.3: "Strike" and "dip" are words used to describe the orientation of rock layers with respect to North/South and horizontal. CrunchyRocks. 2018. CC BY 4.0. https://commons.wikimedia.org/wiki/File:Strike_and_dip_on_bedding.svg

Figure 9.4: Attitude symbol on geologic map (with compass directions for reference) showing strike of N30°E and dip of 45° to the SE. Kindred Grey. 2022. CC BY 4.0. Includes Compass Rose by NAPISAH from Noun Project (Noun Project license).

Figure 9.5: Model of anticline. Speleotherm. 2016. CC BY-SA 4.0. https://commons.wikimedia.org/wiki/File:Anticline.png

Figure 9.6: An Anticline near Bcharre, Lebanon. Not home. 2005. Public domain. https://commons.wikimedia.org/wiki/File:Anticline-lebanon.jpg

Figure 9.7: Monocline at Colorado National Monument. Anky-man. 2007. CC BY-SA 3.0. https://commons.wikimedia.org/wiki/File:Monocline.JPG

Figure 9.8: This prominent circular feature in the Sahara desert of Mauritania has attracted attention since the earliest space missions because it forms a conspicuous bull's-eye in the otherwise rather featureless expanse of the desert. NASA/GSFC/MITI/ERSDAC/JAROS, and U.S./Japan ASTER Science Team. 2000. Public domain. https://commons.wikimedia.org/wiki/File:ASTER_Richat.jpg

Figure 9.9: The Denver Basin is an active sedimentary basin at the eastern extent of the Rocky Mountains. Daniel H. Knepper, Jr. (editor), US Geological Survey. 2002. Public domain. https://commons.wikimedia.org/wiki/File:Denver_Basin_Location_Map.png

Figure 9.10: Common terms used for normal faults. Kindred Grey. 2022. CC BY-SA 3.0. Includes Faults6 by Actualist, 2013 (CC BY-SA 3.0, https://commons.wikimedia.org/wiki/File:Faults6.png).

Figure 9.11: Example of a normal fault in an outcrop of the Pennsylvanian Honaker Trail Formation near Moab, Utah.

James St. John. 2007. CC BY 2.0. https://commons.wikimedia.org/wiki/File:Faults_in_Moenkopi_Formation_Moab_Canyon_Utah_USA_01.jpg

Figure 9.12: Faulting that occurs in the crust under tensional stress. USGS; adapted by Gregors. 2011. Public domain. https://commons.wikimedia.org/wiki/File:Fault-Horst-Graben.svg

Figure 9.13: Simplified block diagram of a reverse fault. Kindred Grey. 2022. CC BY-SA 3.0. Includes Faults6 by Actualist, 2013 (CC BY-SA 3.0, https://commons.wikimedia.org/wiki/File:Faults6.png).

Figure 9.14: Terminology of thrust faults (low-angle reverse faults). Woudloper. 2006. Public domain. https://commons.wikimedia.org/wiki/File:Thrust_system_en.jpg

Figure 9.15: Thrust fault in the North Qilian Mountains (Qilian Shan). Jide. 2006. CC BY-SA 3.0. https://commons.wikimedia.org/wiki/File:Thrust_fault_Qilian_Shan.jpg

Figure 9.16: Flower structures created by strike-slip faults. Mikenorton. 2009. CC BY-SA 3.0. https://commons.wikimedia.org/wiki/File:Flowerstructure1.png

Figure 9.17: Process of elastic rebound: a) Undeformed state, b) accumulation of elastic strain, and c) brittle failure and release of elastic strain. Steven Earle. Unknown date. CC BY 4.0. Figure 11.2 from https://open.maricopa.edu/physicalgeology/chapter/11-1-what-is-an-earthquake/

Figure 9.18: The hypocenter is the point along the fault plane in the subsurface from which seismic energy emanates. Derived from original work by Sam Hocevar. 2014. CC BY-SA 1.0. https://commons.wikimedia.org/wiki/File:Epicenter_Diagram.svg

Figure 9.19: Example of constructive and destructive interference; note red line representing the results of interference. Lookangmany thanks to author of original simulation = Wolfgang Christian and Francisco Esquembre author of Easy Java Simulation = Francisco Esquembre. 2015. CC BY-SA 4.0. https://www.wikiwand.com/en/Wave_interference#Media/File:Waventerference.gif

Figure 9.20: P-waves are compressional. Christophe Dang Ngoc Chan. 2006. CC BY-SA 3.0. https://commons.wikimedia.org/wiki/File:Onde_compression_impulsion_1d_30_petit.gif

Figure 9.21: S waves are shear. Christophe Dang Ngoc Chan. 2006. CC BY-SA 3.0. https://commons.wikimedia.org/wiki/File:Onde_cisaillement_impulsion_1d_30_petit.gif

Figure 9.22: Frequency of earthquakes in the central United States. USGS. 2019. Public domain. https://commons.wikimedia.org/wiki/File:Cumulative_induced_seismicity.png

Figure 9.23: A seismogram showing the arrivals of the P, S, and surface waves. Kindred Grey. 2022. CC BY 4.0. Adapted from USGS (Public domain, https://www.usgs.gov/media/images/seismic-wave-showing-p-wave-and-s-wave-initiation).

Figure 9.24: Global network of seismic stations. USGS. 2022. Public domain. https://www.usgs.gov/media/images/global-seismographic-network-gsn-stations

Figure 9.25: Speed of seismic waves with depth in the earth. Brews ohare. 2010. CC BY-SA 3.0. https://commons.wikimedia.org/wiki/File:Speeds_of_seismic_waves.PNG

Figure 9.26: Simplified and interpreted P- and S-wave velocity variations in the mantle across southern North America showing the subducted Farallon Plate. Oilfieldvegetarian. 2016. CC BY-SA 4.0. https://commons.wikimedia.org/wiki/File:FarallonTomoSlice.png

Figure 9.27: Tomographic image of the Farallon plate in the mantle below North America. Stuart A. Snodgrass and Hans-Peter Bunge via NASA. 2002. Public domain. https://commons.wikimedia.org/wiki/File:Farallon_Plate.jpg

Figure 9.28: Example of a shake map. USGS. 2012. Public domain. https://en.wikipedia.org/wiki/File:USGS_Shakemap_-_1979_Imperial_Valley_earthquake.jpg

Figure 9.29: Example of devastation on unreinforced masonry by seismic motion. M. Mehrain, Dames and Moore via NOAA/NGDC. 2012. Public domain. https://commons.wikimedia.org/wiki/File:Collapse_of_Unreinforced_Masonry_Buildings,_Iran_(Persia)_-_1990_Manjil_Roudbar_Earthquake.jpg

Figure 9.30: Fault trench near Teton Fault. Jaime Delano via USGS. 2017. Public domain. https://www.usgs.gov/media/images/teton-fault-4

Figure 9.31: High density of earthquakes in the New Madrid seismic zone. Kbh3rd. 2011. CC BY-SA 3.0. https://commons.wikimedia.org/wiki/File:New_Madrid_Seismic_Zone_activity_1974-2011.svg

Figure 9.32: Buildings toppled from liquefaction during a 7.5 magnitude earthquake in Japan. Ungtss. 1964. Public domain. https://commons.wikimedia.org/wiki/File:Liquefaction_at_Niigata.JPG

Figure 9.33: As the ocean depth becomes shallower, the wave slows down and pile up on top of itself, making large, high-amplitude waves. Régis Lachaume. 2005. CC BY-SA 3.0. https://commons.wikimedia.org/wiki/File:Propagation_du_tsunami_en_profondeur_variable.gif

Figure 9.34: Schoolhouse in Thistle, Utah destroyed by a landslide. Jenny Bauman. 2006. CC BY-SA 2.0. https://commons.wikimedia.org/wiki/File:Thistle-School_house.jpg

Figure 9.35: Remains of San Francisco after the 1906 earthquake and fire. Lester C. Guernsey. 1906. Public domain. https://commons.wikimedia.org/wiki/File:San_Francisco_1906_earthquake_Panoramic_View.jpg

10.

MASS WASTING

This chapter discusses the fundamental processes driving mass-wasting, types of **mass wasting**, examples and lessons learned from famous mass-wasting events, how **mass wasting** can be predicted, and how people can be protected from this potential hazard. **Mass wasting** is the downhill movement of rock and **soil** material due to gravity. The term **landslide** is often used as a synonym for **mass wasting**, but **mass wasting** is a much broader term referring to all movement downslope. Geologically, **landslide** is a general term for **mass wasting** that involves fast-moving geologic material. Loose material along with overlying **soils** are what typically move during a mass-wasting event. Moving blocks of **bedrock** are called rock topples, rock slides, or rock **falls**, depending on the dominant motion of the blocks. Movements of dominantly liquid material are called flows. Movement by **mass wasting** can be slow or rapid. Rapid movement can be dangerous, such as during **debris flows**. Areas with steep topography and rapid rainfall, such as the California coast, Rocky Mountain Region, and Pacific Northwest, are particularly susceptible to hazardous mass-wasting events.

10.1 Slope Strength

Mass wasting occurs when a slope fails. A slope fails when it is too steep and unstable for existing materials and conditions. Slope stability is ultimately determined by two principal factors: the slope angle and the strength of the underlying material. Force of gravity, which plays a part in **mass wasting**, is constant on the Earth's surface for the most part, although small variations exist depending on the elevation and density of the underlying rock. In the figure, a block of rock situated on a slope is pulled down toward the Earth's center by the force of gravity (fg). The gravitational force acting on a slope can be divided into two components: the **shear** or **driving force** (fs) pushing the block down the slope, and the normal or **resisting force** (fn) pushing into the slope, which produces friction. The relationship between **shear force** and **normal force** is called **shear strength**. When the normal force, i.e., friction, is greater than the **shear force**,

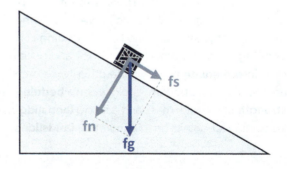

Figure 10.1: Forces on a block on an inclined plane (fg = force of gravity, fn = normal force; fs = shear force).

then the block does *not* move downslope. However, if the slope angle becomes steeper or if the earth material is weakened, **shear force** exceeds **normal force**, compromising **shear strength**, and downslope movement occurs.

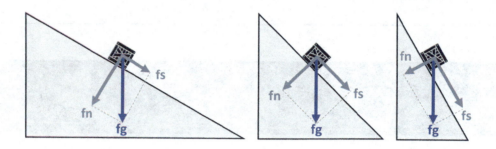

Figure 10.2: As slope increases, the force of gravity (fg) stays the same and the normal force decreases while the shear force proportionately increases.

In the figure, the force vectors change as the slope angle increases. The gravitational force doesn't change, but the **shear force** increases while the **normal force** decreases. The steepest angle at which rock and **soil** material is stable and will *not* move downslope is called the **angle of repose**. The **angle of repose** is measured relative from the horizontal. When a slope is at the **angle of repose**, the **shear force** is in equilibrium with the **normal force**. If the slope becomes just slightly steeper, the **shear force** exceeds the **normal force**, and the material starts to move downhill. The **angle of repose** varies for all material and slopes depending on many factors such as **grain size**, grain **composition**, and water content. The figure shows the **angle of repose** for sand that is poured into a pile on a flat surface. The sand grains cascade down the sides of the pile until coming to rest at the **angle of repose**. At that angle, the base and height of the pile continue to increase, but the angle of the sides remains the same.

Figure 10.3: Angle of repose in a pile of sand.

Water is a common factor that can significantly change the **shear strength** of a particular slope. Water is located in **pore** spaces, which are empty air spaces in **sediments** or rocks between the grains. For example, assume a dry sand pile has an **angle of repose** of 30 degrees. If water is added to the sand, the **angle of repose** will increase, possibly to 60 degrees or even 90 degrees, such as a sandcastle being built at a beach. But if too much water is added to the pore spaces of the sandcastle, the water decreases the **shear strength**, lowers the **angle of repose**, and the sandcastle collapses.

Another factor influencing **shear strength** are planes of weakness in sedimentary rocks. **Bedding** planes (see chapter 5) can act as significant planes of weakness when they are parallel to the slope but less so if they are perpendicular to the slope. At locations A and B, the **bedding** is nearly perpendicular to the slope and relatively stable. At location D, the **bedding** is nearly parallel to the slope and quite unstable. At location C, the **bedding** is nearly horizontal, and the stability is **intermediate** between the other two extremes. Additionally, if clay **minerals** form along **bedding** planes, they can absorb water and become slick. When a **bedding** plane of **shale** (clay and silt) becomes **saturated**, it can lower the **shear strength** of the rock mass and cause a **landslide**, such as at the 1925 Gros Ventre, Wyoming rock slide. See the case studies section for details on this and other **landslides**.

Figure 10.4: Locations A and B have bedding nearly perpendicular to the slope, making for a relatively stable slope. Location D has bedding nearly parallel to the slope, increasing the risk of slope failure. Location C has bedding nearly horizontal and the stability is relatively intermediate.

Take this quiz to check your comprehension of this section.

If you are using an offline version of this text, access the quiz for section 10.1 via the QR code.

An interactive H5P element has been excluded from this version of the text. You can view it online here: https://pressbooks.lib.vt.edu/introearthscience/?p=698#h5p-67

10.2 Mass-Wasting Triggers & Mitigation

Mass-wasting events often have a **trigger**: something changes that causes a **landslide** to occur at a specific time. It could be rapid snowmelt, intense rainfall, earthquake shaking, **volcanic** eruption, storm waves, rapid-stream **erosion**, or human activities, such as **grading** a new road. Increased water content within the slope is the most common mass-wasting **trigger**. Water content can increase due to rapidly melting snow or ice or an intense rain event. Intense rain events can occur more often during El Niño years. Then, the west coast of North America receives more **precipitation** than normal, and **landslides** become more common. Changes in surface-water conditions resulting from earthquakes, previous slope failures that dam up **streams**, or human structures that interfere with **runoff**, such as buildings, roads, or parking lots can provide additional water to a slope. In the case of the 1959 Hebgen Lake rock slide, Madison Canyon, Montana, the **shear strength** of the slope may have been weakened by earthquake shaking. Most **landslide** mitigation diverts and drains water away from slide areas. Tarps and plastic sheeting are often used to drain water off of slide bodies and prevent **infiltration** into the slide. Drains are used to dewater **landslides** and shallow wells are used to monitor the water content of some active **landslides**.

An **oversteepened** slope may also **trigger landslides**. Slopes can be made excessively steep by natural processes of **erosion** or when humans modify the landscape for building construction. An example of how a slope may be **oversteepened** during development occurs where the bottom of the slope is cut into, perhaps to build a road or level a building lot, and the top of the slope is modified by depositing excavated material from below. If done carefully, this practice can be very useful in land development, but in some cases, this can result in devastating consequences. For example, this might have been a contributing factor in the 2014 North Salt Lake City, Utah **landslide**. A former gravel pit was regraded to provide a road and several building lots. These activities may have **oversteepened** the slope, which resulted in a slow moving **landslide** that destroyed one home at the bottom of the slope. Natural processes such as excessive **stream erosion** from a flood or coastal erosion during a storm can also **oversteepen** slopes. For example, natural undercutting of the riverbank was proposed as part of the **trigger** for the famous 1925 Gros Ventre, Wyoming rock slide.

Slope reinforcement can help prevent and mitigate **landslides**. For **rockfall**-prone areas, sometimes it is economical to use long steel bolts. Bolts, drilled a few meters into a rock face, can secure loose pieces of material that could pose a hazard. Shockcrete, a reinforced spray-on form of concrete, can strengthen a slope face when applied properly. Buttressing a slide by adding weight at the toe of the slide and removing weight from the head of the slide, can stabilize a **landslide**. Terracing, which creates a stairstep topography, can be applied to help with slope stabilization, but it must be applied at the proper scale to be effective.

A different approach in reducing **landslide** hazard is to **shield**, catch, and divert the runout material. Sometimes the most economical way to deal with a **landslide** hazard is to divert and slow the falling material. Special stretchable fencing can be applied in areas where **rockfall** is common to protect pedestrians and vehicles. Runout channels, diversion structures, and check dams can be used to slow **debris flows** and divert them around structures. Some highways have special tunnels that divert **landslides** over the highway. In all of these cases the shielding has to be engineered to a scale that is greater than the slide, or catastrophic loss in property and life could result.

Take this quiz to check your comprehension of this section.

If you are using an offline version of this text, access the quiz for section 10.2 via the QR code.

 An interactive H5P element has been excluded from this version of the text. You can view it online here:
https://pressbooks.lib.vt.edu/introearthscience/?p=698#h5p-68

10.3 Landslide Classification & Identification

Mass-wasting events are classified by type of movement and type of material, and there are several ways to classify these events. The figure and table show terms used. In addition, mass-wasting types often share common morphological features observed on the surface, such as the head scarp—commonly seen as crescent shapes on a cliff face; hummocky or uneven surfaces; accumulations of **talus**—loose rocky material falling from above; and toe of slope, which covers existing surface material.

10.3.1 Types of Mass Wasting

The most common mass-wasting types are **falls**, rotational and **translational slides**, flows, and **creep**. **Falls** are abrupt rock movements that detach from steep slopes or cliffs. Rocks separate along existing natural breaks such as **fractures** or **bedding** planes. Movement occurs as free-falling, bouncing, and rolling. **Falls** are strongly influenced by gravity, **mechanical weathering**, and water. **Rotational slides** commonly show slow movement along a curved rupture surface. **Translational slides** often are rapid movements along a plane of distinct weakness between the overlying slide material and more stable underlying material. Slides can be further subdivided into rock slides, debris slides, or earth slides depending on the type of the material involved (see table).

Type of movement	Primary material type and common name of slide		
	Bedrock	Soil: mostly coarse-grained	Soil: mostly fine-grained
Falls	Rock fall		
Rock avalanche	Rock avalanche		
Rotational slide (slump)		Rotational debris slide (slump)	Rotational Earth slide (slump)
Transitional slide	Transitional rock slide	Transitional debris slide	Transitional Earth slide
Flows		Debris flow	Earth flow
Soil creep		Creep	Creep

Table 10.1: Mass wasting types.

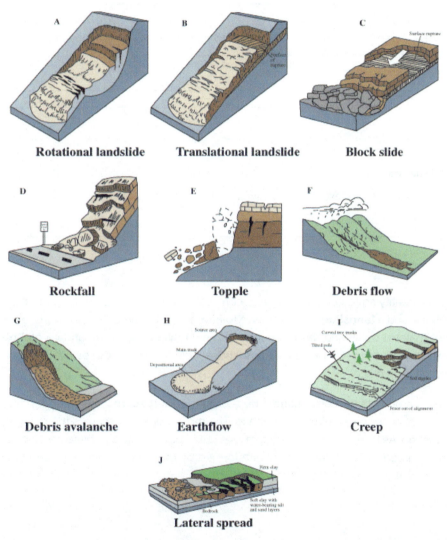

Figure 10.5: Examples of some of the types of landslides.

Flows are rapidly moving mass-wasting events in which the loose material is typically mixed with abundant water, creating long runouts at the slope base. Flows are commonly separated into **debris flow** (coarse material) and **earthflow** (fine material) depending on the type of material involved and the amount of water. Some of the largest and fastest flows on land are called **sturzstroms**, or long runout **landslides**. They are still poorly understood, but are known to travel for long distances, even in places without significant atmospheres like the Moon.

Creep is the imperceptibly slow downward movement of material caused by a regular cycle of nighttime freezing followed by daytime thawing in unconsolidated material such as **soil**. During the freeze, expansion of ice pushes **soil** particles out away from the slope, while the next day following the thaw, gravity pulls them directly downward. The net effect is a gradual movement of surface **soil** particles downhill. **Creep** is indicated by curved tree trunks, bent fences or retaining walls, tilted poles or fences, and small **soil ripples** or ridges. A special type of **soil creep** is solifluction, which is the slow movement of **soil** lobes on low-angle slopes due to **soil** seasonally freezing and thawing in high-**latitude**, typically sub-Arctic, Arctic, and Antarctic locations.

 One or more interactive elements has been excluded from this version of the text. You can view them online here: https://www.youtube.com/watch?v=MVwSpGVfWVo

Video 10.1: Landslide hazards.

If you are using an offline version of this text, access this YouTube video via the QR code.

10.3.2 Parts of a Landslide

Landslides have several identifying features that can be common across the different types of **mass wasting**. Note that there are many exceptions, and a **landslide** does not have to have these features. Displacement of material by **landslides** causes the absence of material uphill and the **deposition** of new material downhill, and careful **observation** can identify the evidence of that displacement. Other signs of **landslides** include tilted or **offset** structures or natural features that would normally be vertical or in place.

Many **landslides** have escarpments or scarps. **Landslide** scarps, like **fault scarps**, are steep terrain created when movement of the adjacent land exposes a part of the subsurface. The most prominent scarp is the main scarp, which marks the uphill extent of the **landslide**. As the disturbed material moves out of place, a step slope forms and develops a new hillside escarpment for the undisturbed material. Main scarps are formed by movement of the displaced material away from the undisturbed ground and are the visible part of slide rupture surface.

Complete this interactive activity to check your understanding.

If you are using an offline version of this text, access this interactive activity via the QR code.

 An interactive H5P element has been excluded from this version of the text. You can view it online here: https://pressbooks.lib.vt.edu/introearthscience/?p=698#h5p-69

The slide rupture surface is the boundary of the body of movement of the **landslide**. The geologic material below the slide surface does not move, and is marked on the sides by the flanks of the **landslide** and at the end by the toe of the **landslide**.

The toe of the **landslide** marks the end of the moving material. The toe marks the runout, or maximum distance traveled, of the **landslide**. In rotational **landslides**, the toe is often a large, disturbed mound of geologic material, forming as the **landslide** moves past its original rupture surface.

Rotational and translational **landslides** often have **extensional** cracks, sag ponds, hummocky terrain and pressure ridges. **Extensional** cracks form when a **landslide**'s toe moves forward faster than the rest of **landslide**, resulting in **tensional** forces. Sag ponds are small bodies of water filling depressions formed where **landslide** movement has impounded **drainage**. Hummocky terrain is undulating and uneven topography that results from the ground being disturbed. Pressure ridges develop on the margins of the **landslide** where material is forced upward into a ridge structure.

Take this quiz to check your comprehension of this section.

If you are using an offline version of this text, access the quiz for section 10.3 via the QR code.

 An interactive H5P element has been excluded from this version of the text. You can view it online here: https://pressbooks.lib.vt.edu/introearthscience/?p=698#h5p-70

10.4 Examples of Landslides

10.4.1 Landslides in United States

Gros Ventre, Wyoming (1925): On June 23, 1925, a 38 million cubic meter (50 million cu yd) translational rock slide occurred next to the Gros Ventre River (pronounced "grow vont") near Jackson Hole, Wyoming. Large boulders dammed the Gros Ventre River and ran up the opposite side of the valley several hundred vertical feet. The dammed **river** created Slide Lake, and two years later in 1927, lake levels rose high enough to destabilize the dam. The dam failed and caused a catastrophic flood that killed six people in the small downstream community of Kelly, Wyoming.

Figure 10.6: Scar of the Gros Ventre landslide in background with landslide deposits in the foreground.

Figure 10.7: Lower Slide Lake was created on June 23, 1925, when the Gros Ventre landslide dammed the Gros Ventre River. It is located in Bridger-Teton National Forest, in the U.S. state of Wyoming.

A combination of three factors caused the rock slide: 1) heavy rains and rapidly melting snow **saturated** the Tensleep sandstone causing the underlying **shale** of the Amsden Formation to lose its **shear strength**, 2) the Gros Ventre River cut through the **sandstone** creating an **oversteepened** slope, and 3) soil on top of the mountain became saturated with water due to poor **drainage**. The cross-section diagram shows how the parallel **bedding** planes between the Tensleep sandstone and Amsden Formation offered little friction against the slope surface as the **river** undercut the **sandstone**. Lastly, the rockslide may have been triggered by an earthquake.

Madison Canyon, Montana (1959): In 1959, the largest earthquake in Rocky Mountain recorded history, **magnitude** 7.5, struck the Hebgen Lake, Montana area, causing a destructive seiche on the lake (see chapter 9). The earthquake caused a rock avalanche that dammed the Madison River, creating Quake Lake, and ran up the other side of the valley hundreds of vertical feet. Today, there are still house-sized boulders visible on the slope opposite their starting point. The slide moved at a velocity of up to 160.9 kph (100 mph), creating an incredible air blast that swept through the Rock Creek Campground. The slide killed 28 people, most of whom were in the campground and remain buried there. In a manner like the Gros Ventre slide, **foliation** planes of weakness in **metamorphic rock** outcrops were parallel with the surface, compromising **shear strength**.

Mount Saint Helens, Washington (1980): On May 18, 1980 a 5.1-**magnitude** earthquake triggered the largest **landslide** observed in the historical record. This **landslide** was followed by the lateral eruption of Mount Saint Helens **volcano**, and the eruption was followed by **volcanic debris flows** known as **lahars**. The volume of material moved by the **landslide** was 2.8 cubic kilometers (0.67 mi^3).

La Conchita, California (1995 and 2005): On March 4, 1995, a fast-moving **earthflow** damaged nine houses in the southern California coastal community of La Conchita. A week later, a **debris flow** in the same location damaged five more houses. Surface-**tension** cracks at the top of the slide gave early warning signs in the summer of 1994. During the rainy winter season of 1994/1995, the cracks grew larger. The likely **trigger** of the 1995 event was unusually

Figure 10.8: Road damage from the August 1959 Hebgen Lake (Montana-Yellowstone) earthquake.

heavy rainfall during the winter of 1994/1995 and rising **groundwater** levels. Ten years later, in 2005, a rapid-**debris flow** occurred at the end of a 15-day **period** of near-record rainfall in southern California. Vegetation remained relatively intact as it was rafted on the surface of the rapid flow, indicating that much of the **landslide** mass simply was being carried on a presumably much more **saturated** and fluidized layer beneath. The 2005 slide damaged 36 houses and killed 10 people.

Figure 10.9: Oblique LIDAR image of La Conchita after the 2005 landslide. Outline of 1995 (blue) and 2005 (yellow) landslides shown; arrows show examples of other landslides in the area; red line outlines main scarp of an ancient landslide for the entire bluff.

Figure 10.10: 1995 La Conchita slide.

Figure 10.11: 2014 Oso slide in Washington killed 43 people and buried many homes.

Oso Landslide, Washington (2014): On March 22, 2014, a **landslide** of approximately 18 million tons (10 million yd^3) traveled at 64 kph (40 mph), extended for nearly a 1.6 km (1 m), and dammed the North Fork of the Stillaguamish River. The **landslide** covered 40 homes and killed 43 people in the Steelhead Haven community near Oso, Washington. It produced a volume of material equivalent to 600 football fields covered in material 3 m (10 ft) deep. The winter of 2013-2014 was unusually wet with almost double the average amount of **precipitation**. The **landslide** occurred in an area of the Stillaguamish River Valley historically active with many **landslides**, but previous events had been small.

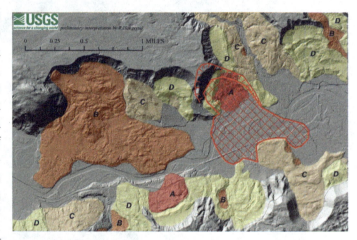

Figure 10.12: Annotated Lidar map of 2014 Oso slide in Washington.

Yosemite National Park Rock Falls: The steep cliffs of Yosemite National Park cause frequent rock **falls**. **Fractures** created to **tectonic stresses** and **exfoliation** and expanded by **frost wedging** can cause house-sized blocks of **granite** to detach from the cliff-faces of Yosemite National Park. The park models potential runout, the distance **landslide** material travels, to better assess the risk posed to the millions of park visitors.

 One or more interactive elements has been excluded from this version of the text. You can view them online here: https://www.youtube.com/watch?v=H0YhlqP1BgE

Video 10.2: Rock fall in Yosemite.

If you are using an offline version of this text, access this YouTube video via the QR code.

10.4.2 Landslides in Utah

Markagunt Gravity Slide: About 21–22 million years ago, one of the biggest land-based **landslides** yet discovered in the geologic record displaced more than 1,700 cu km (408 cu mi) of material in one relatively fast event. Evidence for this slide includes breccia **conglomerates** (see chapter 5), glassy pseudotachylytes, (see chapter 6), slip surfaces (similar to **faults**) see chapter 9), and **dikes** (see chapter 7). The **landslide** is estimated to encompass an area the size of Rhode Island and to extend from near Cedar City, Utah to Panguitch, Utah. This **landslide** was likely the result of material released from the side of a growing **laccolith** (a type of **igneous** intrusion) see chapter 4), after being triggered by an eruption-related earthquake.

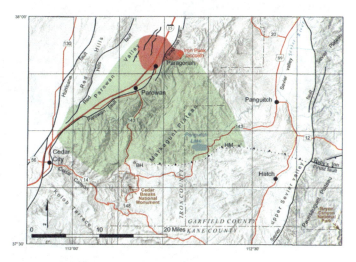

Figure 10.13: Approximate extent of Markagunt gravity slide.

Figure 10.14: The 1983 Thistle landslide (foreground) dammed the Spanish Fork river creating a lake.

Thistle Slide (1983): Starting in April of 1983 and continuing into May of that year, a slow-moving **landslide** traveled 305 m (1,000 ft) downhill and blocked Spanish Fork Canyon with an **earthflow** dam 61 m (200 ft) high. This caused disastrous flooding upstream in the Soldier Creek and Thistle Creek valleys, submerging the town of Thistle. As part of the emergency response, a spillway was constructed to prevent the newly formed lake from breaching the dam. Later, a tunnel was constructed to drain the lake, and currently the **river** continues to flow through this tunnel. The rail line and US-6 highway had to be relocated at a cost of more than $200 million.

Figure 10.15: House before and after destruction from 2013 Rockville rockfall.

Rockville Rock Fall (2013): Rockville, Utah is a small community near the entrance to Zion National Park. In December of 2013, a 2,700 ton (1,400 yd³) block of Shinarump **Conglomerate** fell from the Rockville Bench cliff, landed on the steep 35-degree slope below, and shattered into several large pieces that continued downslope at a high speed. These boulders completely destroyed a house located 375 feet below the cliff (see the before and after photographs) and killed two people inside the home. The topographic map shows other rock **falls** in the area prior to this catastrophic event.

North Salt Lake Slide (2014): In August 2014 after a particularly wet **period**, a slow moving rotational **landslide** destroyed one home and damaged nearby tennis courts.

Reports from residents suggested that ground cracks had been seen near the top of the slope at least a year prior to the catastrophic movement. The presence of easily-drained sands and gravels overlying more impermeable clays weathered from **volcanic ash**, along with recent regrading of the slope, may have been contributing causes of this slide. Local heavy rains seem to have provided the **trigger**. In the two years after the **landslide**, the slope has been partially regraded to increase its stability. Unfortunately, in January 2017, parts of the slope have shown reactivation movement. Similarly, in 1996 residents in a nearby subdivision started reporting distress to their homes. This distress continued until 2012 when 18 homes became uninhabitable due to extensive damage and were removed. A geologic park was constructed in the now vacant area.

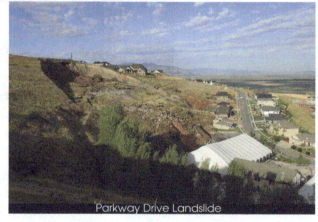

Figure 10.16: Scarp and displaced material from the North Salt Lake (Parkview) slide of 2014.

One or more interactive elements has been excluded from this version of the text. You can view them online here:
https://www.youtube.com/watch?v=NQs_OWgNshg

Video 10.3: North Salt Lake landslide.

If you are using an offline version of this text, access this YouTube video via the QR code.

Bingham Canyon Copper Mine Landslide, Utah (2013): At 9:30 pm on April 10, 2013, more than 65 million cubic meters of steep terraced **mine** wall slid down into the engineered pit of Bingham Canyon **mine**, making it one of the largest historic **landslides** not associated with **volcanoes**. Radar systems maintained by the **mine** operator warned of movement of the wall, preventing the loss of life and limiting the loss of property.

Complete this interactive activity to check your understanding.

If you are using an offline version of this text, access this interactive activity via the QR code.

An interactive H5P element has been excluded from this version of the text. You can view it online here:
https://pressbooks.lib.vt.edu/introearthscience/?p=698#h5p-71

Take this quiz to check your comprehension of this section.

If you are using an offline version of this text, access the quiz for section 10.4 via the QR code.

An interactive H5P element has been excluded from this version of the text. You can view it online here:
https://pressbooks.lib.vt.edu/introearthscience/?p=698#h5p-72

Summary

Mass wasting is a geologic term describing all downhill rock and **soil** movement due to gravity. **Mass wasting** occurs when a slope is too steep to remain stable with existing material and conditions. Loose rock and **soil**, called **regolith**, are what typically move during a mass-wasting event. Slope stability is determined by two factors: the angle of the slope and the **shear strength** of the accumulated materials. Mass-wasting events are triggered by changes that **oversteepen** slope angles and weaken slope stability, such as rapid snow melt, intense rainfall, earthquake shaking, **volcanic** eruption, storm waves, **stream erosion**, and human activities. Excessive **precipitation** is the most common trigger. Mass-wasting events are classified by their type of movement and material, and they share common morphological surface features. The most common types of mass-wasting events are **rockfalls**, slides, flows, and **creep**.

Mass-wasting movement ranges from slow to dangerously rapid. Areas with steep topography and rapid rainfall, such as the California coast, Rocky Mountain Region, and Pacific Northwest, are particularly susceptible to hazardous mass-wasting events. By examining examples and lessons learned from famous mass-wasting events, scientists have a better understanding of how mass-wasting occurs. This knowledge has brought them closer to predicting where and how these potentially hazardous events may occur and how people can be protected.

Take this quiz to check your comprehension of this chapter.

If you are using an offline version of this text, access the quiz for chapter 10 via the QR code.

An interactive H5P element has been excluded from this version of the text. You can view it online here:
https://pressbooks.lib.vt.edu/introearthscience/?p=698#h5p-73

Text References

1. Haugerud, R.A., 2014, Preliminary interpretation of pre-2014 landslide deposits in the vicinity of Oso, Washington: US Geological Survey.
2. Highland, L., 2004, Landslide types and processes: pubs.er.usgs.gov.
3. Highland, L.M., and Bobrowsky, P., 2008, The Landslide Handbook – A Guide to Understanding Landslide: U.S. Geological Survey USGS Numbered Series 1325, 147 p.
4. Highland, L.M., and Schuster, R.L., 2000, Significant landslide events in the United States: United States Geological Survey.
5. Hildenbrand, T.G., and Hendricks, J.D., 1995, Geophysical setting of the Reelfoot rift and relations between rift structures and the New Madrid seismic zone: U.S. Geological Survey Professional Paper 1538-E, 36 p.
6. Hungr, O., Leroueil, S., and Picarelli, L., 2013, The Varnes classification of landslide types, an update: Landslides, v. 11, no. 2, p. 167–194.
7. Jibson, R.W., 2005, Landslide hazards at La Conchita, California: United States Geological Survey Open-File Report 2005-1067.
8. Lipman, P.W., and Mullineaux, D.R., 1981, The 1980 eruptions of Mount St. Helens, Washington: US Geological Survey USGS Numbered Series 1250, 844 p., doi: 10.3133/pp1250.
9. Lund, W.R., Knudsen, T.R., and Bowman, S.D., 2014, Investigation of the December 12, 2013, Fatal Rock Fall at 368 West Main Street, Rockville, Utah: Utah Geological Survey 273, 24 p.

10. United States Forest Service, 2016, A Brief History of the Gros Ventre Slide Geological Site: United States Forest Service.

Figure References

Figure 10.1: Forces on a block on an inclined plane (fg = force of gravity; fn = normal force; fs = shear force). Kindred Grey. 2022. CC BY 4.0. Includes crate by Andrew Doane from Noun Project (Noun Project license).

Figure 10.2: As slope increases, the force of gravity (fg) stays the same and the normal force decreases while the shear force proportionately increases. Kindred Grey. 2022. CC BY 4.0. Includes crate by Andrew Doane from Noun Project (Noun Project license).

Figure 10.3: Angle of repose in a pile of sand. Captain Sprite. 2007. CC BY-SA 2.5. https://en.wikipedia.org/wiki/File:Angle-ofrepose.png

Figure 10.4: Locations A and B have bedding nearly perpendicular to the slope, making for a relatively stable slope. Location D has bedding nearly parallel to the slope, increasing the risk of slope failure. Location C has bedding nearly horizontal and the stability is relatively intermediate. Kindred Grey. 2022. CC BY 4.0.

Figure 10.5: Examples of some of the types of landslides. R.L. Schuster, U.S. Geological Survey. 2004. Public domain. https://pubs.usgs.gov/fs/2004/3072/fs-2004-3072.html

Figure 10.6: Scar of the Gros Ventre landslide in background with landslide deposits in the foreground. Daniel Mayer. 2006. CC BY-SA 3.0. https://commons.wikimedia.org/wiki/File:Gros_Venture_Slide.JPG

Figure 10.7: Grand Teton National Park showing sedimentary layers parallel with the surface and undercutting (over-steepening) of the slope by the river. Daniel Mayer. 2006. CC BY-SA 3.0. https://commons.wikimedia.org/wiki/File:Lower_Slide_Lake.JPG

Figure 10.8: Road damage from the August 1959 Hebgen Lake (Montana-Yellowstone) earthquake. R.B. Colton via USGS. 1959. Public domain. https://commons.wikimedia.org/wiki/File:Roaddamage59quake.JPG

Figure 10.9: Oblique LIDAR image of La Conchita after the 2005 landslide. Todd Stennett via USGS. 2016. Public domain. https://www.usgs.gov/media/images/la-conchita

Figure 10.10: 1995 La Conchita slide. USGS. 2005. Public domain. https://commons.wikimedia.org/wiki/File:Laconchita1995landslide.jpg

Figure 10.11: 2014 Oso slide in Washington killed 43 people and buried many homes. Mark Reid, USGS. 2014. Public domain. https://commons.wikimedia.org/wiki/File:Oso_Landslide_aerial.jpg

Figure 10.12: Annotated Lidar map of 2014 Oso slide in Washington. USGS. 2014. Public domain. https://commons.wikimedia.org/wiki/File:Oso_landslide_geomorphology_map.png

Figure 10.13: Approximate extent of Markagunt gravity slide. Used under fair use from THE EARLY MIOCENE MARKAGUNT MEGABRECCIA: UTAH'S LARGEST CATASTROPHIC LANDSLIDE by Robert F. Biek. https://geology.utah.gov/map-pub/survey-notes/the-early-miocene-markagunt-megabreccia/

Figure 10.14: The 1983 Thistle landslide (foreground) dammed the Spanish Fork river creating a lake. R.L. Schuster, U.S. Geological Survey. 1983. Public domain. https://en.wikipedia.org/wiki/File:Thistlelandslideusgs.jpg

Figure 10.15: House before and after destruction from 2013 Rockville rockfall. Used under fair use from INVESTIGATION OF THE DECEMBER 12, 2013,FATAL ROCK FALL AT 368 WEST MAIN STREET, ROCKVILLE, UTAH by William R. Lund, Tyler R. Knudsen, and Steve D. Bowman. https://ugspub.nr.utah.gov/publications/reports_of_investigations/ri-273.pdf

Figure 10.16: Scarp and displaced material from the North Salt Lake (Parkview) slide of 2014. Used under fair use from PARK-WAY DRIVE LANDSLIDE, NORTH SALT LAKE. https://geology.utah.gov/hazards/landslides/parkway_drive_landslide/

WATER

By the end of this chapter, students should be able to:

- Describe the processesof the water cycle.
- Describe **drainage** basins, **watershed** protection, and water budget.
- Describe reasons for water laws, who controls them, and how water is shared in the western U.S.
- Describe zone of transport, zone of **sediment** production, zone of **deposition**, and equilibrium.
- Describe **stream** landforms: channel types, **alluvial** fans, floodplains, natural levees, deltas, **entrenched meanders**, and **terraces**.
- Describe the properties required for a good **aquifer**; define **confining layer water table**.
- Describe three major groups of water contamination and three types of **remediation**.
- Describe**karst** topography, how it is created, and the landforms that characterize it.

All life on Earth requires water. The **hydrosphere** (Earth's water) is an important agent of geologic change. Water shapes our planet by depositing **minerals**, aiding **lithification**, and altering rocks after they are lithified. Water carried by **subducted oceanic plates** causes **flux melting** of upper **mantle** material. Water is among the **volatiles** in **magma** and emerges at the surface as steam in **volcanoes**.

Figure 11.1: Example of a Roman aqueduct in Segovia, Spain.

Figure 11.2: Chac mask in Mexico.

Humans rely on suitable water sources for consumption, agriculture, power generation, and many other purposes. In pre-industrial civilizations, the powerful controlled water resources. As shown in the figures, two thousand year old Roman aqueducts still grace European, Middle Eastern, and North African skylines. Ancient Mayan architecture depicts water imagery such as frogs, water-lilies, water fowl to illustrate the importance of water in their societies. In the drier lowlands of the Yucatan Peninsula, mask facades of the hooked-nosed rain god, *Chac* (or *Chaac*) are prominent on Mayan buildings such as the Kodz Poop (Temple of the Masks, sometimes spelled Coodz Poop) at the ceremonial site of Kabah. To this day government controlled water continues to be an integral part of most modern societies.

11.1 Water Cycle

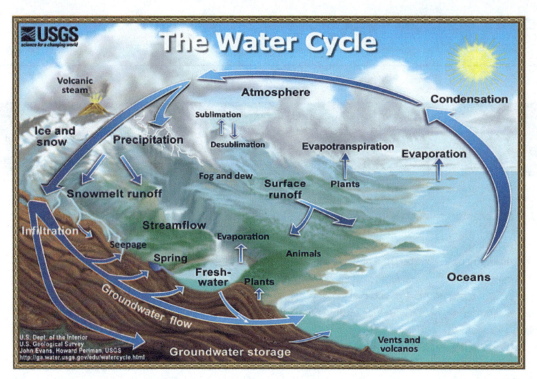

Figure 11.3: The water cycle.

The water cycle is the continuous circulation of water in the Earth's **atmosphere**. During circulation, water changes between solid, liquid, and gas (water vapor) and changes location. The processes involved in the water cycle are evaporation, transpiration, condensation, **precipitation**, and **runoff**.

Evaporation is the process by which a liquid is converted to a gas. Water evaporates when solar energy warms the water sufficiently to excite the water molecules to the point of vaporization. Evaporation occurs from oceans, lakes, and **streams** and the land surface. Plants contribute significant amounts of water vapor as a byproduct of photosynthesis called transpiration that occurs through the minute **pores** of plant leaves. The term **evapotranspiration** refers to these two sources of water entering the **atmosphere** and is commonly used by geologists.

Water vapor is invisible. Condensation is the process of water vapor transitioning to a liquid. Winds carrywater vapor in the **atmosphere** long distances. When water vapor cools or when air masses of different temperatures mix,water vapor may condense back into droplets of liquid water. These water droplets usually form around a microscopic piece of dust or salt called condensation nuclei. Thesesmall droplets of liquid water suspended in the **atmosphere** becomevisible as in a cloud. Water droplets inside clouds collide and stick together, growing into largerdroplets. Once the water droplets become big enough, they fall to Earth as rain, snow, hail, or sleet.

Once **precipitation** has reached the Earth's surface, it can evaporate or flow as **runoff** into **streams**, lakes, and eventually back to the oceans. Water in **streams** and lakes is called surface water. Or water can also **infiltrate** into the **soil** and fill the **pore** spaces in the rock or **sediment** underground to become **groundwater**. **Groundwater** slowly moves through rock and unconsolidated materials. Some **groundwater** may reach the surface again, where it discharges as springs, **streams**, lakes, and the ocean. Also, surface water in **streams** and lakes can **infiltrate** again to **recharge groundwater**. Therefore, the surface water and **groundwater** systems are connected.

One or more interactive elements has been excluded from this version of the text. You can view them online here:
https://www.youtube.com/watch?v=al-do-HGulk

Video 11.1: Water cycle.

If you are using an offline version of this text, access this YouTube video via the QR code.

Take this quiz to check your comprehension of this section.

If you are using an offline version of this text, access the quiz for section 11.1 via the QR code.

 An interactive H5P element has been excluded from this version of the text. You can view it online here:
https://pressbooks.lib.vt.edu/introearthscience/?p=765#h5p-74

11.2 Water Basins and Budgets

The basic unit of division of the landscape is the **drainage basin**, also known as a **catchment** or **watershed**. It is the area of land that captures **precipitation** and contributes **runoff** to a **stream** or **stream** segment. **Drainage** divides are local topographic high points that separate one **drainage basin** from another. Water that **falls** on one side of the divide goes to one **stream**, and water that **falls** on the other side of the divide goes to a different **stream**. Each **stream**, **tributary** and streamlet has its own **drainage basin**. In areas with flatter topography, **drainage** divides are not as easily identified but they still exist.

Figure 11.4: Map view of a drainage basin with main trunk streams and many tributaries with drainage divide in dashed red line.

Latorița River, tributary of the Lotru River
(Drainage basin)

Figure 11.5: Oblique view of the drainage basin and divide of the Latorita River, Romania.

The headwater is where the **stream** begins. Smaller **tributary streams** combine downhill to make the larger trunk of the **stream**. The **mouth** is where the stream finally reaches its end. The mouth of most streams is at the ocean. However, a rare number of **streams** do not flow to the ocean, but rather end in a **closed basin** (or **endorheic basin**) where the only outlet is evaporation. Most **streams** in the Great Basin of Western North America end in **endorheic basins**. For example, in Salt Lake County, Utah, Little Cottonwood Creek and the Jordan River flow into the endorheic Great Salt Lake where the water evaporates.

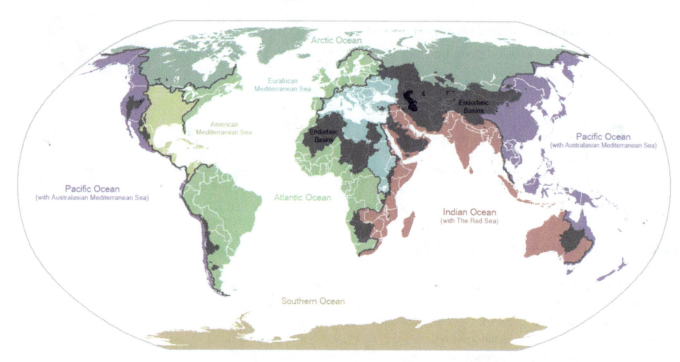

Figure 11.6: Major drainage basins color coded to match the related ocean. Closed basins (or endorheic basins) are shown in gray.

Perennial **streams** flow all year round. Perennial **streams** occur in humid or temperate climates where there is sufficient rainfall and low evaporation rates. Water levels rise and fall with the seasons, depending on the **discharge**. **Ephemeral streams** flow only during rain events or the wet season. In arid climates, like Utah, many **streams** are **ephemeral**. These **streams** occur in dry climates with low amounts of rainfall and high evaporation rates. Their channels are often dry washes or **arroyos** for much of the year and their sudden flow causes **flash floods**.

Along Utah's Wasatch Front, the urban area extending north to south from Brigham City to Provo, there are several **watersheds** that are designated as "**watershed** protection areas" that limit the type of use allowed in those **drainages** in order to protect culinary water. Dogs and swimming are limited in those **watersheds** because of the possibility of contamination by harmful bacteria and substances to the drinking supply of Salt Lake City and surrounding municipalities.

Water in the water cycle is very much like money in a personal budget. Income includes **precipitation** and **stream** and **groundwater** inflow. Expenses include **groundwater** withdrawal, evaporation, and **stream** and **groundwater** outflow. If the expenses outweigh the income, the water budget is not balanced. In this case, water is removed from savings, i.e. water storage, if available. **Reservoirs**, snow, ice, **soil** moisture, and **aquifers** all serve as storage in a water budget. In dry regions, the water is critical for sustaining human activities. Understanding and managing the water budget is an ongoing political and social challenge.

Hydrologists create **groundwater** budgets within any designated area, but they are generally made for **watershed** (**basin**) boundaries, because **groundwater** and surface water are easier to account for within these boundaries. Water budgets can be created for state, county, or aquifer extent boundaries as well. The groundwater budget is an essential component of the hydrologic model; hydrologists use measured data with a conceptual workflow of the model to better understand the water **system**.

Take this quiz to check your comprehension of this section.

If you are using an offline version of this text, access the quiz for section 11.2 via the QR code.

An interactive H5P element has been excluded from this version of the text. You can view it online here:
https://pressbooks.lib.vt.edu/introearthscience/?p=765#h5p-75

11.3 Water Use and Distribution

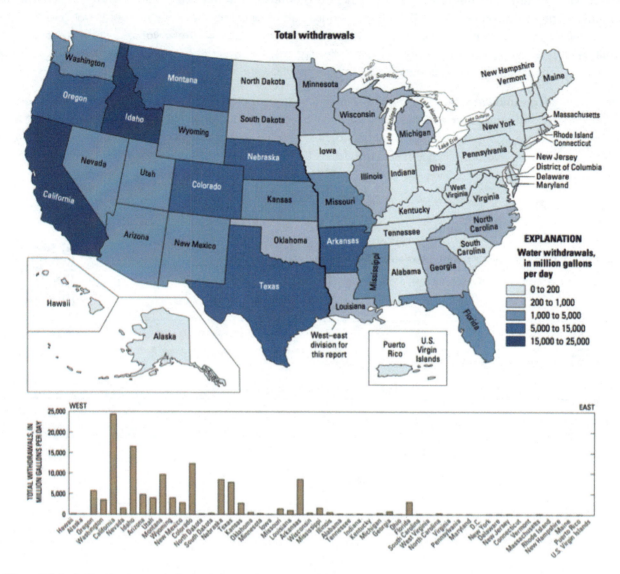

Figure 11.7: Agricultural water use in the United States by state.

In the United States, 1,344 billionliters (355 billion gallons) of ground and surface water are used each day, of which 288 billionliters (76 billion gallons) are fresh **groundwater**. The state of California uses 16% of national **groundwater**.

Utah is the second driest state in the United States. Nevada, having a mean statewide **precipitation** of 31 cm (12.2 inches) per year, is the driest. Utah also has the second highest per capita rate of total domestic water use of 632.16 liters (167 gallons) per person per day. With the combination of relatively high demand and limited quantity, Utah is at risk for water budget deficits.

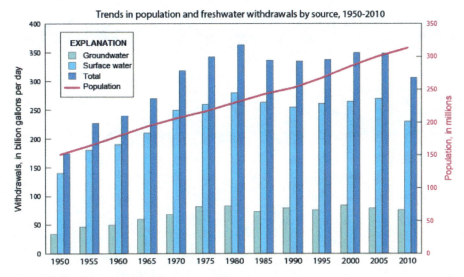

Figure 11.8: Trends in water use by source.

11.3.1 Surface Water Distribution

Fresh water is a precious resource and should not be taken for granted, especially in dry **climates**. Surface water makes up only 1.2% of the fresh water available on the planet, and 69% of that surface water is trapped in ground ice and **permafrost**. **Stream** water accounts for only 0.006% of all freshwater and lakes contain only 0.26% of the world's fresh water.

Global circulation patterns are the most important factor in distributing surface water through **precipitation**. Due to the Coriolis effect and the uneven heating of the Earth, air rises near the equator and near latitudes 60° north and south. Air sinks at the poles and latitudes 30° north and south (see chapter 13). Land masses near rising air are more prone to humid and wet climates. Land masses near sinking air, which inhibits **precipitation**, are prone to dry conditions. Prevailing winds, ocean circulation patterns such as the Gulf Stream's effects on eastern North America, rain shadows (the dry leeward sides of mountains), and even the proximity of bodies of water can affect local **climate** patterns. When this moist air collides with the nearby mountains causing it to rise and cool, the moisture may fall out as snow or rain on nearby areas in a phenomenon known as "lake-effect **precipitation**."

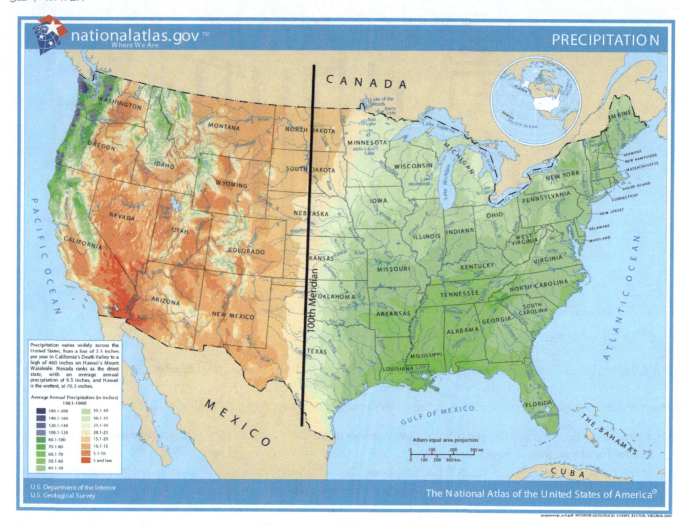

Figure 11.9: Distribution of precipitation in the United States. The 100th Meridian is approximately where the average precipitation transitions from relatively wet to dry. (Source: U.S. Geological Survey)

In the United States, the 100th meridian roughly marks the boundary between the humid and arid parts of the country. Growing crops west of the 100th meridian requires irrigation. In the west, surface water is stored in **reservoirs** and mountain snowpacks, then strategically released through a **system** of canals during times of high water use.

Some of the driest parts of the western United States are in the **Basin and Range** Province. The **Basin and Range** has multiple mountain ranges that are oriented north to south. Most of the **basin** valleys in the **Basin and Range** are dry, receiving less than 30 cm (12 inches) of **precipitation** per year. However, some of the mountain ranges can receive more than 1.52 m (60 inches) of water as snow or snow-water-equivalent. The snow-water equivalent is the amount of water that would result if the snow were melted, as the snowpack is generally much thicker than the equivalent amount of water that it would produce.

11.3.2 Groundwater Distribution

Water source	Water volume (cubic miles)	Fresh water (%)	Total water (%)
Oceans, seas, and bays	321,000,000	—	96.5%
Ica caps, glaciers, and permanent snow	5,773,000	68.7%	1.74%
Groundwater (total)	5,614,000	—	1.69%
Groundwater (fresh)	2,526,000	30.1%	0.76%
Groundwater (saline)	3,088,000	—	0.93%
Soil moisture	3.959	0.05%	0.001%
Ground ice and permafrost	71,970	0.86%	0.022%
Lakes (total)	42,320	—	0.013%
Lakes (fresh)	21,830	0.26%	0.007%
Lakes (saline)	20,490	—	0.006%
Atmosphere	3,095	0.04%	0.001%
Swamp water	2,752	0.03%	0.0008%
Rivers	509	0.006%	0.0002%
Biological water	269	0.003%	0.0001%

Table 11.1: Groundwater distribution. Source: Igor Shiklomanov's chapter "World fresh water resources" in Peter H. Gleick (editor), 1993, Water in Crisis: A Guide to the World's Fresh Water Resources (Oxford University Press, New York).

Groundwater makes up 30.1% of the fresh water on the planet, making it the most abundant **reservoir** of fresh water accessible to most humans. The majority of freshwater, 68.7%, is stored in **glaciers** and ice caps as ice. As the **glaciers** and ice caps melt due to global warming, this fresh water is lost as it flows into the oceans.

Take this quiz to check your comprehension of this section.

If you are using an offline version of this text, access the quiz for section 11.3 via the QR code.

 An interactive H5P element has been excluded from this version of the text. You can view it online here: https://pressbooks.lib.vt.edu/introearthscience/?p=765#h5p-76

11.4 Water Law

Federal and state governments have put laws in place to ensure the fair and equitable use of water. In the United States, the states are tasked with creating a fair and legal **system** for sharing water.

11.4.1 Water Rights

Because of the limited supply of water, especially in the western United States, states disperse a **system** of legal **water rights** defined as a claim to a portion or all of a water source, such as a **spring**, **stream**, well, or lake. Federal law mandates that states control water rights, with the special exception of federally reserved water rights, such as those associated with national parks and Native American tribes, and navigation servitude that maintains navigable water bodies. Each state in the United States has a different way to disperse and manage water rights.

A person, entity, company, or organization, must have a **water right** to legally extract or use surface or **groundwater** in their state. Water rights in some western states are dictated by the concept of prior appropriation, or "first in time, first in right," where the person with the oldest **water right** gets priority water use during times when there is not enough water to fulfill every **water right**.

The Colorado River and its tributaries pass through a desert region, including seven states (Wyoming, Colorado, Utah, New Mexico, Arizona, Nevada, California), Native American reservations, and Mexico. As the western United States became more populated and while California was becoming a key agricultural producer, the states along the Colorado River realized that the **river** was important to sustaining life in the West.

To guarantee certain *perceived* water rights, these western states recognized that a water budget was necessary for the Colorado River Basin. Thus was enacted the Colorado River Compact in 1922 to ensure that each state got a fair share of the **river** water. The Compact granted each state a specific volume of water based on the total measured flow at the time. However, in 1922, the flow of the **river** was higher than its long-term average flow, consequently, more water was allocated to each state than is typically available in the **river**.

Over the next several decades, lawmakers have made many other agreements and modifications regarding the Colorado River Compact, including those agreements that brought about the Hoover Dam (formerly Boulder Dam), and Glen Canyon Dam,and a treaty between the American and Mexican governments. Collectively, the agreements are referred to as "The Law of the River" by the United States Bureau of Reclamation. Despite adjustments to the Colorado River Compact, many believe that the Colorado River is still over-allocated, as the Colorado River flow no longer reaches the Pacific Ocean, its original terminus (**base level**). Dams along the Colorado River have caused water to divert and evaporate, creating serious water budget concerns in the Colorado River Basin. Predicted drought associated with global warming is causing additional concerns about over-allocating the Colorado River flow in the future.

The Law of the River highlights the complex and prolonged nature of interstate water rights agreements, as well as the importance of water.

One or more interactive elements has been excluded from this version of the text. You can view them online here:
https://www.youtube.com/watch?v=MZrKW-Q9X8E

Video 11.2: The Colorado River Compact of 1992.

If you are using an offline version of this text, access this YouTube video via the QR code.

The Snake Valley straddles the border of Utah and Nevada withmore of the irrigable land area lying on the Utah side of the border.In 1989, the Southern Nevada Water Authority (SNWA) submitted applications for water rights to pipe up to 191,189,707 cu m (155,000 ac-ft) of water per year (an acre-foot of water is one acre covered with water one foot deep) from Spring, Snake, Delamar, Dry Lake, and Cave valleys to southern Nevada, mostly for Las Vegas. Nevada and Utah have attempted a comprehensive agreement, but negotiations have not yet been settled.

Complete this interactive activity to check your understanding.

If you are using an offline version of this text, access this interactive activity via the QR code, or by visiting https://www.arcgis.com/apps/MapJournal/index.html?appid=79199afd183e459596e6e21315159354.

NPR story on Snake Valley

 One or more interactive elements has been excluded from this version of the text. You can view them online here: https://www.npr.org/player/embed/10953190/10956967#

If you are using an offline version of this text, access this NPR story via the QR code, or by visiting https://www.npr.org/player/embed/10953190/10956967#.

SNWA History

Dean Baker Story

 One or more interactive elements has been excluded from this version of the text. You can view them online here: https://www.youtube.com/watch?v=eCZ8KLrmUXo

Video 11.3: Transporting Snake Valley water to satisfy a thirsty Las Vegas.

If you are using an offline version of this text, access this YouTube video via the QR code.

11.4.2 Water Quality and Protection

Two major federal laws that protect water quality in the United States are the Clean Water Act and the Safe Drinking Water Act. The Clean Water Act, an amendment of the Federal Water Pollution Control Act, protects navigable waters from dumping and point-source pollution. The Safe Drinking Water Act ensures that water that is provided by public water suppliers, like cities and towns, is safe to drink.

The U.S. Environmental Protection Agency Superfund program ensures the cleanup of hazardous contamination, and can be applied to situations of surface water and **groundwater** contamination. It is part of the Comprehensive Environmental Response, Compensation, and Liability Act of 1980. Under this act, state governments and the U.S. Environmental Protection Agency can use the **superfund** to pay for remediation of a contaminated site and then file a lawsuit against the polluter to recoup the costs. Or to avoid being sued, the polluter that caused the contamination may take direct action or provide funds to remediatethe contamination.

Take this quiz to check your comprehension of this section.

If you are using an offline version of this text, access the quiz for section 11.4 via the QR code.

An interactive H5P element has been excluded from this version of the text. You can view it online here:
https://pressbooks.lib.vt.edu/introearthscience/?p=765#h5p-77

11.5 Surface Water

Geologically, a **stream** is a body of flowing surface water confined to a channel. Terms such as **river**, creek and brook are social terms not used in geology. **Streams** erode and transport **sediments**, making them the most important agents of the earth's surface, along with wave action (see chapter 12) in eroding and transporting **sediments**. They create much of the surface topography and are an important water resource.

Several factors cause **streams** to erode and transport **sediment**, but the two main factors are **stream**–channel **gradient** and velocity. **Stream**–channel **gradient** is the slope of the**stream** usually expressed in meters per kilometer or feet per mile. A steeper channel gradient promotes **erosion**. When **tectonic** forces elevate a mountain, the **stream gradient** increases, causing themountain **stream** to erode downward and deepen its channel eventually forming a valley. **Stream**–channel velocity is the speed at which channel water flows. Factors affecting channel velocityinclude channel **gradient** which decreases downstream, **discharge** and channel size which increaseas tributaries coalesce, and channel roughness which decreases as **sediment** lining the channel walls decreases in size thus reducing friction. The combined effect of these factors is that channel velocity actually increases from mountain brooks to the **mouth** of the **stream**.

11.5.1 Discharge

Stream size is measured in terms of **discharge**, the volume of water flowing past a point in the **stream** over a defined time interval. Volume is commonly measured in cubic units (length x width x depth), shown as feet^3 (ft^3) or meter^3 (m^3). Therefore, the units of **discharge** are cubic feet per second (ft$_3$/sec or cfs). Therefore, the units of **discharge** are cubic meters per second, (m^3/s or cms, or cubic feet per second (ft^3/sec or cfs). **Stream discharge** increases downstream. Smaller **streams**

have less **discharge** than larger **streams**. For example, the Mississippi **River** is the largest **river** in North America, with an average flow of about 16,990.11 cms (600,000 cfs). For comparison, the average **discharge** of the Jordan **River** at Utah Lake is about 16.25 cms (574 cfs) and for the annual **discharge** of the Amazon River, (the world's largest river), annual discharge is about 175,565 cms (6,200,000 cfs).

Discharge can be expressed by the following equation:

Q = V A

- Q = **discharge** cms (or ft^3/sec),
- A = cross-sectional area of the **stream** channel [width times average depth] as m^2 (or in^2 or ft^2),
- V = average channel velocity m/s (or ft/sec)

At a given location along the **stream**, velocity varies with **stream** width, shape, and depth within the **stream** channel as well. When the stream channel narrows but **discharge** remains constant, the same volume of water must flows through a narrower space causing the velocity to increase, similar to putting a thumb over the end of a backyard water hose. In addition, during rain storms or heavy snow melt, **runoff** increases, which increases **stream discharge** and velocity.

When the **stream** channel curves, the highest velocity will be on the outside of the bend. When the **stream** channel is straight and uniformly deep, the highest velocity is in the channel center at the top of the water where it is the farthest from frictional contact with the **stream** channel bottom and sides. In hydrology, the ***thalweg*** of a river is the line drawn that shows its natural progression and deepest channel, as is shown in the diagram.

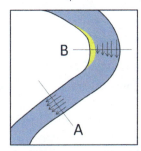 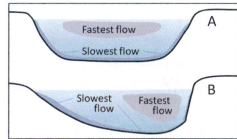

Figure 11.10: Thalweg of a river. In a river bend, the fastest moving water is on the outside of the bend, near the cutbank. Stream velocity is higher on the outside bend and the water surface which is farthest from the friction of the stream bed. Longer arrows indicate faster velocity (Earle 2015).

11.5.2 Runoff versus Infiltration

Factors that dictate whether water will **infiltrate** into the ground or run off over the land include the amount, type, and intensity of **precipitation**; the type and amount of vegetation cover; the slope of the land; the **temperature** and aspect of the land; preexisting conditions; and the type of **soil** in the infiltrated area. High– intensity rain will cause more **runoff** than the same amount of rain spread out over a longer duration. If the rain **falls** faster than the **soil**'s properties allow it to **infiltrate**, then the water that cannot **infiltrate** becomes **runoff**. Dense vegetation can increase **infiltration**, as the vegetative cover slows the water particle's overland flow giving them more time to **infiltrate**. If a parcel of land has more direct solar radiation or higher seasonal temperatures, there will be less **infiltration** and **runoff**, as **evapotranspiration** rates will be higher. As the land's slope increases, so does **runoff**, because the water is more inclined to move downslope than **infiltrate** into the ground. Extreme examples are a **basin** and a cliff, where water **infiltrates** much quicker into a **basin** than a cliff that has the same **soil** properties. Because saturated soil does not have the capacity to take more water, runoff is generally greater over saturated soil. Clay-rich **soil** cannot accept **infiltration** as quickly as gravel-rich **soil**.

11.5.3 Drainage Patterns

The pattern of tributaries within a region is called **drainage pattern**. They depend largely on the type of rock beneath, and on structures within that rock (such as **folds** and **faults**). The main types of **drainage patterns** are dendritic, trellis, **rectangular**, radial, and deranged. **Dendritic patterns** are the most common and develop in areas where the underlying rock or **sediments** are uniform in character, mostly flat lying, and can be eroded equally easily in all directions. Examples are **alluvial sediments** or flat lying sedimentary rocks. **Trellis patterns** typically develop where sedimentary rocks have been folded or tilted and then eroded to varying degrees depending on their strength. The Appalachian Mountains in eastern United States have many good examples of **trellis drainage**. **Rectangular** patterns develop in areas that have very little topography and a **system** of **bedding** planes, **joints**, or **faults** that form a rectangular network. A **radial pattern** forms when **streams** flow away from a central high point such as a mountain top or **volcano**, with the individual **streams** typically having **dendritic drainage** patterns. In places with extensive **limestone** deposits, **streams** can disappear into the **groundwater** via caves and subterranean **drainage** and this creates a **deranged pattern**.

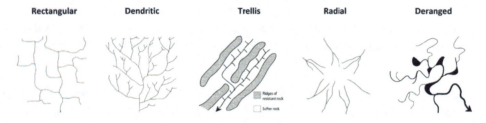

Figure 11.11: Various stream drainage patterns.

11.5.4 Fluvial Processes

Fluvial processes dictate how a **stream** behaves and include factors controlling **fluvial sediment** production, transport, and **deposition**. **Fluvial** processes include velocity, slope and **gradient**, **erosion**, transportation, **deposition**, **stream** equilibrium, and **base level**.

Streams can be divided into three main zones: the many smaller tributaries in the source area, the main trunk **stream** in the **floodplain** and the distributaries at the **mouth** of the **stream**. Major **stream** systems like the Mississippi are composed of many source areas, many tributaries and trunk **streams**, all coalescing into the one main **stream** draining the region. The zones of a **stream** are defined as 1) the zone of **sediment** production (erosion), 2) the zone of transport, and 3) the zone of deposition. The zone of sediment production is located in the **headwaters** of the **stream**. In the zone of **sediment** transport, there is a general balance between **erosion** of the finer **sediment** in its channel and transport of **sediment** across the **floodplain**. **Streams** eventually flow into the ocean or end in quiet water with a **delta** which is a zone of **sediment deposition** located at the **mouth** of a **stream**. The **longitudinal profile** of a **stream** is a plot of the elevation of the **stream** channel at all points along its course and illustrates the location of the three zones.

Zone of Sediment Production

The zone of **sediment** production is located in the **headwaters** of a **stream** where rills and gullies erode **sediment** and contribute to larger **tributary streams**. These tributaries carry **sediment** and water further downstream to the main trunk of the **stream**. Tributaries at the **headwaters** have the steepest **gradient**; **erosion** there produces considerable **sediment** carried b the **stream**. Headwater **streams** tend to be narrow and straight with small or non-existent floodplains adjacent to the channel. Since the zone of **sediment** production is generally the steepest part of the **stream**, **headwaters** are generally located in relatively high elevations. The Rocky Mountains of Wyoming and Colorado west of the Continental Divide contain much of the **headwaters** for the Colorado River which then flows from Colorado through Utah and Arizona to Mexico. **Headwaters** of the Mississippi river system lie east of the Continental Divide in the Rocky Mountains and west of the Appalachian Divide.

Zone of Sediment Transport

Streams transport **sediment** great distances from the **headwaters** to the ocean, the ultimate depositional basins. **Sediment** transportation is directly related to **stream gradient** and velocity. Faster and steeper **streams** can transport larger **sediment** grains. When velocity slows down, larger **sediments** settle to the channel bottom. When the velocity increases, those larger **sediments** are entrained and move again.

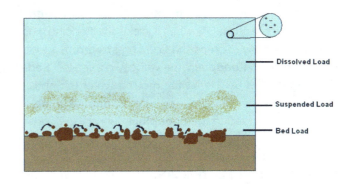

Figure 11.12: A stream carries dissolved load, suspended load, and bedload.

Transported **sediments** are grouped into **bedload**, **suspended load**, and **dissolved load** as illustrated in the above image. **Sediments** moved along the channel bottom are the **bedload** that typically consists of the largest and densest particles. **Bedload** is moved by **saltation** (bouncing) and traction (being pushed or rolled along by the force of the flow). Smaller particles are picked up by flowing water and carried in suspension as **suspended load**. The particle size that is carried in suspended and **bedload** depends on the flow velocity of the **stream**. **Dissolved load** in a **stream** is the total of the ions in **solution** from **chemical weathering**, including such common ions such as bicarbonate ($-HCO_3-$), calcium (Ca_{+2}), chloride (Cl_{-1}), potassium (K_{+1}), and sodium (Na_{+1}). The amounts of these ions are not affected by flow velocity.

 One or more interactive elements has been excluded from this version of the text. You can view them online here:
https://www.youtube.com/watch?v=is-qcxrKKBI

Video 11.4: Bed load sediment transport.

If you are using an offline version of this text, access this YouTube video via the QR code.

A **floodplain** is the flat area of land adjacent to a **stream** channel inundated with flood water on a regular basis. **Stream** flooding is a natural process that adds **sediment** to floodplains. A **stream** typically reaches its greatest velocity when it is close to flooding, known as the **bankfull stage**. As soon as the flooding **stream** overtops its banks and flows onto its **floodplain**, the velocity decreases. **Sediment** that was being carried by the swiftly moving water is deposited at the edge of the channel, forming a low ridge or **natural levée**. In addition, **sediments** are added to the **floodplain** during this flooding process contributing to fertile **soils**.

Zone of Sediment Deposition

Deposition occurs when **bedload** and **suspended load** come to rest on the bottom of the **stream** channel, lake, or ocean due to decrease in **stream gradient** and reduction in velocity. While both **deposition** and **erosion** occur in the zone of transport such as on point bars and **cut banks**, ultimate **deposition** where the **stream** reaches a lake or ocean. Landforms called deltas form where the **stream** enters quiet water composed of the finest **sediment** such as fine sand, silt, and clay.

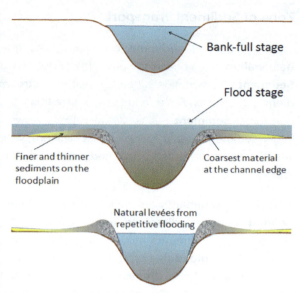

Figure 11.13: Profile of stream channel at bankfull stage, flood stage, and deposition of natural levee.

Equilibrium and Base Level

All three **stream** zones are present in the typical **longitudinal profile** of a **stream** which plots the elevation of the channel at all points along its course (see figure 11.14). All **streams** have a long profile. The long profile shows the **stream gradient** from headwater to **mouth**. All **streams** attempt to achieve an energetic balance among **erosion**, transport, **gradient**, velocity, discharge, and channel characteristics along the **stream**'s profile. This balance is called equilibrium, a state called **grade**.

Another factor influencing equilibrium is **base level**, the elevation of the **stream**'s **mouth** representing the lowest level to which a **stream** can erode. The ultimate **base level** is, of course, sea-level. A lake or **reservoir** may also represent **base level** for a **stream** entering it. The Great Basin of western Utah, Nevada, and parts of some surrounding

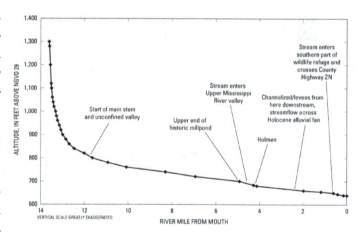

Figure 11.14: Example of a longitudinal profile of a stream; Halfway Creek, Indiana.

states contains no outlets to the sea and provides internal base levels for **streams** within it. **Base level** for a **stream** entering the ocean changes if sea-level rises or **falls**. **Base level** also changes if a natural or human-made dam is added along a **stream**'s profile. When **base level** is lowered, a **stream** will cut down and deepen its channel. When **base level** rises, **deposition** increases as the **stream** adjusts attempting to establish a new state of equilibrium. A **stream** that has approximately achieved equilibrium is called a **graded stream**.

11.5.5 Fluvial Landforms

Stream landforms are the land features formed on the surface by either **erosion** or **deposition**. The **stream**-related landforms described here are primarily related to channel types.

Channel Types

Figure 11.15: The braided Waimakariri river in New Zealand.

Stream channels can be straight, **braided**, **meandering**, or entrenched. The **gradient**, **sediment** load, **discharge**, and location of **base level** all influence channel type. **Straight channels** are relatively straight, located near the **headwaters**, have steep gradients, low **discharge**, and narrow V-shaped valleys. Examples of these are located in mountainous areas.

Figure 11.16: Air photo of the meandering river, Río Cauto, Cuba.

Braided streams have multiple channels splitting and recombining around numerous mid-channel bars. These are found in floodplains with low gradients in areas with near sources of coarse **sediment** such as trunk **streams** draining mountains or in front of **glaciers**.

Meandering streams have a single channel that curves back and forth like a snake within its **floodplain** where it emerges from its **headwaters** into the zone of transport. **Meandering streams** are dynamic creating a wide **floodplain** by eroding and extending meander loops side-to-side. The highest velocity water is located on the outside of a meander bend. **Erosion** of the outside of the curve creates a feature called a **cut bank** and the meander extends its loop wider by this **erosion**.

The **thalweg** of the **stream** is the deepest part of the **stream** channel. In the straight parts of the channel, the **thalweg** and highest velocity are in the center of the channel. But at the bend of a **meandering stream**, the **thalweg** shifts toward the **cut bank**. Opposite the **cutbank** on the inside bend of the channel is the lowest **stream** velocity and is an area of **deposition** called a **point bar**.

In areas of **tectonic** uplift such as on the Colorado Plateau, **meandering streams** that once flowed on the plateau surface have become entrenched or incised as uplift occurred and the **stream** cut its **meandering channel** down into **bedrock**. Over the past several million years, the Colorado River and its tributaries have incised into the flat lying rocks of the plateau by hundreds, even thousands of feet creating deep canyons including the Grand Canyon in Arizona.

Figure 11.17: Point bar and cut bank on the Cirque de la Madeleine in France.

Figure 11.18: An entrenched meander on the Colorado River in the eastern entrance to the Grand Canyon.

Figure 11.19: Panoramic view of incised meanders of the San Juan River at Gooseneck State Park, Utah.

Figure 11.20: The Rincon is an abandoned meaner loop on the entrenched Colorado River in Lake Powell.

Many **fluvial** landforms occur on a **floodplain** associated with a **meandering stream**. Meander activity and regular flooding contribute to widening the **floodplain** by eroding adjacent uplands. The **stream** channels are confined by natural levees that have been built up over many years of regular flooding. Natural levees can isolate and direct flow from **tributary** channels on the **floodplain** from immediately reaching the main channel. These isolated streams are called yazoo streams and flow parallel to the main trunk stream until there is an opening in the levee to allow for a belated confluence.

Figure 11.21: Landsat image of Zambezi Flood Plain, Namibia.

 One or more interactive elements has been excluded from this version of the text. You can view them online here: https://www.youtube.com/watch?v=persGpc6-Dw

Video 11.5: How is a levee formed?

If you are using an offline version of this text, access this YouTube video via the QR code.

To limit flooding, humans build artificial levees on flood plains. **Sediment** that breaches the levees during flood stage is called **crevasse splays** and delivers silt and clay onto the **floodplain**. These deposits are rich in nutrients and often make good farm land. When floodwaters crest over human-made levees, the levees quickly erode with potentially catastrophic impacts. Because of the good **soils**, farmers regularly return after floods and rebuild year after year.

Through **erosion** on the outsides of the meanders and **deposition** on the insides, the channels of **meandering streams** move back and forth across their **floodplain** over time. On very broad floodplains with very low gradients, the meander bends can become so extreme that they cut across themselves at a narrow neck (see figure 11.22) called a cutoff. The former channel becomes isolated and forms an **oxbow** lake seen on the right of the figure. Eventually the **oxbow** lake fills in with **sediment** and becomes a wetland and eventually a **meander scar**. Stream meanders can migrate and form **oxbow** lakes in a relatively short amount of time. Where **stream** channels form geographic and political boundaries, this shifting of channels can cause conflicts.

Figure 11.22: Meander nearing cutoff on the Nowitna River in Alaska.

One or more interactive elements has been excluded from this version of the text. You can view them online here:
https://www.youtube.com/watch?v=8a3r-cG8Wic

Video 11.6: Why do rivers curve?

If you are using an offline version of this text, access this YouTube video via the QR code.

Alluvial fans are a depositional landform created where **streams** emerge from mountain canyons into a valley. The channel that had been confined by the canyon walls is no longer confined, slows down and spreads out, dropping its **bedload** of all sizes, forming a **delta** in the air of the valley. As distributary channels fill with **sediment**, the **stream** is diverted laterally, and the **alluvial** fan develops into a cone shaped landform with distributaries radiating from the canyon **mouth**. **Alluvial** fans are common in the dry climates of the West where **ephemeral streams** emerge from canyons in the ranges of the **Basin and Range**.

Figure 11.23: Alluvial fan in Iraq seen by NASA satellite. A stream emerges from the canyon and creates this cone-shaped deposit.

Complete this interactive activity to check your understanding.

If you are using an offline version of this text, access this interactive activity via the QR code.

An interactive H5P element has been excluded from this version of the text. You can view it online here:
https://pressbooks.lib.vt.edu/introearthscience/?p=765#h5p-78

A **delta** is formed when a **stream** reaches a quieter body of water such as a lake or the ocean and the **bedload** and **suspended load** is deposited. If wave **erosion** from the water body is greater than **deposition** from the **river**, a **delta** will not form. The largest and most famous **delta** in the United States is the Mississippi River **delta** formed where the Mississippi River flows into the Gulf of Mexico. The Mississippi River **drainage basin** is the largest in North America, draining 41% of the contiguous United States. Because of the large **drainage** area, the **river** carries a large amount of **sediment**. The Mississippi River is a major shipping route and human engineering has ensured that the channel has been artificially straightened and remains fixed within the **floodplain**. The **river** is now 229 km shorter than it was before humans began engineering it. Because of these restraints, the **delta** is now focused on one trunk channel and has created a "bird's foot" pattern. The two NASA images below of the **delta** show how the **shoreline** has retreated and land was inundated with water while **deposition** of **sediment** was focused at end of the distributaries. These images have changed over a 25 year **period**

Figure 11.24: Location of the Mississippi River drainage basin and Mississippi River delta.

from 1976 to 2001. These are stark changes illustrating sea-level rise and land **subsidence** from the **compaction** of peat due to the lack of **sediment** resupply.

Complete this interactive activity to check your understanding.

If you are using an offline version of this text, access this interactive activity via the QR code.

 An interactive H5P element has been excluded from this version of the text. You can view it online here:
https://pressbooks.lib.vt.edu/introearthscience/?p=765#h5p-79

The **formation** of the Mississippi River **delta** started about 7500 years ago when postglacial sea level stopped rising. In the past 7000 years, prior to **anthropogenic** modifications, the Mississippi River **delta** formed several sequential lobes. The **river** abandoned each lobe for a more preferred route to the Gulf of Mexico. These **delta** lobes were reworked by the ocean waves of the Gulf of Mexico. After each lobe was abandoned by the **river**, isostatic depression and **compaction** of the **sediments** caused **basin subsidence** and the land to sink.

A clear example of how deltas form came from an earthquake. During the 1959 Madison Canyon 7.5 **magnitude** earthquake in Montana, a large **landslide** dammed the Madison River forming Quake Lake still there today. A small **tributary stream** that once flowed into the Madison River, now flows into Quake Lake forming a **delta** composed of coarse **sediment** actively eroded from the mountainous upthrown block to the north.

Deltas can be further categorized as wave-dominated or **tide**-dominated. Wave-dominated deltas occur where the tides are small and wave energy dominates. An example is the Nile River **delta** in the Mediterranean Sea that has the classic shape of the Greek character (Δ) from which the landform is named. A **tide**-dominated **delta** forms when ocean tides are powerful and influence the shape of the **delta**. For example, Ganges-Brahmaputra Delta in the Bay of Bengal (near India and Bangladesh) is the world's largest **delta** and mangrove swamp called the Sundarban.

Figure 11.25: Delta in Quake Lake Montana. Deposition of this delta began in 1959, when the Madison river was dammed by the landslide caused by the 7.5 magnitude earthquake.

At the Sundarban Delta in Bangladesh, tidal forces create linear intrusions of seawater into the **delta**. This **delta** also holds the world's largest mangrove swamp.

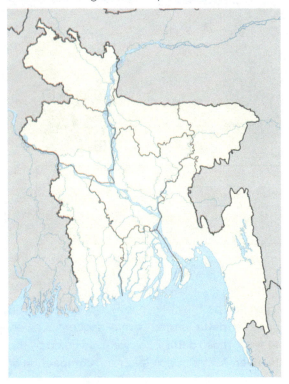

Figure 11.26: Sundarban Delta in Bangladesh, a tide-dominated delta of the Ganges River.

Figure 11.27: Nile Delta showing its classic "delta" shape.

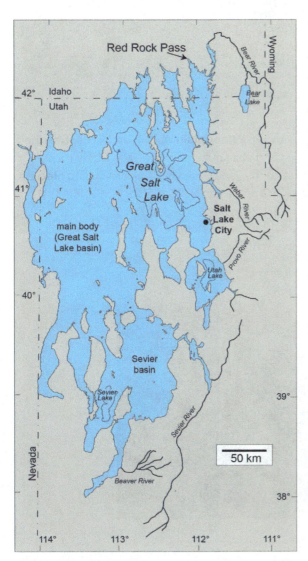

Figure 11.28: Map of Lake Bonneville, showing the outline of the Bonneville shoreline, the highest level of the lake.

Lake Bonneville was a large, **pluvial lake** that occupied the western half of Utah and parts of eastern Nevada from about 30,000 to 12,000 years ago. The lake filled to a maximum elevation as great as approximately 5100 feet above mean sea level, filling the basins, leaving the mountains exposed, many as islands. The presence of the lake allowed for **deposition** of both fine grained lake mud and silt and coarse gravels from the mountains. Variations in lake level were controlled by regional **climate** and a catastrophic failure of Lake Bonneville's main outlet, Red Rock Pass. During extended **periods** of time in which the lake level remained stable, wave-cut **terraces** were produced that can be seen today on the flanks of many mountains in the region. Significant deltas formed at the mouths of major canyons in Salt Lake, Cache, and other Utah valleys. The Great Salt Lake is the remnant of Lake Bonneville and cities have built up on these **delta** deposits.

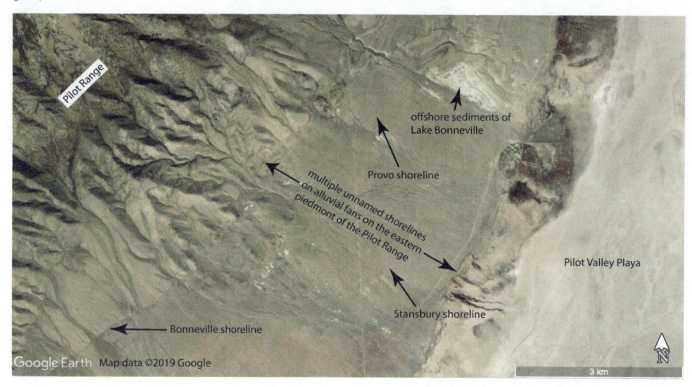

Figure 11.29: Deltaic deposits of Lake Bonneville near Logan, Utah; wave cut terraces can be seen on the mountain slope.

Stream terraces are remnants of older floodplains located above the existing **floodplain** and **river**. Like **entrenched meanders**, **stream terraces** form when uplift occurs or **base level** drops and **streams** erode downward, their meanders widening a new **flood plain**. **Stream terraces** can also form from extreme flood events associated with retreating **glaciers**. A classic example of multiple **stream terraces** are along the Snake **River** in Grand Teton National Park in Wyoming.

Figure 11.30: Terraces along the Snake River, Wyoming.

Take this quiz to check your comprehension of this section.

If you are using an offline version of this text, access the quiz for section 11.5 via the QR code.

 An interactive H5P element has been excluded from this version of the text. You can view it online here:
https://pressbooks.lib.vt.edu/introearthscience/?p=765#h5p-80

11.6 Groundwater

Groundwater is an important source of freshwater. It can be found at varying depths in all places under the ground, but is limited by extractable quantity and quality.

 One or more interactive elements has been excluded from this version of the text. You can view them online here:
https://www.youtube.com/watch?v=g7R0yLX0V9E

Video 11.7: What is an aquifer?

If you are using an offline version of this text, access this YouTube video via the QR code.

11.6.1 Porosity and Permeability

An **aquifer** is a rock unit that contains extractable ground water. A good **aquifer** must be both porous and permeable. Porosity is the space between grains that can hold water, expressed as the percentage of open space in the total volume of the rock. **Permeability** comes from connectivity of the spaces that allows water to move in the **aquifer**. **Porosity** can occur as primary **porosity**, as space between sand grains or vesicles in **volcanic** rocks, or secondary **porosity** as **fractures** or **dissolved** spaces in rock). **Compaction** and **cementation** during **lithification** of **sediments** reduces **porosity** (see section 5.3).

A combination of a place to contain water (**porosity**) and the ability to move water (**permeability**) makes a good **aquifer**—a rock unit or **sediment** that allows extraction of **groundwater**. Well-sorted **sediments** have higher **porosity** because there are not smaller **sediment** particles filling in the spaces between the larger particles. **Shales** made of clays generally have high **porosity**, but the **pores** are poorly connected, thereby causing low **permeability**.

One or more interactive elements has been excluded from this version of the text. You can view them online here:
https://www.youtube.com/watch?v=8mfBomrw0rs

Video 11.8: Porosity and permeability.

If you are using an offline version of this text, access this YouTube video via the QR code.

While **permeability** is an important measure of a porous material's ability to transmit water, **hydraulic conductivity** is more commonly used by geologists to measure how easily a fluid is transmitted. **Hydraulic conductivity** measures both the permeability of the porous material and the properties of the water, or whatever fluid is being transmitted like **oil** or gas. Because **hydraulic conductivity** also measures the properties of the fluid, such as **viscosity**, it is used by both **petro-leum** geologists and hydrogeologists to describe both the production capability of **oil reservoirs** and of **aquifers**. High **hydraulic conductivity** indicates that fluid transmits rapidly through an **aquifer**.

11.6.2 Aquifers

Aquifers are rock layers with sufficient **porosity** and **permeability** to allow water to be both contained and move within them. For rock or **sediment** to be considered an **aquifer**, its **pores** must be at least partially filled with water and it must be permeable enough to transmit water. Drinking water **aquifers** must also contain potable water. **Aquifers** can vary dramatically in scale, from spanning several **formations** covering large regions to being a local **formation** in a limited area. **Aquifers** adequate for water supply are both permeable, porous, and potable.

11.6.3 Groundwater Flow

When surface water **infiltrates** or seeps into the ground, it usually enters the unsaturated zone also called the **vadose zone**, or zone of aeration. The **vadose zone** is the volume of geologic material between the land surface and the **zone of saturation** where the **pore** spaces are not completely filled with water. Plant roots inhabit the upper **vadose zone** and fluid pressure in the **pores** is less than atmospheric pressure. Below the **vadose zone** is the capillary fringe. Capillary fringe is the usually thin zone below the **vadose zone** where the **pores** are completely filled with water (**saturation**), but the fluid pressure is less than atmospheric pressure. The **pores** in the capillary fringe are filled because of capillary action, which occurs because of a combination of **adhesion** and **cohesion**. Below the capillary fringe is the **saturated** zone or phreatic zone, where the **pores** are completely **saturated** and the fluid in the **pores** is at or above atmospheric pressure. The interface between the capillary fringe and the **saturated** zone marks the location of the **water table**.

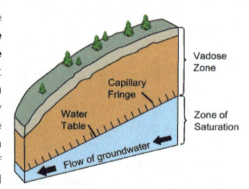

Figure 11.31: Zone of saturation.

Wells are conduits that extend into the ground with openings to the **aquifers**, to extract from, measure, and sometimes add water to the **aquifer**. Wells are generally the way that geologists and hydrologist measure the depth to **groundwater** from the land surface as well as withdraw water from **aquifers**.

Water is found throughout the **pore** spaces in **sediments** and **bedrock**. The **water table** is the area below which the **pores** are fully **saturated** with water. The simplest case of a **water table** is when the **aquifer** is unconfined, meaning it does not have a **confining layer** above it. **Confining layers** can pressurize **aquifers** by trapping water that is **recharged** at a higher elevation underneath the **confining layer**, allowing for a **potentiometric surface** higher than the top of the **aquifer**, and sometimes higher than the land surface.

A **confining layer** is a low **permeability** layer above and/ or below an **aquifer** that restricts the water from moving in and out of the **aquifer**. **Confining layers** include **aquicludes**, which are so impermeable that no water travels through them, and **aquitards**, which significantly decrease the speed at which water travels through them. The **potentiometric surface** represents the height that water would rise in a well penetrating the pressurized **aquifer system**. Breaches in the pressurized **aquifer system**, like **faults** or wells, can cause springs or **flowing wells**, also known as **artesian** wells.

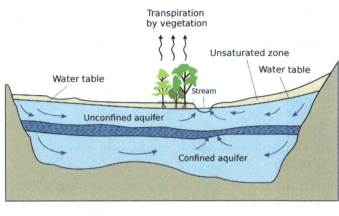

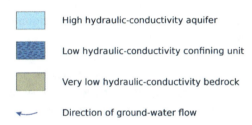

The **water table** will generally mirror surface topography, though more subdued, because hydrostatic pressure is equal to atmospheric pressure along the surface of the **water table**. If the **water table** intersects the ground surface the result will be water at the surface in the form of a gaining **stream**, **spring**, lake, or wetland. The **water table** intersects the channel for gaining **streams** which then gains water from the **water table**. The channels for losing **streams** lie below the **water table**, thus losing **streams** lose water to the **water table**. Losing **streams** may be seasonal during a dry season or **ephemeral** in dry climates

Figure 11.32: An aquifer cross-section. This diagram shows two aquifers with one aquitard (a confining or impermeable layer) between them, surrounded by the bedrock aquiclude, which is in contact with a gaining stream (typical in humid regions).

where they may normally be dry and carry water only after rain storms. **Ephemeral streams** pose a serious danger of **flash flooding** in dry climates.

 One or more interactive elements has been excluded from this version of the text. You can view them online here:
https://www.youtube.com/watch?v=UfgyJkmZgK8

Video 11.9: Where is the water table?

If you are using an offline version of this text, access this YouTube video via the QR code.

Mentioned in the video is the USGS Groundwater Watch site.

Using wells, geologists measure the **water table**'s height and the **potentiometric surface**. Graphs of the depth to **groundwater** over time, are known as hydrographs and show changes in the **water table** over time. Well-water level is controlled by many factors and can change very frequently, even every minute, seasonally, and over longer **periods** of time.

In 1856, French engineer Henry Darcy developed a **hypothesis** to show how **discharge** through a porous medium is controlled by **permeability**, pressure, and cross–sectional area. To prove this relationship, Darcy experimented withtubes of packed **sediment** with water running through them. The results of his experiments empirically established a **quantitative** measure of **hydraulic conductivity** and **discharge** that is known as Darcy's law. The relationships described by Darcy's Law have close similarities to Fourier's law in the field of heat conduction, Ohm's law in the field of electrical networks, or Fick's law in diffusion **theory**.

Q=KA(Δh/L)

- · Q = flow (volume/time)
- · K = **hydraulic conductivity** (length/time)
- · A = cross-sectional area of flow (area)
- · Δh = change in pressure head (pressure difference)
- · L = distance between pressure (h) measurements (length)
- · Δh/L is commonly referred to as the hydraulic **gradient**

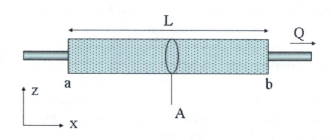

Pumping water from an unconfined **aquifer** lowers the **water table**. Pumping water from a confined **aquifer** lowers the pressure and/or **potentiometric surface** around the well. In an unconfined **aquifer**, the **water table** is lowered as water is removed from the **aquifer** near the well producing drawdown and a **cone of depression** (see figure 11.34). In a confined **aquifer**, pumping on an **artesian well** reduces the pressure or **potentiometric surface** around the well.

Figure 11.33: Pipe showing apparatus that would demonstrate Darcy's Law. Δh would be measured across L from a to b.

When one **cone of depression** intersects another **cone of depression** or a barrier feature like an impermeable mountain block, drawdown is intensified. When a **cone of depression** intersects a recharge zone, the **cone of depression** is lessened.

11.6.4 Recharge

The **recharge** area is where surface water enters an **aquifer** through the process of **infiltration**. **Recharge** areas are generally topographically high locations of an **aquifer**. They are characterized by losing **streams** and permeable rock that allows **infiltration** into the **aquifer**. **Recharge** areas mark the beginning of **groundwater** flow paths.

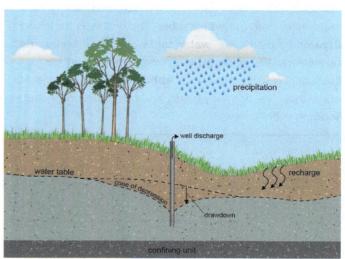

Figure 11.34: Cones of depression.

In the **Basin** and **Range** Province, **recharge** areas for the unconsolidated **aquifers** of the valleys are along mountain foothills. In the foothills of Salt Lake Valley, losing **streams** contribute water to the gravel-rich deltaic deposits of ancient Lake Bonneville, in some cases feeding **artesian** wells in the Salt Lake Valley.

An **aquifer** management practice is to induce **recharge** through storage and recovery. Geologists and hydrologists can increase the recharge rate into an **aquifer system** using injection wells and **infiltration** galleries or basins. Injection wells pump water into an **aquifer** where it can be stored. Injection wells are regulated by state and federal governments to ensure that the injected water is not negatively impacting the quality or supply of the existing **groundwater** in the **aquifer**. Some **aquifers** can store significant quantities of water, allowing water managers to use the **aquifer system** like a surface **reservoir**. Water is stored in the **aquifer** during **periods** of low water demand and high water supply and later extracted during times of high water demand and low water supply.

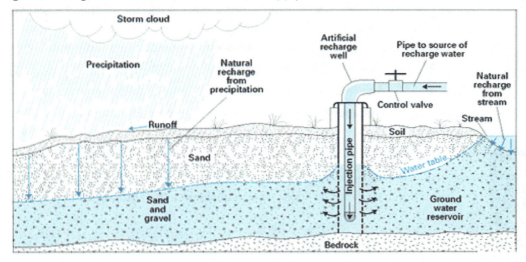

Figure 11.35: Different ways an aquifer can be recharged.

11.6.5 Discharge

Discharge areas are where the **water table** or **potentiometric surface** intersects the land surface. **Discharge** areas mark the end of **groundwater** flow paths. These areas are characterized by springs, flowing (**artesian**) wells, gaining **streams**, and **playas** in the dry valley basins of the **Basin and Range** Province of the western United States.

11.6.6 Groundwater Mining and Subsidence

Like other **natural resources** on our planet, the quantity of fresh and potable water is finite. The only natural source of water on land is from the sky in the form of **precipitation**. In many places, **groundwater** is being extracted faster than it is being replenished. When **groundwater** is extracted faster than it is recharged, **groundwater** levels and potentiometric surfaces decline, and **discharge** areas diminish or dry up completely. Regional pumping-induced **groundwater** decline is known as **groundwater mining** or **groundwater** overdraft. **Groundwater mining** is a serious situation and can lead to dry wells, reduced spring and **stream** flow, and **subsidence**. Groundwater **mining** is happeningis places where more water is extracted by pumping than is being replenished by **precipitation**, and the **water table** is continually lowered. In these situations, **groundwater** must be viewed as a **ore** body and in its depletion, the possibility of producing ghost towns.

 One or more interactive elements has been excluded from this version of the text. You can view them online here: https://www.youtube.com/watch?v=_DuwZjb3_DA

Video 11.10: Groundwater.

If you are using an offline version of this text, access this YouTube video via the QR code.

Figure 11.36: Evidence of land subsidence from pumping of groundwater shown by dates on a pole.

In many places, water actually helps hold up an **aquifer**'s skeleton by the water pressure exerted on the grains in an **aquifer**. This pressure is called **pore** pressure and comes from the weight of overlying water. If **pore** pressure decreases because of **groundwater mining**, the **aquifer** can compact, causing the surface of the ground to sink. Areas especially susceptible to this effect are **aquifers** made of unconsolidated **sediments**. Unconsolidated **sediments** with multiple layers of clay and other fine-grained material are at higher risk because when water is drained, clay compactsconsiderably.

Subsidence from **groundwater mining** has been documented in southwestern Utah, notably Cedar Valley, Iron County, Utah. **Groundwater** levels have declined more than 100 feet in certain parts of Cedar Valley, causing earth fissures and measurable amounts of land **subsidence**.

This photo shows documentation of **subsidence** from pumping of **groundwater** for irrigation in the Central Valley in California. The pole shows **subsidence** from **groundwater** pumping over a **period** of time.

Take this quiz to check your comprehension of this section.

If you are using an offline version of this text, access the quiz for section 11.6 via the QR code.

 An interactive H5P element has been excluded from this version of the text. You can view it online here:
https://pressbooks.lib.vt.edu/introearthscience/?p=765#h5p-81

11.7 Water Contamination and Remediation

Water can be contaminated by natural features like **mineral**-rich geologic **formations** and by human activities such as agriculture, industrial operations, landfills, animal operations, and sewage treatment processes, among many other things. As water runs over the land or **infiltrates** into the ground, it dissolves material left behind by these potential contaminant sources. There are three major groups of contamination: organic and inorganic chemicals and biological agents. Small **sediments** that cloud water, causing turbidity, is also an issue with some wells, but it is not considered contamination. The risks and type of **remediation** for a contaminant depends on the type of chemicals present.

Contamination occurs as point–source and nonpoint–source pollution. **Point source** pollution can be attributed to a single, definable source, while **nonpoint source** pollution is from multiple dispersed sources. **Point sources** include waste disposal sites, storage tanks, sewage treatment plants, and chemical spills. Nonpoint sources are dispersed and indiscreet, where the whole of the contribution of pollutants is harmful, but the individual components do not have harmful concentrations of pollutants. A good example of nonpoint pollution is residential areas, where lawn fertilizer on one person's yard may not contribute much pollution to the **system**, but the combined effect of many residents using fertilizer can lead to significant nonpoint pollution. Other nonpoint sources include nutrients (nitrate and **phosphate**), herbicides, pesticides contributed by farming, nitrate contributed by animal operations, and nitrate contributed by septic systems.

Organic chemicals are common pollutants. They consist of strands and rings of carbon atoms, usually connected by covalent **bonds**. Other types of atoms, like chlorine, and molecules, like hydroxide (OH-), are attached to the strands and rings. The number and arrangement of atoms will decide how the chemical behaves in the environment, its danger to humans or ecosystems, and where the chemical ends up in the environment. The different arrangements of carbon allow for tens of thousands of organic chemicals, many of which have never been studied for negative effects on human health or the environment. Common organic pollutants are herbicides and pesticides, pharmaceuticals, fuel, and industrial solvents and cleansers.

Organic chemicals include surfactants such as cleaning agents and synthetic hormones associated with pharmaceuticals, which can act as endocrine disruptors. Endocrine disruptors mimic hormones, and can cause long-term effects in developing sexual reproduction systems in developing animals. Only very small quantities of endocrine disruptors are needed to cause significant changes in animal populations.

An example of organic chemical contamination is the Love Canal, Niagara Falls, New York. From 1942 to 1952, the Hooker Chemical Company disposed of over 21,337 mt (21,000 t) of chemical waste, including chlorinated hydrocarbons, into a canal and covered it with a thin layer of clay. Chlorinated hydrocarbons are a large group of organic chemicals that have chlorine functional groups, most of which are toxic and carcinogenic to humans. The company sold the land to the New York School Board, who developed it into a neighborhood. After residents began to suffer from serious health ailments and pools of oily fluid started rising into residents' basements, the neighborhood had to be evacuated. This site became a U.S. Environmental Protection Agency **Superfund site**, a site with federal funding and oversight to ensure its cleanup.

Inorganic chemicals are another set of chemical pollutants. They can contain carbon atoms, but not in long strands or links. Inorganic contaminants include chloride, arsenic, and nitrate (NO_3). Nutrients can be from geologic material, like phosphorous-rich rock, but are most often sourced from fertilizer and animal and human waste. Untreated sewage and agricultural **runoff** contain concentrates of nitrogen and phosphorus which are essential for the growth of microorganisms. Nutrients like nitrate and **phosphate** in surface water can promote growth of microbes, like blue-green algae (cyanobacteria), which in turn use oxygen and create toxins (microcystins and anatoxins) in lakes. This process is known as eutrophication.

Metals are common inorganic contaminants. Lead, mercury, and arsenic are some of the more problematic inorganic **groundwater** contaminants. Bangladesh has a well documented case of arsenic contamination from natural geologic material dissolving into the **groundwater**. Acid–**mine drainage** can also cause significant inorganic contamination (see chapter 16).

Salt, typically sodium chloride, is a common inorganic contaminant. It can be introduced into **groundwater** from natural sources, such as **evaporite** deposits like the Arapien Shale of Utah, or from **anthropogenic** sources like the salts applied to roads in the winter to keep ice from forming. Salt contamination can also occur near ocean coasts from saltwater intruding into the cones of depression around fresh **groundwater** pumping, inducing the encroachment of saltwater into the freshwater body.

Biological agents are another common **groundwater** contaminant which includes harmful bacteria and viruses. A common bacteria contaminant is *Escherichia coli* (*E. coli*). Generally, harmful bacteria are not present in **groundwater** unless the **groundwater** source is closely connected with a contaminated surface source, such as a septic **system**. **Karst**, landforms created from **dissolved limestone**, is especially susceptible to this form of contamination, because water moves relatively quickly through theconduits of **dissolved limestone**. Bacteria can also be used for **remediation**.

View USGS lables on contaminants found in groundwater.

Remediation is the act of cleaning contamination. Hydrologists use three types of **remediation**: biological, chemical, and physical. Biological **remediation** uses specific **strains** of bacteria to break down a contaminant into safer chemicals. This type of **remediation** is usually used on organic chemicals, but also works on reducing or oxidizing inorganic chemicals like nitrate. Phytoremediation is a type of bioremediation that uses plants to absorb the chemicals over time.

Chemical **remediation** uses chemicals to remove the contaminant or make it less harmful. One example is to use a reactive barrier, a permeable wall in the ground or at a **discharge** point that chemically reacts with contaminants in the water. Reactive barriers made of **limestone** can increase the pH of **acid mine drainage**, making the water less acidic and more basic, which removes **dissolved** contaminants by **precipitation** into solid form.

Physical **remediation** consists of removing the contaminated water and either treating it with filtration, called pump–and–treat, or disposing of it. All of these options are technically complex, expensive, and difficult, with physical **remediation** typically being the most costly.

Take this quiz to check your comprehension of this section.

If you are using an offline version of this text, access the quiz for section 11.7 via the QR code.

 An interactive H5P element has been excluded from this version of the text. You can view it online here: https://pressbooks.lib.vt.edu/introearthscience/?p=765#h5p-82

11.8 Karst

Karst refers to landscapes and hydrologic features created by the dissolving of **limestone**. **Karst** can be found anywhere there is **limestone** and other soluble subterranean substances like salt deposits. Dissolving of limestone creates features like sinkholes, caverns, disappearing **streams**, and towers.

Figure 11.37: Steep karst towers in China left as remnants as limestone is dissolved away by acidic rain and groundwater.

Figure 11.38: Sinkholes of the McCauley Sink in Northern Arizona, produced by collapse of Kaibab Limestone into caverns caused by solution of underlying salt deposits.

Dissolving of underlying salt deposits has caused sinkholes to form in the Kaibab Limestone on the Colorado Plateau in Arizona.

Figure 11.39: This sinkhole from collapse of surface into a underground cavern appeared in the front yard of this home in Florida.

Collapse of the surface into an underground cavern caused this sinkhole in the front yard of a home in Florida.

CO_2 in the **atmosphere** dissolves readily in the water droplets that form clouds from which **precipitation** comes in the form of rain and snow. This **precipitation** is slightly acidic with **carbonic acid**. Karst forms when carbonic acid dissolves **calcite** (calcium **carbonate**) in **limestone**.

$$H_2O + CO_2 = H_2CO_3$$

Water + Carbon Dioxide Gas equals Carbonic Acid in Water

$$CaCO_3 + H_2CO_3 = Ca^{2+} + 2HCO_3^{-1}$$

Solid Calcite + Carbonic Acid in Water Dissolved equals Calcium Ion + Dissolved Bicarbonate Ion

After the slightly acidic water dissolves the **calcite**, changes in **temperature** or gas content in the water can cause the water to redeposit the **calcite** in a different place as **tufa** (**travertine**), often deposited by a **spring** or in a cave. Speleothems are secondary deposits, typically made of **travertine**, deposited in a cave. **Travertine** speleothems form by water dripping through cracks and **dissolved** openings in caves and evaporating, leaving behind the **travertine** deposits. Speleothems commonly occur in the form of stalactites, when extending from the ceiling, and stalagmites, when standing up from the floor.

Figure 11.40: Mammoth hot springs, Yellowstone National Park.

Figure 11.41: Varieties of speleotherms.

Surface water enters the **karst system** through sinkholes, losing **streams**, and disappearing **streams**. Changes in **base level** can cause **rivers** running over **limestone** to **dissolve** the **limestone** and sink into the ground. As the water continues to **dissolve** its way through the **limestone**, it can leave behind intricate networks of caves and narrow passages. Often **dissolution** will follow and expand **fractures** in the **limestone**. Water exits the **karst system** as springs and rises. In mountainous **terrane**, **dissolution** can extend all the way through the vertical profile of the mountain, with caverns dropping thousands of feet.

Figure 11.42: This stream disappears into a subterranean cavern system to re-emerge a few hundred yards downstream.

Take this quiz to check your comprehension of this section.

If you are using an offline version of this text, access the quiz for section 11.8 via the QR code.

 An interactive H5P element has been excluded from this version of the text. You can view it online here:
https://pressbooks.lib.vt.edu/introearthscience/?p=765#h5p-83

Summary

Water is essential for all living things. It continuously cycles through the **atmosphere**, over land, and through the ground. In much of the United States and other countries, water is managed through a **system** of regional laws and regulations and distributed on paper in a **system** collectively known as "water rights". Surface water follows a **watershed**, which is separate from other areas by its divides (highest ridges). **Groundwater** exists in the **pores** within rocks and **sediment**. It moves predominantly due to pressure and gravitational gradients through the rock. Human and natural causes can make water unsuitable for consumption. There are different ways to deal with this contamination. **Karst** is when **limestone** is **dissolved** by water, forming caves and sinkholes.

Take this quiz to check your comprehension of this chapter.

If you are using an offline version of this text, access the quiz for chapter 11 via the QR code.

 An interactive H5P element has been excluded from this version of the text. You can view it online here:
https://pressbooks.lib.vt.edu/introearthscience/?p=765#h5p-84

URLs Linked Within This Chapter

NPR story on Snake Valley: https://www.npr.org/2007/06/12/10953190/las-vegas-water-battle-crops-vs-craps

SNWA History: https://www.snwa.com/about/mission/index.html#:~:text=In%201991%2C%20seven%20local%20water,Big%20Bend%20Water%20District

Dean Baker Story: The Consequences: Transporting Snake Valley water to satisfy a thirsty Las Vegas. [Video: 25:24] https://www.youtube.com/watch?v=eCZ8KLrmUXo

NASA images: https://earthobservatory.nasa.gov/images/8103/mississippi-river-delta

USGS Groundwater Watch: https://groundwaterwatch.usgs.gov/default.asp

USGS tables on contaminants found in groundwater: https://www.usgs.gov/special-topics/water-science-school/science/contamination-groundwater

Text References

1. Brush, L.M., Jr, 1961, Drainage basins, channels, and flow characteristics of selected streams in central Pennsylvania: pubs.er.usgs.gov.
2. Charlton, R., 2007, Fundamentals of fluvial geomorphology: Taylor & Francis.
3. Cirrus Ecological Solutions, 2009, Jordan River TMDL: Utah State Division of Water Quality.
4. Earle, S., 2015, Physical geology OER textbook: BC Campus OpenEd.
5. EPA, 2009, Water on Tap-What You Need to Know: U.S. Environmental Protection Agency.
6. Fagan, B., 2012, Elixir: A history of water and humankind: Bloomsbury Press.
7. Fairbridge, R.W., 1968, Yazoo rivers or streams, *in* Geomorphology: Springer Berlin Heidelberg Encyclopedia of Earth Science, p. 1238–1239.
8. Freeze, A.R., and Cherry, J.A., 1979, Groundwater: Prentice Hall.
9. Galloway, D., Jones, D.R., and Ingebritsen, S.E., 1999, Land subsidence in the United States: U.S. Geological Survey Circular 1182.
10. Galloway, W.E., Whiteaker, T.L., and Ganey-Curry, P., 2011, History of Cenozoic North American drainage basin evolution, sediment yield, and accumulation in the Gulf of Mexico basin: Geosphere, v. 7, no. 4, p. 938–973.
11. Gilbert, G.K., 1890, Lake Bonneville: United States Geological Survey, 438 p.
12. Gleick, P.H., 1993, Water in Crisis: A Guide to the World's Fresh Water Resources: Oxford University Press.
13. Hadley, G., 1735, Concerning the cause of the general trade-winds: By Geo. Hadley, Esq; FRS: Philosophical Transactions, v. 39, no. 436–444, p. 58–62.
14. Halvorson, S.F., and James Steenburgh, W., 1999, Climatology of lake-effect snowstorms of the Great Salt Lake: University of Utah.
15. Heath, R.C., 1983, Basic ground-water hydrology: U.S. Geological Survey Water-Supply Paper 2220, 91 p.
16. Hobbs, W.H., and Fisk, H.N., 1947, Geological Investigation of the Alluvial Valley of the Lower Mississippi River: JSTOR.
17. Knudsen, T., Inkenbrandt, P., Lund, W., Lowe, M., and Bowman, S., 2014, Investigation of land subsidence and earth fissures in Cedar Valley, Iron County, Utah: Utah Geological Survey Special Study 150.
18. Lorenz, E.N., 1955, Available potential energy and the maintenance of the general circulation: Tell'Us, v. 7, no. 2, p. 157–167.
19. Marston, R.A., Mills, J.D., Wrazien, D.R., Bassett, B., and Splinter, D.K., 2005, Effects of Jackson lake dam on the Snake River and its floodplain, Grand Teton National Park, Wyoming, USA: Geomorphology, v. 71, no. 1–2, p. 79–98.
20. Maupin, M.A., Kenny, J.F., Hutson, S.S., Lovelace, J.K., Barber, N.L., and Linsey, K.S., 2014, Estimated use of water in the United States in 2010: US Geological Survey.
21. Myers, W.B., and Hamilton, W., 1964, The Hebgen Lake, Montana, earthquake of August 17, 1959: U.S. Geol. Surv. Prof. Pap., v. 435, p. 51.
22. Oviatt, C.G., 2015, Chronology of Lake Bonneville, 30,000 to 10,000 yr B.P: Quat. Sci. Rev., v. 110, p. 166–171.
23. Powell, J.W., 1879, Report on the lands of the arid region of the United States with a more detailed account of the land of Utah with maps: Monograph.
24. Reed, J.C., Love, D., and Pierce, K., 2003, Creation of the Teton landscape: a geologic chronicle of Jackson Hole and the Teton Range: pubs.er.usgs.gov.
25. Reese, R.S., 2014, Review of Aquifer Storage and Recovery in the Floridan Aquifer System of Southern Florida.
26. Schele, L., Miller, M.E., Kerr, J., Coe, M.D., and Sano, E.J., 1992, The Blood of Kings: Dynasty and Ritual in Maya Art: George Braziller Inc.
27. Seaber, P.R., Kapinos, F.P., and Knapp, G.L., 1987, Hydrologic unit maps.
28. Solomon, S., 2011, Water: The Epic Struggle for Wealth, Power, and Civilization: Harper Perennial.
29. Törnqvist, T.E., Wallace, D.J., Storms, J.E.A., Wallinga, J., Van Dam, R.L., Blaauw, M., Derksen, M.S., Klerks, C.J.W., Mei-

jneken, C., and Snijders, E.M.A., 2008, Mississippi Delta subsidence primarily caused by compaction of Holocene strata: Nat. Geosci., v. 1, no. 3, p. 173–176.

30. Turner, R.E., and Rabalais, N.N., 1991, Changes in Mississippi River water quality this century: Bioscience, v. 41, no. 3, p. 140–147.

31. United States Geological Survey, 1967, The Amazon: Measuring a Mighty River: United States Geological Survey O-245-247.

32. U.S. Environmental Protection Agency, 2014, Cyanobacteria/Cyanotoxins.

33. U.S. Geological Survey, 2012, Snowmelt – The Water Cycle, from USGS Water-Science School.

34. Utah/Nevada Draft Snake Valley Agreement, 2013.

Figure References

Figure 11.1: Example of a Roman aqueduct in Segovia, Spain. Bernard Gagnon. 2009. CC BY-SA 3.0. https://en.wikipedia.org/wiki/File:Aqueduct_of_Segovia_08.jpg

Figure 11.2: Chac mask in Mexico. Bernard DUPONT. 1995. CC BY-SA 2.0. https://commons.wikimedia.org/wiki/File:Chac_Mask_(21784027699).jpg

Figure 11.3: The water cycle. John Evans and Howard Periman, USGS. 2013. Public domain. https://commons.wikimedia.org/wiki/File:Watercyclesummary.jpg

Figure 11.4: Map view of a drainage basin with main trunk streams and many tributaries with drainage divide in dashed red line. Zimbres. 2005. CC BY-SA 2.5. https://commons.wikimedia.org/wiki/File:Hydrographic_basin.svg

Figure 11.5: Oblique view of the drainage basin and divide of the Latorita River, Romania. Asybaris01. 2011. Public domain. https://commons.wikimedia.org/wiki/File:EN_Bazinul_hidrografic_al_Raului_Latorita,_Romania.jpg

Figure 11.6: Major drainage basins color coded to match the related ocean. Citynoise. 2007. Public domain. https://commons.wikimedia.org/wiki/File:Ocean_drainage.png

Figure 11.7: Agricultural water use in the United States by state. USGS. 2018. Public domain. https://www.usgs.gov/media/images/map-us-state-showing-total-water-withdrawals-2015

Figure 11.8: Trends in water use by source. USGS. 2018. Public domain. https://www.usgs.gov/media/images/trends-population-and-freshwater-withdrawals-source-1950-2015-0

Figure 11.9: Distribution of precipitation in the United States. United States Department of the Interior. 2006. Public domain. https://commons.wikimedia.org/wiki/File:Average_precipitation_in_the_lower_48_states_of_the_USA.png

Figure 11.10: Thalweg of a river. In a river bend, the fastest moving water is on the outside of the bend, near the cutbank. Steven Earle. 2021. CC BY. Figure 8.1.1 from https://openeducationalberta.ca/practicalgeology/chapter/8-1-stream-erosion-and-deposition/#fig8.1.1

Figure 11.11: Various stream drainage patterns. Kindred Grey. 2022. CC BY-SA 3.0. Includes Rectangular by Zimbres, 2006 (CC BY-SA 2.5, https://commons.wikimedia.org/wiki/File:Rectangular.png). Dendritic by Zimbres, 2006 (CC BY-SA 2.5, https://en.wikipedia.org/wiki/File:Dendritic.png). Trellis drainage pattern by Tshf aee, 2007 (CC BY-SA 3.0, https://en.wikipedia.org/wiki/File:Trellis_drainage_pattern.JPG). Radial by Zimbres, 2006 (CC BY-SA 2.5, https://commons.wikimedia.org/wiki/File:Radial.png). Irregular drainage pattern by Tshf aee, 2007 (CC BY-SA 3.0, https://en.wikipedia.org/wiki/File:Irregular_drainage_pattern.JPG).

Figure 11.12: A stream carries dissolved load, suspended load, and bedload. PSUEnviroDan. 2008. Public domain. https://en.wikipedia.org/wiki/File:Stream_Load.gif

Figure 11.13: Profile of stream channel at bankfull stage, flood stage, and deposition of natural levee. Steven Earle.

2021. CC BY. Figure 8.1.4 from https://openeducationalberta.ca/practicalgeology/chapter/8-1-stream-erosion-and-deposition/#retfig8.1.3

Figure 11.14: Example of a longitudinal profile of a stream; Halfway Creek, Indiana. USGS. 2008. Public domain. https://commons.wikimedia.org/wiki/File:HalfwayCreek_fig02.jpg

Figure 11.15: The braided Waimakariri river in New Zealand. Greg O'Beirne. 2007. CC BY 2.5. https://www.wikiwand.com/simple/Braided_river#Media/File:Waimakariri01_gobeirne.jpg

Figure 11.16: Air photo of the meandering river, Rio Cauto, Cuba. Not home~commonswiki. 2007. Public domain. https://commons.wikimedia.org/wiki/File:Rio-cauto-cuba.JPG

Figure 11.17: Point bar and cut bank on the Cirque de la Madeleine in France. Jean-Christophe BENOIST. 2007. CC BY 2.5. https://www.wikiwand.com/en/Bar_(river_morphology)#Media/File:CirqueMadeleine.jpg

Figure 11.18: An entrenched meander on the Colorado River in the eastern entrance to the Grand Canyon. Paul Hermans. 2012. CC BY-SA 3.0. https://commons.wikimedia.org/wiki/File:Horseshoe_Bend_TC_27-09-2012_15-34-14.jpg

Figure 11.19: Panoramic view of incised meanders of the San Juan River at Gooseneck State Park, Utah. Michael Rissi. 2006. CC BY-SA 3.0. https://commons.wikimedia.org/wiki/File:GooseNeckStateParkPanorama.jpg

Figure 11.20: The Rincon is an abandoned meaner loop on the entrenched Colorado River in Lake Powell. NASA's Earth Observatory. 2012. CC BY 2.0. https://commons.wikimedia.org/wiki/File:Lake_Powell_and_The_Rincon,_Utah_-_NASA_Earth_Observatory.jpg

Figure 11.21: Landsat image of Zambezi Flood Plain, Namibia. Jesse Allen and Robert Simmon via NASA. 2010. Public domain. https://commons.wikimedia.org/wiki/File:Zambezi_Flood_Plain,_Namibia_(EO-1).jpg

Figure 11.22: Meander nearing cutoff on the Nowitna River in Alaska. Oliver Kurmis. 2002. CC BY-SA 2.0 DE. https://www.wikiwand.com/en/Oxbow_lake#Media/File:Nowitna_river.jpg

Figure 11.23: Alluvial fan in Iraq seen by NASA satellite. NASA. 2004. Public domain. https://commons.wikimedia.org/wiki/File:Alluvial_fan_in_Iran.jpg

Figure 11.24: Location of the Mississippi River drainage basin and Mississippi River delta. Shannon1. 2016. CC BY-SA 4.0. https://commons.wikimedia.org/wiki/File:Mississippiriver-new-01.png

Figure 11.25: Delta in Quake Lake Montana. Staplegunther. 2007. CC BY-SA 3.0. https://commons.wikimedia.org/wiki/File:Quakelakemontana.jpg

Figure 11.26: Sundarban Delta in Bangladesh, a tide-dominated delta of the Ganges River. NordNordWest. 2015. CC BY-SA 3.0 DE. https://commons.wikimedia.org/wiki/File:Bangladesh_adm_location_map.svg

Figure 11.27: Nile Delta showing its classic "delta" shape. NASA. 2018. Public domain. https://www.earthdata.nasa.gov/worldview/worldview-image-archive/the-nile-delta-from-space

Figure 11.28: Map of Lake Bonneville, showing the outline of the Bonneville shoreline, the highest level of the lake. Staplini. 2019. CC BY-SA 4.0. https://commons.wikimedia.org/wiki/File:Map_of_Lake_Bonneville.jpg

Figure 11.29: Deltaic deposits of Lake Bonneville near Logan, Utah; wave cut terraces can be seen on the mountain slope. Staplini. 2019. CC BY-SA 4.0. https://commons.wikimedia.org/wiki/File:Image_of_Lake_Bonneville_shorelines.png

Figure 11.30: Terraces along the Snake River, Wyoming. Fredlyfish4. 2008. CC BY-SA 3.0. https://commons.wikimedia.org/wiki/File:Snake_River_Overlook.JPG

Figure 11.31: Zone of saturation. USGS. 2011. Public domain. https://commons.wikimedia.org/wiki/File:Vadose_zone.gif

Figure 11.32: An aquifer cross-section. Hans Hillewaert. 2007. CC BY-SA 3.0. https://commons.wikimedia.org/wiki/File:Aquifer_en.svg

Figure 11.33: Pipe showing apparatus that would demonstrate Darcy's Law. Vectorised by Sushant savla from the work by Peter Kapitola. 2018. CC BY-SA 2.5. https://en.m.wikipedia.org/wiki/File:Darcy%27s_Law.svg

Figure 11.34: Cones of depression. USGS. 2018. Public domain. https://www.usgs.gov/media/images/cone-depression-pumping-a-well-can-cause-water-level-lowering

Figure 11.35: Different ways an aquifer can be recharged. USGS. Unknown date. Public domain. https://www.usgs.gov/media/images/groundwater-can-be-recharged-naturally-and-artificially

Figure 11.36: Evidence of land subsidence from pumping of groundwater shown by dates on a pole. Dr. Joseph F. Poland via USGS. 1977. Public domain. https://www.usgs.gov/media/images/location-maximum-land-subsidence-us-levels-1925-and-1977

Figure 11.37: Steep karst towers in China left as remnants as limestone is dissolved away by acidic rain and groundwater. chensiyuan. 2011. CC BY-SA 4.0. https://commons.wikimedia.org/wiki/File:1_li_jiang_guilin_yangshuo_2011.jpg

Figure 11.38: Sinkholes of the McCauley Sink in Northern Arizona, produced by collapse of Kaibab Limestone into caverns caused by solution of underlying salt deposits. Google Earth. Image retrieved 2022 by Kindred Grey. Public domain.

Figure 11.39: This sinkhole from collapse of surface into a underground cavern appeared in the front yard of this home in Florida. Ann Tihansky via USGS. 2010. Public domain. https://www.usgs.gov/media/images/sinkholes-west-central-florida-freeze-event-2010-2

Figure 11.40: Mammoth hot springs, Yellowstone National Park. Brocken Inaglory. 2008. CC BY-SA 3.0. https://en.wikipedia.org/wiki/File:Dead_trees_at_Mammoth_Hot_Springs.jpg

Figure 11.41: Varieties of speleotherms. Dave Bunnell / Under Earth Images. 2006. CC BY-SA 2.5. https://commons.wikimedia.org/wiki/File:Labeled_speleothems.jpg

Figure 11.42: This stream disappears into a subterranean cavern system to re-emerge a few hundred yards downstream. Martyn Gorman. 2007. CC BY-SA 2.0. https://www.geograph.org.uk/photo/471804

12.

COASTLINES

Learning Objectives

By the end of this chapter, students should be able to:

- Describe how waves occur, move, and carry energy.
- Explain wave behavior approaching the **shoreline**.
- Describe **shoreline** features and zones.
- Describe **wave refraction** and its contribution to **longshore currents** and **longshore drift**.
- Explain how **longshore currents** cause the **formation** of spits and baymouth bars.
- Distinguish between **submergent** and **emergent** coasts and describe coastal features associated with each.
- Describe the relationship between the natural **river** of sand in the **littoral** zone and human attempts to alter it for human convenience.
- Describe the pattern of the main ocean currents and explain the different factors involved in surface currents and deep ocean currents.
- Explain how ocean tides occur and distinguish among diurnal, semidiurnal, and **mixed tide** patterns.

The Earth's surface is 29% land and 71% water. Coastlines are the interfaces between, and as such, the longest visible boundaries on Earth. To understand the processes that occur at these boundaries, it is important to first understand wave energy.

12.1 Waves and Wave Processes

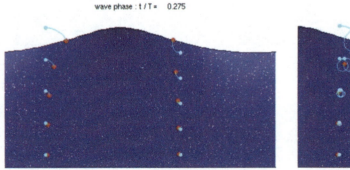
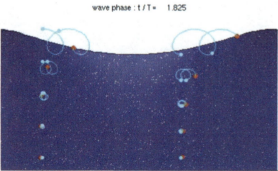

Figure 12.1: Particle motion within a wind-blown wave.

Wind blowing over the surface of water transfers energy to the water through friction. The energy transferred from wind to water causes waves to form. Waves move as individual oscillating particles of water. As the **wave crest** passes, the water is moving forward. As the **wave trough** passes, the water is moving backward. To see wave movement in action, watch a cork or some floating object as a wave passes.

Important terms to understand in the operation of waves include: the **wave crest** is the highest point of the wave; the **trough** is the lowest point of the wave. **Wave height** is the vertical distance from the **trough** to the crest and is determined by wave energy. Wave **amplitude** is half the **wave height**, or the distance from either the crest or trough to the still water line. **Wavelength** is the horizontal distance between consecutive wave crests. **Wave velocity** is the speed at which a **wave crest** moves forward and is related to the wave's energy. **Wave period** is the time interval it takes for adjacent wave crests to pass a given point.

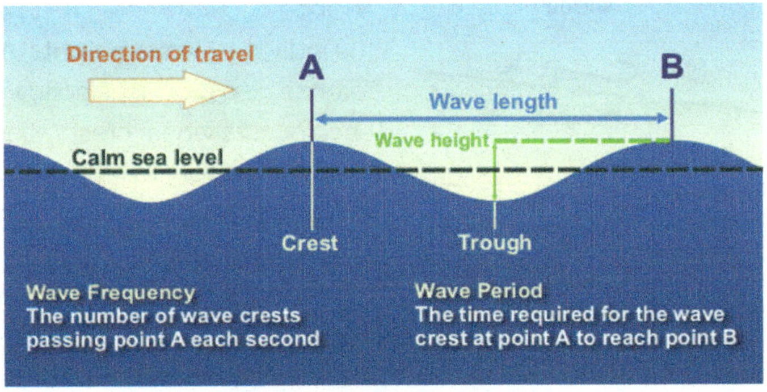

Figure 12.2: Aspects of water waves, labeled.

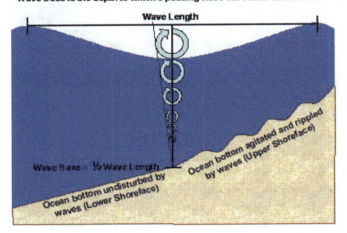

Figure 12.3: Diagram describing wavebase.

The circular motion of water particles diminishes with depth and is negligible at about one-half **wavelength**, an important dimension to remember in connection with waves. **Wave base** is the vertical depth at which water ceases to be disturbed by waves. In water shallower than **wave base**, waves will disturb the bottom and **ripple shore** sand. **Wave base** is measured at a depth of about one-half **wavelength**, where the water particles' circular motion diminishes to zero. If waves approaching a beach have crests at about 6 m (~20 ft) intervals, this wave motion disturbs water to about 3 m (~10 ft) deep. This motion is known as fair-**weather wave base**. In strong storms such as hurricanes, both **wavelength** and **wave base** increase dramatically to a depth known as **storm wave base**, which is approximately 91 m (~300 ft).

Waves are generated by wind blowing across the ocean surface. The amount of energy imparted to the water depends on wind velocity and the distance across which the wind is blowing. This distance is called **fetch**. Waves striking a **shore** are typically generated by storms hundreds of miles from the **coast** and have been traveling across the ocean for days.

Winds blowing in a relatively constant direction generate waves moving in that direction. Such a group of approximately parallel waves traveling together is called a **wave train**. A **wave train** coming from one **fetch** can produce various wavelengths. Longer wavelengths travel at a faster velocity than shorter wavelengths, so they arrive first at a distant **shore**. Thus, there is a **wavelength–sorting** process that takes place during the **wave train**'s travel. This **sorting** process is called wave dispersion.

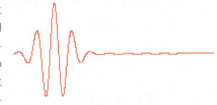

Figure 12.4: Wave train moving with dispersion.

12.1.1 Behavior of Waves Approaching Shore

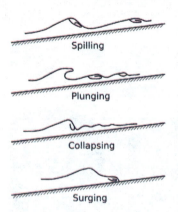

Figure 12.5: Types of breakers.

On the open sea, waves generally appear choppy because wave trains from many directions are interacting with each other, a process called wave interference. Constructive interference occurs where crests align with other crests. The aligned wave height is the sum of the individual **wave heights**, a process referred to as *wave amplification*. Constructive interference also produces hollows where troughs align with other troughs. Destructive interference occurs where crests align with troughs and cancel each other out. As waves approach **shore** and begin to make frictional contact with the sea floor at a depth of about one-half **wavelength** or less, they begin to slow down. However, the energy carried by the wave remains the same, so the waves build up higher. Remember that water moves in a circular motion as a wave passes, and each circle is fed from the **trough** in front of the advancing wave. As the wave encounters shallower water at the **shore**, there is eventually insufficient water in the **trough** in front of the wave to supply a complete circle, so the crest pours over creating a breaker.

A special type of wave is called a **tsunami**, sometimes incorrectly called a "tidal wave." **Tsunamis** are generated by energetic events affecting the sea floor, such as earthquakes, submarine **landslides**, and **volcanic** eruptions (see chapter 9 and chapter 4). During earthquakes for example, **tsunamis** can be produced when the moving crustal rocks below the sea abruptly elevate a portion of the seafloor. Water is suddenly lifted creating a bulge at the surface and a **wave train** spreads out in all directions traveling at tremendous speeds [over 322 kph (200 mph)] and carrying enormous energy. **Tsunamis** may pass unnoticed in the open ocean because they move so fast, the **wavelength** is very long, and the **wave height** is very low. But, as the **wave train** approaches **shore** and each wave

Figure 12.6: All waves, like tsunamis, slow down as they reach shallow water. This causes the wave to increase in hight.

begins to interact with the shallow seafloor, friction increases and the wave slows down. Still carrying its enormous energy, wave height builds up and the wave strikes the shore as a wall of water that can be over 30 m (~100 ft) high. The **massive** wave, called a **tsunami** runup, may sweep inland well beyond the beach destroying structures far inland. **Tsunamis** can deliver a catastrophic blow to people at the beach. As the **trough** water in front of the **tsunami** wave is drawn back, the seafloor is exposed. Curious and unsuspecting people on the beach may run out to see exposed **offshore** sea life only to be overwhelmed when the breaking crest hits.

Take this quiz to check your comprehension of this section.

If you are using an offline version of this text, access the quiz for section 12.1 via the QR code.

 An interactive H5P element has been excluded from this version of the text. You can view it online here:
https://pressbooks.lib.vt.edu/introearthscience/?p=796#h5p-85

12.2 Shoreline Features

Coastlines are dynamic, high energy, and geologically complicated places where many different erosional and depositional features exist (see chapter 5). They include all parts of the land-sea boundary directly affected by the sea, including land far above high **tide** and seafloor well below normal **wave base**. But, the **shoreline** itself is the direct interface between water and land that shifts with the tides. This shifting interface at the **shoreline** is called the **littoral** zone. The combination of waves, currents, **climate**, coastal morphology, and gravity, all act on this land-sea boundary to create **shoreline** features.

12.2.1 Shoreline Zones

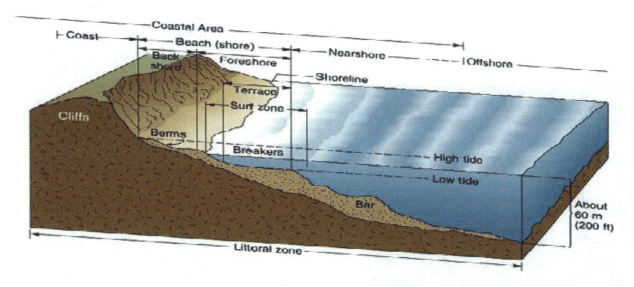

Figure 12.7: Diagram of zones of the shoreline.

Shorelines are divided into five primary zones—**offshore**, **nearshore**, surf, **foreshore**, and **backshore**. The **offshore** zone is below water, but it is still geologically active due to flows of **turbidity currents** that cascade over the **continental slope** and accumulate in the **continental** rise. The **nearshore** zone is the area of the **shore** affected by the waves where water depth is one-half wavelength or less. The width of this zone depends on the maximum **wavelength** of the approaching **wave train** and the slope of the seafloor. The **nearshore** zone includes the **shoreface**, which is where sand is disturbed and deposited. The **shoreface** is broken into two segments: upper and lower **shoreface**. Upper **shoreface** is affected by every-day wave action and consists of finely-laminated and cross-bedded sand. The lower **shoreface** is the only area moved by storm waves and consists of hummocky cross-stratified sand. The **surf zone** is where the waves break.

The **foreshore** zone overlaps the **surf zone** and is periodically wet and dry due to waves and tides. The **foreshore** zone is where planer-laminated, well-sorted sand accumulates. The **beach face** is the part of the **foreshore** zone where the breaking waves swash up and the backwash flows back down. Low ridges above the **beach face** in the **foreshore** zone are called **berms**. During the summer in North America, when most people visit the beach, the zone where people spread their towels and beach umbrellas is the **summer berm**. Wave energy is typically lower in the summer, which allows sand to pile onto the beach. Behind the **summer berm** is a low ridge of sand called the **winter berm**. In winter, higher storm energy moves the **summer berm** sand off the beach and piles it in the **nearshore** zone. The next year, that sand is replaced on the beach and moved back onto the **summer berm**. The **backshore** zone is the area always above sea level in normal conditions. In the **backshore** zone, onshore winds may blow sand behind the beach and the **berms**, creating **dunes**.

12.2.2 Refraction, Longshore Currents, and Longshore Drift

As waves enter shallower water less than one-half **wavelength** depth, they slow down. Waves usually approach the **shoreline** at an angle, with the end of the waves nearest the beach slowing down first. This causes the wave crests to bend, called **wave refraction**. From the **beach face**, this causes it to look like waves are approaching the beach straight on, parallel to the beach. However, as refracted waves actually approach the **shoreline** at a slight angle, they create a slight difference between the swash as it moves up the **beach face** at a slight angle and the backwash as it flows straight back down under gravity. This slight angle between swash and backwash along the beach creates a current called the **longshore current**. Waves stir up sand in the **surf zone** and move it along the **shore**. This movement of sand is called **longshore drift**. **Longshore drift** along both the west and east coasts of North America moves sand north to south on average.

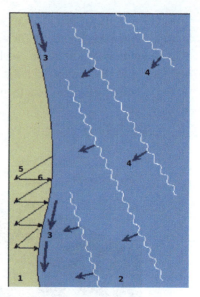

Figure 12.8: Longshore drift. 1=beach, 2=sea, 3=longshore current direction, 4=incoming waves, 5=swash, 6=backwash.

Figure 12.9: Farewell Spit, New Zealand.

Longshore currents can carry **longshore drift** down a **coast** until it reaches a bay or inlet where it will deposit sand in the quieter water (see chapter 11). Here, a **spit** can form. As the **spit** grows, it may extend across the **mouth** of the bay forming a barrier called a **baymouth bar**. Where the bay or inlet serves as boat anchorage, spits and baymouth bars are a severe inconvenience. Often, inconvenienced communities create methods to keep their bays and harbors open.

One way to keep a harbor open is to build a **jetty**, a long concrete or stone barrier constructed to deflect the sand away from a harbor **mouth** or other ocean waterway. If the **jetty** does not deflect the sand far enough out, sand may continue to flow along the **shore**, forming a **spit** around the end of the **jetty**. A more expensive but effective method to keep a bay **mouth** open is to dredge the sand from the growing spit, put it on barges, and deliver it back to the drift downstream of the harbor **mouth**. An even more expensive but more effective option is to install large pumps and pipes to draw in the sand upstream of the harbor, pump it through pipes, and **discharge** it back into the drift downstream of the harbor **mouth**. Because natural processes work continuously, human efforts to mitigate inconvenient spits and baymouth bars require ongoing modifications. For example, the community of Santa Barbara, California, tried several methods to keep their harbor open before settling on pumps and piping.

Figure 12.10: Jetties near Carlsbad, California. Notice the left jetty is loaded with sand, while the right jetty is lacking sand. This is due to the longshore drift going left to right.

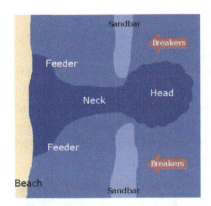

Figure 12.11: Animation of rip currents.

Rip currents are another coastal phenomenon related to **longshore currents**. **Rip currents** occur in the **nearshore** seafloor when wave trains come *straight* onto the **shoreline**. In areas where wave trains push water directly toward the **beach face** or where the shape of the **nearshore** seafloor refracts waves toward a specific point on the beach, the water piles up on **shore**. But this water must find an outlet back to the sea. The outlet is relatively narrow, and **rip currents** carry the water directly away from the beach. Swimmers caught in **rip currents** are carried out to sea. Swimming back to **shore** directly against the strong current is fruitless. A **solution** for good swimmers is to ride out the current to where it dissipates, swim around it, and return to the beach. Another **solution** for average swimmers is to swim parallel to the beach until out of the current, then return to the beach. Where **rip currents** are known to exist, warning signs are often posted. The best **solution** is to understand the nature of **rip currents**, have a plan before entering the water, or watch the signs and avoid them all together.

Like **rip currents**, undertow is a current that moves away from the shore. However, unlike **rip currents**, undertow occurs underneath the approaching waves and is strongest in the **surf zone** where waves are high and water is shallow. Undertow is another return flow for water transported onshore by waves.

12.2.3 Emergent and Submergent Coasts

Figure 12.12: Island Arch, a sea arch in Victoria, Australia.

Emergent coasts occur where sea levels fall relative to land level. **Submergent** coasts occur where sea levels rise relative to land level. **Tectonic** shifts and sea level changes cause the long-term rise and fall of sea level relative to land. Some features associated with **emergent** coasts include high cliffs, headlands, exposed **bedrock**, steep slopes, rocky shores, arches, stacks, tombolos, wave-cut **platforms**, and wave notches.

In **emergent** coasts, wave energy, wind, and gravity erode the **coastline**. The erosional features are elevated relative to the wave zone. Sea cliffs are persistent features as waves cut away at their base and higher rocks calve off by **mass wasting**. Refracted waves that attack **bedrock** at the base of headlands may erode or carve out a sea arch, which can extend below sea level in a sea cave. When a sea arch collapses, it leaves one or more rock columns called stacks.

Figure 12.13: This tombolo, called "Angel Road," connects the stack of Shodo Island, Japan.

Figure 12.14: Wave notches carved by Lake Bonneville, Antelope Island, Utah.

A **stack** or near **shore** island creates a quiet water zone behind it. Sand moving in the **longshore drift** accumulates in this quiet zone forming a **tombolo**: a sand strip that connects the island or **stack** to the **shoreline**. Where sand supply is low, wave energy may erode a wave-cut **platform** across the **surf zone**, exposed as bare rock with tidal pools at low **tide**. This bench-like **terrace** extends to the cliff's base. When wave energy cuts into the base of a sea cliff, it creates a **wave notch**.

Submergent coasts occur where sea levels rise relative to land. This may be due to **tectonic subsidence**—when the Earth's **crust** sinks—or when sea levels rise due to **glacier** melt. Features associated with **submergent** coasts include flooded **river** mouths, **fjords**, **barrier islands**, **lagoons**, **estuaries**, bays, **tidal flats**, and tidal currents. In **submergent** coastlines, **river** mouths are flooded by the rising water, for example Chesapeake Bay. **Fjords** are **glacial** valleys flooded by post-**ice age** sea level rise (see chapter 14). **Barrier islands** are elongated bodies of sand that formed from old beach sands that used to parallel the **shoreline**. Often, **lagoons** lie behind **barrier islands**. **Barrier island formation** is controversial: some scientists believe that they formed when **ice sheets** melted after the last **ice age**, raising sea levels. Another **hypothesis** is that **barrier islands** formed from spits and bars accumulating far **offshore**.

Figure 12.15: Landsat image of Chesapeake Bay, eastern United States. Note the barrier islands parallel to the coastline.

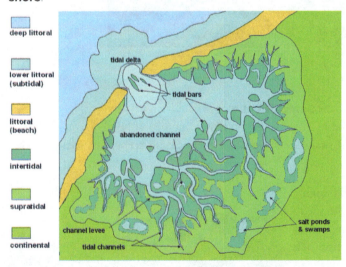

Figure 12.16: General diagram of a tidal flat and associated features.

Tidal flats—or mudflats, form where tides alternately flood and expose low areas along the **coast**. Tidal currents create combinations of symmetrical and asymmetrical **ripple** marks on mudflats, and drying mud creates mud cracks. In the central Wasatch Mountains of Utah, ancient **tidal flat** deposits are exposed in the **Precambrian strata** of the Big Cottonwood Formation. These ancient deposits provide an example of applying Hutton's **principle of uniformitarianism** (see chapter 1). Sedimentary structures common on modern **tidal flats** indicate that these ancient deposits were formed in a similar environment: there were **shorelines**, tides, and shoreline processes acting at that time, yet the ancient age indicates that there were no land plants to hold products of **mechanical weathering** in place (see chapter 5), so **erosion** rates would have been different.

Geologically, **tidal flats** are broken into three different sections: barren zones, marshes, and salt pans. These zones may be present or absent in each individual **tidal flat**. Barren zones are areas with strong flowing water, coarser **sediment**, with **ripple** marks and **cross bedding** common. Marshes are vegetated with sand and mud. Salt pans or flats, less often submerged than the other zones, are the finest-grained parts of **tidal flats**, with silty **sediment** and mud cracks (see chapter 5).

Lagoons are locations where spits, **barrier islands**, or other features partially cut off a body of water from the ocean. **Estuaries** are a vegetated type of **lagoon** where fresh water flows into the area making the water **brackish**—a salinity between salt and fresh water. However, terms like **lagoon, estuary**, and even bay are often loosely used in place of one another. **Lagoons** and **estuaries** are certainly transitional between land and water environments where **littoral**, shallow shorelines; **lacustrine**, lakes or **lagoons**; and **fluvial**, **rivers** or currents can overlap. For more information on **lagoons** and **estuaries**, see chapter 5.

12.2.4 Human Impact on Coastal Beaches

Figure 12.17: Kara-Bogaz Gol lagoon, Turkmenistan.

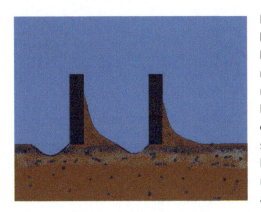

Figure 12.18: Groins gathering sediment from longshore drift.

Humans impact coastal beaches when they build homes, condominiums, hotels, businesses, and harbors—and then again when they try to manage the natural processes of **erosion**. Waves, currents, **longshore drift**, and dams at **river** mouths deplete sand from expensive beachfront property and expose once calm harbors to high-wave energy. To protect their investment, keep sand on their beach, and maintain calm harbors, cities and landowners find ways to mitigate the damage by building **jetties**, **groins**, dams, and breakwaters.

Jetties are large manmade piles of boulders or concrete barriers built at **river** mouths and harbors. A **jetty** is designed to divert the current or **tide**, to keep a channel to the ocean open, and to protect a harbor or beach from wave action. **Groins** are similar but smaller than **jetties**. **Groins** are fences of wire, wood or concrete built across the beach perpendicular to the **shoreline** and downstream of a property. Unlike **jetties**, **groins** are used to preserve sand on a beach rather than to divert it. Sand erodes on the downstream side of the **groin** and collects against the upstream side. Every **groin** on one property thus creates a need for another one on the property downstream. A series of **groins** along a beach develops a scalloped appearance along the **shoreline**.

Inland **streams** and **rivers** flow to the ocean carrying sand to the **longshore current** which distributes it to beaches. When dams are built, they **trap** sand and keep **sediment** from reaching beaches. To replenish beaches, sand may be hauled in from other areas by trucks or barges and dumped on the depleted beach. Unfortunately, this can disrupt the ecosystem that exists along the **shoreline** by exposing native creatures to foreign ecosystems and microorganisms and by introducing foreign objects to humans. For example, visitors to one replenished east coast beach found munitions and metal shards in the sand, which had been dredged from abandoned military test ranges.

Figure 12.19: Groin system on a coast in Virginia.

An approach to protect harbors and moorings from high-energy wave action is to build a **breakwater**—an **offshore** structure against which the waves break, leaving calmer waters behind it. Unfortunately, breakwaters keep waves from reaching the beach and stop sand moving with **longshore drift**. When **longshore drift** is interrupted, sand is deposited in quieter water, and the **shoreline** builds out forming a **tombolo** behind the **breakwater**. The **tombolo** eventually fill in behind the **breakwater** with sand. When the city of Venice, California built a **breakwater** to create a quiet water harbor, **longshore drift** created a **tombolo** behind the **breakwater**, as seen in the image. The **tombolo** now acts as a large **groin** in the beach drift.

12.2.5 Submarine Canyons

Figure 12.20: A tombolo formed behind the breakwater at Venice, CA.

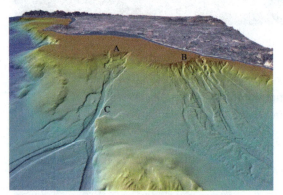

Figure 12.21: Submarine canyons off of Los Angeles. A=San Gabriel Canyon, B=Newport Canyon. At point C, the canyon is 815 m wide and 25 m deep.

Submarine canyons are narrow, deep underwater canyons located on **continental** shelves. **Submarine canyons** typically form at the mouths of large landward **river** systems. They form when **rivers** cut down into the **continental shelf** during low sea level and when material continually slumps or flows down from the **mouth** of a **river** or a **delta**. Underwater currents rich in **sediment** and more dense than sea water, can flow down the canyons, even erode and deepen them, then drain onto the **ocean floor**. Underwater **landslides**, called **turbidity flows**, occur when steep **delta** faces and underwater **sediment** flows are released down the **continental slope**. **Turbidity flows** in submarine canyons can continue to erode the canyon, and eventually, fan-shaped deposits develop at the **mouth** of the canyon on the **continental** rise. See chapter 5 for more information on **turbidity flows**.

Take this quiz to check your comprehension of this section.

If you are using an offline version of this text, access the quiz for section 12.2 via the QR code.

 An interactive H5P element has been excluded from this version of the text. You can view it online here:
https://pressbooks.lib.vt.edu/introearthscience/?p=796#h5p-86

12.3 Currents and Tides

Ocean water moves as waves, currents, and tides. Ocean currents are driven by persistent global winds blowing over the water's surface and by water density. Ocean currents are part of Earth's heat engine in which solar energy is absorbed by ocean water and distributed by ocean currents. Water has another unique property, high specific heat, that relates to ocean currents. Specific heat is the amount of heat necessary to raise a unit volume of a substance one degree. For water it takes one calorie per cubic centimeter to raise its **temperature** one degree Celsius. This means the oceans, covering 71% of the Earth's surface, soak up solar heat with little **temperature** change and distribute that heat around the Earth by ocean currents.

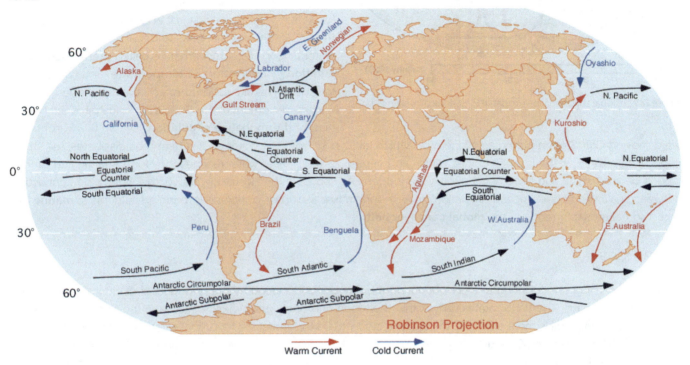

Figure 12.22: World ocean currents.

12.3.1 Surface Currents

The Earth's rotation and the Coriolis effect exert significant influence on ocean currents (see chapter 13). In the figure, the black arrows show global surface currents. Notice the large circular currents in the northern and southern hemispheres in the Atlantic, Pacific, and Indian Oceans. These currents are called **gyres** and are driven by atmospheric circulation—air movement. **Gyres** rotate clockwise in the northern hemisphere and counterclockwise in the southern hemisphere because of the Coriolis Effect. Western boundary currents flow from the equator toward the poles carrying warm water. They are key contributors to local **climate**. Western boundary currents are narrow and move poleward along the east coasts of adjacent continents. The Gulf Stream and the Kuroshio currents in the northern hemisphere and the Brazil, Mozambique, and Australian currents in the southern hemisphere are western boundary currents. Currents returning cold water toward the equator are broad and diffuse along the western coasts of adjacent land masses. These warm western boundary and cold eastern boundary currents affect **climate** of nearby lands making them warmer or colder than other areas at equivalent latitudes. For example, the warm Gulf Stream makes Northern Europe much milder than similar latitudes in northeastern Canada and Greenland. Another example is the cool Humboldt Current, also called the Peru Current, flowing north along the west coast of South America. Cold currents limit evaporation in the ocean, which is one reason the **Atacama Desert** in Chile is cool and arid.

12.3.2 Deep Currents

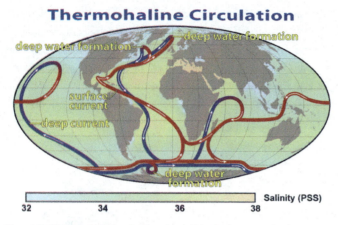

Figure 12.23: Global thermohaline circulation. PSS=practical salinity units.

Whether an ocean current moves horizontally or vertically depends on its density. The density of seawater is determined by **temperature** and salinity.

Evaporation and freshwater influx from **rivers** affect salinity and, therefore, the density of seawater. As the western boundary currents cool at high latitudes and salinity increases due to evaporation and ice **formation** (recall that ice floats; water is densest just above its freezing point). So the cold, denser water sinks to become the ocean's deep waters. Deep-water movement is called **thermohaline circulation**—*thermo* refers to **temperature**, and *haline* refers to salinity. This circulation connects the world's deep ocean waters. Movement of the Gulf Stream illustrates the beginning of **thermohaline circulation**. Heat in the warm poleward moving Gulf Stream promotes evaporation which takes heat from the water and as heat thus dissipates, the water cools. The resulting water is much colder, saltier, and denser. As the denser water reaches the North Atlantic and Greenland, it begins to sink and becomes a deep-water current. As shown in the illustration above, this worldwide connection between shallow and deep-ocean circulation overturns and mixes the entire world ocean, bringing nutrients to **marine** life, and is sometimes referred to as the **global conveyor belt**.

12.3.3 Tides

Tides are the rising and lowering of sea level during the day and are caused by the gravitational effects of the Sun and Moon on the oceans. The Earth rotates daily within the Moon and Sun's gravity fields. Although the Sun is much larger and its gravitational pull is more powerful, the Moon is closer to Earth; hence, the Moon's gravitational influence on tides is dominant. The **magnitude** of the **tide** at a given location and the difference between high and low **tide**—the tidal range, depends primarily on the configuration of the Moon and Sun with respect to the Earth orbit and rotation. **Spring tide** occurs when the Sun, Moon, and Earth line up with each other at the full or new Moon, and the tidal range is at a maximum. **Neap tide** occurs approximately two weeks later when the Moon and Sun are at right angles with the Earth, and the tidal range is lowest.

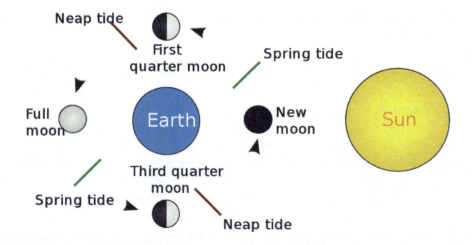

Figure 12.24: The types of tides.

The Earth rotates within a tidal envelope, so tides rise and ebb daily. Tides are measured at coastal locations. These measurements and the tidal predictions based on them are published on the NOAA website. Tides rising and falling create tidal patterns at any given **shore** location. The three types of tidal patterns are *diurnal*, *semidiurnal*, and *mixed*.

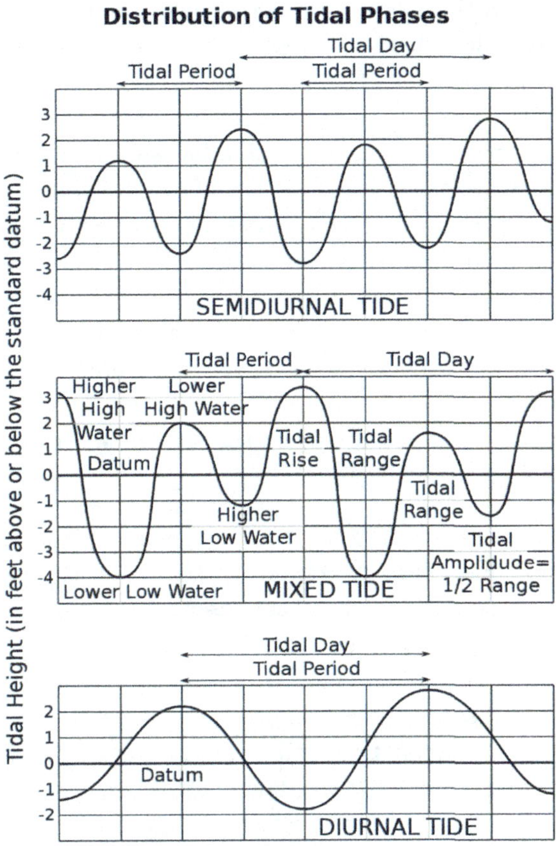

Distribution of Tidal Phases

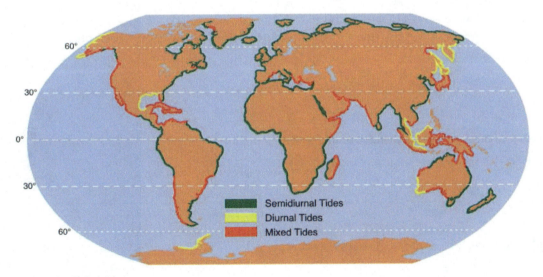

Figure 12.25: Different tide types.

Figure 12.26: Global tide types.

Diurnal tides go through one complete cycle each **tidal day**. A **tidal day** is the amount of time for the Moon to align with a point on the Earth as the Earth rotates, which is slightly longer than 24 hours. Semidiurnal tides go through two complete cycles in each **tidal day**—approximately 12 hours and 50 minutes, with the tidal range typically varying in each cycle. Mixed tides are a combination of diurnal and semidiurnal patterns and show two tidal cycles per **tidal day**, but the relative amplitudes of each cycle and their highs and lows vary during the tidal month. For example, there is a high, high **tide** and a high, low **tide**. The next day, there is a low, high **tide** and a low, low **tide**. Forecasting the tidal pattern and the times tidal phases arrive at a given **shore** location

Figure 12.27: A tidal day lasts slightly longer than 24 hours.

is complicated and can be done for only a few days at a time. Tidal phases are determined by *bathymetry*: the depth of ocean basins and the **continental** obstacles that are in the way of the tidal envelope within which the Earth rotates. Local tidal experts make 48-hour tidal forecasts using tidal charts based on daily observations, as can be seen in the chart of different **tide** types. A typical tidal range is approximately 1 m (3 ft). Extreme tidal ranges occur where the tidal wave enters a narrow restrictive zone that funnels the tidal energy. An example is the English Channel between Great Britain and the European **continent** where the tidal range is 7 to 9.75 m (23 to 32 ft). The Earth's highest tidal ranges occur at the Bay of Fundy, the funnel-like bay between Nova Scotia and New Brunswick, Canada, where the average range is nearly 12 m (40 ft) and the extreme range is around 18 m (60 ft). At extreme tidal range locations, a person who ventures out onto the seafloor exposed during ebb **tide** may not be able to outrun the advancing water during flood **tide**. NOAA has additional information on tides.

Take this quiz to check your comprehension of this section.

If you are using an offline version of this text, access the quiz for section 12.3 via the QR code.

 An interactive H5P element has been excluded from this version of the text. You can view it online here:
https://pressbooks.lib.vt.edu/introearthscience/?p=796#h5p-87

Summary

Shoreline processes are complex, but important for understanding coastal processes. Waves, currents, and tides are the main agents that shape shorelines. Most coastal landforms can be attributed to moving sand via **longshore drift**, and long-term rising or falling sea levels.

The **shoreline** is the interface between water and land and is divided into five zones. Processes at the **shoreline** are called **littoral** processes. Waves approach the beach at an angle, which cause the waves to bend towards the beach. This bending action is called **wave refraction** and is responsible for creating the **longshore current** and **longshore drift**—the process that moves sand along the coasts. When the **longshore current** deposits sand along the **coast** into quieter waters, the sand can accumulate, creating a **spit** or barrier called a **baymouth bar**, which often blocks bays and harbors. Inconvenienced humans create methods to keep their harbors open and preserve sand on their beaches by creating **jetties** and **groins**, which negatively affect natural beach processes.

Emergent coasts are created by sea levels falling, while **submergent** coasts are caused by sea levels rising. Oceans absorb solar energy, which is distributed by currents throughout the world. Circular surface currents, called **gyres**, rotate clockwise in the northern hemisphere and counterclockwise in the southern hemisphere. Thermohaline deep circulation connects the world's deep ocean waters: when shallow poleward moving warm water evaporates, the colder, saltier, and denser water sinks and becomes deep-water currents. The connection between shallow and deep-ocean circulation is called the **global conveyor belt**.

Tides are the rising and lowering of sea level during the day and are caused by the gravitational effects of the Sun and Moon on the oceans. There are three types of tidal patterns: diurnal, semidiurnal, and mixed. Typical tidal ranges are approximately 1 m (3 ft). Extreme tidal ranges are around 18 m (60 ft).

Take this quiz to check your comprehension of this chapter.

If you are using an offline version of this text, access the quiz for chapter 12 via the QR code.

 An interactive H5P element has been excluded from this version of the text. You can view it online here: https://pressbooks.lib.vt.edu/introearthscience/?p=796#h5p-88

URLs Linked Within This Chapter

Tide measurements and predictions: https://tidesandcurrents.noaa.gov/tide_predictions.html

NOAA: https://tidesandcurrents.noaa.gov/

Text References

1. Colling, Angela. 2001. *Ocean Circulation*. Edited by Open University Course Team. Butterworth-Heinemann.
2. Davis, Richard A., Jr., and Duncan M. Fitzgerald. 2009. *Beaches and Coasts*. John Wiley & Sons.
3. Davis, Richard Albert. 1997. *The Evolving Coast*. Scientific American Library New York.
4. Greene, Paul, George Follett, and Clint Henker. 2009. "Munitions and Dredging Experience on the United States Coast." *Marine Technology Society Journal* 43 (4): 127–31.
5. Jackson, Nancy L., Mitchell D. Harley, Clara Armaroli, and Karl F. Nordstrom. 2015. "Beach Morphologies Induced by Breakwaters with Different Orientations." *Geomorphology* 239 (June). Elsevier: 48–57.
6. "Littoral Bypassing and Beach Restoration in the Vicinity of Port Hueneme, California." n.d. In *Coastal Engineering 1966*.
7. Munk, Walter H. 1950. "On the Wind-Driven Ocean Circulation." *Journal of Meteorology* 7 (2): 80–93.
8. Normark, William R., and Paul R. Carlson. 2003. "Giant Submarine Canyons: Is Size Any Clue to Their Importance in the

Rock Record?" *Geological Society of America Special Papers* 370 (January): 175–90.

9. Reineck, H-E, and Indra Bir Singh. 2012. *Depositional Sedimentary Environments: With Reference to Terrigenous Clastics.* Springer Science & Business Media.

10. Rich, John Lyon. 1951. "Three Critical Environments of Deposition, and Criteria for Recognition of Rocks Deposited In Each of Them." *Geological Society of America Bulletin* 62 (1). gsabulletin.gsapubs.org: 1–20.

11. Runyan, Kiki, and Gary Griggs. 2005. "Implications of Harbor Dredging for the Santa Barbara Littoral Cell." In *California and the World Ocean '02*, 121–35. Reston, VA: American Society of Civil Engineers.

12. Schwiderski, Ernst W. 1980. "On Charting Global Ocean Tides." *Reviews of Geophysics* 18 (1): 243–68.

13. Stewart, Robert H. 2008. *Introduction to Physical Oceanography.* Texas A & M University Texas.

14. Stommel, Henry, and A. B. Arons. 2017. "On the Abyssal Circulation of the World ocean—I. Stationary Planetary Flow Patterns on a Sphere – ScienceDirect." Accessed February 26. http://www.sciencedirect.com/science/article/pii/0146631359900656.

15. Stommel, Henry, and A. B. Arons. 2017. "On the Abyssal Circulation of the World ocean—I. Stationary Planetary Flow Patterns on a Sphere – ScienceDirect." Accessed February 26. http://www.sciencedirect.com/science/article/pii/0146631359900656.

Figure References

Figure 12.1: Particle motion within a wind-blown wave. Kraaiennest. 2008. CC BY-SA 4.0. https://commons.wikimedia.org/wiki/File:Deep_water_wave.gif

Figure 12.2: Aspects of water waves, labeled. NOAA. Unknown date. Public domain. https://oceanservice.noaa.gov/education/tutorial_currents/media/supp_cur03a.html

Figure 12.3: Diagram describing wavebase. GregBenson. 2004. CC BY-SA 3.0. https://en.wikipedia.org/wiki/File:Wavebase.jpg

Figure 12.4: Wave train moving with dispersion. Fffred~commonswiki. 2006. Public domain. https://commons.wikimedia.org/wiki/File:Wave_packet_(dispersion).gif

Figure 12.5: Types of breakers. Kraaiennest. 2015. CC BY-SA 3.0. https://commons.wikimedia.org/wiki/File:Breaking_wave_types.svg

Figure 12.6: All waves, like tsunamis, slow down as they reach shallow water. Régis Lachaume. 2005. CC BY-SA 3.0. https://commons.wikimedia.org/wiki/File:Propagation_du_tsunami_en_profondeur_variable.gif

Figure 12.7: Diagram of zones of the shoreline. U.S. Navy. 2006. Public domain. https://commons.wikimedia.org/wiki/File:Littoral_Zones.jpg

Figure 12.8: Longshore drift. 1=beach, 2=sea, 3=longshore current direction, 4=incoming waves, 5=swash, 6=backwash. USGS. 2006. Public domain. https://commons.wikimedia.org/wiki/File:Longshore_i18n.png

Figure 12.9: Farewell Spit, New Zealand. NASA. 2006. Public domain. https://commons.wikimedia.org/wiki/File:Farewell_spit.jpg

Figure 12.10: Jetties near Carlsbad, California. USGS. 2005. Public domain. https://commons.wikimedia.org/wiki/File:Jetty_break2_new(USGS).jpg

Figure 12.11: Animation of rip currents. NOAA. 2008. Public domain. https://www.nws.noaa.gov/mdl/rip_current/

Figure 12.12: Island Arch, a sea arch in Victoria, Australia. Diliff. 2008. CC BY-SA 3.0. https://commons.wikimedia.org/wiki/File:Island_Archway,_Great_Ocean_Rd,_Victoria,_Australia_-_Nov_08.jpg

Figure 12.13: This tombolo, called "Angel Road," connects the stack of Shodo Island, Japan. 663highland. 2012. CC BY-SA 3.0. https://commons.wikimedia.org/wiki/File:Angel_Road_Shodo_Island_Japan01s3.jpg

Figure 12.14: Wave notches carved by Lake Bonneville, Antelope Island, Utah. Wilson44691. 2008. Public domain. https://commons.wikimedia.org/wiki/File:WaveCutPlatformsAntelopeIslandUT.jpg

Figure 12.15: Landsat image of Chesapeake Bay, eastern United States. USGS, NASA. 2009. Public domain. https://www.usgs.gov/media/images/chesapeake-bay-landsat

Figure 12.16: General diagram of a tidal flat and associated features. Foxbat deinos. 2009. Public domain. https://en.m.wikipedia.org/wiki/File:Tidal_flat_general_sketch.png

Figure 12.17: Kara-Bogaz Gol lagoon, Turkmenistan. NASA. 1995. Public domain. https://commons.wikimedia.org/wiki/File:Kara-Bogaz_Gol_from_space,_September_1995.jpg

Figure 12.18: Groins gathering sediment from longshore drift. Archer0630. 2013. CC BY-SA 3.0. https://commons.wikimedia.org/wiki/File:Groin_effect-3.JPG

Figure 12.19: Groin system on a coast in Virginia. Internet Archive Book Images. 1958. Public domain. https://commons.wikimedia.org/wiki/File%3AThe_Annual_bulletin_of_the_Beach_Erosion_Board_(1958)_(18396945096).jpg

Figure 12.20: A tombolo formed behind the breakwater at Venice, CA. Internet Archive Book Images. 1948. Public domain. https://flic.kr/p/wRSfxb

Figure 12.21: Submarine canyons off of Los Angeles. A=San Gabriel Canyon, B=Newport Canyon. At point C, the canyon is 815 m wide and 25 m deep. USGS. 2012. Public domain. https://commons.wikimedia.org/wiki/File:Canyons_off_LA.jpg

Figure 12.22: World ocean currents. Dr. Michael Pidwirny. 2007. Public domain. https://commons.wikimedia.org/wiki/File:Corrientes-oceanicas.png

Figure 12.23: Global thermohaline circulation. NASA. 2008. Public domain. https://commons.wikimedia.org/wiki/File:Thermohaline_Circulation_2.png

Figure 12.24: The types of tides. KVDP. 2009. Public domain. https://commons.wikimedia.org/wiki/File:Tide_schematic.svg

Figure 12.25: Different tide types. Snubcube. 2013. Public domain. https://commons.wikimedia.org/wiki/File:Tide_type.svg

Figure 12.26: Global tide types. KVDP. 2009. Public domain. https://commons.wikimedia.org/wiki/File:Diurnal_tide_types_map.jpg

Figure 12.27: A tidal day lasts slightly longer than 24 hours. NOAA. Unknown date. Public domain. https://oceanservice.noaa.gov/education/tutorial_tides/tides05_lunarday.html

13.

DESERTS

By the end of this chapter, students should be able to:

- Explain the defining characteristic of a desert and distinguish between the three broad categories of deserts.
- Explain how geographic features, **latitude**, atmospheric circulation, and Coriolis Effect influence where deserts are located.
- List the primary desert **weathering** and **erosion** processes.
- Identify desert landforms.
- Explain how desert landforms are formed by **erosion** and **deposition**.
- Describe the main types of sand **dunes** and the conditions that form them.
- Identify the main features of the **Basin and Range** desert (United States).

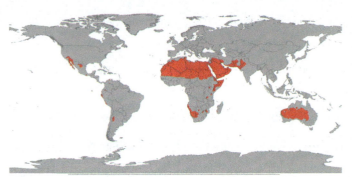

Figure 13.1: World hot deserts (BWh indicated in red).

Approximately 30% of the Earth's **terrestrial** surface consists of deserts, which are defined as locations of low **precipitation**. While **temperature** extremes are often associated with deserts, they do not define them. Deserts exhibit extreme temperatures because of the lack of moisture in the **atmosphere**, including low humidity and scarce cloud cover. Without cloud cover, the Earth's surface absorbs more of the Sun's energy during the day and emits more heat at night.

Deserts are not randomly located on the Earth's surface. Many deserts are located at latitudes between 15° and 30° in both hemispheres and at both the North and South Poles, created by prevailing wind circulation in the **atmosphere**. Sinking, dry air currents occurring at 30° north and south of the equator produce **trade winds** that create deserts like the African Sahara and Australian Outback.

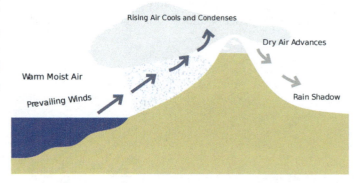

Figure 13.2: Mountainous areas in front of the prevailing winds create a rain shadow.

Another type of desert is found in the rain shadow created from prevailing winds blowing over mountain ranges. As the wind drives air up and over mountains, atmospheric moisture is released as snow or rain. Atmospheric pressure is lower at higher elevations, causing the moisture-laden air to cool. Cool air holds less moisture than hot air, and **precipitation** occurs as the wind rises up the mountain. After releasing its moisture on the windward side of the mountains, the dry air descends on the leeward or downwind side of the mountains to create an arid region with little **precipitation** called a rain shadow. Examples of rain-shadow deserts include the Western Interior Desert of North America and **Atacama Desert** of Chile, which is the Earth's driest, warm desert.

Finally, **polar deserts**, such as vast areas of the Antarctic and Arctic, are created from sinking cold air that is too cold to hold much moisture. Although they are covered with ice and snow, these deserts have very low average annual **precipitation**. As a result, Antarctica is Earth's driest **continent**.

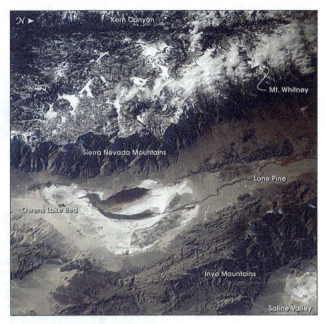

Figure 13.3: In this image from the ISS, the Sierra Nevada Mountains are perpendicular to prevailing westerly winds, creating a rain shadow to the east (down in the image). Note the dramatic decrease in snow on the Inyo Mountains.

 One or more interactive elements has been excluded from this version of the text. You can view them online here:
https://www.youtube.com/watch?v=iMu4dShS74w

Video 13.1: Rain shadow.

If you are using an offline version of this text, access this YouTube video via the QR code.

13.1 The Origin of Deserts

13.1.1 Atmospheric Circulation

Geographic location, atmospheric circulation, and the Earth's rotation are the primary causal factors of deserts. Solar energy converted to heat is the engine that drives the circulation of air in the **atmosphere** and water in the oceans. The strength of the circulation is determined by how much energy is absorbed by the Earth's surface, which in turn is dependent on the average position of the Sun relative to the Earth. In other words, the Earth is heated unevenly depending on **latitude** and **angle of incidence**. **Latitude** is a line circling the Earth parallel to the equator and is measured in degrees. The equator is 0° and the North and South Poles are 90° N and 90° S respectively (see the diagram of generalized atmospheric circulation on Earth). **Angle of incidence** is the angle made by a ray of sunlight shining on the Earth's surface. Tropical zones are located near the equator, where the **latitude** and **angle of incidence** are close to 0°, and receive high amounts of solar energy. The poles, which have latitudes and angles of incidence approaching 90°, receive little or almost no energy.

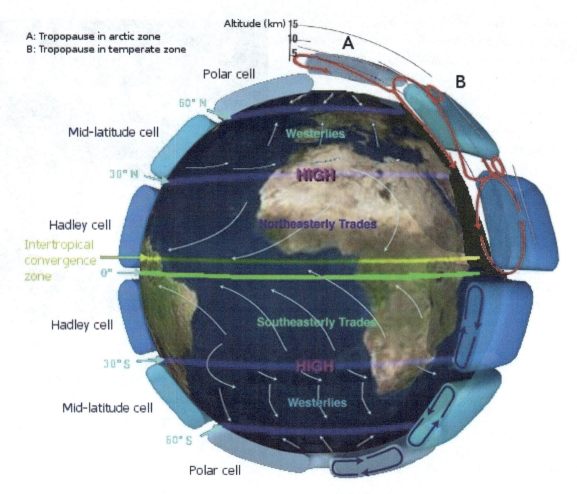

Figure 13.4: Generalized atmospheric circulation.

The figure shows the generalized air circulation within the **atmosphere**. Three cells of circulating air span the space between the equator and poles in both hemispheres, the **Hadley Cell**, the Ferrel or Midlatitude Cell, and the **Polar Cell**. In the **Hadley Cell** located over the tropics and closest to the equatorial belt, the sun heats the air and causes it to rise. The rising air cools and releases its contained moisture as tropical rain. The rising dried air spreads away from the equator and toward the north and south poles, where it collides with dry air in the Ferrel Cell. The combined dry air sinks back to the Earth at 30° **latitude**. This sinking drier air creates belts of predominantly high pressure at approximately 30° north and south of the equator, called the horse latitudes. Arid zones between 15° and 30° north and south of the equator thus exist within which desert conditions predominate. The descending air flowing north and south in the Hadley and Ferrel cells also creates prevailing winds called **trade winds** near the equator, and **westerlies** in the temperate zone. Note the arrows indicating general directions of winds in these zones.

Other deserts, like the **Great Basin Desert** that covers parts of Utah and Nevada, owe at least part of their origin to other atmospheric phenomena. The **Great Basin Desert**, while somewhat affected by sinking air effects from global circulation, is a rain-shadow desert. As westerly moist air from the Pacific rises over the Sierra Nevada and other mountains, it cools and loses moisture as condensation and **precipitation** on the upwind or rainy side of the mountains.

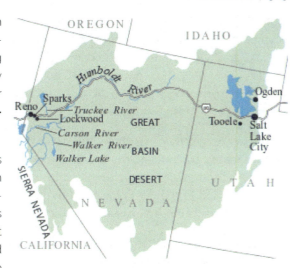

Figure 13.5: USGS Map of the Great Basin Desert.

Figure 13.6: Map of the Atacama desert (yellow) and surrounding related climate areas (orange).

One of the driest places on Earth is the **Atacama Desert** of northern Chile. The **Atacama Desert** occupies a strip of land along Chile's **coast** just north of **latitude** 30°S, at the southern edge of the trade-wind belt. The desert lies west of the Andes Mountains, in the rain shadow created by prevailing **trade winds** blowing west. As this warm moist air crossing the Amazon **basin** meets the eastern edge of the mountains, it rises, cools, and precipitates much of its water out as rain. Once over the mountains, the cool, dry air descends onto the **Atacama desert**. Onshore winds from the Pacific are cooled by the Peru (Humboldt) ocean current. This super-cooled air holds almost no moisture and, with these three factors, some locations in the **Atacama Desert** have received no measured **precipitation** for several years. This desert is the driest, non-polar location on Earth.

Notice in the figure that the polar regions are also areas of predominantly high pressure created by descending cold dry air, the **Polar Cells**. As with the other cells, cold air, which holds much less moisture than warm air, descends to create **polar deserts**. This is why historically, land near the north and south poles has always been so dry.

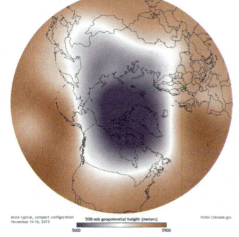

Figure 13.7: The polar vortex of mid-November, 2013. This cold, descending air (shown in purple) is characteristic of polar circulation.

13.1.2 Coriolis Effect

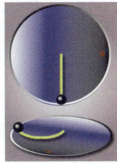

Figure 13.8: In the inertial frame of reference of the top two circles, the ball moves in a straight line. The observer, represented as a red dot, standing in the rotating frame of reference sees the ball following a curved path. This perceived curvature is due to the Coriolis Effect and centrifugal forces.

The Earth rotates toward the east where the sun rises. Think of spinning a weight on a string around your head. The speed of the weight depends on the length of the string. The speed of an object on the rotating Earth depends on its horizontal distance from the Earth's **axis** of rotation. Higher latitudes are a smaller distance from the Earth's rotational **axis**, and therefore do not travel as fast eastward as lower **latitudes** that are closer to the equator. When a fluid like air or water moves from a lower **latitude** to a higher **latitude**, the fluid maintains its momentum from moving at a higher speed, so it will travel relatively faster eastward than the Earth beneath it at the higher latitudes. This factor causes deflection of movements that occur in north-south directions.

Another factor in the Coriolis effect also causes deflection of east-west movement due to the angle between the centripetal effect of Earth's spin and gravity pulling toward the earth's center (see figure 13.9). This produces a net deflection toward the equator. The total Coriolis deflection on a mass moving in any direction on the rotating Earth results from a combination of these two factors.

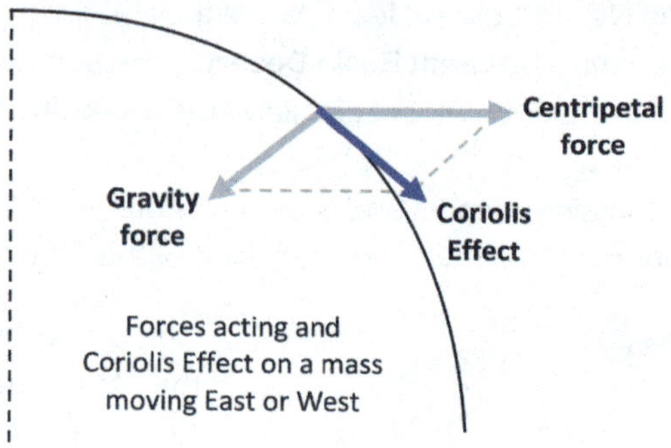

Since each hemisphere has three atmospheric cells moving respectively north and south relative to the Earth beneath them, the Coriolis effect deflects these moving air masses to the right in the Northern Hemisphere and to the left in the Southern Hemisphere. The Coriolis effect also deflects moving masses of water in the ocean currents.

Figure 13.9: Forces acting on a mass moving East–West on the rotating Earth that produce the Coriolis Effect.

For example, in the northern hemisphere **Hadley Cell**, the lower altitude air currents are flowing south towards the equator. These are deflected to the right (or west) by the Coriolis effect. This deflected air generates the prevailing **trade winds** that European sailors used to cross the Atlantic Ocean and reach South America and the Caribbean Islands in their tall ships. This air movement is mirrored in the **Hadley Cell** in the southern hemisphere; the lower altitude air current flowing equatorward is deflected to the left, creating **trade winds** that blow to the northwest.

In the northern Mid-Latitude or Ferrel Cell, surface air currents flow from the horse latitudes (**latitude** 30°) toward the North Pole, and the Coriolis effect deflects them toward the east, or to the right, producing the zone of westerly winds. In the southern hemisphere Mid-Latitude or Ferrel Cell, the poleward flowing surface air is deflected to the left and flows southeast creating the Southern Hemisphere **westerlies**.

Another Coriolis-generated deflection produces the **Polar Cells**. At 60° north and south **latitude**, relatively warmer rising air flows poleward cooling and converging at the poles where it sinks in the polar high. This sinking dry air creates the **polar deserts**, the driest deserts on Earth. Persistence of ice and snow is a result of cold temperatures at these dry locations.

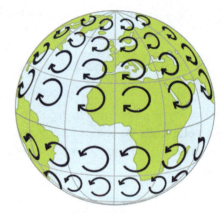

Figure 13.10: Inertia of air masses caused by the Coriolis Effect in the absence of other forces.

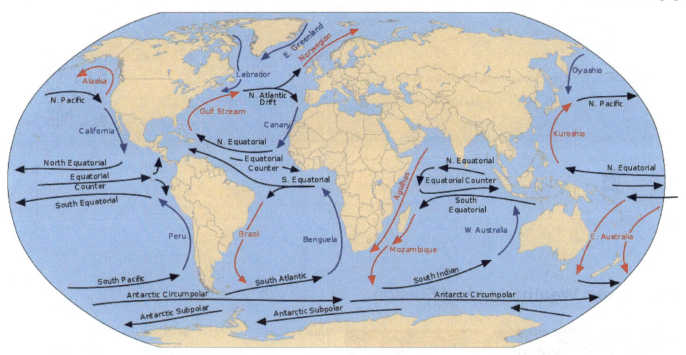

Figure 13.11: Gyres of the Earth's oceans.

The Coriolis effect operates on all motions on the Earth. Artillerymen must take the Coriolis effect into account on ballistic trajectories when making long-distance targeting calculations. Geologists note how its effect on air and **oceanic** currents creates deserts in designated zones around the Earth as well as the surface currents in the ocean. The Coriolis effect causes the ocean **gyres** to turn clockwise in the northern hemisphere and counterclockwise in the southern. It also affects **weather** by creating high-altitude, polar jet **streams** that sometimes push lobes of cold arctic air into the temperate zone, down to as far as **latitude** 30° from the usual 60°. It also causes low pressure systems and intense tropical storms to rotate counter-clockwise in the Northern Hemisphere and clockwise in the Southern Hemisphere.

One or more interactive elements has been excluded from this version of the text. You can view them online here: https://www.youtube.com/watch?v=HIyBpi7B-dE

Video 13.2: The Coriolis effect.

If you are using an offline version of this text, access this YouTube video via the QR code.

Take this quiz to check your comprehension of this section.

If you are using an offline version of this text, access the quiz for section 13.1 via the QR code.

 An interactive H5P element has been excluded from this version of the text. You can view it online here:
https://pressbooks.lib.vt.edu/introearthscience/?p=841#h5p-89

13.2 Desert Weathering and Erosion

Weathering takes place in desert climates by the same means as other climates, only at a slower rate. While higher temperatures typically spur faster **chemical weathering**, water is the main agent of **weathering**, and lack of water slows both mechanical and **chemical weathering**. Low **precipitation** levels also mean less **runoff** as well as **ice wedging**. When **precipitation** does occur in the desert, it is often heavy and may result in **flash floods** in which a lot of material may be dislodged and moved quickly.

Figure 13.12: Weathering and erosion of Canyonlands National Park has created a unique landscape, including arches, cliffs, and spires.

One unique **weathering** product in deserts is **desert varnish**. Also known as desert patina or rock rust, this is thin dark brown layers of clay **minerals** and iron and manganese **oxides** that form on very stable surfaces within arid environments. The exact way this material forms is still unknown, though cosmogenic and biologic mechanisms have been proposed.

Figure 13.13: Newspaper rock, near Canyonlands National Park, has many petroglyphs carved into desert varnish.

Figure 13.14: A dust storm (haboob) hits Texas in 2019.

While water is still the dominant agent of **erosion** in most desert environments, wind is a notable agent of **weathering** and **erosion** in many deserts. This includes suspended **sediment** traveling in **haboobs**, or large dust storms, that frequent deserts. Deposits of windblown dust are called **loess**. **Loess** deposits cover wide areas of the midwestern United States, much of it from rock flour that melted out of the **ice sheets** during the last **ice age**. Loess was also blown from desert regions in the West. Possessing lower energy than water, wind transport nevertheless moves sand, silt, and dust. As noted in chapter 11, the load carried by a fluid (air is a fluid like water) is distributed among **bedload** and **suspended load**. As with water, in wind these components depend on wind velocity.

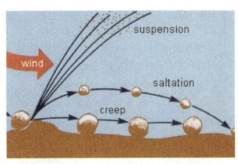

Figure 13.15: Diagram showing the mechanics of saltation.

Figure 13.16: Enlarged image of frosted and rounded windblown sand grains.

Sand size material moves by a process called **saltation** in which sand grains are lifted into the moving air and carried a short distance where they drop and splash into the surface dislodging other sand grains which are then carried a short distance and splash dislodging still others.

Since saltating sand grains are constantly impacting other sand grains, wind blown sand grains are commonly quite well rounded with frosted surfaces. **Saltation** is a cascading effect of sand movement creating a zone of wind blown sand up to a meter or so above the ground. This zone of saltating sand is a powerful **erosive** agent in which **bedrock** features are effectively sandblasted. The fine-grained **suspended load** is effectively sorted from the sand near the surface carrying the silt and dust into **haboobs**. Wind is thus an effective **sorting** agent separating sand and dust sized (≤70 μm) particles (see chapter 5). When wind velocity is high enough to slide or roll materials along the surface, the process is called **creep**.

One extreme version of **sediment** movement was shrouded in mystery for years: **Sliding stones**. Also called **sailing stones** and **sliding rocks**, these are large moving boulders along flat surfaces in deserts, leaving trails. This includes the famous example of the Racetrack **Playa** in Death Valley National Park, California. For years, scientists and enthusiasts attempted to explain their movement, with little definitive results. In recent years, several experimental and observational studies have confirmed that the stones, imbedded in thin layers of ice, are propelled by friction from high winds. These studies include measurements of actual movement, as well as re-creations of the conditions, with resulting movement in the lab.

Figure 13.17: A "sailing stone" at Racetrack Playa in Death Valley National Park, California.

Figure 13.18: Wind-carved ventifact in White Desert National Park, Egypt.

The zone of saltating sand is an effective agent of **erosion** through sand abrasion. A **bedrock** outcrop which has such a sandblasted shape is called a **yardang**. Rocks and boulders lying on the surface may be blasted and polished by saltating sand. When predominant wind directions shift, multiple sandblasted and polished faces may appear. Such wind abraded desert rocks are called **ventifacts**.

Figure 13.19: A yardang near Meadow, Texas.

Figure 13.20: Blowout in Texas.

In places with sand and silt accumulations, clumps of vegetation often anchor **sediment** on the desert surface. Yet, winds may be sufficient to remove materials not anchored by vegetation. The bowl-shaped depression remaining on the surface is called a **blowout**.

Take this quiz to check your comprehension of this section.

If you are using an offline version of this text, access the quiz for section 13.2 via the QR code.

 An interactive H5P element has been excluded from this version of the text. You can view it online here:
https://pressbooks.lib.vt.edu/introearthscience/?p=841#h5p-90

13.3 Desert Landforms

Figure 13.21: Aerial image of alluvial fan in Death Valley.

In the American Southwest, as **streams** emerge into the valleys from the adjacent mountains, they create desert landforms called **alluvial** fans. When the **stream** emerges from the narrow canyon, the flow is no longer constrained by the canyon walls and spreads out. At the lower slope angle, the water slows down and drops its coarser load. As the channel fills with this conglomeratic material, the **stream** is deflected around it. This deposited material deflects the **stream** into a **system** of radial distributary channels in a process similar to how a **delta** is made by a **river** entering a body of water. This process develops a **system** of radial distributaries and constructs a fan shaped feature called an **alluvial** fan.

Alluvial fans continue to grow and may eventually coalesce with neighboring fans to form an apron of **alluvium** along the mountain front called a **bajada**.

As the mountains erode away and their **sediment** accumulates first in **alluvial** fans, then **bajadas**, the mountains eventually are buried in their own erosional debris. Such buried mountain remnants are called **inselbergs**, "island mountains," as first described by the German geologist Wilhelm Bornhardt (1864–1946).

Figure 13.22: Bajada along Frisco Peak in Utah.

Figure 13.23: Inselbergs in the Western Sahara.

Where the desert valley is an enclosed **basin**, i.e. **streams** entering it do not drain out, the water is removed by evaporation and a dry lake **bed** is formed called a **playa**.

Figure 13.25: Dry wash (or ephemeral stream).

Figure 13.24: Satellite image of desert playa surrounded by mountains.

Playas are among the flattest of all landforms. Such a dry lake **bed** may cover a large area and be filled after a heavy thunderstorm to only a few inches deep. **Playa** lakes and desert **streams** that contain water only after rainstorms are called intermittent or **ephemeral**. Because of intense thunderstorms, the volume of water transported by **ephemeral** drainage in arid environments can be substantial during a short **period** of time. Desert soil structures lack organic matter that promotes infiltration by absorbing water. Instead of percolating into the **soil**, the **runoff** compacts the ground surface, making the **soil** hydrophobic or water-repellant. Because of this hardpan surface, **ephemeral streams** may gather water across large areas, suddenly filling with water from storms many miles away.

High-volume **ephemeral** flows, called **flash floods**, may move as sheet flows or **sheetwash**, as well as be channeled through normally dry **arroyos** or canyons. Flash floods are a major factor in desert **deposition**. Dry channels can fill quickly with **ephemeral drainage**, creating a mass of water and debris that charges down the **arroyo**, and even overflowing the banks. **Flash floods** pose a serious hazard for desert travelers because the storm activity feeding the **runoff** may be miles away. People hiking or camping in **arroyos** that have been bone dry for months, or years, have been swept away by sudden **flash floods**.

Figure 13.26: Flash flood in a dry wash.

13.3.1 Sand

The popular concept of a typical desert is a broad expanse of sand. Geologically, deserts are defined by a lack of water and arid regions resembling a sea of sand belong to the category of desert called an **erg**. An **erg** consists of fine-grained, loose sand grains, often blown by wind, or **aeolian** forces, into **dunes**. Probably the best known **erg** is the Rub' al Khali, which means Empty Quarter, of the Arabian Peninsula. Ergs are also found in the Great Sand Dunes National Park (Colorado), Little Sahara Recreation Area (Utah), White Sands National Monument (New Mexico), and parts of Death Valley National Park (California). **Ergs** are not restricted to deserts, but may form anywhere there is a substantial supply of sand, including as far north as 60° N in Saskatchewan, Canada, in the Athabasca Sand Dunes Provincial Park. Coastal **ergs** exist along lakes and oceans as well, and examples are found in Oregon, Michigan, and Indiana.

Figure 13.27: Sahara Desert, a sand sea or erg.

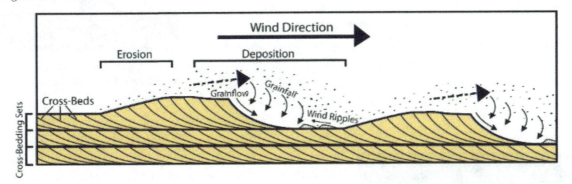

Figure 13.28: Formation of cross bedding in sand dunes.

An internal cross section of a sand **dune** shows a feature called **cross bedding**. As wind blows up the windward side of the **dune**, it carries sand to the **dune** crest depositing layers of sand parallel to the windward (or "stoss") side. The sand builds up the crest of the **dune** and pours over the top until the leeward (downwind or slip) face of the **dune** reaches the **angle of repose**, the maximum angle which will support the slip face. **Dunes** are unstable features and move as the sand erodes from the stoss side and continues to drop down the leeward side covering previous stoss and slip-face layers and creating the cross **beds**. Mostly, these are reworked over and over again, but occasionally, the features are preserved in a depression, then lithified. Shifting wind directions and abundant sand sources create chaotic patterns of cross **beds** like those seen in Zion National Park of Utah.

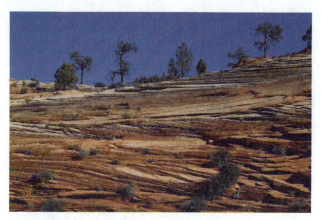

Figure 13.29: Cross beds in the Navajo Sandstone at Zion National Park.

In the **Mesozoic Era**, Utah was covered by a series of **ergs**, with the thickest being in Southern Utah, which lithified into **sandstone** (see chapter 5). Perhaps the best known of these **sandstone formations** is the Navajo Sandstone of **Jurassic** age. This **sandstone** formation consists of dramatic cliffs and spires in Zion National Park and covers a large part of the Colorado Plateau. In Arches National Park, a later series of sand **dunes** covered the Navajo Sandstone and lithified to become the Entrada Formation also during the **Jurassic**. **Erosion** of overlying layers exposed fins of the underlying Entrada Sandstone and carved out weaker parts of the fins forming the arches.

As the cements that hold the grains together in these modern sand cliffs disintegrate and the freed grains gather at the base of the cliffs and move down the washes, sand grains may be recycled and redeposited. These **Mesozoic** sand **ergs** may represent ancient **quartz** sands recycled many times from **igneous** origins in the early **Precambrian**, just passing now through another cycle of **erosion** and **deposition**. An example of this is Coral Pink Sand Dunes State Park in Southwestern Utah, which contains sand that is being eroded from the Navajo Sandstone to form new **dunes**.

Figure 13.30: Enlarged image of frosted and rounded windblown sand grains.

Dune Types

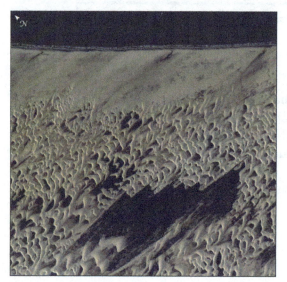

Figure 13.31: NASA image of barchan dune field in coastal Brazil.

Dunes are complex features formed by a combination of wind direction and sand supply, in some cases interacting with vegetation. There are several types of **dunes** representing variables of wind direction, sand supply and vegetative anchoring. Barchan **dunes** or crescent **dunes** form where sand supply is limited and there is a fairly constant wind direction. Barchans move downwind and develop a crescent shape with wings on either side of a **dune** crest. Barchans are known to actually move over homes, even towns.

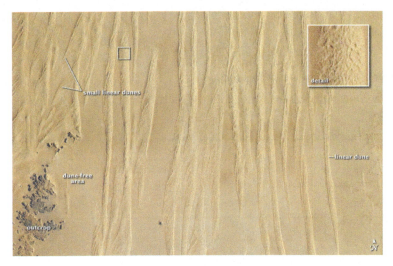

Figure 13.32: Satellite image of longitudinal dunes in Egypt.

Longitudinal dunes or **linear dunes** form where sand supply is greater and the wind blows around a dominant direction, in a back-and-forth manner. They may form ridges tens of meters high lined up with the predominant wind directions.

Parabolic dunes form where vegetation anchors parts of the sand and unanchored parts **blowout**. **Parabolic dune** shape may be similar to barchan **dunes** but usually reversed, and it is determined more by the anchoring vegetation than a strict parabolic form.

Star dunes form where the wind direction is variable in all directions. Sand supply can range from limited to quite abundant. It is the variation in wind direction that forms the star.

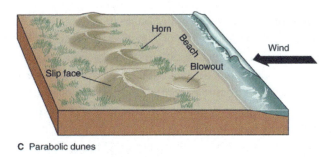

Figure 13.33: Parabolic dunes.

Figure 13.34: Star dune in Namib Desert.

Take this quiz to check your comprehension of this section.

If you are using an offline version of this text, access the quiz for section 13.3 via the QR code.

 An interactive H5P element has been excluded from this version of the text. You can view it online here:
https://pressbooks.lib.vt.edu/introearthscience/?p=841#h5p-91

13.4 The Great Basin and the Basin and Range

The Great Basin is the largest area of interior **drainage** in North America, meaning there is no outlet to the ocean and all **precipitation** remains in the **basin** or is evaporated. It covers western Utah, most of Nevada, and extends into southeastern California, southern Oregon, and southern Idaho. Because there is not outlet to the ocean, **streams** in the Great Basin deliver **runoff** to lakes and **playas** within the **basin**. A subregion within the Great Basin is the **Basin and Range** which extends from the Wasatch Front in Utah west across Nevada to the Sierra Nevada Mountains of California. The basins and ranges referred to in the name are **horsts** and **grabens**, formed by **normal fault** blocks from crustal **extension**, as discussed in chapter 2 and chapter 9. The **lithosphere** of the entire area has stretched by a factor of about 2, meaning from end to end, the distance has doubled over the past 30 million years or so. Valleys without outlets form individual basins, each of which is filled with **alluvial sediments** leading into **playa depositional environments**. During the recent **Ice Age**, The climate was more humid and while glaciers were forming in some of the mountains, **pluvial lakes** formed covering large areas (see section 14.4.3). During the **Ice Age**, valleys in much of western Utah and eastern Nevada were cov-

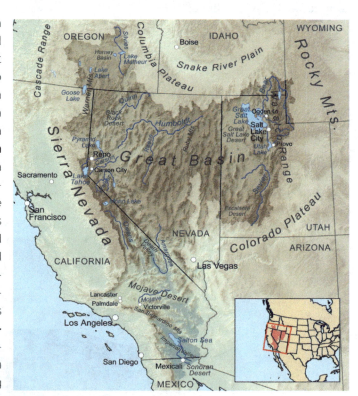

Figure 13.35: The Great Basin.

ered by Lake Bonneville. As the **climate** became arid after the **Ice Age**, Lake Bonneville dried leaving as a remnant the Great Salt Lake in Utah.

Figure 13.36: Typical Basin and Range scene. Ridgecrest, CA sits just east of the southern Sierra Nevada Mountains.

The desert of the **Basin and Range** extends from about 35° to near 40° and results from a rain shadow effect created by westerly winds from the Pacific rising and cooling over the Sierras becoming depleted of moisture by **precipitation** on the western side. The result is relatively dry air descending across Nevada and western Utah.

A journey from the Wasatch Front southwest to the Pacific Ocean will show stages of desert landscape evolution from the **fault** block mountains of Utah with sharp peaks and **alluvial** fans at the mouths of canyons, through land-scapes in Southern Nevada with **bajadas** along the mountain fronts, to the landscapes in the Mojave Desert of California with subdued **inselbergs** sticking up through a sea of coalesced **bajadas**. These landscapes illustrate the evolutionary stages of desert landscape development.

13.4.1 Desertification

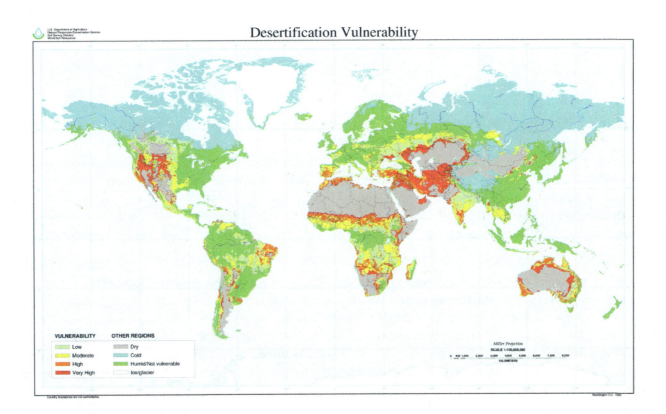

Figure 13.37: World map showing desertification vulnerability.

When previously arable land suitable for agriculture transforms into desert, this process is called **desertification**. Plants and **humus**-rich **soil** (see chapter 5) promote **groundwater infiltration** and water retention. When an area becomes more arid due to changing environmental conditions, the plants and **soil** become less effective in retaining water, creating a **positive feedback** loop of **desertification**. This self-reinforcing loop spirals into increasingly arid conditions and further enlarges the desert regions.

Desertification may be caused by human activities, such as unsustainable crop cultivation practices, overgrazing by livestock, overuse of groundwater, and global **climate** change. Human-caused desertification is a serious worldwide problem. The world map figure above shows what areas are most vulnerable to **desertification**. Note the red and orange areas in the western and midwestern regions of the United States, which also cover large areas of arable land used for raising food crops and animals. The creation of the Dust Bowl in the 1930s (see chapter 5) is a classic example of a high-vulnerability region impacted by human–caused **desertification**. As demonstrated in the Dust Bowl, conflicts may arise between agricultural practices and conservation measures. Mitigating **desertification** while allowing farmers to make a survivable living requires public and individual education to create community support and understanding of sustainable agriculture alternatives.

Take this quiz to check your comprehension of this section.

If you are using an offline version of this text, access the quiz for section 13.4 via the QR code.

An interactive H5P element has been excluded from this version of the text. You can view it online here:
https://pressbooks.lib.vt.edu/introearthscience/?p=841#h5p-92

Summary

Approximately 30% of Earth's surface is arid lands, the location of which is determined by **latitude**, atmospheric circulation, and terrain. The arid belts between 15° and 30° north and south latitudes are produced by descending air masses associated with major cells in the **atmosphere** and include the major deserts like the Sahara in Africa and the Middle East. **Rain shadow deserts** lie behind mountain ranges or long land expanses in zones of prevailing winds like the deserts of western North America, the Atacama of South America, and the Gobi of Asia. Dry descending air also creates the **polar deserts** at the poles.

Major atmospheric circulation involves the **Hadley cells**, midlatitude or Ferrel cells, and the **polar cells** in each hemisphere. Warmed and rising air in the **Hadley cells** rains back on the tropics and moves toward the poles as dryer air. That air meets the dryer equatorward moving air of the Ferrel cells. This dry air descends in the arid zones, called the horse latitudes, to produce the arid belts in each hemisphere. Rotation of the Earth creates the Coriolis Effect that deflects these moving air masses to produce zones of prevailing winds, the **Trade Winds** in the subtropics and the **Westerlies** in the midlatitudes. A combination of **latitude**, rain shadow, and cold adjacent ocean currents causes the **Atacama Desert** of northern Chile, the driest desert on Earth.

Weathering in deserts takes place just as in other climes only slower because of less water. **Desert varnish** is a **weathering** product unique to desert environments. As in more humid climes, water is the main agent of **erosion** although wind is a notable agent. Large dust storms called **haboobs** transport large amounts of **sediment** that may accumulate in sand seas called **ergs** or finer grained **loess** deposits. Sand transport occurs mainly by **saltation** in which grain to grain impact causes frosting of grain surfaces. Sandblasting by persistent winds produces stones with polished surfaces called **ventifacts** and sculpted **bedrock** features called **yardangs**.

Landforms produced in desert environments include **alluvial** fans, **bajadas**, **inselbergs**, and **playas**. Windblown sand can accumulate as **dunes**. The forms of **dunes**, like barchans, parabolic, longitudinal, and **star dunes**, relate to the abundance of sand supply and wind direction as well as presence of vegetation. The internal structure of **dunes** shows **cross bedding**. **Fossil dunes** in an ancient desert leave **cross bedding** in places like Zion and Arches national parks in Utah showing shifting wind directions in these ancient environments. **Ephemeral streams** in desert regions may carry water only after storms and pose risk of **flash floods**.

The Great Basin of North America is an enclosed **basin** with no **drainage** to the ocean. The only exit for **precipitation** there is by evaporation. Travels in the Great Basin show stages of development of desert landscapes from **playas** and **alluvial** fans, to **bajadas**, to **inselbergs** which are eroded mountains buried in their own erosional debris.

Poor land management can result in dying vegetation and loss of **soil** moisture producing an accelerating process of **desertification** in which once productive land is degraded into unproductive desert. This is a serious worldwide problem.

Take this quiz to check your comprehension of this chapter.

If you are using an offline version of this text, access the quiz for chapter 13 via the QR code.

 An interactive H5P element has been excluded from this version of the text. You can view it online here:
https://pressbooks.lib.vt.edu/introearthscience/?p=841#h5p-93

Text References

1. Bagnold, R. A. 1941. "The Physics of Blown Sand and Desert Dunes." *Methum, London, UK*, 265.
2. Boggs, Sam. 2006. "Principles of Sedimentology and Stratigraphy." Pearson Prentice Hall. http://agris.fao.org/agris-search/search.do?recordID=US201300110702.
3. Clements, Thomas. 1952. "Wind-Blown Rocks and Trails on Little Bonnie Claire Playa, Nye County, Nevada." *Journal of Sedimentary Research* 22 (3). Society for Sedimentary Geology. http://archives.datapages.com/data/sepm/journals/v01-32/data/022/022003/0182.htm.
4. Collado, Gonzalo A., Moisés A. Valladares, and Marco A. Méndez. "Hidden Diversity in Spring Snails from the Andean Altiplano, the Second Highest Plateau on Earth, and the Atacama Desert, the Driest Place in the World." *Zoological Studies* 52 (1): 50.
5. Easterbrook, Don J. 1999. *Surface Processes and Landforms*. Pearson College Division.
6. Geist, Helmut. 2005. *The Causes and Progression of Desertification*. Ashgate Aldershot.
7. Grayson, Donald K. 1993. *The Desert's Past: A Natural Prehistory of the Great Basin*. Smithsonian Inst Pr.
8. Hadley, Geo. 1735. "Concerning the Cause of the General Trade-Winds: By Geo. Hadley, Esq; FRS." *Philosophical Transactions* 39 (436-444). The Royal Society: 58–62.
9. Hartley, Adrian J., and Guillermo Chong. 2002. "Late Pliocene Age for the Atacama Desert: Implications for the Desertification of Western South America." *Geology* 30 (1). geology.gsapubs.org: 43–46.
10. Hedin, Sven Anders. 1903. *Central Asia and Tibet*. Vol. 1. Hurst and Blackett, limited.
11. Hooke, Roger Leb. 1967. "Processes on Arid-Region Alluvial Fans." *The Journal of Geology* 75 (4). journals.uchicago.edu: 438–60.

12. King, Lester C. 1953. "Canons of Landscape Evolution." *Geological Society of America Bulletin* 64 (7). gsabulletin.gsa-pubs.org: 721–52.

13. Kletetschka, Gunther, Roger Leb Hooke, Andrew Ryan, George Fercana, Emerald McKinney, and Kristopher P. Schwebler. 2013. "Sliding Stones of Racetrack Playa, Death Valley, USA: The Roles of Rock Thermal Conductivity and Fluctuating Water Levels." *Geomorphology* 195 (August). Elsevier: 110–17.

14. Laity, Julie E. 2009. "Landforms, Landscapes, and Processes of Aeolian Erosion." In *Geomorphology of Desert Environments*, edited by Anthony J. Parsons and Athol D. Abrahams, 597–627. Springer Netherlands.

15. Littell, Eliakim, and Robert S. Littell. 1846. *Littell's Living Age*. T.H. Carter & Company.

16. Livingstone, Ian, and Andrew Warren. 1996. *Aeolian Geomorphology: An Introduction*. Longman.

17. Muhs, Daniel R., and E. A. Bettis. 2003. "Quaternary Loess-Paleosol Sequences as Examples of Climate-Driven Sedimentary Extremes." *Special Papers-Geological Society of America*. Boulder, Colo.; Geological Society of America; 1999, 53–74.

18. Norris, Richard D., James M. Norris, Ralph D. Lorenz, Jib Ray, and Brian Jackson. 2014. "Sliding Rocks on Racetrack Playa, Death Valley National Park: First Observation of Rocks in Motion." *PloS One* 9 (8). journals.plos.org: e105948.

19. Shao, Yaping. 2008. *Physics and Modelling of Wind Erosion*. Springer Science & Business Media.

20. Stanley, George M. 1955. "Origin of Playa Stone Tracks, Racetrack Playa, Inyo County, California." *Geological Society of America Bulletin* 66 (11). gsabulletin.gsapubs.org: 1329–50.

21. Walker, Alta S. 1996. "Deserts: Geology and Resources." Government Printing Office. https://pubs.er.usgs.gov/publication/7000004.

22. Wilson, Ian Gordon. 1971. "Desert Sandflow Basins and a Model for the Development of Ergs." *The Geographical Journal* 137 (2). [Wiley, Royal Geographical Society (with the Institute of British Geographers)]: 180–99.

Figure References

Figure 13.1: World hot deserts (Koppen BWh). Peel, M. C., Finlayson, B. L., and McMahon, T. A.; adapted by Me ne frego. 2011. CC BY-SA 3.0. https://en.wikipedia.org/wiki/File:Koppen_World_Map_BWh.png

Figure 13.2: Mountainous areas in front of the prevailing winds create a rain shadow. domdomegg. 2015. CC BY 4.0. https://commons.wikimedia.org/wiki/File:Rainshadow_copy.svg

Figure 13.3: In this image from the ISS, the Sierra Nevada Mountains are perpendicular to prevailing westerly winds, creating a rain shadow to the east (down in the image). NASA. 2003. Public domain. https://earthobservatory.nasa.gov/images/3273/southern-sierra-nevada-and-owens-lake

Figure 13.4: Generalized atmospheric circulation. NASA. 2009. Public domain. https://commons.wikimedia.org/wiki/File:Earth_Global_Circulation.jpg

Figure 13.5: USGS Map of the Great Basin Desert. USGS. 2012. Public domain. https://commons.wikimedia.org/wiki/File:Central_Basin_and_Range_ecoregion.gif

Figure 13.6: Map of the Atacama desert (yellow) and surrounding related climate areas (orange). cobaltcigs. 2010. CC BY-SA 3.0. https://en.wikipedia.org/wiki/File:Atacama_map.svg

Figure 13.7: The polar vortex of mid-November, 2013. NOAA. 2014. Public domain. https://commons.wikimedia.org/wiki/File:November2013_polar_vortex_geopotentialheight_mean_Large.jpg

Figure 13.8: In the inertial frame of reference of the top picture, the ball moves in a straight line. Hubi. 2003. CC BY-SA 3.0. https://commons.wikimedia.org/wiki/File:Corioliskraftanimation.gif

Figure 13.9: Forces acting on a mass moving East–West on the rotating Earth that produce the Coriolis Effect. Kindred Grey. 2022. CC BY 4.0.

Figure 13.10: Inertia of air masses caused by the Coriolis Effect in the absence of other forces. Kes47. 2015. CC BY-SA 3.0. https://commons.wikimedia.org/wiki/File:Coriolis_effect.svg

Figure 13.11: Gyres of the Earth's oceans. Canuckguy; adapted by Shadowxfox and Popadius. 2012. Public domain. https://commons.wikimedia.org/wiki/File:Corrientes-oceanicas-en.svg

Figure 13.12: Weathering and erosion of Canyonlands National Park has created a unique landscape, including arches, cliffs, and spires. Rick. 2005. CC BY 2.0. https://commons.wikimedia.org/wiki/File:Mesa_Arch_Canyonlands_National_Park.jpg

Figure 13.13: Newspaper rock, near Canyonlands National Park, has many petroglyphs carved into desert varnish. Cacophony. 2009. CC BY-SA 3.0. https://commons.wikimedia.org/wiki/File:Newspaper_Rock_Full.jpg

Figure 13.14: A dust storm (haboob) hits Texas in 2019. Jakeorin. 2019. CC BY-SA 4.0. https://commons.wikimedia.org/wiki/File:Haboob_in_Big_Spring,_TX.jpg

Figure 13.15: Diagram showing the mechanics of saltation. NASA. 2002. Public domain. https://commons.wikimedia.org/wiki/File:Saltation-mechanics.gif

Figure 13.16: Enlarged image of frosted and rounded windblown sand grains. Wilson44691. 2008. Public domain. https://commons.wikimedia.org/wiki/File:CoralPinkSandDunesSand.JPG

Figure 13.17: A "sailing stone" at Racetrack Playa in Death Valley National Park, California. Lgcharlot. 2006. CC BY-SA 4.0. https://commons.wikimedia.org/wiki/File:Racetrack_Playa_in_Death_Valley_National_Park.jpg

Figure 13.18: Wind-carved ventifact in White Desert National Park, Egypt. Christine Schultz. 2003. Public domain. https://commons.wikimedia.org/wiki/File:Weisse_W%C3%BCste.jpg

Figure 13.19: A yardang near Meadow, Texas. United States Department of Agriculture. 2000. Public domain. https://commons.wikimedia.org/wiki/File:Yardang_Lea-Yoakum_Dunes.jpg

Figure 13.20: Blowout in Texas. Leaflet. 1996. CC BY-SA 3.0. https://commons.wikimedia.org/wiki/File:Blowout_Earth_TX.jpg

Figure 13.21: Aerial image of alluvial fan in Death Valley. USGS. 2005. Public domain. https://commons.wikimedia.org/wiki/File:Alluvial_Fan.jpg

Figure 13.22: Bajada along Frisco Peak in Utah. GerthMichael. 2010. CC BY-SA 3.0. https://commons.wikimedia.org/wiki/File:FriscoMountainUT.jpg

Figure 13.23: Inselbergs in the Western Sahara. Nick Brooks. 2007. CC BY 2.0. https://commons.wikimedia.org/wiki/File:Breast-Shaped_Hill.jpg

Figure 13.24: Satellite image of desert playa surrounded by mountains. Robert Simmon, using Landsat data from the U.S. Geological Survey and NASA. 2013. Public domain. https://earthobservatory.nasa.gov/images/80913/eye-exam-for-a-satellite

Figure 13.25: Dry wash (or ephemeral stream). Finetooth. 2010. CC BY-SA 3.0. https://commons.wikimedia.org/wiki/File:Dry_Wash_in_PEFO_NP.jpg

Figure 13.26: Flash flood in a dry wash. USGS. 2016. Public domain. https://www.usgs.gov/media/images/a-flooded-river-australia

Figure 13.27: Sahara Desert, a sand sea or erg. Wsx. 2004. CC BY-SA 3.0. https://commons.wikimedia.org/wiki/File:Sahara_Desert_in_Jalu,_Libya.jpeg

Figure 13.28: Formation of cross bedding in sand dunes. David Tarailo, GSA, GeoCorps Program. 2015. Public domain. https://commons.wikimedia.org/wiki/File:Formation_of_cross-bedding.jpg

Figure 13.29: Cross beds in the Navajo Sandstone at Zion National Park. Roy Luck. 2010. CC BY 2.0. https://commons.wiki-media.org/wiki/File:Sandstone_showing_Cross-bedding_Zion_National_Park_Utah_USA.jpg

Figure 13.30: Enlarged image of frosted and rounded windblown sand grains. Wilson44691. 2008. Public domain. https://commons.wikimedia.org/wiki/File:CoralPinkSandDunesSand.JPG

Figure 13.31: NASA image of barchan dune field in coastal Brazil. NASA. 2003. Public domain. https://earthobserva-tory.nasa.gov/images/3889/coastal-dunes-brazil

Figure 13.32: Satellite image of longitudinal dunes in Egypt. NASA. 2012. Public domain. https://earthobservatory.nasa.gov/images/78151/linear-dunes-great-sand-sea-egypt

Figure 13.33: Parabolic dunes. Po ke jung. 2011. CC BY 3.0. https://commons.wikimedia.org/wiki/File:Parabolic_dune.jpg

Figure 13.34: Star dune in Namib Desert. Dave Curtis. 2005. CC BY-NC-ND 2.0. https://flic.kr/p/4N5Fa

Figure 13.35: The Great Basin. Kmusser; adapted by Hike395. 2022. CC BY-SA 3.0. https://commons.wikimedia.org/wiki/File:Great_Basin_map.gif

Figure 13.36: Typical Basin and Range scene. Matt Affolter (QFL247). 2010. CC BY-SA 3.0. https://commons.wikimedia.org/wiki/File:RidgecrestCA.JPG

Figure 13.37: World map showing desertification vulnerability. USDA. 1998. Public domain. https://commons.wikimedia.org/wiki/File:Desertification_map.png

14.

GLACIERS

Learning Objectives

By the end of this chapter, students should be able to:

- Differentiate the different types of **glaciers** and contrast them with sea icebergs.
- Describe how **glaciers** form, move, and create landforms.
- Describe **glacial budget**; describe the zones of accumulation, equilibrium, and melting.
- Identify **glacial** erosional and depositional landforms and interpret their origin; describe **glacial** lakes.
- Describe the history and causes of past **glaciations** and their relationship to **climate**, sea-level changes, and **isostatic rebound**.

The Earth's **cryosphere**, or ice, has a unique set of erosional and depositional features compared to its **hydrosphere**, or liquid water. This ice exists primarily in two forms, **glaciers** and icebergs. **Glaciers** are large accumulations of ice that exist year-round on the land surface. In contrast, masses ice floating on the ocean are icebergs, although they may have had their origin in **glaciers**.

Glaciers cover about 10% of the Earth's surface and are powerful erosional agents that sculpt the planet's surface. These enormous masses of ice usually form in mountainous areas that experience cold temperatures and high precipitation. Glaciers also occur in low lying areas such as Greenland and Antarctica that remain extremely cold year-round.

14.1 Glacier Formation

Glaciers form when repeated annual snowfall accumulates deep layers of snow that are not completely melted in the summer. Thus there is an accumulation of snow that builds up into deep layers. Perennial snow is a snow accumulation that lasts all year. A thin accumulation of perennial snow is a snow field. Over repeated seasons of perennial snow, the snow settles, compacts, and **bonds** with underlying layers. The amount of void space between the snow grains diminishes. As the old snow gets buried by more new snow, the older snow layers compact into **firn**, or **névé**, a granular mass of ice crystals. As the **firn** continues to be buried, compressed, and recrystallizes, the void spaces become smaller and the ice becomes less porous, eventually turning into **glacier**

Figure 14.1: Glacier in the Bernese Alps.

ice. Solid glacial ice still retains a fair amount of void space and that traps air. These small air pockets provide records of the past **atmosphere composition**.

Figure 14.2: Greenland ice sheet.

There are three general types of **glaciers**: alpine or **valley glaciers**, **ice sheets**, and ice caps. Most **alpine glaciers** are located in the world's major mountain ranges such as the Andes, Rockies, Alps, and Himalayas, usually occupying long, narrow valleys. Alpine **glaciers** may also form at lower elevations in areas that receive high annual **precipitation** such as the Olympic Peninsula in Washington state.

Ice sheets, also called **continental glaciers**, form across millions of square kilometers of land and are thousands of meters thick. Earth's largest **ice sheets** are located on Greenland and Antarctica. The Greenland Ice Sheet is the largest ice mass in the Northern Hemisphere with an extensive surface area of over 2 million sq km (1,242,700 sq mi) and an average thickness of up to 1500 meters (5,000 ft, almost a mile).

The Antarctic Ice Sheet is even larger and covers almost the entire **continent**. The thickest parts of the Antarctic **ice sheet** are over 4,000 meters thick (>13,000 ft or 2.5 mi). Its weight depresses the Antarctic **bedrock** to below sea level in many places. The cross-sectional diagram comparing the Greenland and Antarctica **ice sheets** illustrates the size difference between the two.

Figure 14.3: Thickness of Greenland ice sheet in meters.

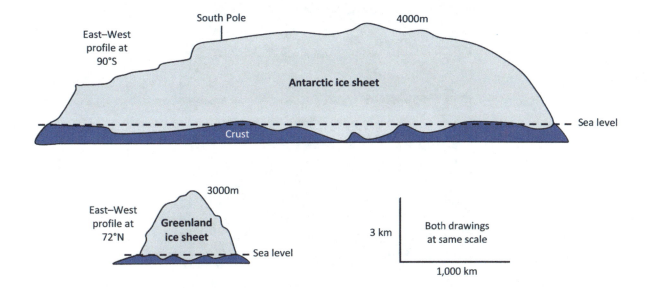

Figure 14.4: Cross-sectional view of both Greenland and Antarctic ice sheets drawn to scale for size comparison.

Ice cap **glaciers** are smaller versions of **ice sheets** that cover less than 50,000 km2 and usually occupy higher elevations and may cover tops of mountains. There are several ice caps on Iceland. A small ice cap called Snow Dome is near Mt. Olympus on the Olympic Peninsula in the state of Washington.

Figure 14.5: Snow Dome ice cap near Mt. Olympus, Washington (left) and Vatnajökull ice cap in Iceland (right).

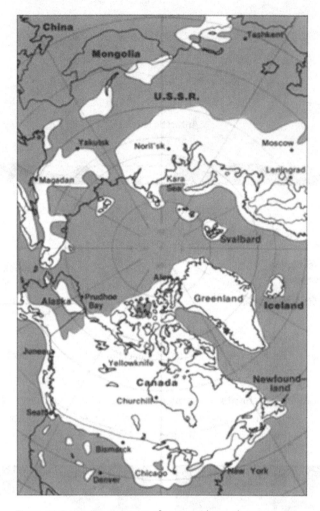

Figure 14.6: Maximum extent of Laurentide ice sheet.

The figure shows the size of the ancient Laurentide Ice Sheet in the Northern Hemisphere. This **ice sheet** was present during the last glacial maximum event, also known as the last Ice Age.

Take this quiz to check your comprehension of this section.

If you are using an offline version of this text, access the quiz for section 14.1 via the QR code.

 An interactive H5P element has been excluded from this version of the text. You can view it online here:
https://pressbooks.lib.vt.edu/introearthscience/?p=3786#h5p-94

14.2 Glacier Movement

Figure 14.7: Glacial crevasses (left) and Cravasse on the Easton Glacier in the North Cascades (right).

As the ice accumulates, it begins to flow downward under its own weight. In 1948, glaciologists installed hollow vertical rods in into the Jungfraufirn Glacier in the Swiss Alps to measure changes in its movement over two years. This study showed that the ice at the surface was fairly rigid and ice within the **glacier** was actually flowing downhill. The cross-sectional diagram of an alpine (valley) **glacier** shows that the rate of ice movement is slow near the bottom, and fastest in the middle with the top ice being carried along on the ice below.

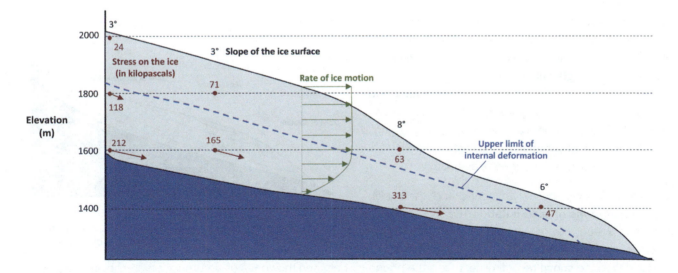

Figure 14.8: Cross-section of a valley glacier showing stress (red numbers) increase with depth under the ice. The ice will deform and flow where the stress is greater than 100 kilopascals, and the relative extent of that deformation is depicted by the red arrows. Down slope movement is shown with blue arrows. The upper ice above the red dashed line does not flow but is pushed along en masse.

One of the unique properties of ice is that it melts under pressure. About half of the overall **glacial** movement was from sliding on a film of meltwater along the **bedrock** surface and half from internal flow. Ice near the surface of the **glacier** is rigid and **brittle** to a depth of about 50 m (165 ft). In this **brittle** zone, large ice cracks called **crevasses** form on the **glacier**'s moving surface. These **crevasses** can be covered and hidden by a snow bridge and are a hazard for **glacier** travelers.

Below the brittle zone, the pressure typically exceeds 100 kilopascals (kPa), which is approximately 100,000 times atmospheric pressure. Under this applied force, the ice no longer breaks, but rather it bends or flows in a zone called the plastic zone. This plastic zone represents the great majority of **glacier** ice. The plastic zone contains a fair amount of **sediment** of various grades from boulders to silt and clay. As the bottom of the **glacier** slides and grinds across the **bedrock** surface, these **sediments** act as grinding agents and create a zone of significant **erosion**.

Take this quiz to check your comprehension of this section.

If you are using an offline version of this text, access the quiz for section 14.2 via the QR code.

 An interactive H5P element has been excluded from this version of the text. You can view it online here:
https://pressbooks.lib.vt.edu/introearthscience/?p=3786#h5p-95

14.3 Glacial Budget

A glacial budget is like a bank account, with the ice being the existing balance. If there is more income (snow accumulating in winter) than expense (snow and ice melting in summer), then the **glacial** budget shows growth. A positive or negative balance of ice in the overall **glacial budget** determines whether a **glacier** advances or retreats, respectively. The area where the ice balance is growing is called the **zone of accumulation**. The area where ice balance is shrinking is called the zone of ablation.

The diagram shows these two zones and the equilibrium line. In the **zone of accumulation**, the snow accumulation rate exceeds the snow melting rate and the ice surface is always covered with snow. The equilibrium

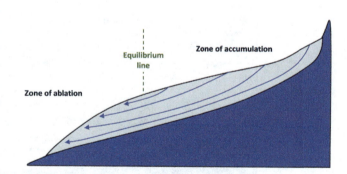

Figure 14.9: Cross-sectional view of an alpine glacier showing internal flow lines, zone of accumulation, snow line, and zone of melting.

line, also called the **snowline** or firnline, marks the boundary between the zones of accumulation and ablation. Below the equilibrium line in the zone of ablation, the melting rate exceeds snow accumulation leaving the bare ice surface exposed. The position of the firnline changes during the season and from year to year as a **reflection** of a positive or negative ice balance in the **glacial budget**. Of the two variables affecting a **glacier**'s budget, winter accumulation and summer melt, summer melt matters most to a **glacier**'s budget. Cool summers promote **glacial** advance and warm summers promote **glacial** retreat.

Figure 14.10: Fjord.

If a handful of warmer summers promote **glacial** retreat, then global **climate** warming over decades and centuries will accelerate glacial melting and retreat even faster. Global warming due to human burning of **fossil fuels** is causing the **ice sheets** to lose in years, an amount of mass that would normally take centuries. Current glacial melting is contributing to rising sea-levels faster than expected based on previous history.

As the Antarctica and Greenland **ice sheets** melt during global warming, they become thinner or deflate. The edges of the **ice sheets** break off and fall into the ocean, a process called **calving**, becoming floating icebergs. A **fjord** is a steep-walled valley flooded with sea water. The narrow shape of a **fjord** has been carved out by a **glacier** during a cooler **climate period**. During a warming trend, **glacial** meltwater may raise the sea level in **fjords** and flood formerly dry valleys. **Glacial** retreat and deflation are well illustrated in the 2009 TED Talk "Time-lapse proof of extreme ice loss" by James Balog.

 One or more interactive elements has been excluded from this version of the text. You can view them online here:
https://www.ted.com/talks/james_balog_time_lapse_proof_of_extreme_ice_loss?language=en

Video 14.1: Time-lapse proof of extreme ice loss, by James Balog.

If you are using an offline version of this text, access this TED Talk via the QR code.

Take this quiz to check your comprehension of this section.

If you are using an offline version of this text, access the quiz for section 14.3 via the QR code.

 An interactive H5P element has been excluded from this version of the text. You can view it online here:
https://pressbooks.lib.vt.edu/introearthscience/?p=3786#h5p-96

14.4 Glacial Landforms

Both alpine and **continental glaciers** create two categories of landforms: erosional and depositional. Erosional landforms are formed by the removal of material. Depositional landforms are formed by the addition of material. Because **glaciers** were first studied by 18th and 19th century geologists in Europe, the terminology applied to **glaciers** and **glacial** features contains many terms derived from European languages.

14.4.1 Erosional Glacial Landforms

Erosional landforms are created when moving masses of **glacial** ice slide and grind over **bedrock**. **Glacial** ice contains large amounts of poorly sorted sand, gravel, and bouldersthat have been plucked and pried from the **bedrock**. As the **glaciers** slide across the **bedrock**, they grind these **sediments** into a fine powder called rock flour. Rock flour acts as fine grit that polishes the surface of the **bedrock** to a smooth finish called **glacial polish**. Larger rock fragments scrape over the surface creating elongated grooves called **glacial striations**.

Figure 14.11: Glacial striations on granite in Whistler, Canada (left) and Glacial striations in Mt. Rainier National Park (right).

Figure 14.12: The U-shape of the Little Cottonwood Canyon, Utah, as it enters into the Salt Lake Valley.

Alpine **glaciers** produce a variety of unique erosional landforms, such as U-shaped valleys, **arêtes**, **cirques**, tarns, **horns**, **cols**, **hanging valleys**, and **truncated spurs**. In contrast, **stream**-carved canyons have a V-shaped profile when viewed in cross-section. Glacial erosion transforms a former V-shaped stream valley into a U-shaped one. **Glaciers** are typically wider than **streams** of similar length, and since **glaciers** tend to erode both at their bases and their sides, they erode V-shaped valleys into relatively flat-bottomed broad valleys with steep sides and a distinctive "U" shape. As seen in the images, Little Cottonwood Canyon near Salt Lake City, Utah was occupied by an **Ice Age glacier** that extended down to the **mouth** of the canyon and into Lake Bonneville. Today, that U-shaped valley hosts many erosional landforms, including polished and striated rock surfaces. In contrast, Big Cottonwood Canyon to the north of Little Cottonwood Canyon has retained the V-shape in its lower portion, indicating that its **glacier** did not extend clear to its **mouth**, but was confined to its upper portion.

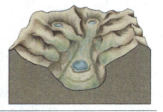

Figure 14.13: Formation of a glacial valley. Glaciers change the shape of the valley from a "V" shape to a "U".

When **glaciers** carve two U-shaped valleys adjacent to each other, the ridge between them tends to be sharpened into a sawtooth feature called an **arête**. At the head of a glacially carved valley is a bowl-shaped feature called a **cirque**. The **cirque** represents where the head of the **glacier** eroded the mountain by plucking rock away from it and the weight of the thick ice eroded out a bowl. After the **glacier** is gone, the bowl at the bottom of the **cirque** often fills with **precipitation** and is occupied by a lake, called a tarn. When three or more mountain **glaciers** erode headward at their **cirques**, they produce **horns**, steep-sided, spire-shaped mountains. Low points along **arêtes** or between **horns** are mountain passes termed **cols**. Where a smaller **tributary glacier** flows into a larger trunk glacier, the smaller **glacier** cuts down less. Once the ice has gone, the **tributary** valley is left as a **hanging valley**, sometimes with a waterfall plunging into the main valley. As the trunk **glacier** straightens and widens a V-shaped valley and erodes the ends of side ridges, a steep triangle-shaped cliff is formed called a **truncated spur**.

Figure 14.14: Cirque with Upper Thornton Lake in the North Cascades National Park, Washington (left). An example of a horn, Kinnerly Peak, Glacier National Park, Montana (center). Bridalveil Falls in Yosemite National Park, California (right) is a good example of a hanging valley.

14.4.2 Depositional Glacial Landforms

Depositional landforms and materials are produced from deposits left behind by a retreating **glacier**. All **glacial** deposits are called drift. These include **till**, **tillites**, **diamictites**, **terminal moraines**, **recessional moraines**, **lateral moraines**, **medial moraines**, **ground moraines**, silt, **outwash plains**, **glacial erratics**, **kettles**, **kettle** lakes, crevasses, **eskers**, kames, and **drumlins**.

Glacial ice carries a lot of **sediment**, which when deposited by a melting **glacier** is called **till**. **Till** is poorly sorted with grain sizes ranging from clay and silt to pebbles and boulders. These clasts may be striated. Many depositional landforms are composed of **till**. The term tillite refers to lithified rock having **glacial** origins. Diamictite refers to a lithified rock that contains a wide range of clast sizes; this includes **glacial till** but is a more **objective** and descriptive term for any rock with a wide range of clast sizes.

Figure 14.15: Boulder of diamictite of the Mineral Fork Formation, Antelope Island, Utah, United States.

Moraines are mounded deposits consisting of **glacial till** carried in the **glacial** ice and rock fragments dislodged by **mass wasting** from the U-shaped valley walls. The **glacier** acts like a conveyor belt, carrying and depositing **sediment** at the end of and along the sides of theice flow. Because the ice in the **glacier** is always flowing downslope, all **glaciers** have **moraines** build up at their terminus, even those not advancing.

Moraines are classified by their location with respect to the **glacier**. A **terminal moraine** is a ridge of **till** located at the end or terminus of the **glacier**. Recessional **moraines** are left as **glaciers** retreat and there are pauses in the retreat. Lateral moraines accumulate along the sides of the **glacier** from material mass wasted from the valley walls. When two **tributary glaciers** merge, the two **lateral moraines** combine to form a **medial moraine** running down the center of the combined glacier. Ground **moraine** is a veneer of **till** left on the land as the **glacier** melts.

Figure 14.16: Lateral moraines of Kaskawulsh Glacier within Kluane National Park in the Canadian territory of Yukon (left) and Medial moraines where tributary glaciers meet. At least seven tributary glaciers from upstream have joined to form the trunk glacier flowing on out of the upper left of the picture (right).

In addition to **moraines**, **glaciers** leave behind other depositional landforms. Silt, sand, and gravel produced by the intense grinding process are carried by streams of water and depositedin front of the glacier in an area called the **outwash plain**. Retreating **glaciers** may leave behind large boulders that don't match the local **bedrock**. These are called **glacial erratics**. When **continental glaciers** retreat, they can leave behind large blocks of ice within the **till**. These ice blocks melt and create a depression in the **till** called a **kettle**. If the depression later fills with water, it is called a **kettle lake**.

If meltwater flowing over the ice surface descends into crevasses in the ice, it may find a channel and continue to flow in sinuous channels within or at the base of the **glacier**. Within or under **continental glaciers**, these **streams** carry **sediments**. When the ice recedes, the accumulated sediment is deposited as a long sinuous ridge known as an **esker**. Meltwater descending down through the ice or over the margins of the ice may deposit mounds of **till** in hills called kames.

Drumlins are common in **continentalglacial** areas of Germany, New York, and Wisconsin, where they typically are found in fields with great numbers. A **drumlin** is an elongated asymmetrical teardrop-shaped hill reflecting ice movement with its steepest side pointing upstream to the flow of ice and its streamlined or low-angled side pointing downstream in the direction of ice movement.

Glacial scientists debate the origins of drumlins. A leading idea is that **drumlins** are created from accumulated **till** being compressed and sculpted under a**glacier** that retreated then advanced again over its own ground moraine. Another idea is that meltwater catastrophically flooded under the glacier and carved the **till** into these streamlined mounds. Still another proposes that the weight of the overlying ice statically deformed the underlying **till**.

Figure 14.17: A small group of Ice Age drumlins in Germany.

Complete this interactive activity to check your understanding.

If you are using an offline version of this text, access this interactive activity via the QR code.

An interactive H5P element has been excluded from this version of the text. You can view it online here:
https://pressbooks.lib.vt.edu/introearthscience/?p=3786#h5p-97

14.4.3 Glacial Lakes

Glacial lakes are commonly found in alpine environments. A lake confined within a **glacial cirque** is called a tarn. A tarn forms when the depression in the **cirque** fills with precipitation after the ice is gone. Examples of tarns include Silver Lake near Brighton Ski resort in Big Cottonwood Canyon, Utah and Avalanche Lake in Glacier National Park, Montana.

Figure 14.18: Tarn in a cirque.

When recessional **moraines** create a series of isolated basins in a glaciated valley, the resulting chain of lakes is called **paternoster lakes**.

Lakes filled by glacial meltwater often looks milky due to finely ground material called rock flour suspended in the water.

Figure 14.19: Paternoster lakes.

Long, glacially carved depressions filled with water are known as **finger lakes**. **Proglacial lakes** form along the edges of all the largest **continental ice sheets**, such as Antarctica and Greenland. The **crust** is depressed isostatically by the overlying ice sheet and these basins fill with **glacial** meltwater. Many such lakes, some of them huge, existed at various times along the southern edge of the Laurentide Ice Sheet. Lake Agassiz, Manitoba, Canada, is a classic example of a **proglacial lake**. Lake Winnipeg serves as the remnant of a much larger proglacial lake.

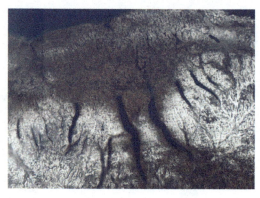

Figure 14.20: Satellite view of Finger Lakes region of New York.

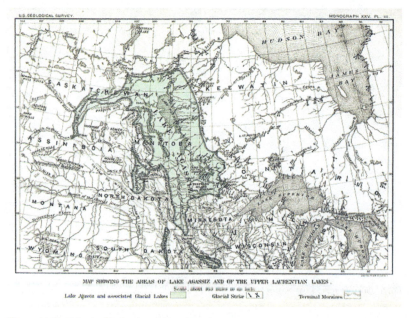

Figure 14.21: Extent of Lake Agassiz.

Other **proglacial lakes** were formed when **glaciers** dammed **rivers** and flooded the **river** valley. A classic example is Lake Missoula, which formed when a lobe of the Laurentide ice sheet blocked the Clark Fork River about 18,000 years ago. Over about 2000 years the ice dam holding back Lake Missoula failed several times. During each breach, the lake emptied across parts of eastern Washington, Oregon, and Idaho into the Columbia River Valley and eventually the Pacific Ocean. After each breach, the dam reformed and the lake refilled. Each breach produced a catastrophic flood over a few days. Scientists estimate that this cycle of ice dam, **proglacial lake**, and torrential **massive** flooding happened at least 25 times over a span of 20 centuries. The rate of each outflow is believed to have equaled the combined **discharge** of all of Earth's current **rivers** combined.

The landscape produced by these massive floods is preserved in the Channeled Scablands of Idaho, Washington, and Oregon.

Figure 14.22: View of Channeled Scablands in central Washington showing huge potholes and massive erosion.

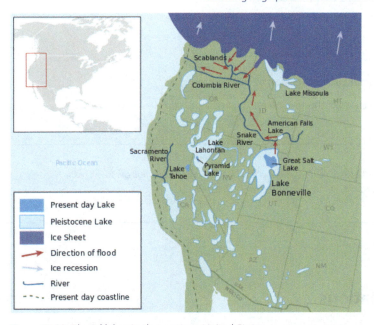

Figure 14.23: Pluvial lakes in the western United States.

Pluvial lakes form in humid environments that experience low temperatures and high **precipitation**. During the last **glaciation**, most of the western United States' **climate** was cooler and more humid than today. Under these low-evaporation conditions, many large lakes, called **pluvial lakes**, formed in the basins of the **Basin and Range** Province. Two of the largest were Lake Bonneville and Lake Lahontan. Lake Lahontan was in northwestern Nevada. The figure illustrates the tremendous size of Lake Bonneville, which occupied much of western Utah and into eastern Nevada. The lake level fluctuated greatly over the centuries leaving several pronounced old shorelines marked by wave-cut **terraces**. These old shorelines can be seen on mountain slopes throughout the western portion of Utah, including

Figure 14.24: The Great Lakes.

the Salt Lake Valley, indicating that the now heavily urbanized valley was once filled with hundreds of feet of water. Lake Bonneville's level peaked around 18,000 years ago when a breach occurred at Red Rock Pass in Idaho and water spilled into the Snake River. The flooding rapidly lowered the lake level and scoured the Idaho landscape across the Pocatello Valley, the Snake River Plain, and Twin Falls. The floodwaters ultimately flowed into the Columbia River across part of the scablands area at an incredible **discharge** rate of about 4,750 cu km/sec (1,140 cumi/sec). For comparison, this **discharge** rate would drain the volume of Lake Michigan completely dry within a few days.

The five Great Lakes in North America's upper Midwest are proglacial lakes that originated during the last **ice age**. The lake basins were originally carved by the encroaching **continental ice sheet**. The basins were later exposed as the ice retreated about 14,000 years ago and filled by precipitation.

Take this quiz to check your comprehension of this section.

If you are using an offline version of this text, access the quiz for section 14.4 via the QR code.

 An interactive H5P element has been excluded from this version of the text. You can view it online here:
https://pressbooks.lib.vt.edu/introearthscience/?p=3786#h5p-98

14.5 Ice Age Glaciations

A **glaciation** (or **ice age**) occurs when the Earth's **climate** becomes cold enough that **continental ice sheets** expand, covering large areas of land. Four major, well-documented **glaciations** have occurred in Earth's history: one during the **Archean**-early **Proterozoic Eon**, ~2.5 billion years ago; another in late **Proterozoic Eon**, ~700 million years ago; another in the Pennsylvanian, 323 to 300 million years ago, and most recently during the Pliocene-Pleistocene **epochs** starting 2.5 million years ago (chapter 8). Some scientists also recognize a minor **glaciation** around 440 million years ago in Africa.

The best-studied **glaciation** is, of course, the most recent. This infographic illustrates the **glacial** and **climate** changes over the last 20,000 years, ending with those caused by human actions since the Industrial Revolution. The Pliocene-Pleistocene **glaciation** was a series of several **glacial** cycles, possibly 18 in total. Antarctic ice-**core** records exhibit especially strong evidence for eight **glacial** advances occurring within the last 420,000 years. The last of these is known in popular media as "The **Ice Age**," but geologists refer to it as the Last Glacial Maximum. The **glacial** advance reached its maximum between 26,500 and 19,000 years ago.

14.5.1 Causes of Glaciations

Glaciations occur due to both long-term and short-term factors. In the geologic sense, long-term means a scale of tens to hundreds of millions of years and short-term means a scale of hundreds to thousands of years.

Long-term causes include **plate tectonics** breaking up the **supercontinents** (see Wilson Cycle, chapter 2), moving land masses to high latitudes near the north or south poles, and changing ocean circulation. For example, the closing of the Panama Strait and isolation of the Pacific and Atlantic oceans may have triggered a change in precipitation cycles, which combined with a cooling **climate** to help expand the **ice sheets**.

Short-term causes of **glacial** fluctuations are attributed to the cycles in the Earth's rotational-**axis** and to variations in the earth's orbit around the Sun which affect the distance between Earth and the Sun. Called **Milankovitch Cycles**, these cycles affect the amount of incoming solar radiation, causing short-term cycles of warming and cooling.

During the **Cenozoic** Era, carbon dioxide levels steadily decreased from a maximum in the Paleocene, causing the **climate** to gradually cool. By the Pliocene, **ice sheets** began to form. The effects of the **Milankovitch Cycles** created short-term cycles of warming and cooling within the larger **glaciation** event.

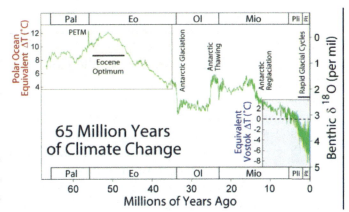

Figure 14.25: Atomospheric CO2 has declined during the Cenozoic from a maximum in the Paleocene–Eocene up to the Industrial Revolution.

Milankovitch Cycles are three orbital changes named after the Serbian astronomer Milutan Milankovitch. The three orbital changes are called **precession**, **obliquity**, and **eccentricity**. Precession is the wobbling of Earth's **axis** with a **period** of about 21,000 years; **obliquity** is changes in the angle of Earth's **axis** with a **period** of about 41,000 years; and **eccentricity** is variations in the Earth's orbit around the sun leading to changes in distance from the sun with a **period** of 93,000 years. These orbital changes created a 41,000-year-long **glacial–interglacial** Milankovitch Cycle from 2.5 to 1.0 million years ago, followed by another longer cycle of about 100,000 years from 1.0 million years ago to today (see Milankovitch Cycles).

Complete this interactive activity to check your understanding.

If you are using an offline version of this text, access this interactive activity via the QR code.

An interactive H5P element has been excluded from this version of the text. You can view it online here:
https://pressbooks.lib.vt.edu/introearthscience/?p=3786#h5p-99

Watch the video to see summaries of the **ice ages**, including their characteristics and causes.

One or more interactive elements has been excluded from this version of the text. You can view them online here:
https://www.youtube.com/watch?v=yNiMhjPHPu0

Video 14.2: Ice ages and climate cycles.

If you are using an offline version of this text, access this YouTube video via the QR code.

14.5.2 Sea-Level Change and Isostatic Rebound

When **glaciers** melt and retreat, two things happen: water runs off into the ocean causing sea levels to rise worldwide, and the land, released from its heavy covering of ice, rises due to **isostatic rebound**. Since the Last Glacial Maximum about 19,000 years ago, sea-level has risen about 125 m (400 ft). A global change in sea level is called **eustatic** sea-level change. During a warming trend, sea-level rises due to more water being added to the ocean and also thermal expansion of sea water. About half of the Earth's **eustatic** sea-level rise during the last century has been the result of **glaciers** melting and about half due to thermal expansion. Thermal expansion describes how a solid, liquid, or gas expands in volume with an increase in **temperature**. This 30 second video demonstrates thermal expansion with the classic brass ball and ring **experiment**.

One or more interactive elements has been excluded from this version of the text. You can view them online here:
https://www.youtube.com/watch?v=QNoE5loRheQ

Video 14.3: Thermal expansion.

If you are using an offline version of this text, access this YouTube video via the QR code.

Relative sea-level change includes vertical movement of both **eustatic** sea-level and continents on **tectonic plates**. In other words, sea-level change is measured relative to land elevation. For example, if the land rises a lot and sea-level rises only a little, then the relative sea-level would appear to drop.

Continents sitting on the **lithosphere** can move vertically upward as a result of two main processes, **tectonic** uplift and **isostatic rebound**. **Tectonic** uplift occurs when **tectonic plates** collide (see chapter 2). Isostatic rebound describes the upward movement of lithospheric **crust** sitting on top of the asthenospheric layer below it. **Continental** crust bearing the weight of **continental** ice sinks into the **asthenosphere** displacing it. After the **ice sheet** melts away, the **asthenosphere** flows back in and continental crust floats back upward. **Erosion** can also create **isostatic rebound** by removing large masses like mountains and transporting the **sediment** away (think of the **Mesozoic** removal of the Alleghanian Mountains and the uplift of the Appalachian plateau; chapter 8), albeit this process occurs more slowly than relatively rapid **glacier** melting.

The isostatic-rebound map below shows rates of vertical crustal movement worldwide. The highest rebound rate is indicated by the blue-to-purple zones (top end of the scale). The orange-to-red zones (bottom end of the scale) surrounding the high-rebound zones indicate isostatic lowering as adjustments in displaced subcrustal material have taken place.

Most **glacial isostatic rebound** is occurring where continental ice sheets rapidly melted about 19,000 years ago, such as in Canada and Scandinavia. Its effects can be seen wherever **Ice Age** ice or water bodies are or were present on **continental** surfaces and in terraces on **river** floodplains that cross these areas. **Isostatic rebound** occurred in Utah when the water from Lake Bonneville drained away. North America's Great Lakes also exhibit emergent **coastline** features caused by **isostatic rebound** since the **continental** ice sheet retreated.

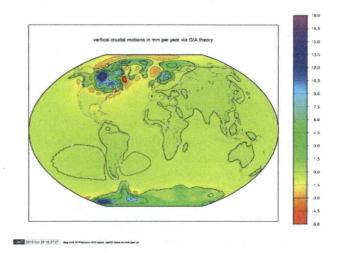

Figure 14.26: Rate of isostatic rebound.

Take this quiz to check your comprehension of this section.

If you are using an offline version of this text, access the quiz for section 14.5 via the QR code.

 An interactive H5P element has been excluded from this version of the text. You can view it online here:
https://pressbooks.lib.vt.edu/introearthscience/?p=3786#h5p-100

Summary

Glaciers form when average annual snowfall exceeds melting and snow compresses into **glacial** ice. There are three types of **glaciers**, alpine or **valley glaciers** that occupy valleys, **ice sheets** that cover **continental** areas, and ice caps that cover smaller areas usually at higher elevations. As the ice accumulates, it begins to flow downslope and outward under its own weight. **Glacial** ice is divided into two zones, the upper rigid or **brittle** zone where the ice cracks into **crevasses** and the lower plastic zone where under the pressure of overlying ice, the ice bends and flows exhibiting **ductile** behavior. Rock material that **falls** onto the ice by **mass wasting** or is plucked and carried by the ice is called **moraine** and acts as grinding agents against the **bedrock** creating significant **erosion**.

Glaciers have a budget of income and expense. The zone of income for the **glacier** is called the **Zone of Accumulation**, where snow is converted into **firn** then ice by **compression** and **recrystallization**, and the zone of expense called the **Zone of Ablation**, where ice melts or sublimes away. The line separating these two zones latest in the year is the Equilibrium or **Firn** Line and can be seen on the **glacier** separating bare ice from snow covered ice. If the **glacial budget** is balanced, even though the ice continues to flow downslope, the end or terminus of the **glacier** remains in a stable position. If income is greater than expense, the position of the terminus moves downslope. If expense is greater than income, a circumstance now affecting **glaciers** and **ice sheets** worldwide due to global warming, the terminus recedes. If this situation continues, the **glaciers** will disappear. An average of cooler summers affects the stability or growth of **glaciers** more than higher snowfall. As the Greenland and Antarctic **ice sheets** flow seaward, the edges calve off forming icebergs.

Glaciers create two kinds of landforms, erosional and depositional. **Alpine glaciers** carve U-shaped valleys and **moraine** carried in the ice polishes and grooves or striates the **bedrock**. Other landscape features produced by **erosion** include **horns**, **arête** ridges, **cirques**, **hanging valleys**, **cols**, and **truncated spurs**. **Cirques** may contain eroded basins that are occupied by post **glacial** lakes called tarns. Depositional features result from deposits left by retreating ice called drift. These include **till**, and **moraine** deposits (terminal, recessional, lateral, medial, and ground), **eskers**, kames, **kettles** and **kettle lakes**, **erratics**, and **drumlins**. A series of **recessional moraines** in glaciated valleys may create basins that are later filled with water to become **paternoster** lakes. **Glacial** meltwater carries fine grained **sediment** onto the **outwash plain**. Lakes containing **glacial** meltwater are milky in color from suspended finely ground rock flour. **Ice Age climate** was more humid and **precipitation** that did not become **glacier** ice filled regional depressions to become **pluvial lakes**. Examples of **pluvial lakes** include Lake Missoula dammed behind an **ice sheet** lobe and Lake Bonneville in Utah whose **shoreline** remnants can be seen on mountainsides. Repeated breaching of the ice lobe allowed Lake Missoula to rapidly drain causing **massive** floods that scoured the Channeled Scablands of Idaho, Washington, and Oregon.

Take this quiz to check your comprehension of this chapter.

If you are using an offline version of this text, access the quiz for chapter 14 via the QR code.

 An interactive H5P element has been excluded from this version of the text. You can view it online here:
https://pressbooks.lib.vt.edu/introearthscience/?p=3786#h5p-101

URLs Listed Within This Chapter

Infographic: Infographic. "A Timeline of Earth's Average Temperature since the Last Ice Age Glaciation" xkcd https://xkcd.com/1732

Milankovitch cycles: https://en.wikipedia.org/wiki/Milankovitch_cycles

Text References

1. Allen, P.A., and Etienne, J.L., 2008, Sedimentary challenge to Snowball Earth: Nat. Geosci., v. 1, no. 12, p. 817–825.
2. Berner, R.A., 1998, The carbon cycle and carbon dioxide over Phanerozoic time: the role of land plants: Philos. Trans. R. Soc. Lond. B Biol. Sci., v. 353, no. 1365, p. 75–82.
3. Cunningham, W.L., Leventer, A., Andrews, J.T., Jennings, A.E., and Licht, K.J., 1999, Late Pleistocene–Holocene marine conditions in the Ross Sea, Antarctica: evidence from the diatom record: The Holocene, v. 9, no. 2, p. 129–139.
4. Deynoux, M., Miller, J.M.G., and Domack, E.W., 2004, Earth's Glacial Record: World and Regional Geology, Cambridge University Press, World and Regional Geology.
5. Earle, S., 2015, Physical geology OER textbook: BC Campus OpenEd.
6. Eyles, N., and Januszczak, N., 2004, "Zipper-rift": a tectonic model for Neoproterozoic glaciations during the breakup of Rodinia after 750 Ma: Earth-Sci. Rev.
7. Francey, R.J., Allison, C.E., Etheridge, D.M., Trudinger, C.M., and others, 1999, A 1000-year high precision record of $\delta13C$ in atmospheric CO2: Tellus B Chem. Phys. Meteorol.
8. Gutro, R., 2005, NASA – What's the Difference Between Weather and Climate? Online, http://www.nasa.gov/mission_pages/noaa-n/climate/climate_weather.html, accessed September 2016.
9. Hoffman, P.F., Kaufman, A.J., Halverson, G.P., and Schrag, D.P., 1998, A neoproterozoic snowball earth: Science, v. 281, no. 5381, p. 1342–1346.
10. Kopp, R.E., Kirschvink, J.L., Hilburn, I.A., and Nash, C.Z., 2005, The Paleoproterozoic snowball Earth: a climate disaster triggered by the evolution of oxygenic photosynthesis: Proc. Natl. Acad. Sci. U. S. A., v. 102, no. 32, p. 11131–11136.
11. Lean, J., Beer, J., and Bradley, R., 1995, Reconstruction of solar irradiance since 1610: Implications for climate change: Geophys. Res. Lett., v. 22, no. 23, p. 3195–3198.
12. Levitus, S., Antonov, J.I., Wang, J., Delworth, T.L., Dixon, K.W., and Broccoli, A.J., 2001, Anthropogenic warming of Earth's climate system: Science, v. 292, no. 5515, p. 267–270.
13. Lindsey, R., 2009, Climate and Earth's Energy Budget: Feature Articles: Online, http://earthobservatory.nasa.gov, accessed September 2016.
14. North Carolina State University, 2013a, Composition of the Atmosphere.
15. North Carolina State University, 2013b, Composition of the Atmosphere: Online, http://climate.ncsu.edu/edu/k12/.AtmComposition, accessed September 2016.
16. Oreskes, N., 2004, The scientific consensus on climate change: Science, v. 306, no. 5702, p. 1686–1686.
17. Pachauri, R.K., Allen, M.R., Barros, V.R., Broome, J., Cramer, W., Christ, R., Church, J.A., Clarke, L., Dahe, Q., Dasgupta, P., Dubash, N.K., Edenhofer, O., Elgizouli, I., Field, C.B., and others, 2014, Climate Change 2014: Synthesis Report. Contribution of Working Groups I, II and III to the Fifth Assessment Report of the Intergovernmental Panel on Climate Change (R. K. Pachauri & L. Meyer, Eds.): Geneva, Switzerland, IPCC, 151 p.
18. Santer, B.D., Mears, C., Wentz, F.J., Taylor, K.E., Gleckler, P.J., Wigley, T.M.L., Barnett, T.P., Boyle, J.S., Brüggemann, W., Gillett, N.P., Klein, S.A., Meehl, G.A., Nozawa, T., Pierce, D.W., and others, 2007, Identification of human-induced changes in atmospheric moisture content: Proc. Natl. Acad. Sci. U. S. A., v. 104, no. 39, p. 15248–15253.
19. Schopf, J.W., and Klein, C., 1992, Late Proterozoic Low-Latitude Global Glaciation: the Snowball Earth, in Schopf, J.W., and Klein, C., editors, The Proterozoic biosphere: a multidisciplinary study: New York, Cambridge University Press, p. 51–52.
20. Webb, T., and Thompson, W., 1986, Is vegetation in equilibrium with climate? How to interpret late-Quaternary pollen data: Vegetatio, v. 67, no. 2, p. 75–91.
21. Weissert, H., 2000, Deciphering methane's fingerprint: Nature, v. 406, no. 6794, p. 356–357.
22. Whitlock, C., and Bartlein, P.J., 1997, Vegetation and climate change in northwest America during the past 125 kyr:

Nature, v. 388, no. 6637, p. 57–61.

23.   Wolpert, S., 2009, New NASA temperature maps provide a 'whole new way of seeing the moon': Online, http://news-room.ucla.edu/releases/new-nasa-temperature-maps-provide-102070, accessed February 2017.

24.   Zachos, J., Pagani, M., Sloan, L., Thomas, E., and Billups, K., 2001, Trends, rhythms, and aberrations in global climate 65 Ma to present: Science, v. 292, no. 5517, p. 686–693.

Figure References

Figure 14.1: Glacier in the Bernese Alps. Dirk Beyer. 2005. CC BY-SA 3.0. https://commons.wikimedia.org/wiki/File:Grosser_Aletschgletscher_3178.JPG

Figure 14.2: Greenland ice sheet. Hannes Grobe. 1995. CC BY-SA 2.5. https://commons.wikimedia.org/wiki/File:Greenland-ice_sheet_hg.jpg

Figure 14.3: Thickness of Greenland ice sheet in meters. Eric Gaba. 2011. CC BY-SA 3.0. https://commons.wikimedia.org/wiki/File:Greenland_ice_sheet_AMSL_thickness_map-en.svg

Figure 14.4: Cross-sectional view of both Greenland and Antarctic ice sheets drawn to scale for size comparison. Kindred Grey. 2022. CC BY 4.0. Adapted from Steve Earle (CC BY 4.0). https://opengeology.org/textbook/14-glaciers/14-2_steve-earle_antarctic-greenland-2-300×128/

Figure 14.5: Snow Dome ice cap near Mt. Olympus, Washington (left) and Vatnajökull ice cap in Iceland (right). Mount Olympus Washington by United States National Park Service, 2004 (Public domain, https://commons.wikimedia.org/wiki/File:Mount_Olympus_Washington.jpg). Vatnajökull by NASA, 2004 (Public domain, https://commons.wikimedia.org/wiki/File:Vatnaj%C3%B6kull.jpeg).

Figure 14.6: Maximum extent of Laurentide ice sheet. USGS. 2005. Public domain. https://commons.wikimedia.org/wiki/File:Pleistocene_north_ice_map.jpg

Figure 14.7: Glacial crevasses (left) and Cravasse on the Easton Glacier in the North Cascades (right). chevron crevasse by Bethan Davies, 2015 (CC BY-NC-SA 3.0, https://www.antarcticglaciers.org/glacier-processes/structural-glaciology/chevron-crevasse/). Glaciereaston by Mauri S. Pelto, 2005 (Public domain, https://en.wikipedia.org/wiki/File:Glaciereaston.jpg).

Figure 14.8: Cross-section of a valley glacier showing stress (red numbers) increase with depth under the ice. Kindred Grey. 2022. CC BY 4.0. Adapted from Steve Earle (CC BY 4.0). https://opengeology.org/textbook/14-glaciers/14-2_steve-earle_ice-flow-and-stress/

Figure 14.9: Cross-sectional view of an alpine glacier showing internal flow lines, zone of accumulation, snow line, and zone of melting. Kindred Grey. 2022. CC BY 4.0. Adapted from Steven Earle (CC BY 4.0). https://opentextbc.ca/geology/chapter/16-2-how-glaciers-work/

Figure 14.10: Fjord. Frédéric de Goldschmidt. 2007. CC BY-SA 3.0. https://sco.wikipedia.org/wiki/File:Geiranger-fjord_(6-2007).jpg

Figure 14.11: Glacial striations on granite in Whistler, Canada (left) and Glacial striations in Mt. Rainier National Park (right). Glacial striations by Amezcackle, 2003 (Public domain, https://en.m.wikipedia.org/wiki/File:Glacial_striations.JPG). Glacial striation 21149 by Walter Siegmund, 2007 (CC BY-SA 3.0, https://commons.wikimedia.org/wiki/File:Glacial_striation_21149.JPG).

Figure 14.12: The U-shape of the Little Cottonwood Canyon, Utah, as it enters into the Salt Lake Valley. Wilson44691. 2008. Public domain. https://commons.wikimedia.org/wiki/File:UshapedValleyUT.JPG

Figure 14.13: Formation of a glacial valley. Cecilia Bernal. 2015. CC BY-SA 4.0. https://commons.wikimedia.org/wiki/File:Glacier_Valley_formation-_Formaci%C3%B3n_Valle_glaciar.gif

Figure 14.14: Cirque with Upper Thornton Lake in the North Cascades National Park, Washington (left). Thornton Lakes 25929 by Walter Siegmund, 2007 (CC BY-SA 3.0, https://commons.wikimedia.org/wiki/File:Thornton_Lakes_25929.JPG). Kinnerly Peak by USGS, 1982 (Public domain, https://uk.wikipedia.org/wiki/%D0%A4%D0%B0%D0%B9%D0%BB:Kinnerly_Peak.jpg). Closeup of Bridalveil Fall seen from Tunnel View in Yosemite NP by Daniel Mayer, 2003 (CC BY-SA 1.0, https://commons.wikimedia.org/wiki/File:Closeup_of_Bridalveil_Fall_seen_from_Tunnel_View_in_Yosemite_NP.JPG).

Figure 14.15: Boulder of diamictite of the Mineral Fork Formation, Antelope Island, Utah, United States. Jstuby. 2002. Public domain. https://commons.wikimedia.org/wiki/File:Diamictite_Mineral_Fork.JPG

Figure 14.16: Lateral moraines of Kaskawulsh Glacier within Kluane National Park in the Canadian territory of Yukon (left) and Medial moraines where tributary glaciers meet. Kluane Icefield 1 by Steffen Schreyer, 2005 (CC BY-SA 2.0 DE, https://commons.wikimedia.org/wiki/File:Kluane_Icefield_1.jpg). Nuussuaq-peninsula-moraines by Algkalv, 2010 (CC BY-SA 3.0, https://commons.wikimedia.org/wiki/File:Nuussuaq-peninsula-moraines.jpg).

Figure 14.17: A small group of Ice Age drumlins in Germany. Martin Groll. 2009. CC BY 3.0 DE. https://commons.wikimedia.org/wiki/File:Drumlin_1789.jpg

Figure 14.18: Tarn in a cirque. Jrmichae. 2012. CC BY-SA 4.0. https://commons.wikimedia.org/wiki/File:Verdi_Leak_in_The_Ruby_Mountains.JPG

Figure 14.19: Paternoster lakes. wetwebwork. 2007. CC BY-SA 2.0. https://en.wikipedia.org/wiki/File:View_from_Forester_Pass.jpg

Figure 14.20: Satellite view of Finger Lakes region of New York. NASA. 2004. Public domain. https://commons.wikimedia.org/wiki/File:New_York%27s_Finger_Lakes.jpg

Figure 14.21: Extent of Lake Agassiz. USGS. 1895. Public domain. https://commons.wikimedia.org/wiki/File:Agassiz.jpg

Figure 14.22: View of Channeled Scablands in central Washington showing huge potholes and massive erosion. DKRKaynor. 2019. CC BY-SA 4.0. https://commons.wikimedia.org/wiki/File:Channeled_Scablands.jpg

Figure 14.23: Pluvial lakes in the western United States. Fallschirmjäger. 2013. CC BY-SA 3.0. https://en.wikipedia.org/wiki/File:Lake_bonneville_map.svg

Figure 14.24: The Great Lakes. SeaWiFS Project, NASA/Goddard Space Flight Center, and ORBIMAGE. 2000. Public domain. https://commons.wikimedia.org/wiki/File:Great_Lakes_from_space.jpg

Figure 14.25: Atmospheric CO_2 has declined during the Cenozoic from a maximum in the Paleocene–Eocene up to the Industrial Revolution. Robert A. Rohde. 2005. CC BY-SA 3.0. https://commons.wikimedia.org/wiki/File:65_Myr_Climate_Change.png

Figure 14.26: Rate of isostatic rebound. NASA. 2010. Public domain. https://commons.wikimedia.org/wiki/File:PGR_Paulson2007_Rate_of_Lithospheric_Uplift_due_to_post-glacial_rebound.png

15.

GLOBAL CLIMATE CHANGE

Learning Objectives

By the end of this chapter, students should be able to:

- Describe the role of greenhouse gases in **climate** change.
- Describe the sources of greenhouse gases.
- Explain Earth's energy budget and global **temperature** changes.
- Explain how positive and **negative feedback** mechanisms can influence **climate**.
- Explain how we know about climates in the geologic past.
- Accurately describe which aspects of the environment are changing due to **anthropogenic climate** change.
- Describe the causes of recent **climate** change, particularly the role of humans in the overall **climate** balance.

This chapter describes the Earth systems involved in **climate** change, the geologic evidence of past **climate** changes, and the human role in today's **climate** change. In science, a **system** is a group of interacting objects and processes. **Earth System Science** is the study of these systems: **geosphere**—rocks; **atmosphere**—gasses; **hydrosphere**—water; **cryosphere**—ice; and **biosphere**—living things. Earth science studies these systems and how they interact and change in response to natural cycles and human-driven, or **anthropogenic** forces. Changes in one Earth **system** affect other systems.

It is critically important for us to be aware of the geologic context of **climate** change processes and how these Earth systems interact, first, for us to understand how and why human activities cause present-day **climate** change and, secondly, to distinguish between natural processes and human processes in the geologic past's **climate** record.

A significant part of this chapter introduces and discusses various processes from these Earth systems, how they influence each other, and how they impact global **climate**. For example, Earth's **temperature** and **climate** largely change based on atmospheric gas **composition**, ocean circulation, and the land-surface characteristics of rocks, **glaciers**, and plants.

Also necessary to understanding **climate** change is to distinguish between **climate** and **weather**. **Weather** is the short-term **temperature** and **precipitation** patterns that occur in days and weeks. **Climate** is the variable range of **temperature** and **precipitation** patterns averaged over the long-term for a particular region (see section 13.1). Thus, a single cold winter does not mean that the entire globe is cooling—indeed, the United States' cold winters of 2013 and 2014 occurred while the rest of the Earth was experiencing record warm-winter temperatures. To avoid these generalizations, many scientists use a 30-year average as a good baseline. Therefore, **climate** change refers to slow **temperature** and **precipitation** changes and trends over the long term for a particular area or the Earth as a whole.

15.1 Earth's Temperature

Without an **atmosphere**, Earth would have huge **temperature** fluctuations between day and night, like the moon. Daytime temperatures would be hundreds of degrees Celsius above normal, and nighttime temperatures would be hundreds of degrees below normal. Because the Moon doesn't have much of an **atmosphere**, its daytime temperatures are around

106 °C (224°F) and nighttime temperatures are around -183°C (-298°F). That is an astonishing 272°C (522°F) degree range between the Moon's light side and dark side. This section describes how Earth's **atmosphere** is involved in regulating the Earth's **temperature**.

15.1.1 Composition of Atmosphere

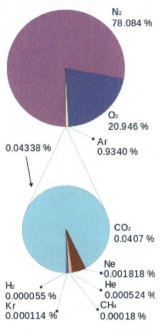

The **atmosphere**'s **composition** is a key component in regulating the planet's **temperature**. The **atmosphere** is 78 percent nitrogen (N_2), 21 percent oxygen (O_2), one percent argon (Ar), and less than one percent trace components, which are all other gases. Trace components include carbon dioxide (CO_2), water vapor (H_2O), neon, helium, and methane. Water vapor is highly variable, mostly based on region, and composes about one percent of the **atmosphere**. Trace component gasses include several important greenhouse gases, which are the gases responsible for warming and cooling the plant. On a geologic scale, **volcanoes** and the **weathering** process, which bury CO_2 in **sediments**, are the **atmosphere**'s CO_2 sources. Biological processes both add and subtract CO_2 from the **atmosphere**.

Greenhouse gases **trap** heat in the **atmosphere** and warm the planet by absorbing some of the longer-wave outgoing infrared radiation that is emitted from Earth, thus keeping heat from being lost to space. More greenhouse gases in the **atmosphere** absorb more longwave heat and make the planet warmer. Greenhouse gasses have little effect on shorter-wave incoming solar radiation.

The most common greenhouse gases are water vapor (H_2O), carbon dioxide (CO_2), methane (CH_4), and nitrous **oxide** (N_2O). Water vapor is the most abundant greenhouse gas, but its atmospheric abundance does not change much over time. Carbon dioxide is much less abundant than water vapor, but carbon dioxide is being added to the **atmosphere** by human activities such as burning **fossil fuels**, land-use changes, and deforestation. Further, natural processes such as **volcanic** eruptions add carbon dioxide, but at an insignificant rate compared to human-caused contributions.

Figure 15.1: Composition of the atmosphere.

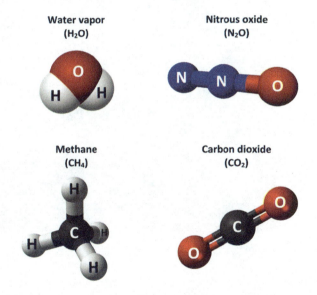

Figure 15.2: Common greenhouse gases.

There are two important reasons why carbon dioxide is the most important greenhouse gas. First, carbon dioxide stays in the **atmosphere** and does not go away for hundreds of years. Second, most of the additional carbon dioxide is "**fossil**" in origin, which means that it is released by burning **fossil fuels**. For example, **coal** and **petroleum** are **fossil fuels**. **Coal** and **oil** are made from long-dead plant material, which was originally created by photosynthesis millions of years ago and stored in the ground. Photosynthesis takes sunlight plus carbon dioxide and creates the substances of plants. This transformation occurs over millions of years as a slow process, accumulating **fossil** carbon in rocks and **sediments**. So, when we burn **coal** and **oil**, we instantaneously release the stored solar energy and **fossil** carbon dioxide that took millions of years to accumulate in the first place. The rate of release is critical to comprehend current **climate** change.

15.1.2 Carbon Cycle

Critical to understanding global **climate** change is to understand the carbon cycle and how Earth's own carbon-balancing **system** is being rapidly thrown off balance by human-driven activities. Earth has two important carbon cycles: the biological and the geological. In the biological cycle, living organisms—mostly plants—consume carbon dioxide from the **atmosphere** to make their tissues and substances through photosynthesis. Then, after the organisms die, and when they decay over years or decades, that carbon is released back into the **atmosphere**. The following is the general equation for photosynthesis.

$$CO_2 + H_2O + sunlight \rightarrow sugars + O_2$$

In the geological carbon cycle, a portion of the biological-cycle carbon becomes part of the geological carbon cycle: plant materials into **coal** and **petroleum**, tiny fragments and molecules into organic-rich **shale**, and the **carbonate** bearing calcareous shells and other parts of **marine** organisms into **limestone**. Such materials become buried and become part of the slow geologic **formation** of **coal** and other sedimentary materials. This cycle actually involves most of Earth's carbon and operates very slowly.

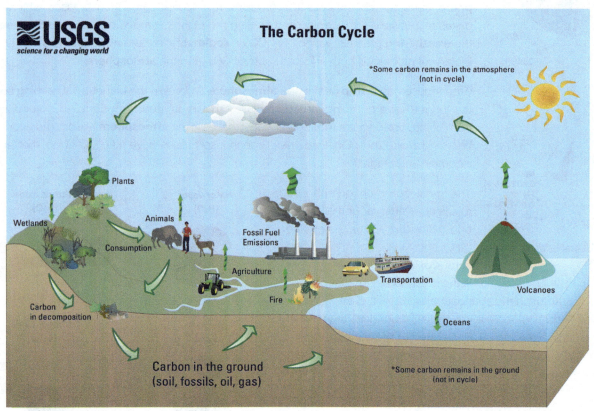

Figure 15.3: Carbon cycle.

The following are geological carbon-cycle storage **reservoirs**:

- Organic matter from plants is stored in peat, **coal**, and **permafrost** for thousands to millions of years.
- **Silicate–mineral weathering** converts atmospheric carbon dioxide to **dissolved** bicarbonate, which is stored in the oceans for thousands to tens of thousands of years.
- **Marine** organisms convert **dissolved** bicarbonate to forms of **calcite**, which is stored in **carbonate** rocks for tens to hundreds of millions of years.
- Carbon compounds are directly stored in **sediments** for tens to hundreds of millions of years; some end up in **petroleum** deposits.
- Carbon-bearing **sediments** are transferred by **subduction** to the **mantle**, where the carbon may be stored for tens of millions to billions of years.
- Carbon dioxide from within the Earth is released back to the **atmosphere** during **volcanic** eruptions, where it is stored for years to decades.

During much of Earth's history, the geological carbon cycle has been balanced by **volcanos** releasing carbon at approximately the same rate that carbon is stored by the other processes. Under these conditions, Earth's **climate** has remained relatively stable. However, in Earth's history, there have been times when that balance has been upset. This can happen during prolonged stretches of above-average **volcanic** activity. One example is the Siberian Traps eruption around 250 million years ago, which contributed to strong **climate** warming over a few million years.

A carbon imbalance is also associated with significant mountain-building events. For example, the Himalayan Range has been forming for about 40 million years, and over that time—and still today—the rate of **weathering** on Earth has been enhanced because those mountains are so huge and the range is so extensive that they present a greater surface area on which weathering takes place. The **weathering** of these rocks—most importantly the **hydrolysis** of **feldspar**—has resulted in consumption of atmospheric carbon dioxide and transfer of the carbon to the oceans and to ocean-floor **carbonate**-rich **sediments**. The steady drop in carbon dioxide levels over the past 40 million years, which contributed to the Pliocene-Pleistocene **glaciations**, is partly attributable to the **formation** of the Himalayan Range.

Another, nongeological form of carbon-cycle imbalance is happening today on a very rapid time scale. In just a few decades, humans have extracted volumes of **fossil fuels**, such as **coal**, **oil**, and gas, which were stored in rocks over the past several hundred million years, and converted these fuels to energy and carbon dioxide. By doing so, we are changing the **climate** faster than has ever happened in the past. Remember, carbon dioxide stays in the **atmosphere** and does not go away for hundreds of years. The more greenhouse gases in the **atmosphere**, the more heat is trapped and the warmer the planet becomes.

15.1.3 Greenhouse Effect

The **greenhouse effect** *is* the reason our global **temperature** is rising, but it's important to understand what this effect is and how it occurs. The **greenhouse effect** occurs because greenhouse gases are present in the **atmosphere**. The **greenhouse effect** is named after a similar process that warms a greenhouse or a car on a hot summer day. Sunlight passes through the glass of the greenhouse or car, reaches the interior, and changes into heat. The heat radiates upward and gets trapped by the glass windows. The **greenhouse effect** for the Earth can be explained in three steps.

Step 1: Solar radiation from the sun is composed of mostly ultraviolet (UV), visible light, and infrared (IR) radiation. Components of solar radiation include parts with a shorter **wavelength** than visible light, like ultraviolet light, and parts of the spectrum with longer wavelengths, like IR and others. Some of the radiation gets absorbed, scattered, or reflected by the atmospheric gases but about half of the solar radiation eventually reaches the Earth's surface.

Step 2: The visible, UV, and IR radiation, that reaches the surface converts to heat energy. Most students have experienced sunlight warming a surface such as pavement, a patio, or deck. When this occurs, the warmer surface then emits thermal radiation, which is a type of IR radiation. So, there is a conversion from visible, UV, and IR to just thermal IR. This thermal IR is what we experience as heat. If you have ever felt heat radiating from a fire or a hot stove top, then you have experienced thermal IR.

Step 3: Thermal IR radiates from the earth's surface back into the **atmosphere**. But since it is thermal IR instead of UV, visible, or regular IR, this thermal IR gets trapped by greenhouse gases. In other words, the sun's energy leaves the Earth at a different **wavelength** than it enters, so the sun's energy is not absorbed in the lower **atmosphere** when energy is coming in, but rather when the energy is

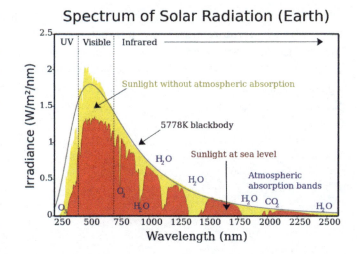

Figure 15.4: Incoming radiation absorbed, scattered, and reflected by atmospheric gases.

going out. The gases that are mostly responsible for this energy blocking on Earth include carbon dioxide, water vapor, methane, and nitrous **oxide**. More greenhouse gases in the **atmosphere** results in more thermal IR being trapped. Explore this external link to an interactive animation on the greenhouse effect from the National Academy of Sciences.

15.1.4 Earth's Energy Budget

The solar radiation that reaches Earth is relatively uniform over time. Earth is warmed, and energy or heat radiates from the Earth's surface and lower **atmosphere** back to space. This flow of incoming and outgoing energy is Earth's energy budget. For Earth's **temperature** to be stable over long stretches of time, incoming energy and outgoing energy have to be equal on average so that the energy budget at the top of the **atmosphere** balances. About 29 percent of the incoming solar energy arriving at the top of the **atmosphere** is reflected back to space by clouds, atmospheric particles, or reflective ground surfaces like sea ice and snow. About 23 percent of incoming solar energy is absorbed in the **atmosphere** by water vapor, dust, and ozone. The remaining 48 percent passes through the **atmosphere** and is absorbed at the surface. Thus, about 71 percent of the total incoming solar energy is absorbed by the Earth **system**.

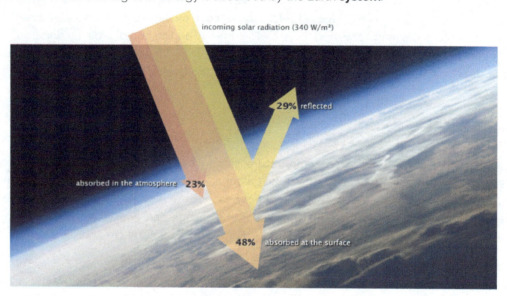

Figure 15.5: Incoming solar radiation filtered by the atmosphere.

When this energy reaches Earth, the atoms and molecules that makeup the **atmosphere** and surface absorb the energy, and Earth's **temperature** increases. If this material *only* absorbed energy, then the **temperature** of the Earth would continue to increase and eventually overheat. For example, if you continuously run a faucet in a stopped-up sink, the water level rises and eventually overflows. However, **temperature** does not infinitely rise because the Earth is not just absorbing sunlight; it is also radiating thermal energy or heat *back* into the **atmosphere**. If the **temperature** of the Earth rises, the planet emits an increasing amount of heat to space, and this is the primary mechanism that prevents Earth from continually heating.

Some of the thermal infrared heat radiating from the surface is absorbed and trapped by greenhouse gasses in the **atmosphere**, which act like a giant canopy over Earth. The more greenhouse gases in the **atmosphere**, the more outgoing heat Earth retains, and the less thermal infrared heat dissipates to space.

Factors that can affect the Earth's energy budget are not limited to greenhouse gases. Increasing solar energy can increase the energy Earth receives. However, these increases are very small over time. In addition, land and water will absorb more sunlight when there is less ice and snow to reflect the sunlight back to the **atmosphere**. For example, the ice covering the Arctic Sea reflects sunlight back to the **atmosphere**; this reflectivity is called **albedo**. Furthermore, aerosols (dust particles) produced from burning **coal**, diesel engines, and **volcanic** eruptions can

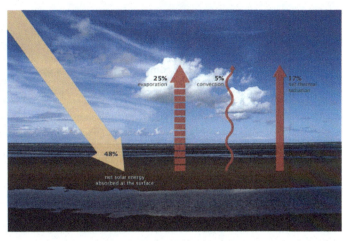

Figure 15.6: Some of the thermal infrared energy (heat) radiated from the surface into the atmosphere is trapped by gasses in the atmosphere.

reflect incoming solar radiation and actually cool the planet. While the effect of **anthropogenic** aerosols on the **climate**'s **system** is weak, the effect of human-produced greenhouse gases is not weak. Thus, the net effect of human activity is warming due to more **anthropogenic** greenhouse gases associated with **fossil fuel** combustion.

An effect that changes the planet can **trigger** feedback mechanisms that amplify or suppress the original effect. A **positive feedback** mechanism occurs when the output or effect of a process *enhances* the original stimulus or cause. Thus, it increases the ongoing effect. For example, the loss of sea ice at the North Pole makes that area less reflective, reducing **albedo**. This allows the surface air and ocean to absorb more energy in an area that was once covered by sea ice. Another example is melting **permafrost**. Permafrost is permanently frozen **soil** located in the high latitudes, mostly in the Northern Hemisphere. As the **climate** warms, more **permafrost** thaws, and the thick deposits of organic matter are exposed to oxygen and begin to decay. This **oxidation** process releases carbon

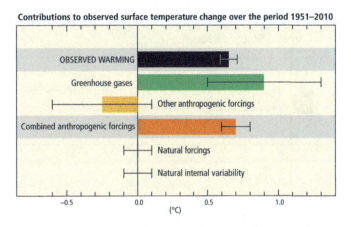

Figure 15.7: Net effect of factors influencing warming.

dioxide and methane, which in turn causes more warming, which melts more **permafrost**, and so on and on.

A **negative feedback** mechanism occurs when the output or effect *reduces* the original stimulus or cause. For example, in the short term, more carbon dioxide (CO_2) is expected to cause forest canopies to grow, which absorb more CO_2. Another example for the long term is that increased carbon dioxide in the **atmosphere** will cause more **carbonic acid** and **chemical weathering**, which results in transporting dissolved bicarbonate and other ions to the oceans, which are then stored in **sediment**.

Global warming is evidence that Earth's energy budget is not balanced. Positive effects on Earth's **temperature** are now greater than negative effects.

Take this quiz to check your comprehension of this section.

If you are using an offline version of this text, access the quiz for section 15.1 via the QR code.

An interactive H5P element has been excluded from this version of the text. You can view it online here:
https://pressbooks.lib.vt.edu/introearthscience/?p=928#h5p-102

15.2 Evidence of Recent Climate Change

While **climate** has changed often in the past due to natural causes (see section 14.5.1 and section 15.3), the scientific consensus is that human activity is causing current very rapid **climate** change. While this seems like a new idea, it was suggested more than 75 years ago. This section describes the evidence of what most scientists agree is **anthropogenic** or **human-caused climate change**. For more information, watch the video below on climate change by two professors at the North Carolina State University.

One or more interactive elements has been excluded from this version of the text. You can view them online here:
https://www.youtube.com/watch?v=dquECwUflQg

Video 15.1: Evidence for climate change.

If you are using an offline version of this text, access this YouTube video via the QR code.

15.2.1 Global Temperature Rise

The land-ocean **temperature** index, 1880 to present, compared to a base reference time of 1951-1980, shows ocean temperatures steadily rising. The solid black line is the global annual mean, and the solid red line is the five-year Lowess smoothing. The blue uncertainty bars (95 percent confidence limit) account only for incomplete spatial sampling.

Since 1880, Earth's surface-temperature average has trended upward with most of that warming occurring since 1970 (see this NASA animation). Surface temperatures include both land and ocean because water absorbs much additional trapped heat. Changes in land-surface or ocean-surface temperatures compared to a reference **period** from 1951 to 1980, where the long-term average remained relatively constant, are called **temperature anomalies**. A temperature anomaly thus represents the difference between the measured **temperature** and the average value during the reference **period**. **Climate** scientists calculate long-term average temperatures over thirty years or more which identified the reference **period** from 1951 to 1980. Another common range is a century, for example, 1900-2000. Therefore, an **anomaly** of 1.25 °C (34.3°F) for

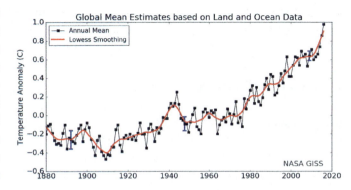

Figure 15.8: Land-ocean temperature index, 1880 to present, with a base time 1951-1980. The solid black line is the global annual mean and the solid red line is the five-year Lowess smoothing. The blue uncertainty bars (95% confidence limit) account only for incomplete spatial sampling. The graph shows that Earth's temperature is rising.

2015 means that the average **temperature** for 2015 was 1.25 °C (34.3°F) greater than the 1900-2000 average. In 1950, the **temperature anomaly** was -0.28 °C (31.5°F), so this is -0.28 °C (31.5°F) lower than the 1900-2000 average. These temperatures are annual average measured surface temperatures.

This video figure of **temperature anomalies** shows worldwide **temperature** changes since 1880. The more blue, the cooler; the more yellow and red, the warmer.

 One or more interactive elements has been excluded from this version of the text. You can view them online here: https://www.youtube.com/watch?v=haBG2IIbwbA

Video 15.2: Earth's long-term warming trend can be seen in this visualization of NASA's global temperature record, which shows how the planet's temperatures are changing over time, compared to a baseline average from 1951 to 1980. The record is shown as a running five-year average.

If you are using an offline version of this text, access this YouTube video via the QR code.

In addition to average land-surface temperatures rising, the ocean has absorbed much heat. Because oceans cover about 70 percent of the Earth's surface and have such a high specific heat value, they provide a large opportunity to absorb energy. The ocean has been absorbing about 80 to 90 percent of human activities' additional heat. As a result, **temperature** in the ocean's top 701.4 m (2,300 ft) has increased by -17.6°C (0.3°F) since 1969 (watch this 3 minute video by NASA JPL on the ocean's heat capacity). The reason the ocean has warmed less than the **atmosphere**, while still taking on most of the heat, is due to water's very high specific heat, which means that water can absorb a lot of heat energy with a small **temperature** increase. In contrast, the lower specific heat of the **atmosphere** means it has a higher **temperature** increase as it absorbs less heat energy.

Some scientists suggest that **anthropogenic** greenhouse gases do not cause global warming because between 1998 and 2013, Earth's surface temperatures did not increase much, despite greenhouse gas concentrations continuing to increase. However, since the oceans are absorbing most of the heat, decade-scale circulation changes in the ocean, similar to La Niña, push warmer water deeper under the surface. Once the ocean's absorption and circulation is accounted for, and this heat is added back into surface temperatures, then **temperature** increases become apparent, as shown in the figure. Also, the ocean's heat storage is temporary, as reflected in the record-breaking warm years of 2014-2016. Indeed, with this temporary ocean-storage effect, the twenty-first century's first 15 of its 16 years were the hottest in recorded history.

15.2.2 Carbon Dioxide

Anthropogenic greenhouse gases, mostly carbon dioxide (CO_2), have increased since the industrial revolution when humans dramatically increased burning **fossil fuels**. These levels are unprecedented in the last 800,000-year Earth history as recorded in geologic sources such as ice cores. Carbon dioxide has increased by 40 percent since 1750, and the rate of increase has been the fastest during the last decade. For example, since 1750, 2040^9 tonnes (2040 gigatons) of CO_2 have been added to the **atmosphere**; about 40 percent has remained in the **atmosphere** while the remaining 60 percent has been absorbed into the land by plants and **soil** or into the oceans. Indeed, during the lifetime of most young adults, the total atmospheric CO_2 has increased by 50 ppm, or 15 percent.

Charles Keeling, an oceanographer with Scripps Institution of Oceanography in San Diego, California, was the first person to regularly measure atmospheric CO_2. Using his methods, scientists at the Mauna Loa Observatory, Hawaii, have constantly measured atmospheric CO_2 since 1957. NASA regularly publishes these measurements at https://keelingcurve.ucsd.edu. Go there now to see the very latest measurement. Keeling's measured values have been posted in a curve of increasing values, called the Keeling Curve. This curve varies up and down in a regular annual cycle, from summer when the plants in the Northern Hemisphere are using CO_2 to winter when the plants are dormant. But the curve shows a steady CO_2 increase over the past several decades. This curve increases exponentially, not linearly, showing that the rate of CO_2 increase is itself increasing!

*Latest CO$_2$ reading: 415.91 ppm

Figure 15.9: Latest CO2 reading as of September 20, 2022.

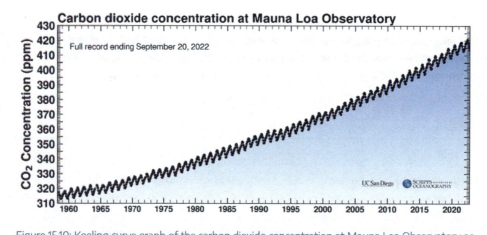

Figure 15.10: Keeling curve graph of the carbon dioxide concentration at Mauna Loa Observatory as of September 20, 2022.

The following Atmospheric CO_2 video shows how atmospheric CO_2 has varied recently and over the last 800,000 years, as determined by an increasing number of CO_2 monitoring stations as shown on the insert map. It is also instructive to watch the video's Keeling portion of how CO_2 varies by **latitude**. This shows that most human CO_2 sources are in the Northern Hemisphere where most of the land is and where most of the developed nations are.

One or more interactive elements has been excluded from this version of the text. You can view them online here:
https://www.youtube.com/watch?v=gH6fQh9eAQE

Video 15.3: History of atmospheric CO₂ from 800,000 years ago until January 2016. Visit https://gml.noaa.gov/ccgg/trends/ for more information.

If you are using an offline version of this text, access this YouTube video via the QR code.

15.2.3 Melting Glaciers and Shrinking Sea Ice

Glaciers are large ice accumulations that exist year round on the land's surface. In contrast, icebergs are masses of floating sea ice, although they may have had their origin in **glaciers** (see chapter 14). **Alpine glaciers**, **ice sheets**, and sea ice are all melting. Explore melting **glaciers** at NASA's interactive Global Ice Viewer). Satellites have recorded that Antarctica is melting at 1189 tonnes (118 gigatons) per year, and Greenland is melting at 2819 tonnes (281 gigatons) per year; 1 metric tonne is 1000 kilograms (1 gigaton is over 2 trillion pounds). Almost all major **alpine glaciers** are shrinking, deflating, and retreating. The ice-mass loss rate is unprecedented—never observed before—since the 1940's when quality records for **glaciers** began.

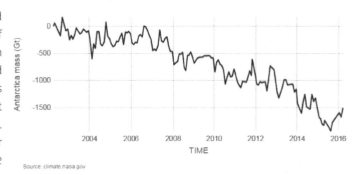

Figure 15.11: Decline of Antarctic ice mass from 2002 to 2016.

Before **anthropogenic** warming, **glacial** activity was variable with some retreating and some advancing. Now, **spring** snow cover is decreasing, and sea ice is shrinking. Most sea ice is at the North Pole, which is only occupied by the Arctic Ocean and sea ice. The NOAA animation shows how perennial sea ice has declined from 1987 to 2015. The oldest ice is white, and the youngest, seasonal ice is dark blue. The amount of old ice has declined from 20 percent in 1985 to 3 percent in 2015.

One or more interactive elements has been excluded from this version of the text. You can view them online here:
https://www.youtube.com/watch?v=Fw7GfNR5PLA

Video 15.4: This animation tracks the relative amount of ice of different ages from 1987 through early November 2015. The oldest ice is white; the youngest (seasonal) ice is dark blue. Key patterns are the export of ice from the Arctic through Fram Strait and the melting of old ice as it passes through the warm waters of the Beaufort Sea. Sea ice age is estimated by tracking of ice parcels using satellite imagery and drifting ocean buoys.

If you are using an offline version of this text, access this YouTube video via the QR code.

15.2.4 Rising Sea-Level

Sea level is rising 3.4 millimeters (0.13 inches) per year and has risen 0.19 meters (7.4 inches) from 1901 to 2010. This is thought largely to be from both **glaciers** melting and thermal expansion of sea water. Thermal expansion means that as objects such as solids, liquids, and gases heat up, they expand in volume.

Classic video demonstration (30 second) on thermal expansion with brass ball and ring (North Carolina School of Science and Mathematics).

 One or more interactive elements has been excluded from this version of the text. You can view them online here:
https://www.youtube.com/watch?v=QNoE5IoRheQ

Video 15.5: Thermal expansion.

If you are using an offline version of this text, access this YouTube video via the QR code.

15.2.5 Ocean Acidification

Since 1750, about 40 percent of new **anthropogenic** carbon dioxide has remained in the **atmosphere**. The remaining 60 percent gets absorbed by the ocean and vegetation. The ocean has absorbed about 30 percent of that carbon dioxide. When carbon dioxide gets absorbed in the ocean, it creates **carbonic acid**. This makes the ocean more acidic, which then has an impact on **marine** organisms that secrete calcium **carbonate** shells. Recall that hydrochloric acid reacts by effervescing with **limestone** rock made of **calcite**, which is calcium **carbonate**. A more acidic ocean is associated with **climate** change and is linked to some sea snails (pteropods) and small protozoan zooplanktons' (foraminifera) thinning **carbonate** shells and to ocean coral **reefs**' declining growth rates. Small animals like protozoan zooplankton are an important component at the base of the **marine** ecosystem. Acidification combined with warmer **temperature** and lower oxygen levels is expected to have severe impacts on **marine** ecosystems and human-harvested fisheries, possibly affecting our ocean-derived food sources.

 One or more interactive elements has been excluded from this version of the text. You can view them online here:
https://www.youtube.com/watch?v=bvg0FajepzU

Video 15.6: Ocean acidification: The other carbon dioxide problem.

If you are using an offline version of this text, access this YouTube video via the QR code.

15.2.6 Extreme Weather Events

Extreme **weather** events such as hurricanes, **precipitation**, and heatwaves are increasing and becoming more intense. Since the 1980's, hurricanes, which are generated from warm ocean water, have increased in frequency, intensity, and duration and are likely connected to a warmer **climate**. Since 1910, average **precipitation** has increased by 10 percent in the contiguous United States, and much of this increase is associated with heavy **precipitation** events. However, the distribution is not even, and more **precipitation** is projected for the northern United States while less **precipitation** is projected for the already dry southwest. Also, heatwaves have increased, and rising temperatures are already affecting crop yields in northern latitudes. Increased heat allows for greater moisture capacity in the **atmosphere**, increasing the potential for more extreme events.

Take this quiz to check your comprehension of this section.

If you are using an offline version of this text, access the quiz for section 15.2 via the QR code.

 An interactive H5P element has been excluded from this version of the text. You can view it online here:
https://pressbooks.lib.vt.edu/introearthscience/?p=928#h5p-103

15.3 Prehistoric Climate Change

Over Earth's history, the **climate** has changed a lot. For example, during the **Mesozoic Era**, the Age of Dinosaurs, the **climate** was much warmer, and carbon dioxide was abundant in the **atmosphere**. However, throughout the **Cenozoic Era**, 65 million years ago to today, the **climate** has been gradually cooling. This section summarizes some of these major past **climate** changes.

15.3.1 Past Glaciations

Through geologic history, **climate** has changed slowly over millions of years. Before the most recent Pliocene-**Quaternary glaciation**, there were other major **glaciations**. The oldest, known as the Huronian, occurred toward the end of the **Archean Eon**-early **Proterozoic Eon**, about 2.5 billion years ago. The event of that time, the **Great Oxygenation Event**, was a major happening (see chapter 8) most commonly associated with causing that **glaciation**. The increased oxygen is thought to have reacted with the potent greenhouse gas methane, causing cooling.

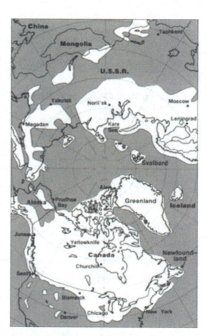

Figure 15.12: Maximum extent of Laurentide Ice Sheet.

The end of the **Proterozoic Eon**, about 700 million years ago, had other **glaciations**. These ancient **Precambrian glaciations** are included in the **Snowball Earth hypothesis**. Widespread global rock sequences from these ancient times contain evidence that **glaciers** existed even in low-latitudes. Two examples are **limestone** rock—usually formed in tropical **marine** environments—and **glacial** deposits—usually formed in cold climates—have been found together from this time in many regions around the world. One example is in Utah. Evidence of **continental glaciation** is seen in interbedded **limestone** and **glacial** deposits (**diamictites**) on Antelope Island in the Great Salt Lake.

The controversial **Snowball Earth hypothesis** suggests that a runaway **albedo** effect—where ice and snow reflect solar radiation and increasingly spread from polar regions toward the equator—caused land and ocean surfaces to completely freeze and biological activity to collapse. Thinking is that because carbon dioxide could not enter the then-frozen ocean, the ice covering Earth could only melt when **volcanoes** emitted high enough carbon dioxide into the **atmosphere** to cause greenhouse heating. Some studies estimate that because of the frozen ocean surface, carbon dioxide 350 times higher than today's concentration was required. Because biological activity did survive, the complete freezing and its extent in the **snowball earth hypothesis** are controversial. A competing **hypothesis** is the Slushball Earth **hypothesis** in which some regions of the equatorial ocean remained open. Differing scientific conclusions about the stability of Earth's magnetic poles, impacts on ancient rock evidence from subsequent **metamorphism**, and alternate interpretations of existing evidence keep the idea of **Snowball Earth** controversial.

Glaciations also occurred in the **Paleozoic Era**, notably the Andean-Saharan **glaciation** in the late **Ordovician**, about 440–460 million years ago, which coincided with a major **extinction** event, and the Karoo **Ice Age** during the Pennsylvanian Period, 323 to 300 million years ago. This **glaciation** was one of the evidences cited by Wegener for his Continental Drift hypothesis as his proposed **Pangea** drifted into south polar latitudes. The Karoo **glaciation** was associated with an increase of oxygen and a subsequent drop in carbon dioxide, most likely produced by the evolution and rise of land plants.

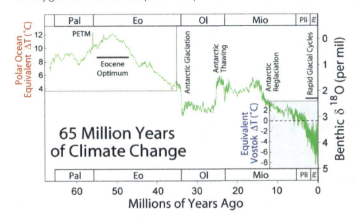

Figure 15.13: Global average surface temperature over the past 65 million years.

About 40 million years ago, the narrow gap between the South American Plate and the Antarctica Plate widened, which opened the Drake Passage. This opening allowed the water around Antarctica—the Antarctic Circumpolar Current—to flow unrestrictedly west-to-east, which effectively isolated the southern ocean from the warmer waters of the Pacific, Atlantic, and Indian Oceans. The region cooled significantly, and by 35 million years ago, during the Oligocene **Epoch**, **glaciers** had started to form on Antarctica.

Around 15 million years ago, **subduction**-related **volcanos** between Central and South America created the Isthmus of Panama, which connected North and South America. This prevented water from flowing between the Pacific and Atlantic Oceans and reduced heat transfer from the tropics to the poles. This reduced heat transfer created a cooler Antarctica and larger Antarctic **glaciers**. As a result, the **ice sheet** expanded on land and water, increased Earth's reflectivity and enhanced the **albedo** effect, which created a **positive feedback** loop: more reflective **glacial** ice, more cooling, more ice, more cooling, and so on.

During the **Cenozoic Era**—the last 65 million years, **climate** started out warm and gradually cooled to today. This warm time is called the **Paleocene-Eocene Thermal Maximum**, and Antarctica and Greenland were ice free during this time. Since the Eocene, **tectonic** events during the **Cenozoic Era** caused the planet to persistently and significantly cool. For example, the Indian Plate and Asian Plate collided, creating the Himalaya Mountains, which increased the rate of **weathering** and **erosion** of **silicate minerals**, especially **feldspar**. Increased **weathering** consumes carbon dioxide from the **atmosphere**, which reduces the **greenhouse effect**, resulting in long-term cooling.

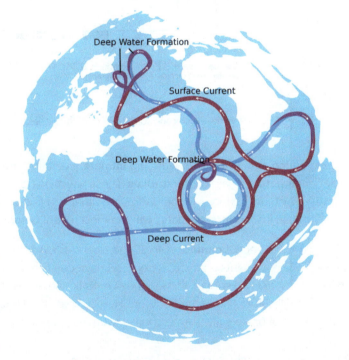

Figure 15.14: The Antarctic Circumpolar Current.

By 5 million years ago, during the Pliocene **Epoch**, **ice sheets** had started to grow in North America and northern Europe. The most intense part of the current **glaciation** is the Pleistocene **Epoch**'s last 1 million years. The Pleistocene's **temperature** varies significantly through a range of almost 10°C (18°F) on time scales of 40,000 to 100,000 years, and **ice sheets** expand and contract correspondingly. These variations are attributed to subtle changes in Earth's orbital parameters, called

Milankovitch cycles (see chapter 14). Over the past million years, the **glaciation** cycles occurred approximately every 100,000 years, with many **glacial** advances occurring in the last 2 million years (Lisiecki and Raymo, 2005).

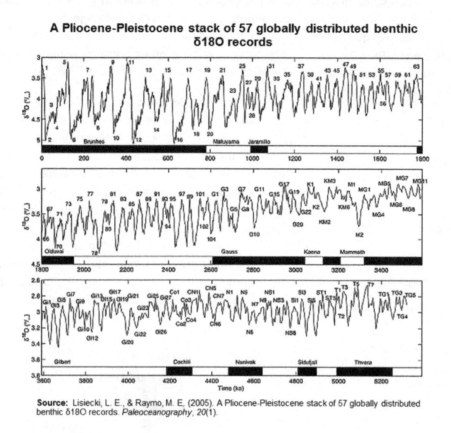

A Pliocene-Pleistocene stack of 57 globally distributed benthic δ18O records

Source: Lisiecki, L. E., & Raymo, M. E. (2005). A Pliocene-Pleistocene stack of 57 globally distributed benthic δ18O records. *Paleoceanography, 20*(1).

Figure 15.15: A Pliocene–Pleistocene stack of 57 globally distributed benthic δ18O records.

During an **ice age**, **periods** of warming **climate** are called **interglacials**; during **interglacials**, very brief **periods** of even warmer **climate** are called **interstadials**. These warming upticks are related to Earth's **climate** variations, like **Milankovitch cycle**s, which are changes to the Earth's orbit that can fluctuate **climate** (see chapter 14). In the last 500,000 years, there have been five or six **interglacials**, with the most recent belonging to our current time, the **Holocene Epoch**.

The two more recent **climate** swings, the Younger Dryas and the **Holocene** Climatic Optimum, demonstrate complex changes. These events are more recent, yet have conflicting information. The Younger Dryas' cooling is widely recognized in the Northern Hemisphere, though the event's timing, about 12,000 years ago, does not appear to be equal everywhere. Also, it is difficult to find in the Southern Hemisphere. The **Holocene** Climatic Optimum is a warming around 6,000 years ago; it was not universally warmer, nor as warm as current warming, and not warm at the same time everywhere.

15.3.2 Proxy Indicators of Past Climates

How do we know about past climates? Geologists use proxy indicators to understand past **climate**. A **proxy indicator** is a biological, chemical, or physical signature preserved in the rock, **sediment**, or ice record that acts like a fingerprint of something in the past. Thus, they are an *indirect* indicator of **climate**. An indirect indicator of ancient **glaciations** from the **Proterozoic Eon** and **Paleozoic Era** is the Mineral Fork Formation in Utah, which contains rock **formations** of **glacial sediments** such as **diamictite** (**tillite**). This dark rock has many fine-grained components plus some large out-sized clasts like a modern **glacial till**.

Deep-sea **sediment** is an indirect indicator of **climate** change during the **Cenozoic Era**, about the last 65 million years. Researchers from the Ocean Drilling Program, an international research collaboration, collect deep-sea **sediment** cores that record continuous **sediment** accumulation. The **sediment** provides detailed chemical records of stable carbon and oxygen **isotopes** obtained from deep-sea benthic foraminifera shells that accumulated on the **ocean floor** over millions of years. The oxygen **isotopes** are a **proxy indicator** of deep-sea temperatures and **continental** ice volume.

Sediment Cores—Stable Oxygen Isotopes

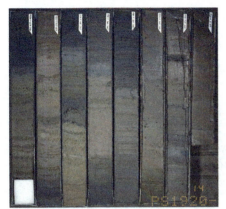

Figure 15.16: Sediment core from the Greenland continental slope.

How do oxygen **isotopes** indicate past **climate**? The two main stable oxygen **isotopes** are ^{16}O and ^{18}O. They both occur in water (H_2O) and in the calcium **carbonate** ($CaCO_3$) shells of foraminifera as both of those substances' oxygen component. The most abundant and lighter **isotope** is ^{16}O. Since it is lighter, it evaporates more readily from the ocean's surface as water vapor, which later turns to clouds and **precipitation** on the ocean and land. This evaporation is enhanced in warmer sea water and slightly increases the concentration of ^{18}O in the surface seawater from which the plankton derives the **carbonate** for its shells. Thus the ratio of ^{16}O and ^{18}O in the fossilized shells in seafloor **sediment** is a **proxy indicator** of the **temperature** and evaporation of seawater.

Keep in mind, it is harder to evaporate the heavier water and easier to condense it. As evaporated water vapor drifts toward the poles and tiny droplets form clouds and **precipitation**, droplets of water with ^{18}O tend to form more readily than droplets of the lighter form and **precipitate** out, leaving the drifting vapor depleted in ^{18}O. During geologic times when the **climate** is cooler, more of this lighter **precipitation** that **falls** on land is locked in the form of **glacial** ice. Consider that the giant **ice sheets** were more than a mile thick and covered a large part of North America during the last **ice age** only 14,000 years ago. During **glaciation**, the **glaciers** thus effectively lock away more ^{16}O, thus the ocean water and foraminifera shells become enriched in ^{18}O. Therefore, the ratio of ^{18}O to ^{16}O ($\diamond^{18}O$) in calcium **carbonate** shells of foraminifera is a **proxy indicator** of past **climate**. The sediment cores from the Ocean Drilling Program record a continuous accumulation of these **fossils** in the **sediment** and provide a record of glacials, **interglacials** and **interstadials**.

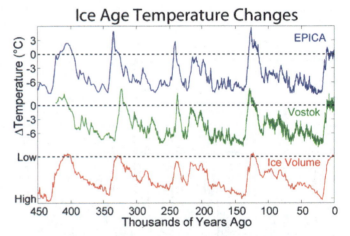

Figure 15.17: Antarctic temperature changes during the last few glaciations compared to global ice volume. The first two curves are based on the deuterium (heavy hydrogen) record from ice cores (EPICA Community Members 2004, Petit et al. 1999). The bottom line is ice volume based on oxygen isotopes from a composite of deep-sea sediment cores (Lisiecki and Raymo 2005).

Sediment Cores—Boron–Isotopes and Acidity

Ocean acidity is affected by **carbonic acid** and is a proxy for past atmospheric CO_2 concentrations. To estimate the ocean's pH (acidity) over the past 60 million years, researchers collected deep-sea **sediment** cores and examined the ancient planktonic foraminifera shells' boron-**isotope** ratios. Boron has two **isotopes**: ^{11}B and ^{10}B. In aqueous compounds of boron, the relative abundance of these two **isotopes** is sensitive to pH (acidity), hence CO_2 concentrations. In the early **Cenozoic**, around 60 million years ago, CO_2 concentrations were over 2,000 ppm, higher pH, and started falling around 55 to 40 million years ago, with noticeable drop in pH, indicated by boron **isotope** ratios. The drop was possibly due to reduced CO_2 outgassing from ocean ridges, **volcanoes** and **metamorphic** belts, and increased carbon burial due to **subduction** and the Himalaya Mountains uplift. By the Miocene **Epoch**, about 24 million years ago, CO_2 levels were below 500 ppm, and by 800,000 years ago, CO^2 levels didn't exceed 300 ppm.

Carbon Dioxide Concentrations in Ice Cores

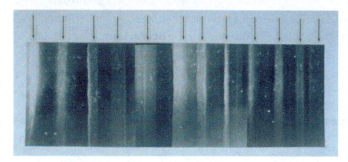

Figure 15.18: 19 cm long section of ice core showing 11 annual layers with summer layers (arrowed) sandwiched between darker winter layers.

For the recent Pleistocene **Epoch**'s **climate**, researchers get a more detailed and direct chemical record of the last 800,000 years by extracting and analyzing ice cores from the Antarctic and Greenland **ice sheets**. Snow accumulates on these **ice sheets** and creates yearly layers. Oxygen **isotopes** are collected from these annual layers, and the ratio of ^{18}O to ^{16}O ($\Diamond^{18}O$) is used to determine **temperature** as discussed above. In addition, the ice contains small bubbles of atmospheric gas as the snow turns to ice. Analysis of these bubbles reveals the **composition** of the **atmosphere** at these previous times.

Small pieces of this ice are crushed, and the ancient air is extracted into a **mass spectrometer** that can detect the ancient **atmosphere**'s chemistry. Carbon dioxide levels are recreated from these measurements. Over the last 800,000 years, the *maximum* carbon dioxide concentration during warm times was about 300 parts per million (ppm), and the minimum was about 170 ppm during cold stretches. Currently, the earth's atmospheric carbon dioxide content is over 410 ppm.

Figure 15.19: Antarctic ice showing hundreds of tiny trapped air bubbles from the atmosphere thousands of years ago.

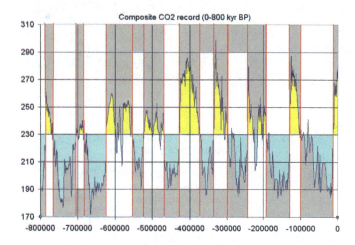

Figure 15.20: Composite carbon dioxide record from last 800,000 years based on ice core data from EPICA Dome C Ice Core.

Oceanic Microfossils

Microfossils, like foraminifera, diatoms, and radiolarians can be used as a proxy to interpret past **climate** record. Different species of microfossils are found in the **sediment core**'s different layers. Microfossil groups are called assemblages and their **composition** differs depending on the climatic conditions when they lived. One assemblage consists of species that lived in cooler ocean water, such as in **glacial** times, and at a different level in the same **sediment core**, another assemblage consists of species that lived in warmer waters.

One or more interactive elements has been excluded from this version of the text. You can view them online here:
https://www.youtube.com/watch?v=Yvu-g8Bkklg

Video 15.7: What sediment cores from the world's oceans reveal about climate patterns.

If you are using an offline version of this text, access this YouTube video via the QR code.

Tree Rings

Tree rings, which form every year as a tree grows, are another past **climate** indicator. Rings that are thicker indicate wetter years, and rings that are thinner and closer together indicate dryer years. Every year, a tree will grow one ring with a light section and a dark section. The rings vary in width. Since trees need much water to survive, narrower rings indicate colder and drier climates. Since some trees are several thousand years old, scientists can use their rings for regional paleoclimatic reconstructions, for example, to reconstruct past **temperature**, **precipitation**, vegetation, streamflow, sea-surface **temperature**, and other climate-dependent conditions. Paleoclimatic study means relating to a distinct past geologic climate. Also, dead trees, such as those found in Puebloan ruins, can be used to extend this **proxy indicator** by showing long-term droughts in the region and possibly explain why villages were abandoned.

Figure 15.21: Tree rings form every year. Rings that are farther apart are from wetter years and rings that are closer together are from dryer years.

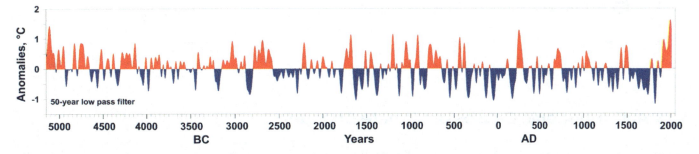

Figure 15.22: Summer temperature anomalies for the past 7,000 years.

Pollen

Pollen is also a proxy **climate** indicator. Flowering plants produce pollen grains. Pollen grains are distinctive when viewed under a microscope. Sometimes, pollen is preserved in lake **sediments** that accumulate in layers every year. Lake-**sediment** cores can reveal ancient pollen. **Fossil**-pollen assemblages are pollen groups from multiple species, such as spruce, pine, and oak. Through time, via the **sediment** cores and radiometric age-dating techniques, the pollen assemblages change, revealing the plants that lived in the area at the time. Thus, the pollen assemblages are a past **climate** indicator, since different plants will prefer different climates. For example, in the Pacific Northwest, east of the Cascades in a region close to grassland and forest borders, scientists tracked pollen over the last 125,000 years, covering the last two **glaciations**. As shown in the figure (Fig. 2 from reference Whitlock and Bartlein 1997), pollen assemblages with more pine tree pollen are found during **glaciations** and pollen assemblages with less pine tree pollen are found during **interglacial** times.

Figure 15.23: Scanning electron microscope image of modern pollen with false color added to distinguish plant species.

Other Proxy Indicators

Paleoclimatologists study many other phenomena to understand past climates, such as human historical accounts, human instrument records from the recent past, lake **sediments**, cave deposits, and corals.

Take this quiz to check your comprehension of this section.

If you are using an offline version of this text, access the quiz for section 15.3 via the QR code.

 An interactive H5P element has been excluded from this version of the text. You can view it online here:
https://pressbooks.lib.vt.edu/introearthscience/?p=928#h5p-104

15.4 Anthropogenic Causes of Climate Change

As shown in the previous section, prehistoric **climate** changes occur slowly over many millions of years. The **climate** changes observed today are rapid and largely human caused. Evidence shows that **climate** is changing, but what is causing that change? Since the late 1800s, scientists have suspected that human-produced, i.e. **anthropogenic** changes in atmospheric greenhouse gases would likely cause **climate** change because changes in these gases have been the case every time in the geologic past. By the middle 1900s, scientists began conducting systematic measurements, which confirmed that human-produced carbon dioxide was accumulating in the **atmosphere** and other earth systems, such as forests and oceans. By the end of the 1900s and into the early 2000s, scientists solidified the Theory of Anthropogenic Climate Change when evidence from thousands of ground-based studies and continuous land and ocean satellite measurements mounted, revealing the expected **temperature** increase. The **Theory** of **Anthropogenic Climate** Change is that humans are causing most of the current climate changes by burning **fossil fuels** such as **coal**, **oil**, and **natural gas**. Theories evolve and **transform** as new data and new techniques become available, and they represent a particular field's state of thinking. This section summarizes the scientific consensus of **anthropogenic climate** change.

15.4.1 Scientific Consensus

The overwhelming majority of **climate** studies indicate that human activity is causing rapid changes to the **climate**, which will cause severe environmental damage. There is strong scientific consensus on the issue. Studies published in peer-reviewed scientific journals show that 97 percent of **climate** scientists agree that **climate** warming is caused from human activities. There is no alternative explanation for the observed link between human-produced greenhouse gas emissions and changing modern **climate**. Most leading scientific organizations endorse this position, including the U.S. National Academy of Science, which was established in 1863 by an act of Congress under President Lincoln. Congress charged the National Academy of Science "with providing independent, **objective** advice to the nation on matters related to science and technology." Therefore, the National Academy of Science is the leading authority when it comes to policy advice related to scientific issues.

One way we know that the increased greenhouse gas emissions are from human activities is with isotopic fingerprints. For example, **fossil fuels**, representing plants that lived millions of years ago, have a stable carbon-13 to carbon-12 ($^{13}C/^{12}C$) ratio that is different from today's atmospheric stable-carbon ratio (**radioactive** ^{14}C is unstable). Isotopic carbon signatures have been used to identify **anthropogenic** carbon in the **atmosphere** since the 1980s. Isotopic records from the Antarctic Ice Sheet show stable isotopic signatures from ~1000 AD to ~1800 AD and a steady isotopic signature gradually changing since 1800, followed by a more rapid change after 1950 as burning of **fossil fuels** dilutes the CO_2 in the **atmosphere**. These changes show the **atmosphere** as having a carbon isotopic signature increasingly more similar to that of **fossil fuels**.

15.4.2 Anthropogenic Sources of Greenhouse Gases

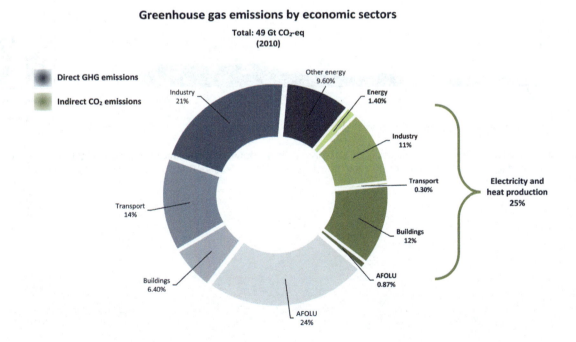

Figure 15.24: Total anthropogenic greenhouse gas emissions (gigatonne of CO2-equivalent per year, GtCO2-eq/yr) from economic sectors in 2010. The circle shows the shares of direct GHG emissions (in % of total anthropogenic GHG emissions) from five economic sectors in 2010. The pull-out shows how shares of indirect CO2 emissions (in % of total anthropogenic GHG emissions) from electricity and heat production are attributed to sectors of final energy use.

Anthropogenic emissions of greenhouse gases have increased since pre-industrial times due to global economic growth and population growth. Atmospheric concentrations of the leading greenhouse gas, carbon dioxide, are at unprecedented levels that haven't been observed in at least the last 800,000 years. Pre-industrial level of carbon dioxide was at about 278 parts per million (ppm). As of 2016, carbon dioxide was, for the first time, above 400 ppm for the entirety of the year. Measurements of atmospheric carbon at the Mauna Loa Carbon Dioxide Observatory show a continuous increase since 1957 when the observatory was established from 315 ppm to over 417 ppm in 2022. The daily reading today can be seen at Daily CO2. Based on the ice **core** record over the past 800,000 years, carbon dioxide ranged from about 185 ppm during **ice ages** to 300 ppm during warm times. View the data-accurate NOAA animation below of carbon dioxide trends over the last 800,000 years.

What is the source of these **anthropogenic** greenhouse gas emissions? **Fossil fuel** combustion and industrial processes contributed 78 percent of all emissions since 1970. The economic sectors responsible for most of this include electricity and heat production (25 percent); agriculture, forestry, and land use (24 percent); industry (21 percent); transportation, including automobiles (14 percent); other energy production (9.6 percent); and buildings (6.4 percent). More than half of greenhouse gas emissions have occurred in the last 40 years, and 40 percent of these emissions have stayed in the **atmosphere**. Unfortunately, despite scientific consensus, efforts to mitigate **climate** change require political action. Despite growing **climate** change concern, mitigation efforts, legislation, and international agreements have reduced emissions in some places, yet the less developed world's continual economic growth has increased global greenhouse gas emissions. In fact, the years 2000 to 2010 saw the largest increases since 1970.

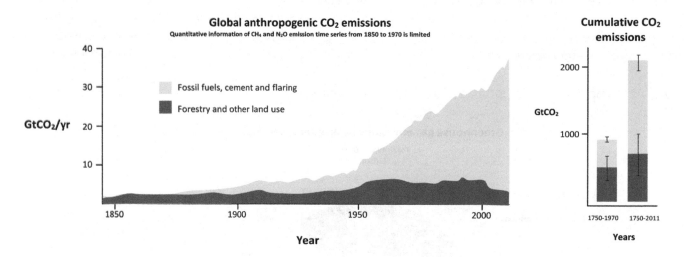

Figure 15.25: Annual global anthropogenic carbon dioxide (CO2) emissions in gigatonne of CO2-equivalent per year (GtCO2/yr) from fossil fuel combustion, cement production and flaring, and forestry and other land use (FOLU), 1750–2011. Cumulative emissions and their uncertainties are shown as bars and whiskers.

Take this quiz to check your comprehension of this section.

If you are using an offline version of this text, access the quiz for section 15.4 via the QR code.

An interactive H5P element has been excluded from this version of the text. You can view it online here:
https://pressbooks.lib.vt.edu/introearthscience/?p=928#h5p-105

Summary

Included in Earth Science is the study of the **system** of processes that affect surface environments and **atmosphere** of the Earth. Recent changes in atmospheric **temperature** and **climate** over intervals of decades have been observed. For Earth's **climate** to be stable, incoming radiation from the sun and outgoing radiation from the sun-warmed Earth must be in balance. Gases in the **atmosphere** called greenhouse gases absorb the infrared thermal radiation from the Earth's surface, trapping that heat and warming the atmosphere, a process called the **Greenhouse Effect**. Thus the energy budget is not now in balance and the Earth is warming. Human activity produces many greenhouse gases that have accelerated **climate** change. CO_2 from **fossil fuel** burning is one of the major ones. While atmospheric **composition** is mostly nitrogen and oxygen, trace components including the greenhouse gases (CO_2 and methane are the major ones and there are others) have the greatest effect on global warming.

A number of **Positive Feedback** Mechanisms, processes whose results reinforce the original process, take place in the Earth **system**. An example of a PFM of great concern is **permafrost** melting which causes decay of melting organic material that produces CO_2 and methane (both powerful greenhouse gases) that warm the **atmosphere** and promote more **permafrost** melting. Two carbon cycles affect Earth's atmospheric CO_2 **composition**, the biologic carbon cycle and the geologic carbon cycle. In the biologic cycle, organisms (mostly plants and also animals that eat them) remove CO_2 from the **atmosphere** for energy and to build their body tissues and return it to the **atmosphere** when they die and decay. The biologic cycle is a rapid cycle. In the geologic cycle, some organic matter is preserved in the form of **petroleum** and **coal** while more is **dissolved** in seawater and captured in **carbonate sediments**, some of which is **subducted** into the **mantle** and returned by **volcanic** activity. The geologic carbon cycle is slow over geologic time.

Measurements of increasing atmospheric **temperature** have been made since the nineteenth century but the upward **temperature** trend itself increased in the mid twentieth century showing the current trend is exponential. Because of the high specific heat of water, the oceans have absorbed most of the added heat. That this is temporary storage is revealed by the record-breaking warm years of the recent decade and the increase in intense storms and hurricanes. In 1957 the Mauna Loa CO_2 Observatory was established in Hawaii providing constant measurements of atmospheric CO_2 since 1958. The initial value was 315 ppm. The Keeling curve, named for the observatory founder, shows that value has steadily increased, exponentially, to over 417 ppm now. Compared to proxy data from atmospheric gases trapped in ice cores that show a maximum value for CO_2 of about 300 ppm over the last 800,000 years, the Keeling increase of over 100 ppm in 50 years is dramatic evidence of human caused CO_2 increase and **climate** change! As Earth's **temperature** rises, **glaciers** and **ice sheets** are shrinking resulting in sea level rise. Atmospheric CO_2 is also absorbed in sea water producing increased concentrations of **carbonic acid** which is raising the pH of the oceans making it harder for **marine** life to extract **carbonate** for their skeletal materials.

Earth's **climate** has changed over geologic time with **periods** of major **glaciations**. There was a high **temperature period** in the **Mesozoic** shown by fossils in high latitudes and the Western Interior Seaway covering what is now the Midwest. However, **climate** has been cooling during the **Cenozoic** culminating in the **Ice Age**. Since the **Ice Age**, several proxy indicators of ancient **climate** show that the rate and amount of current **climate** change is unique in geologic history and can only be attributed to human activity. Those who ignore the consequences of increasing global warming for our planet's future do so at the peril of our posterity!

Take this quiz to check your comprehension of this chapter.

If you are using an offline version of this text, access the quiz for chapter 15 via the QR code.

An interactive H5P element has been excluded from this version of the text. You can view it online here:
https://pressbooks.lib.vt.edu/introearthscience/?p=928#h5p-106

URLs Linked Within This Chapter

Interactive animation on the greenhouse effect: https://climate.nasa.gov/climate_resources/188/graphic-the-greenhouse-effect/

NASA animation: NASA Scientific Visualization Studio (2022), Global Temperature Anomalies from 1880 to 2021 https://svs.gsfc.nasa.gov/4964

3 minute NASA video: NASA JPL, Oceans of Climate Change, https://climate.nasa.gov/climate_resources/40/video-oceans-of-climate-change

Daily CO2: https://www.co2.earth/daily-co2

Text References

1. Allen, P.A., and Etienne, J.L., 2008, Sedimentary challenge to Snowball Earth: Nat. Geosci., v. 1, no. 12, p. 817–825.
2. Berner, R.A., 1998, The carbon cycle and carbon dioxide over Phanerozoic time: the role of land plants: Philos. Trans. R. Soc. Lond. B Biol. Sci., v. 353, no. 1365, p. 75–82.
3. Cunningham, W.L., Leventer, A., Andrews, J.T., Jennings, A.E., and Licht, K.J., 1999, Late Pleistocene–Holocene marine conditions in the Ross Sea, Antarctica: evidence from the diatom record: The Holocene, v. 9, no. 2, p. 129–139.
4. Deynoux, M., Miller, J.M.G., and Domack, E.W., 2004, Earth's Glacial Record: World and Regional Geology, Cambridge University Press, World and Regional Geology.
5. Earle, S., 2015, Physical geology OER textbook: BC Campus OpenEd.
6. Eyles, N., and Januszczak, N., 2004, "Zipper-rift": a tectonic model for Neoproterozoic glaciations during the breakup of Rodinia after 750 Ma: Earth-Sci. Rev.
7. Francey, R.J., Allison, C.E., Etheridge, D.M., Trudinger, C.M., and others, 1999, A 1000-year high precision record of δ13C in atmospheric CO2: Tellus B Chem. Phys. Meteorol.
8. Gutro, R., 2005, NASA – What's the Difference Between Weather and Climate? Online, http://www.nasa.gov/mission_pages/noaa-n/climate/climate_weather.html, accessed September 2016.
9. Hoffman, P.F., Kaufman, A.J., Halverson, G.P., and Schrag, D.P., 1998, A neoproterozoic snowball earth: Science, v. 281,

no. 5381, p. 1342–1346.

10. Kopp, R.E., Kirschvink, J.L., Hilburn, I.A., and Nash, C.Z., 2005, The Paleoproterozoic snowball Earth: a climate disaster triggered by the evolution of oxygenic photosynthesis: Proc. Natl. Acad. Sci. U. S. A., v. 102, no. 32, p. 11131–11136.

11. Lean, J., Beer, J., and Bradley, R., 1995, Reconstruction of solar irradiance since 1610: Implications for climate change: Geophys. Res. Lett., v. 22, no. 23, p. 3195–3198.

12. Levitus, S., Antonov, J.I., Wang, J., Delworth, T.L., Dixon, K.W., and Broccoli, A.J., 2001, Anthropogenic warming of Earth's climate system: Science, v. 292, no. 5515, p. 267–270.

13. Lindsey, R., 2009, Climate and Earth's Energy Budget: Feature Articles: Online, http://earthobservatory.nasa.gov, accessed September 2016.

14. North Carolina State University, 2013a, Composition of the Atmosphere.

15. North Carolina State University, 2013b, Composition of the Atmosphere: Online, http://climate.ncsu.edu/edu/k12/.AtmComposition, accessed September 2016.

16. Oreskes, N., 2004, The scientific consensus on climate change: Science, v. 306, no. 5702, p. 1686–1686.

17. Pachauri, R.K., Allen, M.R., Barros, V.R., Broome, J., Cramer, W., Christ, R., Church, J.A., Clarke, L., Dahe, Q., Dasgupta, P., Dubash, N.K., Edenhofer, O., Elgizouli, I., Field, C.B., and others, 2014, Climate Change 2014: Synthesis Report. Contribution of Working Groups I, II and III to the Fifth Assessment Report of the Intergovernmental Panel on Climate Change (R. K. Pachauri & L. Meyer, Eds.): Geneva, Switzerland, IPCC, 151 p.

18. Santer, B.D., Mears, C., Wentz, F.J., Taylor, K.E., Gleckler, P.J., Wigley, T.M.L., Barnett, T.P., Boyle, J.S., Brüggemann, W., Gillett, N.P., Klein, S.A., Meehl, G.A., Nozawa, T., Pierce, D.W., and others, 2007, Identification of human-induced changes in atmospheric moisture content: Proc. Natl. Acad. Sci. U. S. A., v. 104, no. 39, p. 15248–15253.

19. Schopf, J.W., and Klein, C., 1992, Late Proterozoic Low-Latitude Global Glaciation: the Snowball Earth, in Schopf, J.W., and Klein, C., editors, The Proterozoic biosphere: a multidisciplinary study: New York, Cambridge University Press, p. 51–52.

20. Webb, T., and Thompson, W., 1986, Is vegetation in equilibrium with climate? How to interpret late-Quaternary pollen data: Vegetatio, v. 67, no. 2, p. 75–91.

21. Weissert, H., 2000, Deciphering methane's fingerprint: Nature, v. 406, no. 6794, p. 356–357.

22. Whitlock, C., and Bartlein, P.J., 1997, Vegetation and climate change in northwest America during the past 125 kyr: Nature, v. 388, no. 6637, p. 57–61.

23. Wolpert, S., 2009, New NASA temperature maps provide a 'whole new way of seeing the moon': Online, http://newsroom.ucla.edu/releases/new-nasa-temperature-maps-provide-102070, accessed February 2017.

24. Zachos, J., Pagani, M., Sloan, L., Thomas, E., and Billups, K., 2001, Trends, rhythms, and aberrations in global climate 65 Ma to present: Science, v. 292, no. 5517, p. 686–693.

Figure References

Figure 15.1: Composition of the atmosphere. NASA. 2019. Public domain. https://climate.nasa.gov/news/2915/the-atmosphere-getting-a-handle-on-carbon-dioxide/#:~:text=By%20volume%2C%20the%20dry%20air,methane%2C%20nitrous%20oxide%20and%20ozone.

Figure 15.2: Common greenhouse gases. Kindred Grey. 2022. CC BY 4.0. Water molecule 3D by Dbc334, 2006 (Public domain, https://commons.wikimedia.org/wiki/File:Water_molecule_3D.svg). Nitrous-oxide-dimensions-3D-balls by Ben Mills, 2007 (Public domain, https://commons.wikimedia.org/wiki/File:Nitrous-oxide-dimensions-3D-balls.png). Methane-CRC-MW-3D-balls by Ben Mills, 2009 (Public domain, https://en.m.wikipedia.org/wiki/File:Methane-CRC-MW-3D-balls.png). Carbon dioxide 3D ball by Jynto, 2011 (Public domain, https://commons.wikimedia.org/wiki/File:Carbon_dioxide_3D_ball.png).

Figure 15.3: Carbon cycle. USGS. 2022. Public domain. https://www.usgs.gov/media/images/usgs-carbon-cycle

Figure 15.4: Incoming radiation absorbed, scattered, and reflected by atmospheric gases. Robert A. Rohde. 2013. CC BY SA 3.0. https://commons.wikimedia.org/wiki/File:Solar_spectrum_en.svg

Figure 15.5: Incoming solar radiation filtered by the atmosphere. NASA illustration by Robert Simmon. Astronaut photograph ISS013-E-8948. 2009. Public domain. https://earthobservatory.nasa.gov/features/EnergyBalance/page4.php

Figure 15.6: Some of the thermal infrared energy (heat) radiated from the surface into the atmosphere is trapped by gasses in the atmosphere. NASA illustration by Robert Simmon. Photograph by Cyron, 2006 (CC BY-NC 2.0). https://earthobservatory.nasa.gov/features/EnergyBalance/page5.php

Figure 15.7: Net effect of factors influencing warming. IPCC. 2014. Public domain. https://commons.wikimedia.org/wiki/File:Contributions_to_observed_surface_temperature_change_over_the_period_1951%E2%80%932010.svg

Figure 15.8: Land-ocean temperature index, 1880 to present, with a base time 1951-1980. NASA. 2010. Public domain. https://commons.wikimedia.org/wiki/File:Global_Temperature_Trends.png

Figure 15.9: Latest CO2 reading as of September 20, 2022. Scripps Institution of Oceanography at UC San Diego. 2022. Permissions statement. https://keelingcurve.ucsd.edu/

Figure 15.10: Keeling curve graph of the carbon dioxide concentration at Mauna Loa Observatory as of September 20, 2022. Scripps Institution of Oceanography at UC San Diego. 2022. Permissions statement. https://keelingcurve.ucsd.edu/

Figure 15.11: Decline of Antarctic ice mass from 2002 to 2016. NASA. 2017. Public domain. https://commons.wikimedia.org/wiki/File:ANTARCTICA_MASS_VARIATION_SINCE_2002.png

Figure 15.12: Maximum extent of Laurentide Ice Sheet. USGS. 2005. Public domain. https://commons.wikimedia.org/wiki/File:Pleistocene_north_ice_map.jpg

Figure 15.13: Global average surface temperature over the past 65 million years. Robert A. Rohde. 2005. CC BY-SA 3.0. https://commons.wikimedia.org/wiki/File:65_Myr_Climate_Change.png

Figure 15.14: The Antarctic Circumpolar Current. Avsa. 2009. CC BY-SA 3.0. https://commons.wikimedia.org/wiki/File:Conveyor_belt.svg

Figure 15.15: A Pliocene–Pleistocene stack of 57 globally distributed benthic δ18O records. Used under fair use from A Pliocene-Pleistocene stack of 57 globally distributed benthic δ18O records by Lorraine E. Lisiecki, Maureen E. Raymo (https://doi.org/10.1029/2004PA001071).

Figure 15.16: Sediment core from the Greenland continental slope. Hannes Grobe. 2008. CC BY 3.0. https://en.m.wikipedia.org/wiki/File:PS1920-1_0-750_sediment-core_hg.jpg

Figure 15.17: Antarctic temperature changes during the last few glaciations compared to global ice volume. This figure was produced by Robert A. Rohde from publicly available data and is incorporated into the Global Warming Art project. 2012. CC BY-SA 3.0. https://commons.wikimedia.org/wiki/File:Ice_Age_Temperature.png

Figure 15.18: 19 cm long section of ice core showing 11 annual layers with summer layers (arrowed) sandwiched between darker winter layers. NOAA. 2005. Public domain. https://commons.wikimedia.org/wiki/File:GISP2_1855m_ice_core_layers.png?uselang=en-gb

Figure 15.19: Antarctic ice showing hundreds of tiny trapped air bubbles from the atmosphere thousands of years ago. Atmospheric Research, CSIRO. 2000. CC BY 3.0. https://commons.wikimedia.org/wiki/File:CSIRO_ScienceImage_518_Air_Bubbles_Trapped_in_Ice.jpg

Figure 15.20: Composite carbon dioxide record from last 800,000 years based on ice core data from EPICA Dome C Ice Core. Tomruen. 2011. CC BY-SA 3.0. https://commons.wikimedia.org/w/index.php?curid=16147504

Figure 15.21: Tree rings form every year. Arpingstone. 2005. Public domain. https://commons.wikimedia.org/wiki/File:Tree.ring.arp.jpg

Figure 15.22: Summer temperature anomalies for the past 7,000 years. R.M.Hantemirov. 2010. CC BY 3.0. https://commons.wikimedia.org/wiki/File:Yamal50.gif

Figure 15.23: Scanning electron microscope image of modern pollen with false color added to distinguish plant species. Dartmouth Electron Microscope Facility, Dartmouth College. 2011. Public domain. https://en.wikipedia.org/wiki/File:Misc_pollen_colorized.jpg

Figure 15.24: Total anthropogenic greenhouse gas emissions (gigatonne of CO2-equivalent per year, GtCO2-eq/yr) from economic sectors in 2010. Kindred Grey. 2022. CC BY 4.0. Data from IPCC, 2014: Climate Change 2014: Synthesis Report. Contribution of Working Groups I, II and III to the Fifth Assessment Report of the Intergovernmental Panel on Climate Change [Core Writing Team, R.K. Pachauri and L.A. Meyer (eds.)]. IPCC, Geneva, Switzerland, 151 pp. https://www.ipcc.ch/site/assets/uploads/2018/05/SYR_AR5_FINAL_full_wcover.pdf

Figure 15.25: Annual global anthropogenic carbon dioxide (CO2) emissions in gigatonne of CO2-equivalent per year (GtCO2/yr) from fossil fuel combustion, cement production and flaring, and forestry and other land use (FOLU), 1750–2011. Kindred Grey. 2022. CC BY 4.0. Data from IPCC, 2014: Climate Change 2014: Synthesis Report. Contribution of Working Groups I, II and III to the Fifth Assessment Report of the Intergovernmental Panel on Climate Change [Core Writing Team, R.K. Pachauri and L.A. Meyer (eds.)]. IPCC, Geneva, Switzerland, 151 pp. https://www.ipcc.ch/site/assets/uploads/2018/05/SYR_AR5_FINAL_full_wcover.pdf

16.

ENERGY AND MINERAL RESOURCES

Learning Objectives

By the end of this chapter, students should be able to:

- Describe how a **renewable** resource is different from a **nonrenewable** resource.
- Compare the pros and cons of extracting and using **fossil fuels** and conventional and unconventional **petroleum** sources.
- Describe how **metallic minerals** are formed and extracted.
- Understand how society uses **nonmetallic mineral** resources.

Figure 16.1: A Mode 1 Oldowan tool used for chopping.

This text has previously discussed geology's pioneers, such as scientists James Hutton and Charles Lyell, but the first real "geologists" were the hominids who picked up stones and began the stone age. Maybe stones were first used as curiosity pieces, maybe as weapons, but ultimately, they were used as tools. This was the Paleolithic **Period**, the beginning of geologic study, and it dates back 2.6 million years to east Africa.

In modern times, geologic knowledge is important for locating economically valuable materials for society's use. In fact, all things we use come from only three sources: they are farmed, hunted or fished, or **mined**. At the turn of the twentieth century, speculation was rampant that food supplies would not keep pace with world demand, suggesting the need to develop artificial fertilizers. Sources of fertilizer ingredients are: nitrogen is processed from the **atmosphere**, using the Haber process for the manufacture of ammonia from atmospheric nitrogen and hydrogen; potassium comes from the **hydrosphere**, such as lakes or ocean evaporation; and phosphorus is **mined** from the **lithosphere**, such as minerals like apatite from phosphorite rock, which is found in Florida, North Carolina, Idaho, Utah, and around the world. Thus, without **mining** and processing of natural materials, modern civilization would not exist. Indeed, geologists are essential in this process.

16.1 Mining

Simplified world mining map

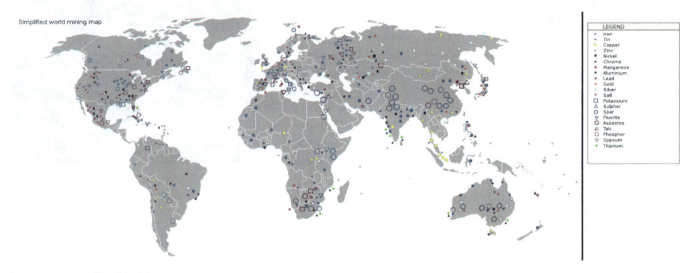

Figure 16.2: Map of world mining areas.

Mining is defined as extracting valuable materials from the Earth for society's use. Usually, these include solid materials such as gold, iron, **coal**, diamond, sand, and gravel, but materials can also include fluid resources such as **oil** and **natural gas**. Modern **mining** has a long relationship with modern society. The oldest **mine** dates back 40,000 years to the Lion Cavern in Swaziland where there is evidence of concentrated digging into the Earth for hematite, an important iron **ore** used as red dye. Resources extracted by **mining** are generally considered to be **nonrenewable**.

16.1.1 Renewable versus Nonrenewable Resources

Resources generally come in two major categories: **renewable** and **nonrenewable**. **Renewable** resources can be reused over and over or their availability replicated over a short human life span; **nonrenewable** resources cannot.

Figure 16.3: Hoover Dam provides hydroelectric energy and stores water for southern Nevada.

Renewable resources are materials present in our environment that can be exploited and replenished. Some common **renewable** energy sources are linked with green energy sources because they are associated with relatively small or easily remediated environmental impact. For example, solar energy comes from **fusion** within the Sun, which radiates electromagnetic energy. This energy reaches the Earth constantly and consistently and should continue to do so for about five billion more years. Wind energy, also related to solar energy, is maybe the oldest **renewable** energy and is used to sail ships and power windmills. Both solar and wind-generated energy are variable on Earth's surface. These limitations are **offset** because we can use energy storing devices, such as batteries or electricity exchanges between producing sites. The Earth's heat, known as geothermal energy, can be viable anywhere that geologists drill deeply enough. In practice, geothermal energy is more useful where heat flow is great, such as **volcanic** zones or regions with a thinner **crust**. Hydroelectric dams provide energy by allowing water to fall through the dam under gravity, which activates turbines that produce the energy. Ocean tides are also a reliable energy source. All of these **renewable** resources provide energy that powers society. Other **renewable** resources are plant and animal matter, which are used for food, clothing, and other necessities, but are being researched as possible energy sources.

Nonrenewable resources cannot be replenished at a sustainable rate. They are finite within human time frames. Many **nonrenewable** resources come from planetary, **tectonic**, or long-term biologic processes and include materials such as gold, lead, copper, diamonds, **marble**, sand, **natural gas**, **oil**, and **coal**. Most **nonrenewable** resources include specific concentrated **elements** listed on the periodic table; some are compounds of those **elements**. For example, if society needs iron (Fe) sources, then an exploration geologist will search for iron-rich deposits that can be economically extracted. **Nonrenewable** resources may be abandoned when other materials become cheaper or serve a better purpose. For example, **coal** is abundantly available in England and other nations, but because **oil** and **natural gas** are available at a lower cost and lower environmental impact, **coal** use has decreased. Economic competition among **nonrenewable** resources is shifting use away from **coal** in many developed countries.

Figure 16.4: Natural, octahedral shape of diamond.

16.1.2 Ore

Earth's materials include the periodic table **elements**. However, it is rare that these **elements** are concentrated to the point where it is profitable to extract and process the material into usable products. Any place where a valuable material is concentrated is a geologic and geochemical **anomaly**. A body of material from which one or more valuable substances can be mined at a profit, is called an **ore** deposit. Typically, the term **ore** is used for only metal-bearing **minerals**, but it can be applied to valuable **nonrenewable** resource concentrations such as **fossil fuels**, building stones, and other nonmetal deposits, even **groundwater**. If a metal-bearing resource is not profitable to **mine**, it is referred to as a **mineral** deposit. The term **natural resource** is more common than the term **ore** for non-metal-bearing materials.

Figure 16.5: Banded-iron formations are an important ore of iron (Fe).

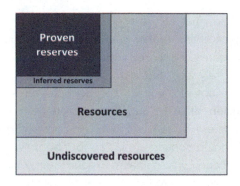

Figure 16.6: Diagram illustrating the relative abundance of proven reserves, inferred reserves, resources, and undiscovered resources.

It is implicit that the technology to **mine** is available, economic conditions are suitable, and political, social and environmental considerations are satisfied in order to classify a **natural resource** deposit as **ore**. Depending on the substance, it can be concentrated in a narrow vein or distributed over a large area as a low-concentration **ore**. Some materials are **mined** directly from bodies of water (e.g. sylvite for potassium; water through desalination) and the **atmosphere** (e.g. nitrogen for fertilizers). These differences lead to various methods of **mining**, and differences in terminology depending on the certainty. **Ore** mineral resource is used for an indication of **ore** that is potentially extractable, and the term **ore mineral** reserve is used for a well defined (proven), profitable amount of extractable **ore**.

Cumulative Production	IDENTIFIED RESOURCES			UNDISCOVERED RESOURCES	
	Demonstrated		Inferred	Probability Range (or)	
	Measured	Indicated		Hypothetical	Speculative
ECONOMIC	Reserves		Inferred Reserves		
MARGINALLY ECONOMIC	Marginal Reserves		Inferred Marginal Reserves		
SUB-ECONOMIC	Demonstrated Subeconomic Resources		Inferred Subeconomic Resources		
Other Occurrences	Includes nonconventional and low-grade materials				

Figure 16.7: McKelvey diagram showing different definitions for different degrees of concentration and understanding of mineral deposits.

16.1.3 Mining Techniques

The **mining** style is determined by technology, social license, and economics. It is in the best interest of the company extracting the resources to do so in a cost-effective way. Fluid resources, such as **oil** and gas, are extracted by drilling wells and pumping. Over the years, drilling has evolved into a complex discipline in which directional drilling can produce multiple bifurcations and curves originating from a single drill collar at the surface. Using geophysical tools like **seismic** imaging, geologists can pinpoint resources and extract efficiently.

Solid resources are extracted by two principal methods of which there are many variants. **Surface mining** is used to remove material from the outermost part of the Earth. **Open pit mining** is used to target shallow, broadly disseminated resources.

Open pit mining requires careful study of the **ore** body through surface mapping and drilling exploratory cores. The pit is progressively deepened through additional **mining** cuts to extract the **ore**. Typically, the pit's walls are as steep as can be safely managed. Once the pit is deepened, widening the top is very expensive. A steep wall is thus an engineering balance between efficient and profitable **mining** (from the company's point of view) and **mass wasting** (**angle of repose** from a safety point of view) so that there is less waste to remove. The waste is called non-valuable rock or overburden and moving it is costly. Occasionally, **landslides** do occur, such as the very large **landslide** in the Kennecott Bingham Canyon **mine**, Utah, in 2013. These events are costly and dangerous. The job of engineering geologists is to carefully monitor the mine; when company management heeds their warnings, there is ample time and action to avoid or prepare for any slide.

Figure 16.8: Bingham Canyon Mine, Utah. This open pit mine is the largest man-made removal of rock in the world.

Figure 16.9: A surface coal mine in Wyoming.

Strip mining and mountaintop mining are surface mining techniques that are used to mine resources that cover large areas, especially layered resources, such as coal. In this method, an entire mountaintop or rock layer is removed to access the ore below. Surface mining's environmental impacts are usually much greater due to the large surface footprint that's disturbed.

Figure 16.10: Underground mining in Estonia of oil shale.

Underground mining is a method often used to mine higher-grade, more localized, or very concentrated resources. For one example, geologists mine some underground ore minerals by introducing chemical agents, which dissolve the target mineral. Then, they bring the solution to the surface where precipitation extracts the material. But more often, a mining shaft tunnel or a large network of these shafts and tunnels is dug to access the material. The decision to mine underground or from Earth's surface is dictated by the ore deposit's concentration, depth, geometry, land-use policies, economics, surrounding rock strength, and physical access to the ore. For example, to use surface mining techniques for deeper deposits might require removing too much material, or the necessary method may be too dangerous or impractical, or removing the entire overburden may be too expensive, or the mining footprint would be too large. These factors may prevent geologists from surface mining materials and cause a project to be mined underground. The mining method and its feasibility depends on the commodity's price and the cost of the technology needed to remove it and deliver it to market. Thus, mines and the towns that support them come and go as the commodity price varies. And, conversely, technological advances and market demands may reopen mines and revive ghost towns.

16.1.4 Concentrating and Refining

All ore minerals occur mixed with less desirable components called gangue. The process of physically separating gangue minerals from ore bearing minerals is called concentrating. Separating a desired element from a host mineral by chemical means, including heating, is called smelting. Finally, taking a metal such as copper and removing other trace metals such as gold or silver is done through the refining process. Typically, refining is done one of three ways: 1. Materials can either be mechanically separated and processed based on the ore mineral's unique physical properties, such as recovering placer gold based on its high density. 2. Materials can be heated to chemically separate desired components, such as refining crude oil into gasoline. 3. Materials can be smelted, in which controlled chemical reactions unbind metals from the minerals they are contained in, such as when copper is taken out of chalcopyrite ($CuFeS_2$). Mining, concentrating, smelting, and refining processes require enormous energy. Continual advances in metallurgy- and mining-practice strive to develop ever more energy efficient and environmentally benign processes and practices.

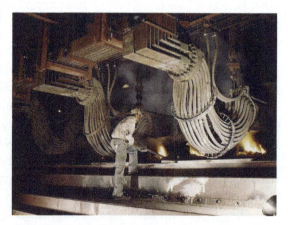

Figure 16.11: A phosphate smelting operation in Alabama, 1942.

Take this quiz to check your comprehension of this section.

If you are using an offline version of this text, access the quiz for section 16.1 via the QR code.

 An interactive H5P element has been excluded from this version of the text. You can view it online here:
https://pressbooks.lib.vt.edu/introearthscience/?p=973#h5p-107

16.2 Fossil Fuels

Figure 16.12: Coal power plant in Helper, Utah.

Fossil fuels are extractable sources of stored energy that were created by ancient ecosystems. The **natural resources** that typically fall under this category are **coal**, **oil**, **petroleum**, and **natural gas**. These resources were originally formed via photosynthesis by living organisms such as plants, phytoplankton, algae, and cyanobacteria. This energy is actually **fossil** solar energy, since the sun's ancient energy was converted by ancient organisms into tissues that preserved the chemical energy within the **fossil fuel**. Of course, as the energy is used, just like photosynthetic respiration that occurs today, carbon enters the **atmosphere** as CO_2, causing **climate** consequences (see chapter 15). Today humanity uses **fossil fuels** for most of the world's energy.

Converting solar energy by living organisms into hydrocarbon **fossil fuels** is a complex process. As organisms die, they decompose slowly, usually due to being buried rapidly, and the chemical energy stored within the organisms' tissues is buried within surrounding geologic materials. All **fossil fuels** contain carbon that was produced in an ancient environment. In environments rich with organic matter such as swamps, coral **reefs**, and planktonic blooms, there is a higher potential for **fossil fuels** to accumulate. Indeed, there is some evidence that over geologic time, organic hydrocarbon **fossil fuel** material was highly produced globally. Lack of oxygen and moderate temperatures in the environment seem to help preserve these organic substances. Also, the heat and pressure applied to organic material after it is buried contribute to transforming it into higher quality materials, such as brown **coal** to anthracite and **oil** to gas. Heat and pressure can also cause mobile materials to migrate to conditions suitable for extraction.

Figure 16.13: Modern coral reefs and other highly-productive shallow marine environments are thought to be the sources of most petroleum resources.

16.2.1 Oil and Gas

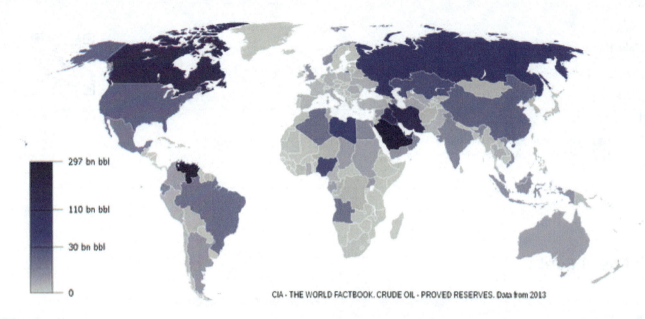

CIA - THE WORLD FACTBOOK, CRUDE OIL - PROVED RESERVES. Data from 2013

Figure 16.14: World oil reserves in 2013. Scale in billions of barrels.

Petroleum is principally derived from organic-rich shallow **marine** sedimentary deposits where the remains of micro-organisms like plankton accumulated in fine grained **sediments**. **Petroleum**'s liquid component is called **oil**, and its gas component is called **natural gas**, which is mostly made up of methane (CH_4). As rocks such as **shale**, **mudstone**, or **limestone** lithify, increasing pressure and **temperature** cause the **oil** and gas to be squeezed out and migrate from the **source rock** to a different rock unit higher in the rock column. Similar to the discussion of good **aquifers** in chapter 11, if that rock is a **sandstone**, **limestone**, or other porous and permeable rock, and involved in a suitable **stratigraphic** or structural trapping process, then that rock can act as an **oil** and gas **reservoir**.

A **trap** is a combination of a subsurface geologic structure, a porous and permeable rock, and an impervious layer that helps block **oil** and gas from moving further, which concentrates it for humans to extract later. A **trap** develops due to many different geologic situations. Examples include an **anticline** or domal structure, an impermeable salt **dome**, or a **fault** bounded **stratigraphic** block, which is porous rock next to nonporous rock. The different **traps** have one thing in common: they pool fluid **fossil fuels** into a configuration in which extracting it is more likely to be profitable. **Oil** or gas in **strata** outside of a **trap** renders it less viable to extract.

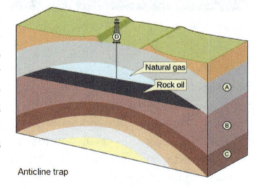

Anticline trap

Figure 16.15: Examples of different forms of hydrocarbon traps: in the core region of anticlines.

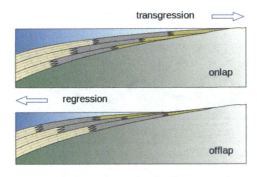

Figure 16.16: The rising sea levels of transgressions create onlapping sediments, regressions create offlapping.

Sequence stratigraphy is a branch of geology that studies sedimentary **facies** both horizontally and vertically and is devoted to understanding how sea level changes create organic-rich shallow **marine** muds, **carbonates**, and sands in areas that are close to each other. For example, **shoreline** environments may have beaches, **lagoons**, **reefs**, **nearshore** and **offshore** deposits, all next to each other. Beach sand, lagoonal and **nearshore** muds, and coral reef layers accumulate into **sediments** that include **sandstones**—good **reservoir** rocks— next to **mudstones**, next to **limestones**, both of which are potential **source rocks**. As sea level either rises or falls, the shoreline's location changes, and the sand, mud, and reef locations shift with it (see the figure). This places oil and gas producing rocks, such as mudstones and limestones next to oil and gas reservoirs, such as sandstones and some limestones. Understanding how the lithology and the facies/stratigraphic relationships interplay is very important in finding new petroleum resources. Using sequence stratigraphy as a model allows geologists to predict favorable locations of the source rock and **reservoir**.

16.2.2 Tar Sands

Conventional **oil** and gas, which is pumped from a **reservoir**, is not the only way to obtain hydrocarbons. There are a few fuel sources known as unconventional **petroleum** sources. However, they are becoming more important as conventional sources become scarce. Tar sands, or **oil sands**, are **sandstones** that contain **petroleum** products that are highly **viscous**, like tar, and thus cannot be drilled and pumped out of the ground readily like conventional **oil**. This unconventional **fossil fuel** is bitumen, which can be pumped as a fluid only at very low recovery rates and only when heated or mixed with solvents. So, using steam and solvent injections or directly **mining** tar sands to process later are ways to extract the tar from the sands. Alberta, Canada is known to have the largest **tar sand** reserves in the world. Note: as with **ores**, an energy resource becomes uneconomic if the total extraction and processing costs exceed the extracted material's sales revenue. Environmental costs may also contribute to a resource becoming uneconomic.

Figure 16.17: Tar sandstone from the Miocene Monterrey Formation of California.

16.2.3 Oil Shale

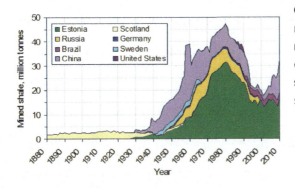

Figure 16.18: Global production of oil shale, 1880-2010.

Oil shale, or **tight oil**, is a fine-grained **sedimentary rock** that has significant **petroleum** or **natural gas** quantities locked tightly in the **sediment**. **Shale** has high **porosity** but very low **permeability** and is a common **fossil fuel source rock**. To extract the **oil** directly from the **shale**, the material has to be **mined** and heated, which, like with tar sands, is expensive and typically has a negative environmental impact.

16.2.4 Fracking

Another process used to extract the **oil** and gas from **shale** and other unconventional tight resources is called **hydraulic fracturing**, better known as **fracking**. In this method, high-pressure water, sand grains, and added chemicals are injected and pumped underground. Under high pressure, this creates and holds open **fractures** in the rocks, which help release the hard-to-access mostly **natural gas** fluids. **Fracking** is more useful in tighter sediments, especially shale, which has a high porosity to store the hydrocarbons but low **permeability** to allow transmission of the hydrocarbons. Fracking has become controversial because its methods contaminate **groundwater** and induce **seismic** activity. This has created much controversy between public concerns, political concerns, and energy value.

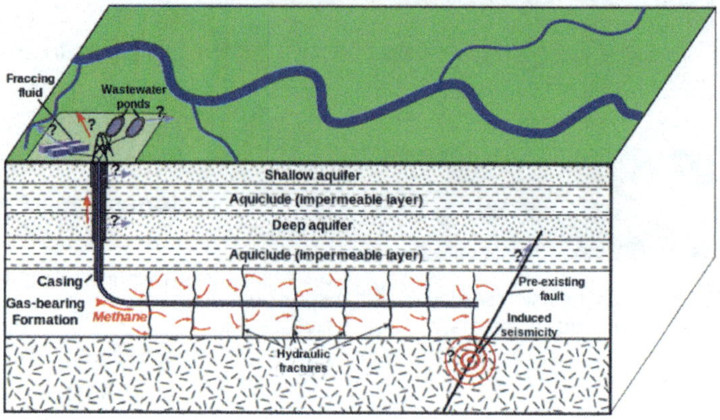

Figure 16.19: Schematic diagram of fracking.

16.2.5 Coal

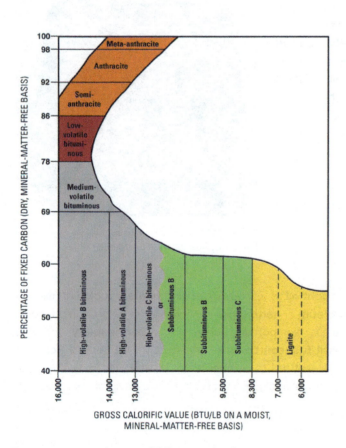

Figure 16.20: USGS diagram of different coal rankings.

Coal comes from fossilized swamps, though some older **coal** deposits that predate **terrestrial** plants are presumed to come from algal buildups. **Coal** is chiefly carbon, hydrogen, nitrogen, sulfur, and oxygen, with minor amounts of other **elements**. As plant material is incorporated into **sediments**, heat and pressure cause several changes that concentrate the fixed carbon, which is the **coal**'s combustible portion. So, the more heat and pressure that **coal** undergoes, the greater is its carbon concentration and fuel value and the more desirable is the **coal**.

This is the general sequence of a swamp progressing through the various stages of **coal formation** and becoming more concentrated in carbon: Swamp => Peat => Lignite => Sub-bituminous => Bituminous => Anthracite => Graphite. As swamp materials collect on the swamp floor and are buried under accumulating materials, they first turn to peat.

Peat itself is an economic fuel in some locations like the British Isles and Scandinavia. As **lithification** occurs, peat turns to lignite. With increasing heat and pressure, lignite turns to sub-bituminous **coal**, bituminous **coal**, and then, in a process like **metamorphism**, anthracite. Anthracite is the highest **metamorphic grade** and most desirable **coal** since it provides the highest energy output. With even more heat and pressure driving out all the **volatiles** and leaving pure carbon, anthracite can become graphite.

Figure 16.21: Peat (also known as turf) consists of partially decayed organic matter. The Irish have long mined peat to be burned as fuel though this practice is now discouraged for environmental reasons.

Figure 16.22: Anthracite coal, the highest grade of coal.

Humans have used **coal** for at least 6,000 years, mainly as a fuel source. **Coal** resources in Wales are often cited as a primary reason for Britain's rise, and later, for the United States' rise during the Industrial Revolution. According to the US Energy Information Administration, US **coal** production has decreased due to competing energy sources' cheaper prices and due to society recognizing its negative environmental impacts, including increased very fine-grained particulate matter as an air pollutant, greenhouse gases, acid rain, and heavy metal pollution. Seen from this perspective, the **coal** industry as a source of **fossil** energy is unlikely to revive.

As the world transitions away from **fossil fuels** including **coal**, and manufacturing seeks strong, flexible, and lighter materials than steel including carbon fiber for many applications, current research is exploring **coal** as a source of this carbon.

Take this quiz to check your comprehension of this section.

If you are using an offline version of this text, access the quiz for section 16.2 via the QR code.

 An interactive H5P element has been excluded from this version of the text. You can view it online here: https://pressbooks.lib.vt.edu/introearthscience/?p=973#h5p-108

16.3 Mineral Resources

Mineral resources, while principally **nonrenewable**, generally placed in two main categories: **metallic**, which contain metals, and **nonmetallic**, which contain other useful materials. Most **mining** has been traditionally focused on extracting **metallic minerals**. Human society has advanced significantly because we've developed the knowledge and technologies to yield metal from the Earth. This knowledge has allowed humans to build the machines, buildings, and monetary systems that dominate our world today. Locating and recovering these metals has been a key facet of geologic study since its inception. Every **element** across the periodic table has specific applications in human civilization. **Metallic mineral mining** is the source of many of these **elements**.

Figure 16.23: Gold-bearing quartz vein from California.

16.3.1 Types of Metallic Mineral Deposits

The various ways in which **minerals** and their associated **elements** concentrate to form **ore** deposits are too complex and numerous to fully review in this text. However, entire careers are built around them. In the following section, we describe some of the more common deposit types along with their associated elemental concentrations and world class occurrences.

Magmatic Processes

When a magmatic body crystallizes and differentiates (see chapter 4), it can cause certain **minerals** and **elements** to concentrate. Layered intrusions, typically **ultramafic** to **mafic**, can host deposits that contain copper, nickel, platinum, palladium, rhodium, and chromium. The Stillwater Complex in Montana is an example of economic quantities of layered **mafic** intrusion. Associated deposit types can contain chromium or titanium-vanadium. The largest magmatic deposits in the world are the chromite deposits in the Bushveld Igneous Complex in South Africa. These rocks have an areal extent larger than the state of Utah. The chromite occurs in layers, which resemble sedimentary layers, except these layers occur within a crystallizing **magma chamber**.

Figure 16.24: Layered intrusion of dark chromium-bearing minerals, Bushveld Complex, South Africa.

Figure 16.25: This pegmatite contains lithium-rich green elbaite (a tourmaline) and purple lepidolite (a mica).

Water and other **volatiles** that are not incorporated into **mineral** crystals when a **magma** crystallizes can become concentrated around the crystallizing **magma**'s margins. Ions in these hot fluids are very mobile and can form exceptionally large crystals. Once crystallized, these large crystal masses are then called **pegmatites**. They form from **magma** fluids that are expelled from the solidifying **magma** when nearly the entire **magma** body has crystallized. In addition to **minerals** that are predominant in the main **igneous** mass, such as **quartz**, **feldspar**, and **mica**, **pegmatite** bodies may also contain very large crystals of unusual **minerals** that contain rare **elements** like beryllium, lithium, tantalum, niobium, and tin, as well as **native elements** like gold. Such **pegmatites** are **ores** of these metals.

An unusual magmatic process is a **kimberlite** pipe, which is a **volcanic conduit** that transports **ultramafic magma** from within the **mantle** to the surface. Diamonds, which are formed at great temperatures and pressures of depth, are transported by a **Kimberlite** pipe to locations where they can be **mined**. The process that created these **kimberlite ultramafic** rocks is no longer common on Earth. Most known deposits are from the **Archean Eon**.

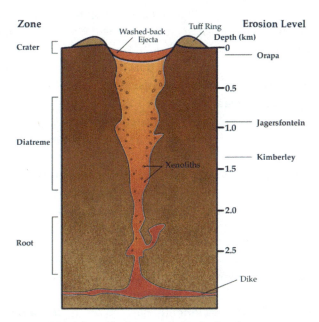

Figure 16.26: Schematic diagram of a kimberlite pipe.

Hydrothermal Processes

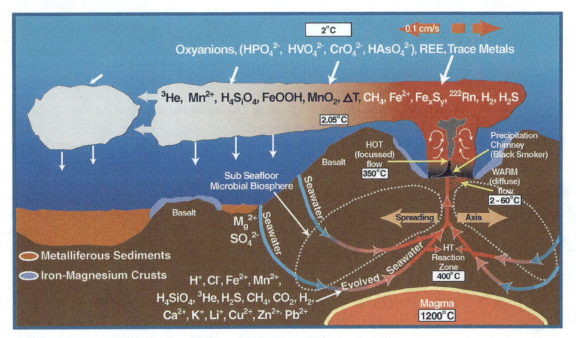

Figure 16.27: The complex chemistry around mid-ocean ridges.

Fluids rising from crystallizing magmatic bodies or that are heated by the **geothermal gradient** cause many geochemical reactions that form various **mineral** deposits. The most active **hydrothermal** process today produces **volcanogenic massive sulfide** (VMS) deposits, which form from **black smoker hydrothermal** chimney activity near **mid-ocean ridges** all over the world. They commonly contain copper, zinc, lead, gold, and silver when found at the surface. Evidence from around 7000 BC in a **period** known as the Chalcolithic shows copper was among the earliest metals smelted by humans as means of obtaining higher temperatures were developed. The largest of these VMS deposits occur in **Precambrian period** rocks. The Jerome deposit in central Arizona is a good example.

Another deposit type that draws on **magma**-heated water is a **porphyry** deposit. This is not to be confused with the **porphyritic igneous** texture, although the name is derived from the **porphyritic texture** that is nearly always present in the **igneous** rocks associated with a **porphyry** deposit. Several types of **porphyry** deposits exist, such as **porphyry** copper, **porphyry** molybdenum, and **porphyry** tin. These deposits contain low-**grade** disseminated **ore minerals** closely associated with **intermediate** and **felsic intrusive** rocks that are present over a very large area. **Porphyry** deposits are typically the largest **mines** on Earth. One of the largest, richest, and possibly best studied **mine** in the world is Utah's Kennecott Bingham Canyon Mine. It's an **open pit mine**, which, for over 100 years, has produced several **elements** including copper, gold, molybdenum, and silver. Underground **carbonate** replacement deposits produce lead, zinc, gold, silver, and copper. In the **mine**'s past, the open pit predominately produced copper and gold from chalcopyrite and bornite. Gold only occurs in minor quantities in the copper-bearing **minerals**, but because the Kennecott Bingham Canyon Mine produces on such a large scale, it is one of the largest gold **mines** in the US. In the future, this **mine** may produce more copper and molybdenum (molybdenite) from deeper **underground mines**.

Most **porphyry** copper deposits owe their high metal content, and hence, their economic value to **weathering** processes called **supergene enrichment** which occurs when the deposit is uplifted, eroded, and exposed to **oxidation**. This process occurred millions of years after the initial **igneous** intrusion and **hydrothermal** expulsion ends. When the deposit's upper pyrite-rich portion is exposed to rain, the pyrite in the oxidizing zone creates an extremely acid condition that dissolves copper out of copper **minerals**, such as chalcopyrite, and converts the chalcopyrite to iron **oxides**, such as hematite or goethite. The copper **minerals** are carried downward in water until they arrive at the **groundwater** table and an environment where the primary copper **minerals** are converted into secondary higher-copper content **minerals**. Chalcopyrite (35% Cu) is converted to bornite (63% Cu), and ultimately, chalcocite (80% Cu). Without this enriched zone, which is two to five times higher in copper content than the main deposit, most **porphyry** copper deposits would not be economic to **mine**.

Figure 16.28: The Morenci porphyry is oxidized toward its top (as seen as red rocks in the wall of the mine), creating supergene enrichment.

Figure 16.29: Garnet-augite skarn from Italy.

If **limestone** or other calcareous sedimentary rocks are near the magmatic body, then another type of **ore** deposit called a **skarn** deposit forms. These **metamorphic** rocks form as **magma**-derived, highly saline metalliferous fluids react with **carbonate** rocks to create calcium-magnesium-**silicate minerals** like **pyroxene**, **amphibole**, and garnet, as well as high-**grade** iron, copper, zinc **minerals**, and gold. Intrusions that are genetically related to the intrusion that made the Kennecott Bingham Canyon deposit have also produced copper-gold skarns, which were **mined** by the early European settlers in Utah. When iron and/or **sulfide** deposits undergo **metamorphism**, the **grain size** commonly increases, which makes separating the **gangue** from the desired **sulfide** or **oxide minerals** much easier.

Figure 16.30: In this rock, a pyrite cube has dissolved (as seen with the negative "corner" impression in the rock), leaving behind small specks of gold.

Sediment-hosted disseminated gold deposits consist of low concentrations of microscopic gold as **inclusions** and disseminated atoms in pyrite crystals. These are formed via low-**grade hydrothermal** reactions, generally in the realm of **diagenesis**, that occur in certain rock types, namely muddy **carbonates** and limey **mudstones**. This **hydrothermal** alteration is generally far removed from a **magma** source, but can be found in rocks situated with a high **geothermal gradient**. The Mercur deposit in Utah's Oquirrh Mountains was this type's earliest locally **mined** deposit. There, almost a million ounces of gold was recovered between 1890 and 1917. In the 1960s, a metallurgical process using cyanide was developed for these low-**grade ore** types. These deposits are also called **Carlin-type** deposits because the disseminated deposit near Carlin, Nevada, is where the new technology was first applied and where the first definitive scientific studies were conducted. Gold was introduced into these deposits by **hydrothermal** fluids that reacted with silty calcareous rocks, removing **carbonate**, creating additional **permeability**, and adding silica and gold-bearing pyrite in the **pore** space between grains. The Betze-Post mine and the Gold Quarry mine on the Carlin Trend are two of the largest disseminated gold deposits in Nevada. Similar deposits, but not as large, have been found in China, Iran, and Macedonia.

Non-magmatic Geochemical Processes

Geochemical processes that occur at or near the surface without **magma**'s aid also concentrate metals, but to a lesser degree than **hydrothermal** processes. One of the main reactions is **redox**, short for reduction/**oxidation** chemistry, which has to do with the amount of available oxygen in a **system**. Places where oxygen is plentiful, as in the **atmosphere** today, are considered oxidizing environments, while oxygen-poor places are considered reducing environments. Uranium deposits are an example of where **redox concentrated** the metal. Uranium is soluble in oxidizing **groundwater** environments and precipitates as uraninite when encountering reducing conditions. Many of the deposits across the Colorado Plateau, such as in Moab, Utah, were formed by this method.

Figure 16.31: Underground uranium mine near Moab, Utah.

Redox reactions are also responsible for creating **banded iron formations** (BIFs), which are interbedded layers of iron **oxide**—hematite and magnetite, **chert**, and **shale beds**. These deposits formed early in the Earth's history as the **atmosphere** was becoming oxygenated. Cycles of oxygenating iron-rich waters initiated **precipitation** of the iron **beds**. Because BIFs are generally **Precambrian** in age, happening at the event of atmospheric oxygenation, they are only found in some of the older exposed rocks in the United States, such as in Michigan's upper peninsula and northeast Minnesota.

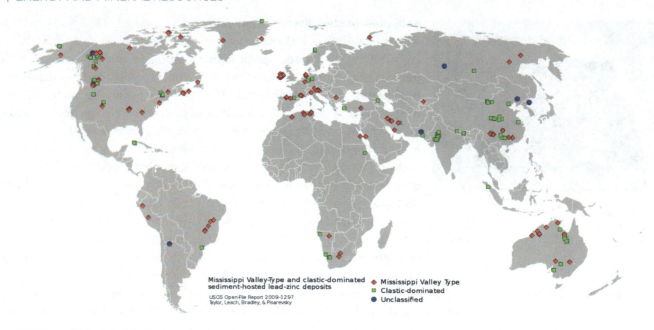

Figure 16.32: Map of Mississippi-Valley type ore deposits.

Deep, saline, **connate fluids** (trapped in **pore** spaces) within **sedimentary basins** may be highly metalliferous. When expelled outward and upward as **basin sediments** compacted, these fluids formed lead and zinc deposits in **limestone** by replacing or filling open spaces, such as caves and **faults**, and in **sandstone** by filling **pore** spaces. The most famous are called **Mississippi Valley-type** deposits. Also known as **carbonate-hosted replacement** deposits, they are large deposits of galena and sphalerite lead and zinc **ores** that form from hot fluids ranging from 100°C to 200°C (212°F to 392°F). Although they are named for occurring along the Mississippi River Valley in the US, they are found worldwide.

Sediment-hosted copper deposits occurring in **sandstones**, **shales**, and marls are enormous, and their contained resources are comparable to **porphyry** copper deposits. These deposits were most likely formed diagenetically by **groundwater** fluids in highly permeable rocks. Well-known examples are the Kupferschiefer in Europe, which has an areal coverage of >500,000 Km², (310,685.596mi) and the Zambian Copper Belt in Africa.

Soils and **mineral** deposits that are exposed at the surface experience deep and intense **weathering**, which can form surficial deposits. **Bauxite**, an aluminum **ore**, is preserved in **karst** topography and laterites, which are **soils** formed in wet tropical environments. **Soils** containing aluminum concentrate **minerals**, such as **feldspar**, and ferromagnesian **minerals** in **igneous** and **metamorphic** rocks, undergo **chemical weathering** processes that concentrate the metals. **Ultramafic** rocks that undergo **weathering** form nickel-rich **soils**, and when the magnetite and hematite in **banded iron formations** undergo **weathering**, it forms goethite, a friable **mineral** that is easily **mined** for its iron content.

Figure 16.33: A sample of bauxite. Note the unweathered igneous rock in the center.

Surficial Physical Processes

At the Earth's surface, **mass wasting** and moving water can cause hydraulic **sorting**, which forces high-density **minerals** to concentrate. When these minerals are concentrated in **streams**, **rivers**, and beaches, they are called **placer** deposits, and occur in modern sands and ancient lithified rocks. **Native** gold, **native** platinum, **zircon**, ilmenite, rutile, magnetite, diamonds, and other gemstones can be found in **placers**. Humans have mimicked this natural process to recover gold manually by gold panning and by mechanized means such as dredging.

Figure 16.34: Lithified heavy mineral sand (dark layers) from a beach deposit in India.

16.3.2 Environmental Impacts of Metallic Mineral Mining

Figure 16.35: Acid mine drainage in the Rio Tinto, Spain.

Metallic mineral mining's primary impact comes from the **mining** itself, including disturbing the land surface, covering landscapes with tailings impoundments, and increasing **mass wasting** by accelerating **erosion**. In addition, many metal deposits contain pyrite, an uneconomic **sulfide mineral**, that when placed on waste dumps, generates **acid rock drainage** (ARD) during **weathering**. In oxygenated water, **sulfides** such as pyrite react and undergo complex reactions to release metal ions and hydrogen ions, which lowers pH to highly acidic levels. **Mining** and processing of **mined** materials typically increase the surface area to volume ratio in the material, causing chemical reactions to occur even faster than would occur naturally. If not managed properly, these reactions lead to acidic **streams** and **groundwater** plumes that carry **dissolved** toxic metals. In **mines** where **limestone** is a waste rock or where **carbonate minerals** like **calcite** or dolomite are present, their acid neutralizing potential helps reduce **acid rock drainage**. Although this is a natural process too, it is very important to isolate **mine** dumps and tailings from oxygenated water, both to prevent the **sulfides** from dissolving and subsequently percolating the **sulfate**-rich water into waterways. Industry has taken great strides to prevent contamination in recent decades, but earlier **mining** projects are still causing problems with local ecosystems.

16.3.3 Nonmetallic Mineral Deposits

While receiving much less attention, **nonmetallic mineral** resources, also known as industrial **minerals**, are just as vital to ancient and modern society as **metallic minerals**. The most basic is building stone. **Limestone**, **travertine**, **granite**, **slate**, and **marble** are common building stones and have been quarried for centuries. Even today, building stones from **slate** roof tiles to **granite** countertops are very popular. Especially pure **limestone** is ground up, processed, and reformed as plaster, cement, and concrete. Some **nonmetallic mineral** resources are not **mineral** specific; nearly any rock or **mineral** can be used. This is generally called aggregate, which is used in concrete, roads, and foundations. Gravel is one of the more common aggregates.

Evaporites

Figure 16.37: Salt-covered plain known as the Bonneville Salt Flats, Utah.

Figure 16.36: Carrara marble quarry in Italy, source to famous sculptures like Michelangelo's David.

Evaporite deposits form in restricted basins where water evaporates faster than it **recharges**, such as the Great Salt Lake in Utah, or the Dead Sea, which borders Israel and Jordan. As the waters evaporate, soluble **minerals** are concentrated and become supersaturated, at which point they **precipitate** from the now highly-saline waters. If these conditions persist for long stretches, thick rock salt, rock **gypsum**, and other **mineral** deposits accumulate (see chapter 5).

Evaporite minerals, such as **halite**, are used in our food as common table salt. Salt was a vitally important food preservative and economic resource before refrigeration was developed. While still used in food, **halite** is now mainly **mined** as a chemical agent, water softener, or road de-icer. **Gypsum** is a common **nonmetallic mineral** used as a building material; it is the main component in dry wall. It is also used as a fertilizer. Other **evaporites** include sylvite—potassium chloride, and bischofite—magnesium chloride, both of which are used in agriculture, medicine, food processing, and other applications. Potash, a group of highly soluble potassium-bearing **evaporite minerals**, is used as a fertilizer. In hyper-arid locations, even more rare and complex **evaporites**, like borax, trona, ulexite, and hanksite are **mined**. They can be found in places such as Searles Dry Lake and Death Valley, California, and in the Green River Formation's ancient **evaporite** deposits in Utah and Wyoming.

Figure 16.38: Hanksite, Na22K(SO4)9(CO3)2Cl, one of the few minerals that is considered a carbonate and a sulfate.

Phosphorus

Phosphorus is an essential **element** that occurs in the **mineral** apatite, which is found in trace amounts in common **igneous** rocks. Phosphorite rock, which is formed in sedimentary environments in the ocean, contains abundant apatite and is **mined** to make fertilizer. Without phosphorus, life as we know it is not possible. Phosphorous is an important component of bone and DNA. Bone **ash** and guano are natural sources of phosphorus.

Figure 16.39: Apatite from Mexico.

Take this quiz to check your comprehension of this section.

If you are using an offline version of this text, access the quiz for section 16.3 via the QR code.

 An interactive H5P element has been excluded from this version of the text. You can view it online here:
https://pressbooks.lib.vt.edu/introearthscience/?p=973#h5p-109

Summary

Energy and **mineral** resources are vital to modern society, and it is the role of the geologist to locate these resources for human benefit. As environmental concerns have become more prominent, the value of the geologist has not decreased, as they are still vital in locating the deposits and identifying the least **intrusive** methods of extraction.

Energy resources are general grouped as being **renewable** or **nonrenewable**. Geologists can aid in locating the best places to exploit **renewable** resources (e.g. locating a dam), but are commonly tasked with finding **nonrenewable fossil fuels**. **Mineral** resources are also grouped in two categories: **metallic** and **nonmetallic**. **Minerals** have a wide variety of processes that concentrate them to economic levels, and are usually **mined** via surface or underground methods.

Take this quiz to check your comprehension of this chapter.

If you are using an offline version of this text, access the quiz for chapter 16 via the QR code.

An interactive H5P element has been excluded from this version of the text. You can view it online here:
https://pressbooks.lib.vt.edu/introearthscience/?p=973#h5p-110

Text References

1. Ague, Jay James, and George H. Brimhall. 1989. "Geochemical Modeling of Steady State Fluid Flow and Chemical Reaction during Supergene Enrichment of Porphyry Copper Deposits." *Economic Geology and the Bulletin of the Society of Economic Geologists* 84 (3). economicgeology.org: 506–28.

2. Arndt, N. T. 1994. "Chapter 1 Archean Komatiites." In *Developments in Precambrian Geology*, edited by K.C. Condie, 11:11–44. Elsevier.

3. Bárdossy, György, and Gerardus Jacobus Johannes Aleva. 1990. *Lateritic Bauxites*. Vol. 27. Elsevier Science Ltd.

4. Barrie, C. T. 1999. "Volcanic-Associated Massive Sulfide Deposits: Processes and Examples in Modern and Ancient Settings." Reviews in Economic Geology, v. 8. https://www.researchgate.net/profile/Michael_Perfit/publication/241276560_Geologic_petrologic_and_geochemical_relationships_between_magmatism_and_massive_sulfide_mineralization_along_the_eastern_Galapagos_Spreading_Center/links/02e7e51c8707bbfe9c000000.pdf.

5. Barrie, L. A., and R. M. Hoff. 1984. "The Oxidation Rate and Residence Time of Sulphur Dioxide in the Arctic Atmosphere." *Atmospheric Environment* 18 (12). Elsevier: 2711–22.

6. Bauquis, Pierre-René. 1998. "What Future for Extra Heavy Oil and Bitumen: The Orinoco Case." In *Paper Presented by TOTAL at the World Energy Congress*, 13:18.

7. Belloc, H. 1913. *The Servile State*. T.N. Foulis.

8. Blander, M., S. Sinha, A. Pelton, and G. Eriksson. 2011. "Calculations of the Influence of Additives on Coal Combustion Deposits." *Argonne National Laboratory, Lemont, Illinois*. enersol.pk, 315.

9. Boudreau, Alan E. 2016. "The Stillwater Complex, Montana–Overview and the Significance of Volatiles." *Mineralogical Magazine* 80 (4). Mineralogical Society: 585–637.

10. Bromfield, C. S., A. J. Erickson, M. A. Haddadin, and H. H. Mehnert. 1977. "Potassium-Argon Ages of Intrusion, Extrusion, and Associated Ore Deposits, Park City Mining District, Utah." *Economic Geology and the Bulletin of the Society of Economic Geologists* 72 (5). economicgeology.org: 837–48.

11. Brown, Valerie J. 2007. "Industry Issues: Putting the Heat on Gas." Environmental Health Perspectives 115 (2). ncbi.nlm.nih.gov: A76.

12. Cabri, Louis J., Donald C. Harris, and Thorolf W. Weiser. 1996. "Mineralogy and Distribution of Platinum-Group Mineral (PGM) Placer Deposits of the World." *Exploration and Mining Geology* 2 (5). infona.pl: 73–167.

13. Crutzen, Paul J., and Jos Lelieveld. 2001. "Human Impacts on Atmospheric Chemistry." *Annual Review of Earth and Planetary Sciences* 29 (1). Annual Reviews 4139 El Camino Way, PO Box 10139, Palo Alto, CA 94303-0139, USA: 17–45.

14. Delaney, M. L. 1998. "Phosphorus Accumulation in Marine Sediments and the Oceanic Phosphorus Cycle." *Global Biogeochemical Cycles* 12 (4). Wiley Online Library: 563–72.

15. Demaison, G. J., and G. T. Moore. 1980. "Anoxic Environments and Oil Source Bed Genesis." Organic Geochemistry 2 (1). Elsevier: 9–31.

16. Dott, Robert H., and Merrill J. Reynolds. 1969. "Sourcebook for Petroleum Geology." American Association of Petro-

leum Geologists Tulsa, Okla. http://archives.datapages.com/data/specpubs/method01/data/a072/a072/0001/0000/vi.htm.

17. Duffield, Wendell A. 2005. "Volcanoes, Geothermal Energy, and the Environment." *Volcanoes and the Environment*. Cambridge University Press, 304.

18. Einaudi, Marco T., and Donald M. Burt. 1982. "Introduction; Terminology, Classification, and Composition of Skarn Deposits." *Economic Geology and the Bulletin of the Society of Economic Geologists* 77 (4). economicgeology.org: 745–54.

19. Gandossi, Luca. 2013. "An Overview of Hydraulic Fracturing and Other Formation Stimulation Technologies for Shale Gas Production." *Eur. Commisison Jt. Res. Cent. Tech. Reports*. skalunudujos.lt. http://skalunudujos.lt/wp-content/uploads/an-overview-of-hydraulic-fracturing-and-other-stimulation-technologies.pdf.

20. Gordon, Mackenzie, Jr, Joshua I. Tracey Jr, and Miller W. Ellis. 1958. "Geology of the Arkansas Bauxite Region." pubs.er.usgs.gov. https://pubs.er.usgs.gov/publication/pp299.

21. Gordon, W. Anthony. 1975. "Distribution by Latitude of Phanerozoic Evaporite Deposits." *The Journal of Geology* 83 (6). journals.uchicago.edu: 671–84.

22. Haber, Fritz. 2002. "The Synthesis of Ammonia from Its Elements Nobel Lecture, June 2, 1920." *Resonance* 7 (9). Springer India: 86–94.

23. Hawley, Charles Caldwell. 2014. *A Kennecott Story: Three Mines, Four Men, and One Hundred Years, 1887-1997*. University of Utah Press.

24. Hirsch, Robert L., Roger Bezdek, and Robert Wendling. 2006. "Peaking of World Oil Production and Its Mitigation." *AIChE Journal. American Institute of Chemical Engineers* 52 (1). Wiley Subscription Services, Inc., A Wiley Company: 2–8.

25. Hitzman, M., R. Kirkham, D. Broughton, J. Thorson, and D. Selley. 2005. "The Sediment-Hosted Stratiform Copper Ore System." *Economic Geology and the Bulletin of the Society of Economic Geologists* 100th . eprints.utas.edu.au. http://eprints.utas.edu.au/705/.

26. Hofstra, Albert H., and Jean S. Cline. 2000. "Characteristics and Models for Carlin-Type Gold Deposits." *Reviews in Economic Geology* 13. Society of Economic Geologists: 163–220.

27. James, L. P. 1979. *Geology, Ore Deposits, and History of the Big Cottonwood Mining District, Salt Lake County, Utah*. Bulletin (Utah Geological and Mineral Survey). Utah Geological and Mineral Survey, Utah Department of Natural Resources.

28. Kim, Won-Young. 2013. "Induced Seismicity Associated with Fluid Injection into a Deep Well in Youngstown, Ohio." *Journal of Geophysical Research, [Solid Earth]* 118 (7). Wiley Online Library: 3506–18.

29. Klein, Cornelis. 2005. "Some Precambrian Banded Iron-Formations (BIFs) from around the World: Their Age, Geologic Setting, Mineralogy, Metamorphism, Geochemistry, and Origins." *The American Mineralogist* 90 (10). Mineralogical Society of America: 1473–99.

30. Laylin, James K. 1993. *Nobel Laureates in Chemistry, 1901-1992*. Chemical Heritage Foundation.

31. Leach, D. L., and D. F. Sangster. 1993. "Mississippi Valley-Type Lead-Zinc Deposits." *Mineral Deposit Modeling: Geological*. researchgate.net. https://www.researchgate.net/profile/Elisabeth_Rowan/publication/252527999_Genetic_link_between_Ouachita_foldbelt_tectonism_and_the_Mississippi_Valley-type_Lead-zinc_deposits_of_the_Ozarks/links/00b7d53c97ac2d6fe7000000.pdf.

32. Lehmann, Bernd. 2008. "Uranium Ore Deposits." *Rev. Econ. Geol. AMS Online 2008*. kenanaonline.com: 16–26.

33. London, David, and Daniel J. Kontak. 2012. "Granitic Pegmatites: Scientific Wonders and Economic Bonanzas." *Elements* 8 (4). GeoScienceWorld: 257–61.

34. Mancuso, Joseph J., and Ronald E. Seavoy. 1981. "Precambrian Coal or Anthraxolite; a Source for Graphite in High-Grade Schists and Gneisses." *Economic Geology and the Bulletin of the Society of Economic Geologists* 76 (4). economicgeology.org: 951–54.

35. McKenzie, Hermione, and Barrington Moore. 1970. "Social Origins of Dictatorship and Democracy." JSTOR. http://www.jstor.org/stable/27856441.

36. Needham, Joseph, Ling Wang, and Gwei Djen Lu. 1963. *Science and Civilisation in China*. Vol. 5. Cambridge University Press Cambridge.

37. Nuss, Philip, and Matthew J. Eckelman. 2014. "Life Cycle Assessment of Metals: A Scientific Synthesis." *PloS One* 9 (7). journals.plos.org: e101298.

38. Orton, E. 1889. *The Trenton Limestone as a Source of Petroleum and Inflammable Gas in Ohio and Indiana*. U.S. Government Printing Office.

39. Palmer, M. A., E. S. Bernhardt, W. H. Schlesinger, K. N. Eshleman, E. Foufoula-Georgiou, M. S. Hendryx, A. D. Lemly, et al. 2010. "Science and Regulation. Mountaintop Mining Consequences." *Science* 327 (5962). science.sciencemag.org: 148–49.

40. Pratt, Wallace Everette. 1942. *Oil in the Earth*. University of Kansas Press.

41. Quéré, C. Le, Robert Joseph Andres, T. Boden, T. Conway, R. A. Houghton, Joanna I. House, Gregg Marland, et al. 2013. "The Global Carbon Budget 1959–2011." *Earth System Science Data* 5 (1). Copernicus GmbH: 165–85.

42. Richards, J. P. 2003. "Tectono-Magmatic Precursors for Porphyry Cu-(Mo-Au) Deposit Formation." *Economic Geology and the Bulletin of the Society of Economic Geologists* 98 (8). economicgeology.org: 1515–33.

43. Rui-Zhong, Hu, Su Wen-Chao, Bi Xian-Wu, Tu Guang-Zhi, and Albert H. Hofstra. 2002. "Geology and Geochemistry of Carlin-Type Gold Deposits in China." *Mineralium Deposita* 37 (3-4). Springer-Verlag: 378–92.

44. Schröder, K-P, and Robert Connon Smith. 2008. "Distant Future of the Sun and Earth Revisited." *Monthly Notices of the Royal Astronomical Society* 386 (1). mnras.oxfordjournals.org: 155–63.

45. Semaw, Sileshi, Michael J. Rogers, Jay Quade, Paul R. Renne, Robert F. Butler, Manuel Dominguez-Rodrigo, Dietrich Stout, William S. Hart, Travis Pickering, and Scott W. Simpson. 2003. "2.6-Million-Year-Old Stone Tools and Associated Bones from OGS-6 and OGS-7, Gona, Afar, Ethiopia." *Journal of Human Evolution* 45 (2). Academic Press: 169–77.

46. Tappan, Helen, and Alfred R. Loeblich. 1970. "Geobiologic Implications of Fossil Phytoplankton Evolution and Time-Space Distribution." *Geological Society of America Special Papers* 127 (January). specialpapers.gsapubs.org: 247–340.

47. Taylor, E. L., T. N. Taylor, and M. Krings. 2009. *Paleobotany: The Biology and Evolution of Fossil Plants*. Elsevier Science.

48. Tissot, B. 1979. "Effects on Prolific Petroleum Source Rocks and Major Coal Deposits Caused by Sea-Level Changes." *Nature* 277. adsabs.harvard.edu: 463–65.

49. Vail, P. R., R. M. Mitchum Jr, S. Thompson III, R. G. Todd, J. B. Sangree, J. M. Widmier, J. N. Bubb, and W. G. Hatelid. 1977. "Seismic Stratigraphy and Global Sea Level Changes." *Seismic Stratigraphy-Applications to Hydrocarbon Exploration, Edited by Payton, CE, Tulsa, American Association of Petroleum Geologists Memoir* 26: 49–212.

50. Vogel, J. C. 1970. "Groningen Radiocarbon Dates IX." *Radiocarbon* 12 (2). journals.uair.arizona.edu: 444–71.

51. Willemse, J. 1969. "The Geology of the Bushveld Igneous Complex, the Largest Repository of Magmatic Ore Deposits in the World." *Economic Geology Monograph* 4: 1–22.

52. Wrigley, E. A. 1990. *Continuity, Chance and Change: The Character of the Industrial Revolution in England. Ellen McArthur Lectures ; 1987*. Cambridge University Press.

53. Youngquist, Walter. 1998. "Shale Oil–The Elusive Energy." *Hubbert Center Newsletter* 4.

Figure References

Figure 16.1: A Mode 1 Oldowan tool used for chopping. José-Manuel Benito Álvarez. 2007. Public domain. https://commons.wikimedia.org/wiki/File:Canto_tallado_2-Guelmim-Es_Semara.jpg

Figure 16.2: Map of world mining areas. KVDP. 2009. Public domain. https://commons.wikimedia.org/wiki/File:Simplified_world_mining_map_1.png

Figure 16.3: Hoover Dam provides hydroelectric energy and stores water for southern Nevada. Ubergirl. 2012. CC BY-SA 3.0. https://commons.wikimedia.org/wiki/File:Hoover_Dam,_Colorado_River.JPG

Figure 16.4: Natural, octahedral shape of diamond. USGS. 2003. Public domain. https://commons.wikimedia.org/wiki/File:Rough_diamond.jpg

Figure 16.5: Banded-iron formations are an important ore of iron (Fe). Wilson44691. 2008. Public domain. https://commons.wikimedia.org/wiki/File:MichiganBIF.jpg

Figure 16.6: Diagram illustrating the relative abundance of proven reserves, inferred reserves, resources, and undiscovered resources. Kindred Grey. 2022. CC BY 4.0.

Figure 16.7: McKelvey diagram showing different definitions for different degrees of concentration and understanding of mineral deposits. USGS. 1980. Public domain. https://commons.wikimedia.org/wiki/File:McKelveyDiagram.jpg

Figure 16.8: Bingham Canyon Mine, Utah. Doc Searls. 2016. CC BY 2.0. https://commons.wikimedia.org/wiki/File:Bingham_Canyon_mine_2016.jpg

Figure 16.9: A surface coal mine in Wyoming. Bureau of Land Management. Unknown date. Public domain. https://www.usgs.gov/news/science-snippet/earthword-thermal-maturity

Figure 16.10: Underground mining in Estonia of oil shale. Kaupo Kikkas. 2011. CC BY-SA 4.0. https://commons.wikimedia.org/wiki/File:VKG_Ojamaa_kaevandus.jpg

Figure 16.11: A phosphate smelting operation in Alabama, 1942. Alfred T. Palmer. 1942. Public domain. https://commons.wikimedia.org/wiki/File:TVA_phosphate_smelting_furnace.jpg

Figure 16.12: Coal power plant in Helper, Utah. David Jolley. 2007. CC BY-SA 3.0. https://commons.wikimedia.org/wiki/File:Castle_Gate_Power_Plant,_Utah_2007.jpg

Figure 16.13: Modern coral reefs and other highly-productive shallow marine environments are thought to be the sources of most petroleum resources. Toby Hudson. 2010. CC BY-SA 3.0. https://commons.wikimedia.org/wiki/File:Coral_Outcrop_Flynn_Reef.jpg

Figure 16.14: World oil reserves in 2013. GunnMap; generated with settings from Emilfaro and A5b. 2014. CC BY-SA 1.0. https://commons.wikimedia.org/wiki/File:Oil_Reserves.png

Figure 16.15: Examples of different forms of hydrocarbon traps: in the core region of anticlines. MagentaGreen. 2014. CC BY-SA 4.0. https://commons.wikimedia.org/wiki/File:Anticlinal_Oil_trap.png

Figure 16.16: The rising sea levels of transgressions create onlapping sediments, regressions create offlapping. Woudloper. 2009. CC BY-SA 1.0. https://commons.wikimedia.org/wiki/File:Offlap_%26_onlap_EN.svg

Figure 16.17: Tar sandstone from the Miocene Monterrey Formation of California. James St. John. 2015. CC BY 2.0. https://flic.kr/p/rECMTD

Figure 16.18: Global production of oil shale, 1880-2010. USGS. 2011. Public domain. https://commons.wikimedia.org/wiki/File:Production_of_oil_shale.png

Figure 16.19: Schematic diagram of fracking. Mikenorton. 2012. CC BY-SA 3.0. https://en.wikipedia.org/wiki/File:HydroFrac.png

Figure 16.20: USGS diagram of different coal rankings. USGS. 2009. Public domain. https://commons.wikimedia.org/wiki/File:Coal_Rank_USGS.png

Figure 16.21: Peat (also known as turf) consists of partially decayed organic matter. David Stanley. 2019. CC BY 2.0. https://commons.m.wikimedia.org/wiki/File:Peat_(49302157252).jpg

Figure 16.22: Anthracite coal, the highest grade of coal. USGS. 2007. Public domain. https://commons.wikimedia.org/wiki/File:Coal_anthracite.jpg

Figure 16.23: Gold-bearing quartz vein from California. James St. John. 2014. CC BY 2.0. https://commons.wikimedia.org/wiki/File:Mother_Lode_Gold_OreHarvard_mine_quartz-gold_vein.jpg

Figure 16.24: Layered intrusion of dark chromium-bearing minerals, Bushveld Complex, South Africa. kevinzim / Kevin Walsh. 2006. CC BY 2.0. https://commons.wikimedia.org/wiki/File:Chromitite_Bushveld_South_Africa.jpg

ORIGIN OF THE UNIVERSE AND OUR SOLAR SYSTEM

Learning Objectives

By the end of this chapter, students should be able to:

- Explain the formation of the **universe** and how we observe it.
- Understand the origin of our **solar system**.
- Describe how the objects in our **solar system** are identified, explored, and characterized.
- Describe the types of small bodies in our solar system, their locations, and how they formed.
- Describe the characteristics of the giant **planets**, terrestrial **planets**, and small bodies in the solar system.
- Explain what influences the **temperature** of a planet's surface.
- Explain why there is geological activity on some planets and not on others.
- Describe different methods for dating planets and the age of the solar system.
- Describe how the characteristics of **extrasolar** systems help us to model our own solar system.

The **universe** began 13.77 billion years ago when energy, matter, and space expanded from a single point. Evidence for the **big bang** is the cosmic "afterglow" from when the universe was still very dense, and red-shifted light from distant galaxies, which tell us the universe is still expanding.

The **big bang** produced hydrogen, helium, and lithium, but heavier elements come from nuclear **fusion** reactions in stars. Large stars make elements such as silicon, iron, and magnesium, which are important in forming terrestrial planets. Large stars explode as supernovae and scatter the elements into space.

Planetary systems begin with the collapse of a cloud of gas and dust. Material drawn to the center forms a star, and the remainder forms a disk around the star. Material within the disk clumps together to form planets. In our **solar system**, rocky planets are closer to the Sun, and ice and gas giants are farther away. This is because temperatures near the Sun were too high for ice to form, but silicate minerals and metals could solidify.

Early Earth was heated by **radioactive** decay, collisions with bodies from space, and gravitational compression. Heating melted Earth, causing molten metal to sink to Earth's center and form a core, and silicate melt to float to the surface and form the mantle and crust. A collision with a planet the size of Mars knocked debris into orbit around Earth, and the debris coalesced into the moon. Earth's atmosphere is the result of volcanic degassing, contributions by comets and meteorites, and photosynthesis.

The search for **exoplanets** has identified 12 planets that are similar in size to Earth and within the habitable zone of their stars. These are thought to be rocky worlds like Earth, but the compositions of these planets are not known for certain.

17.1 The Big Bang

According to the **big bang theory**, the **universe** blinked violently into existence 13.77 billion years ago. The big bang is often described as an explosion, but imagining it as an enormous fireball isn't accurate. The big bang involved a sudden expansion of matter, energy, and space from a single point. The kind of Hollywood explosion that might come to mind involves expansion of matter and energy *within* space, but during the big bang, space *itself* was created.

At the start of the big bang, the universe was too hot and dense to be anything but a sizzle of particles smaller than atoms, but as it expanded, it also cooled. Eventually some of the particles collided and stuck together. Those collisions produced hydrogen and helium, the most common **elements** in the universe, along with a small amount of lithium.

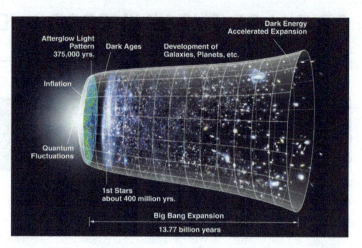

Figure 17.1: The big bang. The universe began 13.77 billion years ago with a sudden expansion of space, matter, and energy, and it continues to expand today.

You may wonder how a universe can be created out of nothing, or how we can know that the big bang happened at all. Creating a universe out of nothing is mostly beyond the scope of this chapter, but there is a way to think about it. The particles that make up the **universe** have opposites that cancel each other out, similar to the way that we can add the numbers 1 and –1 to get zero (also known as "nothing"). As far as the math goes, having zero is exactly the same as having a 1 and a –1. It is also exactly the same as having a 2 and a –2, a 3 and a –3, two –1s and a 2, and so on. In other words, *nothing* is really the potential for *something* if you divide it into its opposite parts. As for how we can know that the big bang happened at all, there are very good reasons to accept that it is indeed how our universe came to be.

17.1.1 Looking Back to the Early Stages of the Big Bang

The notion of seeing the past is often used metaphorically when we talk about ancient events, but in this case it is meant literally. In our everyday experience, when we watch an event take place, we perceive that we are watching it as it unfolds in real time. In fact, this isn't true. To see the event, light from that event must travel to our eyes. Light travels very rapidly, but it does not travel instantly. If we were watching a digital clock 1 m away from us change from 11:59 a.m. to 12:00 p.m., we would actually see it turn to 12:00 p.m. three billionths of a second after it happened. This isn't enough of a delay to cause us to be late for an appointment, but the **universe** is a very big place, and the "digital clock" in question is often much, much farther away. In fact, the universe is so big that it is convenient to describe distances in terms of **light years**, or the distance light travels in one year. What this means is that light from distant objects takes so long to get to us that we see those objects as they were at some considerable time in the past. For example, the star Proxima Centauri is 4.24 light years from the sun. If you viewed Proxima Centauri from Earth on January 1, 2015, you would actually see it as it appeared in early October 2010.

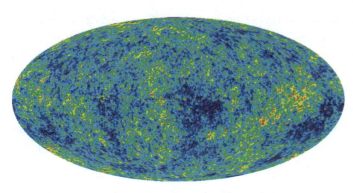

Figure 17.2: Cosmic microwave background (CMB) map of the sky, a baby picture of the universe. The CMB is light from 375,000 years after the big bang. The colors reveal variations in density. Red patches have the highest density and blue patches have the lowest density. Regions of higher density eventually formed the stars, planets, and other objects we see in space today.

We now have tools that are powerful enough to look deep into space and see the arrival of light from early in the universe's history. Astronomers can detect light from approximately 375,000 years after the big bang is thought to have occurred. Physicists tell us that if the big bang happened, then particles within the universe would still be very close together at this time. They would be so close that light wouldn't be able to travel far without bumping into another particle and getting scattered in another direction. The effect would be to fill the sky with glowing fog, the "afterglow" from the formation of the universe.

In fact, this is exactly what we see when we look at light from 375,000 years after the big bang. The fog is referred to as the **cosmic microwave background** (or CMB), and it has been carefully mapped throughout the sky. The map displays the cosmic microwave background as temperature variations, but these variations translate to differences in the density of matter in the early universe. The red patches are the highest density regions and the blue patches are the lowest density. Higher density regions represent the eventual beginnings of stars and planets. The map has been likened to a baby picture of the universe.

17.1.2 The Big Bang is Still Happening, and We Can See the Universe Expanding

Figure 17.3: Doppler effect. The ripples made in the direction the car is moving are closer together than the ripples behind the car.

The expansion that started with the big bang never stopped. It continues today, and we can see it happening by observing **galaxies** that large clusters of billions of stars, called galaxies, are moving away from us. (The exception is the Andromeda galaxy with which we are on a collision course.) The astronomer Edwin Hubble came to this conclusion when he observed that the light from other galaxies was red-shifted. The **red shift** is a consequence of the Doppler effect. This refers to how we see waves when the object that is creating the waves is moving toward us or away from us.

Sun

BAS11

Figure 17.4: Red shift in light from the supercluster BAS11 compared to the sun's light. Black lines represent wavelengths absorbed by atoms (mostly H and He). For BAS11 the black lines are shifted toward the red end of the spectrum compared to the Sun.

Before we get to the Doppler effect as it pertains to the red shift, let's see how it works on something more tangible. The swimming duckling is generating waves as it moves through the water. It is generating waves that move forward as well as back, but notice that the ripples ahead of the duckling are closer to each other than the ripples behind the duckling. The distance from one ripple to the next is called the **wavelength**. The wavelength is shorter in the direction that the duckling is moving, and longer as the duckling moves away.

When waves are in air as sound waves rather than in water as ripples, the different wavelengths manifest as sounds with different pitches—the short wavelengths have a higher pitch, and the long wavelengths have a lower pitch. This is why the pitch of a car's engine changes as the car races past you.

Take this quiz to check your comprehension of this section.

If you are using an offline version of this text, access the quiz for section 17.1 via the QR code.

An interactive H5P element has been excluded from this version of the text. You can view it online here:
https://pressbooks.lib.vt.edu/introearthscience/?p=1809#h5p-111

17.2 Overview of Our Planetary System[1]

The **solar system** consists of the Sun and many smaller objects: the planets, their **moons** and rings, and such "debris" as **asteroids**, **comets**, and dust. Decades of observation and spacecraft exploration have revealed that most of these objects formed together with the Sun about 4.5 billion years ago. They represent clumps of material that condensed from an enormous cloud of gas and dust. The central part of this cloud became the Sun, and a small fraction of the material in the outer parts eventually formed the other objects.

1. Source: OpenStax Astronomy (CC BY). Access for free at https://openstax.org/books/astronomy-2e/pages/
    7-1-overview-of-our-planetary-system

During the past 50 years, we have learned more about the solar system than anyone imagined before the space age. In addition to gathering information with powerful new telescopes, we have sent spacecraft directly to many members of the **planetary system**. (Planetary astronomy is the only branch of our science in which we can, at least vicariously, travel to the objects we want to study.) With evocative names such as *Voyager*, *Pioneer*, *Curiosity*, and *Pathfinder*, our robot explorers have flown past, orbited, or landed on every planet, returning images and data that have dazzled both astronomers and the public. In the process, we have also investigated two **dwarf planets**, hundreds of fascinating moons, four ring systems, a dozen asteroids, and several comets (smaller members of our solar system that we will discuss later).

Our probes have penetrated the **atmosphere** of Jupiter and landed on the surfaces of Venus, Mars, our Moon, Saturn's moon Titan, the asteroids Eros, Itokawa, Ryugu, and Bennu, and the Comet Churyumov-Gerasimenko (usually referred to as 67P). Humans have set foot on the Moon and returned samples of its surface soil for laboratory analysis. We have flown a helicopter drone on Mars. We have even discovered other places in our solar system that might be able to support some kind of life.

17.2.1 An Inventory

The Sun, a star that is brighter than about 80% of the stars in the Galaxy, is by far the most massive member of the solar system. It is an enormous ball about 1.4 million kilometers in diameter, with surface layers of incandescent gas and an interior temperature of millions of degrees. The Sun will be discussed in later chapters as our first, and best-studied, example of a star.

Figure 17.5: Astronauts on the Moon. The lunar lander and surface rover from the Apollo 15 mission are seen in this view of the one place beyond Earth that has been explored directly by humans.

Object	Percentage of total mass of solar system
Sun	99.8%
Jupiter	0.1%
Comets	0.0005-0.03% (estimate)
All other planets and dwarf planets	0.04%
Moons and rings	0.00005%
Asteroids	0.000002% (estimate)
Cosmic dust	0.0000001% (estimate)

Table 17.1: Mass of members of the solar system. Note that the Sun is by far the most massive member of the solar system.

Most of the material of the planets in the solar system is actually concentrated in the largest one, Jupiter, which is more massive than all the rest of the planets combined. Astronomers were able to determine the masses of the planets centuries ago using Kepler's laws of planetary motion and Newton's law of gravity to measure the planets' gravitational effects on one another or on moons that orbit them. Today, we make even more precise measurements of their masses by tracking their gravitational effects on the motion of spacecraft that pass near them.

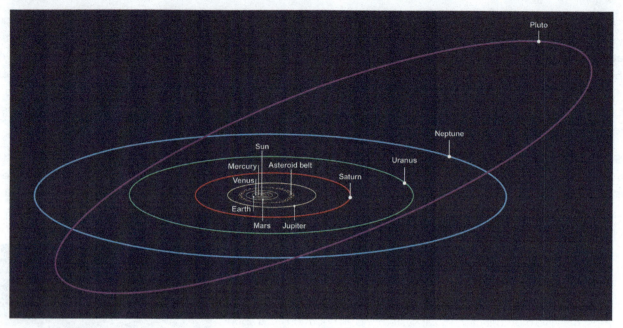

Figure 17.6: Orbits of the planets. All eight major planets orbit the Sun in roughly the same plane. Note that Pluto's orbit is not in the plane of the planets.

Beside Earth, five other planets were known to the ancients—Mercury, Venus, Mars, Jupiter, and Saturn—and two were discovered after the invention of the telescope: Uranus and Neptune. The eight planets all **revolve** in the same direction around the Sun. They orbit in approximately the same plane, like cars traveling on concentric tracks on a giant, flat race-course. Each planet stays in its own "traffic lane," following a nearly circular orbit about the Sun and obeying the "traffic" laws discovered by Galileo, Kepler, and Newton. Besides these planets, we have also been discovering smaller worlds beyond Neptune that are called trans-Neptunian objects or TNOs. The first to be found, in 1930, was Pluto, but others have been discovered during the twenty-first century. One of them, Eris, is about the same size as Pluto and has at least one moon (Pluto has five known moons). The largest TNOs are also classed as *dwarf planets*, as is the largest asteroid, Ceres. To date, more than 2600 of these TNOs have been discovered, and one, called Arrokoth, was explored by the New Horizons spacecraft.

Each of the **planets** and **dwarf planets** also **rotates** (spins) about an axis running through it, and in most cases the direction of rotation is the same as the direction of revolution about the Sun. The exceptions are Venus, which rotates backward very slowly (that is, in a retrograde direction), and Uranus and Pluto, which also have strange rotations, each spinning about an axis tipped nearly on its side. We do not yet know the spin orientations of Eris, Haumea, and Makemake.

The four planets closest to the Sun (Mercury through Mars) are called the inner or terrestrial planets. Often, the Moon is also discussed as a part of this group, bringing the total of terrestrial *objects* to five (we generally call Earth's satellite "the Moon," with a capital M, and the other satellites "moons," with lowercase m's). The terrestrial planets are relatively small worlds, composed primarily of rock and metal. All of them have solid surfaces that bear the records of their geological history in the forms of **craters**, **mountains**, and **volcanoes**.

Figure 17.7: Surface of Mercury. The pockmarked face of the terrestrial world of Mercury is more typical of the inner planets than the watery surface of Earth. This image shows Caravaggio, a double-ring impact basin (approximately 160 kilometers in diameter), with another large impact crater on its south-south-western side.

Figure 17.8: The four giant planets. This montage shows the four giant planets: Jupiter, Saturn, Uranus, and Neptune. Below them, Earth is shown to scale in terms of size. Distance is not to scale.

The next four planets (Jupiter through Neptune) are much larger and are composed primarily of lighter ices, liquids, and gases. We call these four the Jovian planets (after "Jove," another name for Jupiter in mythology) or giant planets—a name they richly deserve. About 1,300 Earths could fit inside Jupiter, for example. These planets do not have solid surfaces on which future explorers might land. They are more like vast, spherical oceans with much smaller, dense cores.

Near the outer edge of the system lies Pluto, which was the first of the distant icy worlds to be discovered beyond Neptune (Pluto was visited by a spacecraft, the NASA New Horizons mission, in 2015).

Figure 17.9: This intriguing image from the New Horizons spacecraft, taken when it flew by the dwarf planet Pluto in July 2015, shows some of its complex surface features. The rounded white area is called the Sputnik Plain, after humanity's first spacecraft.

Name	Distance from Sun (AU)	Revolution period (y)	Diameter (km)	Mass (10^{23} kg)	Density (g/cm^3)
Mercury	0.39	0.24	4,878	3.3	5.4
Venus	0.72	0.62	12,120	48.7	5.2
Earth	1.00	1.00	12,756	59.8	5.5
Mars	1.52	1.88	6,787	6.4	3.9
Jupiter	5.20	11.86	142,984	18,991	1.3
Saturn	9.54	29.46	120,536	5,686	0.7
Uranus	19.18	84.07	51,118	866	1.3
Neptune	30.06	164.82	49,660	1,030	1.6

Table 17.2: The planets.

The outermost part of the solar system is known as the **Kuiper belt**, which is a scattering of rocky and icy bodies. Beyond that is the **Oort cloud**, a zone filled with small and dispersed ice traces. These two locations are where most **comets** form and continue to orbit, and objects found here have relatively irregular orbits compared to the rest of the solar system. Pluto, formerly the ninth planet, is located in this region of space. The XXVIth General Assembly of the International Astronomical Union (IAU) stripped Pluto of planetary status in 2006 because scientists discovered an object more massive than Pluto, which they named Eris. The IAU decided against including Eris as a planet, and therefore, excluded Pluto as well.

The IAU narrowed the definition of a planet to three criteria: 1) it must orbit a star (in our cosmic neighborhood, the Sun), 2) it must be big enough to have enough gravity to force it into a spherical shape, and 3) it must be big enough that its gravity cleared away any other objects of a similar size near its orbit around the Sun. Pluto passed the first two parts of the definition, but not the third. Pluto and Eris are currently classified as dwarf planets.

17.2.2 Smaller Members of the Solar System

Most of the planets are accompanied by one or more **moons**; only Mercury and Venus move through space alone. There are more than 210 known moons orbiting planets and dwarf planets, and undoubtedly many other small ones remain undiscovered. The largest of the moons are as big as small planets and just as interesting. In addition to our Moon, they include the four largest moons of Jupiter (called the Galilean moons, after their discoverer) and the largest moons of Saturn and Neptune (confusingly named Titan and Triton).

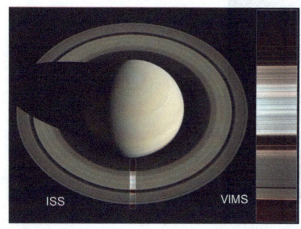

Figure 17.10: Saturn and its A, B, and C rings in visible and (inset) infrared light. In the false-color IR view, greater water ice content and larger grain size lead to blue-green color, while greater non-ice content and smaller grain size yield a reddish hue.

Each of the giant planets also has rings made up of countless small bodies ranging in size from mountains to mere grains of dust, all in orbit about the equator of the planet. The bright rings of Saturn are, by far, the easiest to see. They are among the most beautiful sights in the solar system. But, all four ring systems are interesting to scientists because of their complicated forms, influenced by the pull of the moons that also orbit these giant planets.

The solar system has many other less-conspicuous members. Another group is the **asteroids**, rocky bodies that orbit the Sun like miniature planets, mostly in the space between Mars and Jupiter (although some do cross the orbits of planets like Earth). Most asteroids are remnants of the initial population of the solar system that existed before the planets themselves formed. Some of the smallest moons of the planets, such as the moons of Mars, are very likely captured asteroids.

Another class of small bodies is composed mostly of ice, made of frozen gases such as water, carbon dioxide, and carbon monoxide; these objects are called **comets**. Comets also are remnants from the formation of the solar system, but they were formed and continue (with rare exceptions) to orbit the Sun in distant, cooler regions—stored in a sort of cosmic deep freeze. This is also the realm of the larger icy worlds, called dwarf planets.

Finally, there are countless grains of broken rock, which we call cosmic dust, scattered throughout the solar system. When these particles enter Earth's **atmosphere** (as millions do each day), they burn up, producing a brief flash of light in the night sky known as a **meteor** (meteors are often referred to as shooting stars). Occasionally, some larger chunk of rocky or metallic material survives its passage through the atmosphere and lands on Earth. Any piece that strikes the ground is known as a **meteorite**. You can see meteorites on display in many natural history museums and can sometimes even purchase pieces of them from gem and mineral dealers.

Figure 17.11: Asteroid Eros. This small Earth-crossing asteroid image was taken by the NEAR-Shoemaker spacecraft from an altitude of about 100 kilometers. This view of the heavily cratered surface is about 10 kilometers wide. The spacecraft orbited Eros for a year before landing gently on its surface.

Figure 17.12: Comet Churyumov-Gerasimenko (67P). This approximately true color image of the comet was taken by the Rosetta spacecraft on August 6, 2014, at a distance of 120 kilometers. There is surprisingly little color variation across the surface of the comet.

17.2.3 A Scale Model of the Solar System

Astronomy often deals with dimensions and distances that far exceed our ordinary experience. What does 1.4 billion kilometers—the distance from the Sun to Saturn—really mean to anyone? It can be helpful to visualize such large systems in terms of a scale model.

In our imaginations, let us build a scale model of the solar system, adopting a scale factor of 1 billion (10^9)—that is, reducing the actual solar system by dividing every dimension by a factor of 10^9. Earth, then, has a **diameter** of 1.3 centimeters, about the size of a grape. The Moon is a pea orbiting this at a distance of 40 centimeters, or a little more than a foot away. The Earth-Moon system fits into a standard backpack.

In this model, the Sun is nearly 1.5 meters in diameter, about the average height of an adult, and our Earth is at a distance of 150 meters—about one city block—from the Sun. Jupiter is five blocks away from the Sun, and its diameter is 15 centimeters, about the size of a very large grapefruit. Saturn is 10 blocks from the Sun; Uranus, 20 blocks; and Neptune, 30 blocks. Pluto, with a distance that varies quite a bit during its 249-year orbit, is currently just beyond 30 blocks and getting farther with time. Most of the moons of the outer solar system are the sizes of various kinds of seeds orbiting the grapefruit, oranges, and lemons that represent the outer planets.

In our scale model, a human is reduced to the dimensions of a single atom, and cars and spacecraft to the size of molecules. Sending the Voyager spacecraft to Neptune involves navigating a single molecule from the Earth–grape toward a lemon 5 kilometers away with an accuracy equivalent to the width of a thread in a spider's web.

If that model represents the solar system, where would the nearest stars be? If we keep the same scale, the closest stars would be tens of thousands of kilometers away. If you built this scale model in the city where you live, you would have to place the representations of those stars on the other side of Earth or beyond.

By the way, model solar systems like the one we just presented have been built in cities throughout the world. In Sweden, for example, Stockholm's huge Globe Arena has become a model for the Sun, and Pluto is represented by a 12-centimeter sculpture in the small town of Delsbo, 300 kilometers away. Another model solar system is in Washington on the Mall between the White House and Congress (perhaps proving they are worlds apart?).

Take this quiz to check your comprehension of this section.

If you are using an offline version of this text, access the quiz for section 17.2 via the QR code.

 An interactive H5P element has been excluded from this version of the text. You can view it online here:
https://pressbooks.lib.vt.edu/introearthscience/?p=1809#h5p-112

17.3 Composition and Structure of Planets[2]

The fact that there are two distinct kinds of planets—the rocky terrestrial planets and the gas-rich Jovian planets—leads us to believe that they formed under different conditions. Certainly their compositions are dominated by different **elements**. Let us look at each type in more detail.

17.3.1 The Giant Planets

The two largest planets, Jupiter and Saturn, have nearly the same chemical makeup as the Sun; they are composed primarily of the two **elements** hydrogen and helium, with 75% of their mass being hydrogen and 25% helium. On Earth, both hydrogen and helium are gases, so Jupiter and Saturn are sometimes called gas planets. But, this name is misleading. Jupiter and Saturn are so large that the gas is compressed in their interior until the hydrogen becomes a liquid. Because the bulk of both planets consists of compressed, liquefied hydrogen, we should really call them liquid planets.

2. Source: OpenStax Astronomy (CC BY). Access for free at https://openstax.org/books/astronomy/pages/7-2-composition-and-structure-of-planets

Under the force of gravity, the heavier elements sink toward the inner parts of a liquid or gaseous planet. Both Jupiter and Saturn, therefore, have cores composed of heavier rock, metal, and ice, but we cannot see these regions directly. In fact, when we look down from above, all we see is the atmosphere with its swirling clouds. We must infer the existence of the denser core inside these planets from studies of each planet's gravity.

Uranus and Neptune are much smaller than Jupiter and Saturn, but each also has a core of rock, metal, and ice. Uranus and Neptune were less efficient at attracting hydrogen and helium gas, so they have much smaller atmospheres in proportion to their cores.

Chemically, each giant planet is dominated by hydrogen and its many compounds. Nearly all the oxygen present is combined chemically with hydrogen to form water (H_2O). Chemists call such a hydrogen-dominated composition ***reduced***. Throughout the outer solar system, we find abundant water (mostly in the form of ice) and reducing chemistry.

Figure 17.13: Jupiter with its moon Europa on the left. Earth's diameter is 11 times smaller than Jupiter, and 4 times larger than Europa.

17.3.2 The Terrestrial Planets

The terrestrial planets are quite different from the giants. In addition to being much smaller, they are composed primarily of rocks and metals. These, in turn, are made of **elements** that are less common in the **universe** as a whole. The most abundant rocks, called **silicates**, are made of silicon and oxygen, and the most common **metal** is iron. We can tell from their densities that Mercury has the greatest proportion of metals (which are denser) and the Moon has the lowest. Earth, Venus, and Mars all have roughly similar bulk compositions: about one third of their mass consists of iron-nickel or iron-sulfur combinations; two thirds is made of silicates. Because these planets are largely composed of oxygen compounds (such as the silicate minerals of their crusts), their chemistry is said to be ***oxidized***.

When we look at the internal structure of each of the terrestrial planets, we find that the densest metals are in a central core, with the lighter silicates near the surface. If these planets were liquid, like the giant planets, we could understand this effect as the result the sinking of heavier elements due to the pull of gravity. This leads us to conclude that, although the terrestrial planets are solid today, at one time they must have been hot enough to melt.

Differentiation is the process by which gravity helps separate a planet's interior into layers of different compositions and densities. The heavier metals sink to form a core, while the lightest minerals float to the surface to form a crust. Later, when the planet cools, this layered structure is preserved. In order for a rocky planet to differentiate, it must be heated to the melting point of rocks, which is typically more than 1300 K.

17.3.3 Moons, Asteroids, and Comets

Chemically and structurally, Earth's Moon is like the terrestrial planets, but most moons are in the outer solar system, and they have compositions similar to the cores of the giant planets around which they orbit. The three largest moons—Ganymede and Callisto in the Jovian system, and Titan in the Saturnian system—are composed half of frozen water, and half of rocks and metals. Most of these moons **differentiated** during formation, and today they have cores of rock and metal, with upper layers and crusts of very cold and—thus very hard—ice.

Most of the **asteroids** and **comets**, as well as the smallest **moons**, were probably never heated to the melting point. However, some of the largest asteroids, such as Vesta, appear to be differentiated; others are fragments from differentiated bodies. Many of the smaller objects seem to be fragments or rubble piles that are the result of collisions. Because most asteroids and comets retain their original composition, they represent relatively unmodified material dating back to the time of the formation of the solar system. In a sense, they act as chemical fossils, helping us to learn about a time long ago whose traces have been erased on larger worlds.

17.3.4 Temperatures: Going to Extremes

Generally speaking, the farther a planet or moon is from the Sun, the cooler its surface. The planets are heated by the radiant energy of the Sun, which gets weaker with the square of the distance. You know how rapidly the heating effect of a fireplace or an outdoor radiant heater diminishes as you walk away from it; the same effect applies to the Sun. Mercury, the closest planet to the Sun, has a blistering surface temperature that ranges from 280–430 °C on its sunlit side, whereas the surface temperature on Pluto is only about –220 °C, colder than liquid air.

Figure 17.14: Jupiter's moon Ganymede. The brownish gray color of the surface indicates a dusty mixture of rocky material and ice. The bright spots are places where recent impacts have uncovered fresh ice from underneath.

Mathematically, the **temperatures** decrease approximately in proportion to the square root of the distance from the Sun. Pluto is about 30 **AU** at its closest to the Sun (or 100 times the distance of Mercury) and about 49 AU at its farthest from the Sun. Thus, Pluto's temperature is less than that of Mercury by the square root of 100, or a factor of 10: from 500 **K** to 50 K.

In addition to its distance from the Sun, the surface temperature of a planet can be influenced strongly by its **atmosphere**. Without our atmospheric insulation (the greenhouse effect, which keeps the heat in), the oceans of Earth would be permanently frozen. Conversely, if Mars once had a larger atmosphere in the past, it could have supported a more temperate climate than it has today. Venus is an even more extreme example, where its thick atmosphere of carbon dioxide acts as insulation, reducing the escape of heat built up at the surface, resulting in temperatures greater than those on Mercury. Today, Earth is the only planet where surface temperatures generally lie between the freezing and boiling points of water. As far as we know, Earth is the only planet to support life.

17.3.5 Dating Planetary Surfaces[3]

How do we know the age of the surfaces we see on planets and moons? If a world has a surface (as opposed to being mostly gas and liquid), astronomers have developed some techniques for estimating how long ago that surface solidified. Note that the age of these surfaces is not necessarily the age of the planet as a whole. On geologically active objects (including Earth), vast outpourings of molten rock or the erosive effects of water and ice, which we call planet **weathering**, have erased evidence of earlier epochs and present us with only a relatively young surface for investigation.

One way to estimate the age of a surface is by counting the number of impact **craters**. This technique works because the rate at which impacts have occurred in the solar system has been roughly constant for several billion years. Thus, in the absence of forces to eliminate craters, the number of craters is simply proportional to the length of time the surface has been exposed. This technique has been applied successfully to many solid **planets** and **moons**.

---

3. Source: OpenStax Astronomy (CC BY). Access for free at https://openstax.org/books/astronomy/pages/ 7-3-dating-planetary-surfaces

Bear in mind that crater counts can tell us only the time since the surface experienced a major change that could modify or erase preexisting craters. Estimating ages from crater counts is a little like walking along a sidewalk in a snowstorm after the snow has been falling steadily for a day or more. You may notice that in front of one house the snow is deep, while next door the sidewalk may be almost clear. Do you conclude that less snow has fallen in front of Ms. Jones' house than Mr. Smith's? More likely, you conclude that Jones has recently swept the walk clean and Smith has not. Similarly, the numbers of craters indicate how long it has been since a planetary surface was last "swept clean" by ongoing lava flows or by molten materials ejected when a large impact happened nearby.

Still, astronomers can use the numbers of craters on different parts of the same world to provide important clues about how regions on that world evolved. On a given planet or moon, the more heavily cratered terrain will generally be older (that is, more time will have elapsed there since something swept the region clean).

Figure 17.15: Our cratered Moon. This composite image of the Moon's surface was made from many smaller images taken between November 2009 and February 2011 by the Lunar Reconnaissance Orbiter (LRO) and shows craters of many different sizes.

17.3.6 Radioactive Rocks

Another way to trace the history of a solid world is to measure the age of individual rocks. After samples were brought back from the Moon by Apollo astronauts, the techniques that had been developed to date rocks on Earth were applied to rock samples from the Moon to establish a geological chronology for the Moon. Furthermore, a few samples of material from the Moon, Mars, and the large asteroid Vesta have fallen to Earth as **meteorites** and can be examined directly.

Scientists measure the age of rocks using the properties of natural **radioactivity**. Around the beginning of the twentieth century, physicists began to understand that some atomic nuclei are not stable but can split apart (decay) spontaneously into smaller nuclei. The process of radioactive decay involves the emission of particles such as **electrons**, or of radiation in the form of **gamma rays**.

For any one radioactive nucleus, it is not possible to predict when the decay process will happen. Such decay is random in nature, like the throw of dice: as gamblers have found all too often, it is impossible to say just when the dice will come up 7 or 11. But, for a very large number of dice tosses, we can calculate the odds that 7 or 11 will come up. Similarly, if we have a very large number of radioactive atoms of one type (say, uranium), there is a specific time period, called its half-life, during which the chances are fifty-fifty that decay will occur for any of the nuclei.

A particular nucleus may last a shorter or longer time than its half-life, but in a large sample, almost exactly half of the nuclei will have decayed after a time equal to one half-life. Half of the remaining nuclei will have decayed after two half-lives pass, leaving only one half of a half—or one quarter—of the original sample.

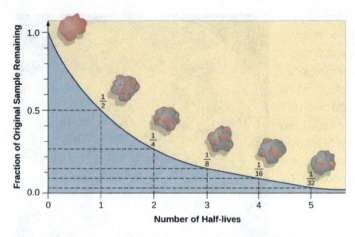

Figure 17.16: Radioactive decay. This graph shows (in pink) the amount of a radioactive sample that remains after several half-lives have passed. After one half-life, half the sample is left; after two half-lives, one half of the remainder (or one quarter) is left; and after three half-lives, one half of that (or one eighth) is left. Note that, in reality, the decay of radioactive elements in a rock sample would not cause any visible change in the appearance of the rock; the splashes of color are shown here for conceptual purposes only.

If you had 1 gram of pure radioactive nuclei with a half-life of 100 years, then after 100 years you would have 1/2 gram; after 200 years, 1/4 gram; after 300 years, only 1/8 gram; and so forth. However, the material does not disappear. Instead, the radioactive atoms are replaced with their decay products. Sometimes the radioactive atoms are called **parents** and the decay products are called **daughter** elements.

In this way, radioactive elements with half-lives we have determined can provide accurate nuclear clocks. By comparing how much of a radioactive **parent** element is left in a rock to how much of its **daughter** products have accumulated, we can learn how long the decay process has been going on and hence how long ago the rock formed. The following table summarizes the decay reactions used most often to date lunar and terrestrial rocks.

Parent	Daughter	Half-life (billions of years)
Samarium-147	Neodymium-143	106
Rubidium-87	Strontium-87	48.8
Thorium-232	Lead-208	14
Uranium-238	Lead-206	4.47
Potassium-40	Argon-40	1.31

Table 17.3: Radioactive decay reaction used to date rocks. The number after each element is its atomic weight, equal to the number of protons plus neutrons in its nucleus. This specifies the isotope of the element, different isotopes of the same element differ in the number of neutrons.

When astronauts first flew to the Moon, one of their most important tasks was to bring back lunar rocks for radioactive age-dating. Until then, astronomers and geologists had no reliable way to measure the age of the lunar surface. Counting **craters** had let us calculate relative ages (for example, the heavily cratered lunar highlands were older than the dark lava plains), but scientists could not measure the actual age in years. Some thought that the ages were as young as those of Earth's surface, which has been resurfaced by many geological events. For the Moon's surface to be so young would imply active geology on our satellite. Only in 1969, when the first Apollo samples were dated, did we learn that the Moon is an ancient, geologically dead world. Using such dating techniques, we have been able to determine the ages of both Earth and the Moon: each was formed about 4.5 billion years ago (although, as we shall see, Earth probably formed earlier than the Moon).

We should also note that the decay of radioactive nuclei generally releases energy in the form of heat. Although the energy from a single nucleus is not very large (in human terms), the enormous numbers of radioactive nuclei in a planet or moon (especially early in its existence) can be a significant source of internal energy for that world. Geologists estimate that about half of Earth's current internal heat budget comes from the decay of radioactive isotopes in its interior.

Take this quiz to check your comprehension of this section.

If you are using an offline version of this text, access the quiz for section 17.3 via the QR code.

An interactive H5P element has been excluded from this version of the text. You can view it online here:
https://pressbooks.lib.vt.edu/introearthscience/?p=1809#h5p-113

17.4 Origin of the Solar System[4]

Much of **astronomy** is motivated by a desire to understand the origin of things: to find at least partial answers to age-old questions of where the **universe**, the Sun, Earth, and we ourselves came from. Each planet and moon is a fascinating place that may stimulate our imagination as we try to picture what it would be like to visit. Taken together, the members of the **solar system** preserve patterns that can tell us about the formation of the entire system. As we begin our exploration of the planets, we want to introduce our modern picture of how the solar system formed.

The recent discovery of thousands of planets in orbit around other stars has shown astronomers that many **exoplanetary** systems can be quite different from our own solar system. For example, it is common for these systems to include planets intermediate in size between our terrestrial and giant planets. These are often called *superearths*. Some **exoplanet** systems even have giant planets close to the star, reversing the order we see in our system.

17.4.1 Looking for Patterns

One way to approach our question of origin is to look for regularities among the planets. We found, for example, that all the planets lie in nearly the same plane and **revolve** in the same direction around the Sun. The Sun also spins in the same direction about its own axis. Astronomers interpret this pattern as evidence that the Sun and planets formed together from a spinning cloud of gas and dust that we call the **solar nebula**.

---

4. Source: OpenStax Astronomy (CC BY). Access for free at https://openstax.org/books/astronomy/pages/7-4-origin-of-the-solar-system

The composition of the planets gives another clue about origins. Spectroscopic analysis allows us to determine which **elements** are present in the Sun and the planets. The Sun has the same hydrogen-dominated composition as Jupiter and Saturn, and therefore appears to have been formed from the same reservoir of material. In comparison, the terrestrial planets and our Moon are relatively deficient in the light gases and the various ices that form from the common elements oxygen, carbon, and nitrogen. Instead, on Earth and its neighbors, we see mostly the rarer heavy elements such as iron and silicon. This pattern suggests that the processes that led to planet formation in the inner solar system must somehow have excluded much of the lighter materials that are common elsewhere. These lighter materials must have escaped, leaving a residue of heavy stuff.

Figure 17.17: NASA artist's conception of various planet formation processes, including exocomets and other planetesimals, around Beta Pictoris, a very young type A V star.

The reason for this is not hard to guess, bearing in mind the heat of the Sun. The inner planets and most of the **asteroids** are made of rock and metal, which can survive heat, but they contain very little ice or gas, which **evaporate** when temperatures are high (to see what we mean, just compare how long a rock and an ice cube survive when they are placed in the sunlight). In the outer solar system, where it has always been cooler, the planets and their moons, as well as icy **dwarf planets** and **comets**, are composed mostly of ice and gas.

17.4.2 The Evidence from Far Away

A second approach to understanding the origins of the **solar system** is to look outward for evidence that other systems of planets are forming elsewhere. We cannot look back in time to the formation of our own system, but many stars in space are much younger than the Sun. In these systems, the processes of planet formation might still be accessible to direct observation. We observe that there are many other "**solar nebulas**" or *circumstellar disks*—flattened, spinning clouds of gas and dust surrounding young stars. These disks resemble our own solar system's initial stages of formation billions of years ago.

17.4.3 Building Planets

Circumstellar disks are a common occurrence around very young stars, suggesting that disks and stars form together. Astronomers can use theoretical calculations to see how solid bodies might form from the gas and dust in these disks as they cool. These models show that material begins to coalesce first by forming smaller objects, precursors of the planets, which we call **planetesimals**.

Today's fast computers can simulate the way millions of planetesimals, probably no larger than 100 kilometers in diameter, might gather together under their mutual gravity to form the planets we see today. We are beginning to understand that this process was a violent one, with planetesimals crashing into each other and sometimes even disrupting the growing planets themselves. As a consequence of those violent impacts (and the heat from radioactive elements in them), all the planets were heated until they were liquid and gas, and therefore differentiated, which helps explain their present internal structures.

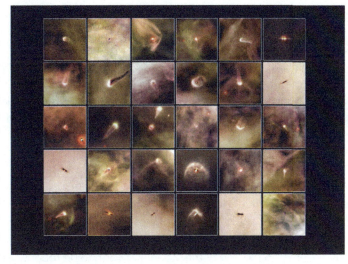

Figure 17.18: Atlas of Planetary Nurseries. These Hubble Space Telescope photos show sections of the Orion Nebula, a relatively close-by region where stars are currently forming. Each image shows an embedded circumstellar disk orbiting a very young star. Seen from different angles, some are energized to glow by the light of a nearby star while others are dark and seen in silhouette against the bright glowing gas of the Orion Nebula. Each is a contemporary analog of our own solar nebula—a location where planets are probably being formed today.

The process of impacts and collisions in the early solar system was complex and, apparently, often random. The **solar nebula** model can explain many of the regularities we find in the solar system, but the random collisions of massive collections of planetesimals could be the reason for some exceptions to the "rules" of solar system behavior. For example, why do the planets Uranus and Pluto spin on their sides? Why does Venus spin slowly and in the opposite direction from the other planets? Why does the composition of the Moon resemble Earth in many ways and yet exhibit substantial differences? The answers to such questions probably lie in enormous collisions that took place in the solar system long before life on Earth began.

Today, some 4.5 billion years after its origin, the solar system is—thank goodness—a much less violent place. However, some planetesimals have continued to interact and collide, and their fragments move about the solar system as roving "transients" that can make trouble for the established members of the Sun's family, such as our own Earth.

Take this quiz to check your comprehension of this section.

If you are using an offline version of this text, access the quiz for section 17.4 via the QR code.

An interactive H5P element has been excluded from this version of the text. You can view it online here:
https://pressbooks.lib.vt.edu/introearthscience/?p=1809#h5p-114

Summary

Our **solar system** currently consists of the Sun, eight planets, five dwarf planets, nearly 200 known **moons**, and a host of smaller objects. The planets can be divided into two groups: the inner terrestrial planets and the outer giant planets. Smaller members of the solar system include **asteroids** (including the dwarf planet Ceres), which are rocky and metallic objects found mostly between Mars and Jupiter; **comets**, which are made mostly of frozen gases and generally orbit far from the Sun; and countless smaller grains of cosmic dust. When a **meteor** survives its passage through our atmosphere and falls to Earth, we call it a **meteorite**.

The ages of the surfaces of objects in the solar system can be estimated by counting craters: on a given world, a more heavily cratered region will generally be older than one that is less cratered. We can also use samples of rocks with **radioactive** elements in them to obtain the time since the layer in which the rock formed last solidified. The **half-life** of a radioactive element is the time it takes for half the sample to decay; we determine how many half-lives have passed by how much of a sample remains the radioactive element and how much has become the decay product. In this way, we have estimated the age of the Moon and Earth to be roughly 4.5 billion years.

Regularities among the planets have led astronomers to hypothesize that the Sun and the planets formed together in a giant, spinning cloud of gas and dust called the **solar nebula**. Astronomical observations show tantalizingly similar circumstellar disks around other stars. Within the solar nebula, material first coalesced into **planetesimals**; many of these gathered together to make the planets and moons. The remainder can still be seen as comets and asteroids. Probably all planetary systems have formed in similar ways, but many **exoplanet** systems have evolved along quite different paths.

Take this quiz to check your comprehension of this chapter.

If you are using an offline version of this text, access the quiz for chapter 17 via the QR code.

An interactive H5P element has been excluded from this version of the text. You can view it online here:
https://pressbooks.lib.vt.edu/introearthscience/?p=1809#h5p-115

Text References

Parts of this chapter are from OpenStax's Astronomy (chapter 7). 2016. CC BY 4.0.

Chapter 17 Origin of Earth and the Solar System (CC BY 4.0) by Karla Panchuk was added from Earle, Steven (2019) Physical Geology, 2nd edition. BC Campus https://opentextbc.ca/physicalgeology2ed/chapter/22-1-starting-with-a-big-bang

Figure References

Figure 17.1: The big bang. NASA/WMAP Science Team. 2006. Public domain. https://en.wikipedia.org/wiki/File:CMB_Timeline300_no_WMAP.jpg

Figure 17.2: Cosmic microwave background (CMB) map of the sky, a baby picture of the universe. NASA / WMAP Science Team. 2012. Public domain. https://commons.wikimedia.org/wiki/File:Ilc_9yr_moll4096.png

Figure 17.3: Doppler effect. Charly Whisky. 2007. CC BY-SA 3.0. https://commons.wikimedia.org/wiki/File%3ADopplerfrequenz.gif

Figure 17.4: Red shift in light from the supercluster BAS11 compared to the sun's light. Kindred Grey. 2022. CC BY 4.0. Includes Duck by parkjisun from Noun Project (Noun Project license).

Figure 17.5: Astronauts on the Moon. NASA Johnson Space Center; Restored by Bammesk. 1971. Public domain. https://en.wikipedia.org/wiki/File:AS15-88-11866_-_Apollo_15_flag,_rover,_LM,_Irwin_-_restoration1.jpg

Figure 17.6: Orbits of the planets. Arabik4892. 2022. CC BY-SA 4.0. https://commons.wikimedia.org/wiki/File:Planet_Orbits.jpg

Figure 17.7: Surface of Mercury. NASA/Johns Hopkins University Applied Physics Laboratory/Carnegie Institution of Washington. 2009. Public domain. https://commons.wikimedia.org/wiki/File:Mercury_Double-Ring_Impact_Basin.png

Figure 17.8: The four giant planets. Solar System Exploration, NASA. 2008. Public domain. https://commons.wikimedia.org/wiki/File:Gas_planet_size_comparisons.jpg

Figure 17.9: This intriguing image from the New Horizons spacecraft, taken when it flew by the dwarf planet Pluto in July 2015, shows some of its complex surface features. NASA / Johns Hopkins University Applied Physics Laboratory / Southwest Research Institute. 2015. Public domain. https://en.wikipedia.org/wiki/File:Pluto-01_Stern_03_Pluto_Color_TXT.jpg

Figure 17.10: Saturn and its A, B, and C rings in visible and (inset) infrared light. NASA/JPL-Caltech/Space Science Institute/ G. Ugarkovic (ISS), NASA/JPL-Caltech/University of Arizona/CNRS/LPG-Nantes (VIMS). 2019. Public domain. https://commons.wikimedia.org/wiki/File:PIA23170-Saturn-Rings-IR-Map-20190613.jpg

Figure 17.11: Asteroid Eros. NASA/JPL/JHUAPL. 2000. Public domain. https://commons.wikimedia.org/wiki/File:Eros_-_PIA02923_(color).jpg

Figure 17.12: Comet Churyumov-Gerasimenko (67P). ESA/Rosetta/MPS for OSIRIS Team MPS/UPD/LAM/IAA/SSO/INTA/ UPM/DASP/IDA. 2014. CC BY-SA 4.0. https://commons.wikimedia.org/wiki/File:Comet_67P_True_color.jpg

Figure 17.13: Jupiter with its moon Europa on the left. NASA, ESA, STScI, A. Simon (Goddard Space Flight Center), and M.H. Wong (University of California, Berkeley) and the OPAL team. 2020. Public domain. https://commons.wikimedia.org/wiki/File:Jupiter_and_Europa_2020.tiff

Figure 17.14: Jupiter's moon Ganymede. NOAA. 2009. Public domain. https://commons.wikimedia.org/wiki/File:Moon_Ganymede_by_NOAA.jpg

Figure 17.15: Our cratered Moon. NASA/Goddard/Arizona State University. 2011. Public domain. https://www.nasa.gov/mission_pages/LRO/news/lro-farside.html

Figure 17.16: Radioactive decay. Andrew Fraknoi, David Morrison, and Sidney Wolff. 2015. CC BY 4.0. https://en.wikipedia.org/wiki/File:OSC_Astro_07_03_Decay_(1).jpg

Figure 17.17: NASA artist's conception of various planet formation processes, including exocomets and other planetesimals, around Beta Pictoris, a very young type A V star. NASA/FUSE/Lynette Cook. 2007. Public domain. https://commons.wikimedia.org/wiki/File:NASA-ExocometsAroundBetaPictoris-ArtistView.jpg

Figure 17.18: Atlas of Planetary Nurseries. NASA/ESA and L. Ricci (ESO). 2009. CC BY 4.0. https://esahubble.org/copyright/

GLOSSARY

---

aa

A blocky, stubby, rubble-like lava.

absolute dating

Quantitative method of dating a geologic substance or event to a specific amount of time in the past.

abyssal

The deep, flat part of the ocean. Also known as the ocean floor.

abyssal plain

Relatively flat ocean floor, which accumulates very fine grained detrital and chemical sediments.

accretionary wedge

Mix of sediments that form as a subducting plate descends and the overriding plate scrapes material and material is added.

acid rock drainage

Toxic waters rich in heavy metals and often of low pH that come from unregulated mining districts.

active margin

A convergent boundary between continental and oceanic plates.

actual preservation

Unchanged materials preserved in the fossil record. This is rare, and is exceedingly less likely with soft materials and older materials.

adhesion

Forces that cause one substance to stick to another.

aeolian

Deposition with wind-blown sediment.

aftershock

Earthquake(s) that occur after the mainshock, usually decreasing in amount and magnitude over time.

albedo

The amount of light that is reflected off of an object. It is measured on a scale of 0 (absorbs all light) to 1 (reflects all light).

alluvial

Depositional environments that are associated with running water.

alluvium

Loose sediment deposited from running water.

alpha decay

Radioactive decay where two protons and two neutrons leave the isotope.

alpine glacier

A glacier that forms on a mountain.

amphibole

A group of chain silicate minerals that form needlelike or prismatic crystals. Can be many colors but the most common form, hornblende, is dark brown to black. Has oblique cleavages at 54° and 126°. Common in many igneous rocks and some metamorphic rocks.

amphoteric

Possesses properties of both acid and base.

amplitude

Height or depth of a wave from the middle point.

andesite

General name of an intermediate igneous rock that is extrusive. Generally has a gray groundmass color.

angle of incidence

Angle from perpendicular to the ground surface at which light rays hit the ground. If the sun is directly above a point and light rays are hitting the ground directly, then the angle of incidence is 0.

angle of repose

Slope angle where shear forces and normal forces are equal.

angular unconformity

Angular discordance between two sets of rock layers. Caused when sedimentary strata are tilted and eroded, followed by new deposition of horizontal strata above.

anhedral

A mineral that shows no crystal habit, either because it is not prone to have a habit, or because it grew in a way that it was confined so it could not grow with its normal habit.

anion

A negatively-charged ion. In geology, this commonly includes elements and molecules like $SiO_4{-4}$, $S-2$, $SO_4{-4}$, and $O-2$.

anomaly

> Data which is out of the ordinary and does not fit previous trends.

Anthropocene

> A newly-proposed time segment (an epoch) that would be representative of time since humans have changed (and left evidence behind within) the geologic record.

anthropogenic

> Made or influenced by humans.

anthropogenic climate change

> Climate change caused by human activity, namely, the burning of fossil fuels.

anticline

> Downward-facing fold, that has older rock in its core.

antidune

> Similar to dunes, in that they are ridges of sand that form perpendicular to flow, but internally, the sediments dip up stream. Forms in the upper part of the upper flow regime.

aphanitic

> Small, microscopic, hard-to-see crystals (i.e. no visible crystals) within an igneous rock. This is common in extrusive rocks.

aquiclude

> A layer with so little porosity and/or permeability that fluids essentially cannot flow through them and only flow around them.

aquifer

> A rock or sediment that has good permeability and porosity, and allows water to move easily, making it possible to get water for human use.

aquitard

> A layer with lower porosity and/or permeability which allows only minimal and/or slow fluid flow.

arc

> A chain of volcanic activity, typically in a curved pattern, rising from a subduction zone. The arc is on the overriding plate, typically a few hundred kilometers from the trench, but parallel to the trench.

Archean

> Eon defined as the time between 4 billion years ago to 2.5 billion years ago. Most of the oldest rocks on Earth, including large portions of the continents, formed at this time.

arête

A ridge that is carved between two glacial valleys.

arkose

A sandstone rich in feldspar.

arroyo

Dry riverbed in an arid region.

artesian well

A well which allows pressurized water to reach the surface.

aseismic

Fault, or movement along a fault, that does not have earthquake activity.

ash

Volcanic tephra that is less than 2 mm in diameter.

assimilation

Bedrock around the magma chamber being incorporated into the magma, sometimes changing the composition of the magma.

asteroid

A small rocky body orbiting the sun.

asthenosphere

A ductile physical layer of the Earth, below the lithosphere. Movement within the asthenosphere is the main driver of plate motion, as the overriding lithosphere is pushed by this.

Astronomy

The branch of science which deals with celestial objects, space, and the physical universe as a whole.

Atacama Desert

Driest nonpolar desert on Earth, located in west-central South America.

atmosphere

The gases that are part of the Earth, which are mainly nitrogen and oxygen.

atmosphere (astronomy definition)

The layers of gases surrounding a planet or other celestial body.

AU

An AU (or astronomical unit) is the average distance from Earth to the Sun.

augen

> Strong crystals that do not deform as easily under ductile deformation, and form lens-shaped porphyroblasts.

aulacogen

> A depression that occurs in an area that was subject to earlier rifting.

aureole

> A zone of contact metamorphism that surrounds an intrusion. Since intrusions are typically somewhat round in cross section, the pattern of metamorphism is concentric about the intrusion.

authigenic mineralization

> Specialized mineralization around organic material which produces highly precise molds and casts.

axial plane

> Dividing two-dimensional line between the two sides of a fold.

back-arc

> Area behind the arc, which can be subject to compressional (causing thrusted mountain belts) or extensional (causing back-arc basins) forces.

back-arc basin

> Depression formed behind an arc, where extension has caused a basin, typically with seafloor spreading.

backshore

> Area of the shoreline that is always entirely above normal wave action.

bajada

> A group of several alluvial fans that have come together and formed a single surface.

banded iron formation

> A sedimentary rock that formed long ago as free oxygen changed the solubility of iron, causing layers of iron rich and iron-poor sediments to form in thin layers, or bands.

banding

> A separation of light (felsic) and dark (mafic) minerals in higher grade metamorphic rocks like gneiss.

bankfull stage

> Largest amount of flow a river can hold before flooding.

barchan dune

> Crescent-shaped dune formed by consistent wind and limited sediment.

barrier island

> Ridges of sand, made from former beach sediments, that form parallel to the shoreline.

basalt

General name of a mafic rock that is extrusive. Generally has a black groundmass color.

base level

Elevation of the mouth of a river.

basin

A down-warped feature in the crust.

basin and range

Term for the extensional tectonic province that extends from California's Sierra Nevada Mountains in the west, to Utah's Wasatch Mountains to the east, to southern Oregon and Idaho to the north, to northern Mexico to the south. Known as a wide rift, each graben (basin) is bounded by horsts (ranges).

batholith

Used to describe a large mass or chain of many plutons and intrusive rocks.

bauxite

A highly weathered soil deposit that consists of aluminum ores.

baymouth bar

A place where a spit extends out and covers a bay.

beach face

Active area of crashing waves.

beach replenishment

Adding sediment to a beach system in order to replace lost sediment due to longshore drift.

bed

A specific layer of rock with identifiable properties.

bedding

Discernible layers of rock, typically from a sedimentary rock.

bedform

A specific type of sedimentary structure (ripples, plane beds, etc.) linked to a specific flow regime.

bedload

Sediment that large and dense, typically sits on the bottom of stream channels, and is only moved with higher-speed flows.

bedrock

Term for the underlying lithified rocks that make up the geologic record in an area. This term can sometimes refer to only the deeper, crystalline (non-layered) rocks.

berm

Ridge of sand built above the beach face.

beta decay

A radioactive decay process where a neutron changes into a proton, releasing an electron.

Big Bang Theory

The theory that the Universe started with a expansive explosion. Shortly after, elements were created (mostly hydrogen) and galaxies started to form.

biochemical

Chemical sedimentary rocks that have a biologic component to their origin. Many limestones are biochemical.

biosphere

The living things that inhabit the Earth.

biostratigraphic correlation

A type of stratigraphic correlation in which fossils are used to match different rock layers.

bioturbation

Sedimentary layering disturbed by movement of organisms.

black smoker

Mineral chimneys that form at hydrothermal vents.

blowout

A depression in dune sediment formed because of a lack of anchoring vegetation.

blueschist

A metamorphic facies of low temperature, high pressure rocks, typified by the rock blueschist, a metamorphic rock containing a blue amphibole called glaucophane.

body wave

Seismic waves that travel through the Earth, mainly P waves and S waves.

bolide

A large extraterrestrial object, such as a meteor or asteroid, that hits the surface of the Earth.

bomb

Large volcanic tephra greater than 64 mm in diameter.

bond

Two or more atoms or ions that are connected chemically.

Bouma sequence

Predictable sequence of fining upward sediments, caused by turbidity flows.

Bowen's Reaction Series

A series of mineral formation temperatures that can explain the minerals that form in specific igneous rocks. For example, pyroxene will form with olivine and amphibole, but not quartz.

brackish

Water that is a mixture of sea water and fresh water.

braided channel

Channel type with many switching channels, common with large sediment volumes.

breakwater

Offshore durable structure designed to lessen wave action and reduce longshore drift.

brittle

A property of solids in which a force applied to an object causes the object to fracture, break, or snap. Most rocks, at low temperatures, are brittle.

brittle deformation

A style of strain in which an object suddenly breaks, fractures, or otherwise fails in a different way than ductile deformation.

burial metamorphism

Metamorphism that is caused by confining pressure and heat, both increasing with depth.

calcite

$CaCO_3$. Pure form is clear, but can take on many different colors with impurities. It is soft, fizzes in acid, and has three cleavages that are not at 90°.

calcite compensation depth

Also known as the CCD, it is the point in the depths of the ocean where calcite start to dissolve, leaving only siliceous ooze behind.

caldera

Hole left behind after a large volume of material erupts out of a volcano. This depression is often turned into a valley or lake after the eruption is over.

calving

A process where ice from the ends of glaciers falls off into the ocean.

Cambrian

The first period of the Paleozoic, 541 million years ago-485 million years ago.

Cambrian Explosion

A period of time in the early Cambrian (about 541-516 million years ago) in which a large diversification of life forms was found in the fossil record. Many of the modern phyla of organisms evolved in this time span.

carbonate

Mineral group in which the carbonate ion, CO_3^{-2}, is the building block. This can also refer to the rocks that are made from these minerals, namely limestone and dolomite (dolostone).

carbonatite

An igneous composition or rock containing more than 50% carbonate minerals (e.g. calcite). Magma of this composition is very low temperature (500-600 C) relative to other magmas.

carbonic acid

An acid that forms from carbon dioxide and water. It is a large contributor to chemical weathering.

Carboniferous

The fifth (second to last) period in the Paleozoic, 359-299 million years ago. In North America, the Carboniferous is split into two different periods, the Mississippian (359-323 million years ago) and the Pennsylvanian (323-299 million years ago).

Carbonization

A type of fossilization where only a carbon-rich film is preserved, common in plants.

cast

Material filling in a cavity left by a organism that has dissolved away.

cataclasite

A type of breccia that forms in a brittle way within fault zones.

Catastrophism

The idea that large, damaging events are the cause of most geologic events.

cation

A positively-charged ion. In geology, this commonly includes ions of the elements Ca^{+2}, Na^{+1}, K^{+1}, $Fe^{+2,+3}$, Al^{+3}, and Mg^{+2}.

cementation

Sediment being "glued" together via mineralization, typically calcite and quartz from groundwater fluids.

Cenozoic

The last (and current) era of the Phanerozoic eon, starting 66 million years ago and spanning through the present.

chalk

A limestone made of coccolithophore shells, a type of single-celled algae.

chemical sedimentary

Sedimentary rocks that are precipitated, from solution.

chemical weathering

Breaking down of mineral material via chemical methods, like dissolution and oxidation.

chemosynthesis

A biologic process of gaining energy from chemicals from within the Earth, similar to using the energy of the sun in photosynthesis.

chert

A very fine grained version of silica deposited with or without microfossils.

Chicxulub Crater

A 180 kilometer (110 mile) crater that exists near Chicxulub, Mexico, on the Yucatan Peninsula. Widely accepted to have caused the K-T extinction.

Chordata

Organisms that possess vertebrate or some form of a spinal column, including humans.

chronostratigraphic correlation

A type of stratigraphic correlation which is based on similar ages.

cinder

A type of tephra which forms as blobs of magma splatter out of a volcanic vent (e.g. cinder cone) and cool and harden quickly.

cinder cone

Volcano formed from piles of cinders and tephra. Forms with low viscosity lava with high volatile content.

cirque

Glacially-carved, bowl-shaped valley.

clastic

Sedimentary rocks that are made of sediment, weathered pieces of bedrock.

claystone

A rock made primarily of clay.

cleavage

A weakness within the atomic structure of a mineral, which allows the mineral to break more easily along that plane.

Minerals can have one, two, three, or more cleavages. Cleavage can also refer to the alignment of features within metamorphic rocks., though they are unrelated.

climate

Long term averages and variations within the conditions of the atmosphere.

closed basin

An internally draining watershed, whose waters do not flow to the ocean.

coal

Former swamp-derived (plant) material that is part of the rock record.

coastline

The entire area which is related to land-sea interactions.

cohesion

Forces that hold a substance together.

col

Low point within an arête.

collision

When two continents crash, with no subduction (and thus little to no volcanism), since each continent is too buoyant. Many of the largest mountain ranges and broadest zones of seismic activity come from collisions.

comet

a celestial object consisting of a nucleus of ice and dust and, when near the Sun, a "tail" of gas and dust particles pointing away from the Sun

compaction

Sediment being squeezed together into a coherent mass.

composition

The mineral makeup of a rock, i.e. which minerals are found within a rock.

compression

Stresses that push objects together into a smaller surface area or volume; contracting forces.

concentrator

A mechanical process which takes ore and separates it from gangue material.

conchoidal

Fractures that have a circular appearance.

conduit

Pipe that connects the magma chamber to the volcanic vent.

cone of depression

Area with a lower water table due to water pumping from a well.

confining

Non-directional pressure resulting from burial.

confining layer

A layer that has lower permeability and porosity and does not allow fluid flow as easily.

conglomerate

A sedimentary rock with rounded, larger (≥2 mm) clasts.

connate water

Original water trapped inside a forming rock.

contact metamorphism

Metamorphism that occurs when rocks are next to a hot intrusion of magma.

continental crust

The layers of igneous, sedimentary, and metamorphic rocks that form the continents. Continental crust is much thicker than oceanic crust. Continental crust is defined as having higher concentrations of very light elements like K, Na, and Ca, and is the lowest density rocky layer of Earth. Its average composition is similar to granite.

continental glacier

A body of ice covering large stretches of land over a continent (mainly found in Antarctica).

continental shelf

Submerged part of the continental mass, with a gentle slope.

continental slope

Steep part of an ocean basin that is the transition between the continental mass and the ocean floor.

convection

The property of unevenly-heated (heated from one direction) fluids (like water, air, ductile solids) in which warmer, less dense parts within the fluid rise while cooler, denser parts sink. This typically creates convection cells: round loops of rising and sinking material.

convergent

Place where two plates come together, casing subduction or collision.

coquina

Limestone made of shell fragments cemented together.

core

The innermost chemical layer of the Earth, made chiefly of iron and nickel. It has both liquid and solid components.

correlation

Matching rocks of similar ages, types, etc.

cosmic microwave background

Radiation left over from the an early stage in the development of the universe at the time when protons and neutrons were recombining to form atoms.

cosmic microwave background radiation

Trace amounts of energy found throughout the Universe.

crater

A bowl-shaped depression, or hollowed-out area, produced by the impact of a meteorite, volcanic activity, or an explosion.

craton

The stable interior part of a continent, typically more than a billion years old, and sometimes as old as 2.5-3 billion years. When exposed on the surface, a craton is called a shield.

creep

A slow and steady movement. Can occur as part of faults, mass wasting, and grain movement.

Cretaceous

The last period of the Mesozoic, 145-66 million years ago.

Cretaceous Interior Seaway

A waterway that existed in North America around 100 million years ago. Western North America was separated from eastern North America.

crevasse

Cracks that form with glacial movement in the upper, brittle part of the glacier.

crevasse splay

Sediment that breaks through a levée and deposits in a floodplain during a flood event

cross bed

A sedimentary structure that forms in the lower flow regime, where ridges of sediment form perpendicular to flow direction, but within the ridges, sediment layers and dips toward flow direction. Found in ripples and dunes. Can be tabular, sinuous, or trough shaped.

cross bedding

A sedimentary structure that has inclined layers within an overall layer. Forms commonly in dunes, larger in eolian dunes.

crust

The outermost chemical layer of the Earth, defined by its low density and higher concentrations of lighter elements. The crust has two types: continental, which is the thick, more ductile, and lowest density, and oceanic, which is higher density, more brittle, and thinner.

cryosphere

The part of the hydrosphere (water) that is frozen, found mainly at the poles.

crystal habit

The typical form or forms a crystal takes when it grows.

crystallization

The process of liquid rock solidifying into solid rock. Because liquid rock is made of many components, the process is complex as different components solidify at different temperatures.

cut bank

Erosional part of a meandering channel.

daughter isotope

The atom that is made after a radioactive decay.

debris flow

A mixture of coarse material and water, channeled and flowing downhill rapidly.

decay chain

A series of several radioactive decays which eventually leads to a stable isotope.

Deccan Trapps

Large flood basalt province in India that occurred around the same time as the K-T Extinction, 66 million years ago.

Decompression melting

Melting that occurs as material is moved upward and pressure is released, typically found at divergent plate boundaries or hot spots.

deductive reasoning

Taking known truths in order to develop new truths.

deformation

A strain that occurs in a substance in which the item changes shape due to a stress.

delta

Place where rivers enter a large body of water, forming a triangular shape as the river deposits sediment and switches course.

dendritic drainage

A common branching style of drainage pattern that resembles the pattern of tree roots.

Density

We give densities in units where the density of water is 1 g/cm3. To get densities in units of kg/m3, multiply the given value by 1000.

deposition

Sediment gathering together and collecting, typically in a topographic low point.

depositional environment

An interpretation of the rock record which describes the cause of sedimentation (i.e. ancient beach, river, swamp, etc.).

deranged pattern

Drainages that are erratic and disappearing, typically in karst environments.

desert varnish

Dark mineralization that forms on rocks in desert environments.

desertification

The process that turns non-desert land into desert.

detachment fault

A style of low-angle, high extension normal faulting.

detrital

Sedimentary rocks made of mineral grains weathered as mechanical detritus of previous rocks, e.g. sand, gravel, etc.

Devonian

Known as the "Age of Fishes," the 4th period of the Paleozoic, about 419-359 million years ago.

dextral

Movement in a transform or strike-slip setting which is toward the right across the fault. As viewed across the fault, objects will move to the right.

diagenesis

Changes in sedimentary rocks due to increased (but low when compared to metamorphism) temperatures and pressures. This can include deposition of new minerals (e.g. limestone converting to dolomite) or dissolution of existing minerals.

diameter

A straight line passing from side to side through the center of a body or figure, especially a circle or sphere.

diamictite

A sedimentary rock containing two distinct grain sizes, typically cobbles (or larger) mixed with mud.

diapir

A ductile material that moves toward the surface of Earth. Can be used to describe salt domes and intrusions.

Differentiation

In planetary science, differentiation is the process of separating out the different components within a planetary body as a consequence of their physical or chemical behavior (e.g. density and chemical affinities).

dike

A narrow igneous intrusion that cuts through existing rock, not along bedding planes.

diorite

General name of an intermediate rock that is intrusive. Has about the same amount of felsic minerals and mafic minerals.

dip

A measure of a plane's (maximum) angle with respect to horizontal, where a perfectly horizontal plane has a dip of zero and a vertical plane has a dip of 90°.

dip slip

Faulting that occurs with a vertical motion.

directed stress

Stress that has a strong directional component (unequal), typically creating elongated or flattened features.

directivity

Increased intensity due to being along the path of fault propagation.

discharge

Amount of water that leaves a system, such as a river or aquifer.

disconformity

Two layered rocks that may seem conformable, but an erosional surface exists between them.

dissolution

The process in which solids (like minerals) are disassociated and the ionic components are dispersed in a liquid (usually water).

dissolved load

Amount of material dissolved in stream water.

diurnal tide

Areas that have two clear high and low tides per tidal day.

divergent

Place where two plates are moving apart, creating either a rift (continental lithosphere) or a mid-ocean ridge (oceanic lithosphere).

dome

A rock up-warping of symmetrical anticlines.

Doppler Effect

A change in wavelength and frequency of a wave due to the source of a wave moving relative to the observer of a wave.

drainage basin

The area within a topographic basin or drainage divide in which water collects.

drainage divide

Topographic prominence which sheds water into a specific drainage basin.

drainage pattern

The shape or form of a river and/or tributary drainage system.

drumlin

Ridge of sediment that forms under a glacier, with a steep uphill (with respect to the glacier) side and gentle downhill side.

ductile

A property of a solid, such that when a force is applied, the solid flows, stretches, or bends along with the force, instead of cracking or breaking. For example, many plastics are ductile.

ductile deformation

A bending, squishing, or stretching style of deformation where an object changes shape smoothly.

dune

A large pile of sediment, deposited perpendicular to flow. Internal bedding in dunes dips toward flow direction (i.e. cross bedding). Formed in the upper part of the lower flow regime.

dwarf planet

A small planetary-mass object that is in direct orbit of the Sun – something smaller than any of the eight classical planets, but still a world in its own right.

Earth System Science

The study of the interaction of the spheres within the system that is the Earth, mainly the study of the hydrosphere, atmosphere, geosphere (lithosphere), and biosphere.

earthflow

Plastic moving, fine-grained type of flow.

eccentricity

The measure of the amount of circular or elliptical nature of the Earth's orbit.

Ediacaran fauna

A group of relatively complex organisms that existed at the end of the Proterozoic.

elastic deformation

A type of deformation that reverses when the stress is removed.

elastic rebound

A theory of building energy that is released during an earthquake.

electromagnetic spectrum

Visible light and its related energetic waves, including X-rays, UV rays, and radio waves.

electron

A stable subatomic particle with a charge of negative electricity, found in all atoms and acting as the primary carrier of electricity in solids.

electron capture

A type of radioactive decay where an electron combines with a proton, making a neutron.

element

A group of all atoms with a specific number of protons, having specific, universal, and unique properties.

emergent coastline

Features of a coastline where relative sea level is falling.

entrenched channel

A channel that carves into existing bedrock, preserving its original shape and character.

eon

The largest span of time recognized by geologists, larger than an era. We are currently in the Phanerozoic eon. Rocks of a specific eon are called eonotherms.

ephemeral stream

A stream or river that can be wet or dry depending on the season.

epicenter

The location at the surface directly above the focus of an earthquake, typically associated with strong damage.

epoch

A unit of geological time recognized by geologists; smaller than a period. We are currently in the Holocene epoch.

equant

Stubby, not long in any direction.

era

The second largest span of time recognized by geologists; smaller than an eon, larger than a period. We are currently in the Cenozoic era.

erg

A vast stretch of sand dunes.

erosion

The transport and movement of weathered sediments.

esker

Ridge of sediment that forms under a glacier by meltwater which forms a river.

estuary

Lagoon with brackish water, typically with abundant biologic factors.

euhedral

A mineral that perfectly shows its true crystal habit.

eukaryote

A type of organism in with a cell or cells that contains a nucleus.

eustatic

An overall global sea level change, either due to climate or seafloor spreading rate.

evaporate

Turn from liquid into vapor.

evaporite

A chemical sedimentary rock that forms as water evaporates.

evapotranspiration

A combination of evaporation and transpiration from plants, which is a measure of water entering the atmosphere.

exfoliation

A type of mechanical weathering in which outer layers of rock, approximately parallel to the surface, fracture off.

exoplanet

Any planet beyond our solar system.

exoplanetary

Of or pertaining to an exoplanet, a planet outside the solar system.

experiment

A test of an idea in which new information can be gathered to either accept or reject a hypothesis.

extinct

When a species no longer exists.

extrasolar

Originating or existing outside the solar system.

extrusive

Igneous rock cooling, and thus forming, outside of the Earth, i.e. on the surface.

facies

A specific set of features that are tied together in an interpretive group. Facies can be based on mineralogy, biologic factors, fossils, rock types, etc.

failed rift arm

A section of a rift that starts but does not complete. This typically occurs at 120° angles to the active rift.

fair weather wave base

The depth normal, non-storm waves reach.

falsifiable

The idea that any claim in science can be proved wrong with proper evidence.

fault

Planar feature where two blocks of bedrock move past each other via earthquakes.

fault scarp

Place where fault movement cuts the surface of the Earth.

feldspar

Consisting of three end members: potassium feldspar (K-spar, $KAlSi_3O_8$), plagioclase with calcium ($CaAl_2Si_2O_8$, called anorthite), and plagioclase with sodium ($NaAlSi_3O_8$, called albite). Commonly blocky, with two cleavages at ~90°. Plagioclase is typically more dull white and gray, and K-spar is more vibrant white, orange, or red.

felsic

Can refer to a volcanic rock with higher silica composition, or the minerals that make up those rocks, namely quartz, feldspar, and muscovite mica. Felsic rocks are lighter in color and contain more minerals that are light in color. Primary felsic rocks are rhyolite (extrusive) and granite (intrusive).

fetch

Distance wind has been building a wave.

finger lake

Lake that fills a glacial valley.

firn

Snow which has been compressed, and is starting to turn into ice.

fissile

Easily split along bedding planes, a characteristic of shale.

fjord

Glacial valley filled by ocean water.

flash flood

Dangerous flooding that occurs in arid regions.

flood basalt

Rare very low viscosity eruption that covers vast areas. None have been observed in human history.

floodplain

Flat area around a river channel that is filled with water during flooding events.

flow regime

A qualitative measure of the speed of a fluid flow, with different amounts of flow corresponding to different sedimentary structures, called bedforms. Typically, it is split into upper and lower flow regimes, with upper being a more rapid flow.

flower structure

A small area along a strike-slip or transform fault with branching structures of transpression/transtension, causing local hills or valleys.

fluvial

Deposition that has to do with rivers.

flux melting

The process in which volatiles enter the mantle wedge, and the volatiles lower the melting temperature, causing volcanism.

focus

Initiation point of an earthquake or fault movement.

fold

A rock layer that has been bent in a ductile way instead of breaking (as with faulting).

foliation

A planar alignment of minerals and textures within a rock.

footwall

On a dipping fault, the part of the block that is below the fault. Moves down in normal faulting, up in reverse faulting.

forearc

Area in front of the arc, between the arc and the trench. Often marked by an accretionary wedge or a forearc basin.

forearc basin

Any depression formed between the arc and the trench, commonly between the arc and the accretionary wedge.

foreshock

An earthquake that sometimes occurs before the larger mainshock.

foreshore

Area between high tide and low tide.

formation

An extensive, distinct, and mapped set of geologic layers.

fossil

Any evidence of ancient life.

fossil fuel

Energy resources (typically hydrocarbons) derived from ancient chemical energy preserved in the geologic record. Includes coal, oil, and natural gas.

fossiliferous

Adjective for a rock filled with fossils, most commonly with limestones.

fracking

A process of injecting pressurized fluids into the ground to aid in hydrocarbon migration.

fractionation

The process of a magma changing from mafic to felsic via cooling. As the magma cools, higher temperature, mafic minerals crystalize, and a more felsic magma is left.

fracture

A break within a rock that has no relative movement between the sides. Caused by cooling, pressure release, tectonic forces, etc.

fracture zone

Faults along mid-ocean ridges that have a transform motion but do not produce earthquakes. These faults accommodate different amounts of movement along the mid-ocean ridge.

frost wedging

A process where water freezes inside cracks in rocks, causing expansion and mechanical weathering.

fumerole

Gas expulsions from the subsurface, usually related to volcanic activity.

fusion

A process inside stars where smaller atoms combine and form larger atoms.

gabbro

General name of a mafic rock that is intrusive. Has more mafic minerals than felsic minerals.

galaxies

A gravitationally-bound system of stars and interstellar matter.

gamma ray

A penetrating form of electromagnetic radiation arising from the radioactive decay of atomic nuclei. It consists of the shortest wavelength electromagnetic waves, typically shorter than those of X-rays.

gangue

Material found around ore which is less valuable and needs to be removed in order to obtain ore.

geopetal structure

A feature in a rock that allows the observer to determine which direction was up in the past.

geosphere

The solid, rocky parts of the Earth, including the crust, mantle, and core. Also referred to as the lithosphere.

geothermal gradient

The average change in temperature that is experienced as material moves into the Earth. Near the surface, this rate is about 25°C/km.

Giant Impact Hypothesis

Idea that a large body struck the Earth, sprayed material into space, and that material eventually collected to form the Moon.

glacial

Deposition and erosion tied to glacier movement.

glacial budget

The net gain or loss of ice within a glacier.

glacial erratic

Large sediment (e.g. boulder) carried and then dropped by a glacier.

glacial polish

Smooth surface carved in harder rocks by glacial action.

glacial striation

Grooves scratched in rock by glacial action.

glaciation

A period of cooler temperatures on Earth in which ice sheets can grow on continents.

glacier

A body of ice that moves downhill under its own mass.

glaciers

A body of ice that moves downhill under its own mass.

Glossary

[glossary]

gneiss

A very high grade metamorphic rock, higher grade than schist, with a separation of light and dark minerals.

Goldich Dissolution Series

Working opposite of Bowen's reaction series, it states that minerals that are formed at conditions more dissimilar to the surface are more quickly prone to chemical weathering.

graben

A valley formed by normal faulting.

grade

A qualitative measure of the amount of metamorphism that has occurred or the amount of a resource present in an ore.

gradient

Slope of a stream channel.

Grading

A sequence of layers in which the sediment changes linearly in size, either getting coarser or finer.

grain size

The average diameter of a grain of sediment, ranging from small, also known as fine-grained (e.g. clay, silt) to large, also known as coarse-grained (e.g. boulder).

granite

General name of a felsic rock that is intrusive. Has more felsic minerals than mafic minerals.

Great Basin Desert

Desert area stretching from California to the west, Utah to the east, and Idaho/Oregon to the north. Partially caused by latitude, partially caused by rain shadow.

Great Oxygenation Event

A period of the early Proterozoic (around 2.5-2 billion years ago) where atmospheric oxygen levels dramatically increased, killing many non-oxygen-breathing organisms and allowing oxygen-breathing organisms to thrive.

greenhouse effect

The ability for the atmosphere to absorb heat that is emitted by a planet's surface.

greywacke

A sandstone with a significant mud component OR a sandstone with a significant lithic fragment component.

groin

A hard stabilization structure built perpendicular to the shoreline to help control longshore drift.

ground moraine

Moraine that forms beneath a glacier.

groundmass

General term for the fine-grained, not discernible part of a rock. In igneous rocks, this is the part of the rock that is not phenocrysts, and can help in determining the composition of extrusive rocks. In sedimentary rocks, it typically refers to the fine-grained components, namely mud. In metamorphic rocks, it is usually referring to material between porphyroblasts or a low-grade rock with only microscopic mineralization.

groundwater

Water that is below the surface.

Groundwater mining

When discharge exceeds recharge, and the groundwater is withdrawn at a rate that depletes groundwater storage.

gypsum

An evaporite mineral, $CaSO_4 \cdot 2H_2O$. Has one cleavage, hardness of 2. Typically clear or white.

gyre

Large circular ocean currents formed by global atmospheric circulation patters.

haboob

Dust storms that occur in desert areas.

Hadean

Eon that represents the time from Earth's formation to 4 billion years ago. Noted for high levels of volcanism, impacts, and very low preservation.

Hadley Cell

A part of the global circulation system that rises at the equator and sinks at 30°.

half graben

A valley formed by normal faulting on just one side.

half life

The calculated amount of time that half of the mass of an original (parent) radioactive isotope breaks down into a new (daughter) isotope.

halide

Minerals based on bonds to column 17 halogens, such as chlorine and fluorine.

halite

Also known as rock salt, or table salt. 3 cleavages at 90°, cubic crystal habit. Typically clear or white, hardness of 3.

hanging valley

A feature formed by a tributary glacier going into a main glacier, forming a tributary valley floor higher in elevation than the main valley floor.

hanging wall

On a dipping fault, the side that is on top of the fault plane. Moves down in normal faulting, up in reverse faulting.

hardness

The ease or difficulty in scratching a mineral, measured by the qualitative Mohs hardness scale, which ranges from soft talc (#1 on the scale) to hard diamond (#10 on the scale).

headwaters

The source or beginning of a river.

Holocene

The most recent epoch of geologic time, from 11,700 years ago to present.

hopper crystal

Evaporites (like salt) which form cavities within rocks, which mimic the shape of the crystal.

horn

Steep spire carved by several glaciers.

hornfels

A dense, hard metamorphic rock, typically derived from contact metamorphism.

horst

Uplifted mountain block caused by normal faulting.

hot spot

Rising stationary magma, forming a succession of volcanism. This is reflected as islands on oceanic plates, and volcanic mountains or craters on land.

hummocky cross stratification

A special type of cross bedding that forms when strong storms produce mounds and divots of cross-bedded sand in deeper water.

humus

Organic rich material found in soil.

hydraulic

Relating to movement brought about by water.

hydraulic conductivity

The measure of how well a fluid flows through an object.

hydrogen bond

A weak chemical bond which attracts hydrogen to a negative part of a molecule. Many of water's properties are due to hydrogen bonds.

hydrolysis

Water breaking into ions and replacing ions in minerals; a major type of chemical weathering in silicates.

hydrosphere

The water part of the Earth, as a solid, liquid, or gas.

hydrothermal

Metamorphism which occurs with hot fluids going within rocks, altering and changing the rocks.

hypotheses

Proposed explanations for an observation that can be tested.

hypothesis

A proposed explanation for an observation that can be tested.

ice sheet

Thick glaciers that cover continents during ice ages.

igneous

Relating to molten rock.

Igneous rocks

Rocks that are formed from liquid rock, i.e. from volcanic processes.

imbrication

Stacked cobbles in the direction of flow.

inclusion

A piece of a rock that is caught up inside of another rock.

index fossil

A fossil with a wide geographic reach but short geologic time span used to match rock layers to a specific time period.

index mineral

Minerals that form at a specific range of temperatures and pressures. Using a collection of index minerals narrows down the conditions of rock formation.

induced seismicity

Earthquakes that occur due to human activity.

inductive reasoning

Establishing evidence (including new observations) to infer a possible truth.

infiltration

Water that works its way down into the subsurface.

inner core

The innermost physical layer of the Earth, which is solid.

inselberg

Isolated piece of bedrock which sticks above an alluvial surface.

interglacial

Period of warming within a glacial or ice age cycle.

intermediate

A volcanic rock with medium silica composition, equally rich in felsic minerals (feldspar) and mafic minerals (amphibole, biotite, pyroxene). Intermediate rocks are grey in color and contain somewhat equal amounts of minerals that are light and dark in color. Primary intermediate rocks are andesite (extrusive) and diorite (intrusive).

interplate

Activity that occurs at the boundaries between plates.

interstadial

A very brief period of warming, even warmer than a interglacial, within a glacial or ice age cycle.

intraplate

Activities that occur within plates, away from plate boundaries.

intrusive

Igneous rock cooling, and thus forming, inside of the Earth, i.e. under the surface.

ion

An atom or molecule that has a charge (positive or negative) due to the loss or gain of electrons.

island arc

Place where oceanic-oceanic subduction causes volcanoes to form on an overriding oceanic plate, making a chain of active volcanoes.

isostasy

Relative balance of an object based on how it floats.

isostatic rebound

An upwards movement of the lithosphere when weight is removed, such as water or ice.

isotope

An atom that has different number of neutrons but the same number of protons. While most properties are based on the number of protons in an element, isotopes can have subtle changes between them, including temperature fractionation and radioactivity.

jetty

Artificial device (typically a wall of concrete or rocks) placed to stop or slow longshore drift.

Jurassic

The middle period of the Mesozoic era, 201-145 million years ago.

K

The kelvin, symbol K, is the SI base unit of temperature. Absolute zero is 0 K, the equivalent of –273.15°C.

K-T Extinction

The most recent mass extinction, which killed the non-avian dinosaurs and paved the way for the diversification of mammals. Occurred when a bolide hit near Chicxulub, Mexico, 66 million years ago.

karst

Carbonate rocks which dissolve, leaving behind caverns and holes which affect the landscape.

kettle

Depression formed by ice resting in sediment, then preserved after the ice melts and the sediment lithifies.

kettle lake

Lake that forms in a kettle.

kimberlite

An ultramafic rock from deep volcanic vents that can contain diamonds.

Kuiper belt

A circumstellar disc in the outer Solar System, extending from the orbit of Neptune at 30 astronomical units (AU) to approximately 50 AU from the Sun.

laccolith

Large igneous intrusion that is wedged between sedimentary layers, bulging upwards. Called a lopolith if bulging downward.

lacustrine

Deposition in and around lakes.

lagerstätte

An exceptionally-well preserved fossil locality, often including soft tissues.

lagoon

Interior body of ocean water, at least partially cut off from the main ocean water.

lahar

A type of volcanic mudslide, in which rain or snowmelt accumulates volcanic ash of the slopes of steep volcanoes or other mountains and then wash downhill, causing damaging flooding.

laminae

Thin (less than 1 cm) beds of rock.

landslide

General term for sudden material falling (sliding) down a slope due to gravity.

lapilli

Volcanic tephra that has a diameter between 2 mm and 64 mm. Many cinders are within the category of lapilli.

late heavy bombardment

A hypothesis that states that movement of Jupiter and Saturn about 4 billion years ago caused a destabilization of orbits in the Asteroid and Kuiper Belts, which then caused a spike in impacts throughout of solar system.

lateral moraine

Moraines that form at the sides of glaciers.

latitude

The measure of degrees north or south from the equator, which has a latitude of 0 degrees. The Earth's north and south poles have latitudes of 90 degrees north and south, respectively.

Laurentia

Geologic name for the craton that makes up North America.

lava

Liquid rock on the surface of the Earth.

Lava dome

Very steep sided volcanic feature formed by higher viscosity, higher-silica lava.

layered intrusion

Metallic mineral deposit consisting of mafic plutonic rocks, typically containing platinum-group elements, chromium, copper, nickel, etc.

light years

The distance that light can travel through space in a year. One light year is 9.4607×10^{12} km.

limestone

A chemical or biochemical rock made of mainly calcite.

linear dune

Dunes that are much longer than wide, forming from wind that varies in two opposite directions.

lineation

Linear alignment of minerals within a rock.

liquefaction

Process of saturated sediments becoming internally weak (like quicksand) and destabilizing foundations.

lithification

The process of turning sediment into sedimentary rock, including deposition, compaction, and cementation.

lithosphere

> The outermost physical layer of the Earth, made of the entire crust and upper mantle. It is brittle and broken into a series of plates, and these plates move in various ways (relative to one another), causing the features of the theory of plate tectonics.

lithostratigraphic correlation

> A type of stratigraphic correlation in which the physical characteristics of rocks are used to correlate.

littoral

> The beach (shoreline) zone, where waves are crashing.

loess

> Wind-blown silt, mainly formed from glacial processes.

longitudinal profile

> Illustration of the topography of the base of a stream, showing zones of sediment production, transport, and deposition.

longshore current

> A net movement that occurs as waves intersect the shoreline at non-perpendicular angles.

longshore drift

> Sediment that moves via a longshore current.

Love wave

> Surface waves that have a side-to-side motion.

luster

> The shine a mineral takes on, based on the way light reflects off of a mineral. This is typically divided into two main categories: metallic (metal-like shine) and nonmetallic (non metal-like shine).

mafic

> Can refer to a volcanic rock with lower silica composition, or the minerals that make up those rocks, namely olivine, pyroxene, amphibole, and biotite. Mafic rocks are darker in color and contain more minerals that are dark in color, but can contain some plagioclase feldspar. Primary mafic rocks are basalt (extrusive) and gabbro (intrusive).

magma

> Liquid rock within the Earth.

magma chamber

> A reservoir of magma below a volcano.

magmatic differentiation

> The process of changing a magma's composition, usually through assimilation or fractionation.

magnetic striping

Symmetric (about the ridge) patterns of magnetism created by ocean floor rocks recording changes in Earth's magnetic field.

magnitude

A measure of earthquake strength. Scales include Richter and Moment.

mainshock

Largest earthquake in an earthquake sequence.

mantle

Middle chemical layer of the Earth, made of mainly iron and magnesium silicates. It is generally denser than the crust (except for older oceanic crust) and less dense than the core.

mantle plume

Rising material and heat derived from the mantle. These may be responsible for hot spots.

mantle wedge

The area of the mantle where volatiles rise from the slab, causing flux melting and volcanism.

marble

A metamorphosed limestone.

marine

Places that are under ocean water at all times.

mass extinction

A pronounced increase in the extinction rate, typically caused by significant environmental change. There have been 5 mass extinctions in geologic history, and a sixth that has been suggested to be currently occurring.

mass spectrometer

A device that can determine the amounts of different isotopes in a substance.

mass wasting

Any downhill movement of material, caused by gravity.

massive

A feature with no internal structure, habit, or layering.

meander scar

Silted-in oxbow which still has a topographic expression.

meandering channel

Low-gradient channel where rivers sweep across broad flood plains.

mechanical weathering

> The physical breakdown (weathering) of bedrock by processes such as pressure, ice expansion, etc.

medial moraine

> A place where two or more glaciers combine, and the lateral moraines combine to form a moraine within the glacier.

megathrust

> Term for faulting that occurs in subduction.

mesosphere

> Also called lower mantle, a solid, more brittle physical layer of the Earth, below the asthenosphere.

Mesozoic

> Meaning "middle life," it is the middle era of the Phanerozoic, starting at 252 million years ago and ending 66 million years ago. Known as the Age of Reptiles.

metal

> A solid material that is typically hard, shiny, malleable, fusible, and ductile, with good electrical and thermal conductivity.

metallic

> Minerals with a luster similar to metal and contain metals, including valuable elements like lead, zinc, copper, tin, etc.

metamorphic

> Rocks and minerals that change within the Earth are called metamorphic, changed by heat and pressure. Metamorphism is the name of the process.

metamorphic facies

> A specific set of index minerals tied to specific styles of metamorphism. When these minerals are present, it allows a history of metamorphism to be determined.

metamorphic rock

> Rocks formed via heat and/or pressure which change the minerals within the rock.

meteor

> A small body of matter from outer space that enters the Earth's atmosphere, becoming incandescent as a result of friction and appearing as a streak of light.

meteorite

> A stoney and/or metallic object from our solar system which was never incorporated into a planet and has fallen onto Earth. Meteorite is used for the rock on Earth, meteoroid for the object in space, and meteor as the object travels in Earth's atmosphere.

mica

X1A2-3Z4O10(OH, F)2, where commonly X=K, Na, Ca; A=Al, Mg, Fe; Z=Si, Al. Has two more-common occurrences, light-colored (translucent and pearly tan) muscovite, and dark-colored biotite. Has one strong cleavage, and is typically seen as sheets, in stacks or "books." Common in many igneous and metamorphic rocks. Structure is two-dimensional sheets of silica tetrahedra in a hexagonal network.

micrite

Limestone made of primarily fine-grained calcite mud. Microscopic fossils are commonly present.

mid-ocean ridge

A divergent boundary within an oceanic plate, where new lithosphere and crust is created as the two plates spread apart. Mid-ocean ridge and spreading center are synonyms.

migmatite

A rock transitional between metamorphic and igneous rock, i.e. rocks so metamorphosed that they begin the process of melting.

Milankovitch Cycles

A series of changes in the Earth's orbit/position in relation to the Sun which can fluctuate climate over varying period-icities.

mine

Place where material is extracted from the Earth for human use.

mineral

A natural substance that is typically solid, has a crystalline structure, and is typically formed by inorganic processes. Minerals are the building blocks of most rocks.

mineraloid

A mineral-like substance that does not meet all the criteria as a true mineral. Examples include glass, coal, opal, and obsidian.

minerals

A natural substance that is typically solid, has a crystalline structure, and is typically formed by inorganic processes. Minerals are the building blocks of most rocks.

Mississippi Valley-type

Metallic mineral deposit of mainly lead and zinc from groundwater movements within sedimentary rocks.

mixed tide

Areas with an irregular sequence of tides over the course of a month.

Modified Mercalli Intensity Scale

A qualitative earthquake scale, from I-XII, of the degree of shaking in an earthquake.

moho

Short for Mohorovičić Discontinuity, it is the seismically-recognized layer within the Earth in which the crust ends and the mantle begins. Because the crust is very different in composition to the mantle, the moho is easy to find, since seismic waves travel differently through the two materials.

mold

Organic material making a preserved impression in a rock.

Moment magnitude

A magnitude scale based on calculation of the energy released in an earthquake.

monocline

A one-sided fold-like structure in which layers of rock warp upwards or downwards.

moon

An object that orbits a planet or something else that is not a star. Besides planets, moons can circle dwarf planets, large asteroids, and other bodies.

moons

An object that orbits a planet or something else that is not a star. Besides planets, moons can circle dwarf planets, large asteroids, and other bodies.

moraine

Accumulation of sediment at the margins of glaciers, including the base, sides, and end.

mountain

A landform that rises above its surrounding area.

mouth

The end of a river.

mud chip

Pieces of mudcracks that are incorporated into a sedimentary rock.

mudcrack

Polygonal cracking that occurs with shrinking clays. Indicative of mud submerged underwater and then exposed to air.

mudstone

A rock made of primarily mud, i.e. particles smaller than sand (≤0.064 mm).

mylonite

Fault-formed rock via ductile deformation, deeper within the Earth.

native element minerals

Minerals made from just a single element, bonded to itself. Examples include gold, silver, copper, and diamond, which is a native version of carbon.

natural gas

Gaseous fossil fuel derived from petroleum, mostly made of methane.

natural hazard

A significant and dangerous event that is part of a natural process.

natural levée

Built-up area around a river channel which can hold river flow within a channel.

natural resources

Items that are found within Earth that are valuable and limited. Examples include coal, water, and gold.

neap tide

Lowest low tide of the month.

nearshore

Shore area between low tide and storm wave base. Upper part is dominated by fair weather wave base, lower part is dominated by storm wave base.

nebula

A cloud of gas and dust in space that can form a new star/solar system if it collapses.

nebular hypothesis

The idea that a nebula can collapse and form a star with planets.

negative feedback

A system which reverts back to a baseline when it deviates.

non-foliated

Metamorphic textures that do not have a directional component of its minerals.

nonconformity

Layered rocks on top of a non-layered rock, such as crystalline basement.

nonmetallic

Minerals that have a luster that is not similar to metal. Divided into subtypes based on the way light reflects (or doesn't), including glassy/vitreous, greasy, pearly, dull, etc.

nonpoint source

Pollution that does not come from one specific, known place, but instead, comes from a wide, broad zone.

nonrenewable

A resource that is not able to be replaced on human time scales.

normal fault

A dip-slip fault in which the hanging wall drops relative to the footwall, caused by extensional forces.

normal force

Component of the gravitational force which holds material on a slope.

obduction

Process which allows a continental plate to bring up oceanic plate, frequently occurring in collision zones.

objective

An observation that is completely free of bias, i.e. anyone and everyone would make the same observation.

obliquity

The angle of the Earth's axis with respect to the plane of rotation.

observation

The act of gathering new information from the senses or from a scientific instrument.

obsidian

Dark colored volcanic glass, with extremely small microscopic crystals or no crystals. Typically form from felsic volcanism.

oceanic crust

The thin, outer layer of the Earth which makes up the rocky bottom of the ocean basins. Oceanic crust is much thinner (but denser) than continental crust. Oceanic crust is made of rocks similar to basalt and as it cools, becomes more dense.

oceanic-continental subduction

Where an ocean plate subducts beneath a continental plate, causing a volcanic arc to form.

oceanic-oceanic subduction

Where a dense oceanic plate subducts beneath a less dense oceanic plate, causing an island arc to form on the overriding plate.

octet rule

A rule that says the outer valence shell of electrons is complete when it contains 8 electrons.

offset

Amount of movement during a faulting event.

offshore

The part of the coastline which is below any wave base action.

oil

A dark liquid fossil fuel derived from petroleum.

oil shale

Oil which is found in low-permeability, high-porosity rocks such as shale.

olivine

$(Fe,Mg)_2SiO_4$. Typically translucent olive green and equant, with no cleavage. Common in mafic igneous rocks and in the mantle, but easily weathered in surface conditions. Structure is isolated silica tetrahedra. Known as peridot when a gem.

ooid

Spheres of calcite that form in saline waters with slight wave agitation. Ooid refers to the sphere, oolite the rock with the spheres.

Oort cloud

A spherical layer of icy objects surrounding our Sun; likely occupies space at a distance between about 2,000 and 100,000 astronomical units (AU) from the Sun.

open pit mine

Large surface mine with opening carved into the ground.

ophiolite

Rocks of the ocean floor, such as mid-ocean ridge rocks, which are brought to the surface.

Ordovician

The second period of the Paleozoic era, 485-444 million years ago.

ore

Valuable material in the Earth, typically used for metallic mineral resources.

ore mineral reserve

A proven commodity of profitable material that could be mined.

ore mineral resource

Potentially extractible and valuable material, but unproven.

orogeny

The process of uplifting mountains and creating mountain belts, primarily via tectonic movement. Orogenic belts are the mountain belts that result from these movements, and orogenesis is the name for the process of forming mountain belts.

outer core

The outer physical layer of the core, which is liquid. Movement within the outer core is believed to be responsible for Earth's magnetic field and flips of the magnetic field.

outwash plain

Accumulation of fine-grained sediment formed downhill of the terminal moraine.

oversteepen

A slope, that by natural or human activity, becomes steeper than the angle of repose.

oxbow

Abandoned meanders that are cut off from the main channel.

oxidation

Certain metallic elements (like iron) take in oxygen, causing reactions like rust.

oxide

Minerals in which ions are bonded to oxygen, such as hematite (Fe_2O_3).

oxidized

Oxidation is the loss of electrons or an increase in the oxidation state of a chemical or atoms within it.

p wave

The fastest seismic wave that occurs after an earthquake, compressional in nature.

pahoehoe

Rope-like, flowing basaltic lava.

Paleocene-Eocene Thermal Maximum

A warm climate spike, the warmest in the recent geologic past, occurring about 55.5 million years ago. Commonly abbreviated as the PETM.

paleocurrent

Direction of flow preserved in the rock record.

Paleomagnetism

As a rock cools, the iron minerals within the rock align with the current magnetic field. Since the magnetic field changes (by where you are on Earth, by flips where "north" and "south" switch, and by migration of the magnetic north pole), scientists use the magnetic alignment within rocks to determine past movement or the magnetic field itself, along with the movement of rocks and plates via plate tectonics.

Paleozoic

Meaning "ancient life," the era that started 541 million years ago and ending 252 million years ago. Vertebrates (including fish, amphibians, and reptiles) and arthropods (including insects) evolved and diversified throughout the Paleozoic. Pangea formed toward the end of the Paleozoic.

paludal

Deposition in swamps.

Pangea

The most recent supercontinent, which formed over 300 million years ago and started breaking apart less than 200 million years ago. Africa and South America, as well as Europe and North America, bordered each other.

parabolic dune

Dunes that form semicircular shapes due to anchoring vegetation.

parasitic cone

Small side vent of a stratovolcano where secondary eruption can occur.

parent isotope

A radioactive atom that can and will decay.

partial melt

The process of some material being derived from a heterogenous mixture when melting (e.g. rocks). Because all rocks are made of many different components, they have many different melting points. As they are heated, certain easy-to-melt components will be melted first.

parting lineation

Subtle ridges formed in the upper flow regime on top of plane beds in the direction of flow.

passive margin

A boundary between continental and oceanic plates that has no relative movement, making it a place where an oceanic plate is connected to a continental plate, but it is not a plate boundary.

paternoster lake

Series of lakes between moraines within an alpine glacier basin, typically a cirque.

peer review

A process where experts in a field review and comment on a newly-introduced work, typically a part of publication.

pegmatite

A rock (or texture within a rock) with unusually-large crystals, minerals with rare trace element concentrations, and/or unusual minerals, typically forming in veins as the last dredges of magma crystallize.

peridotite

An intrusive ultramafic rock, which is the main component of the mantle. The minerals in peridotite are typically olivine with some pyroxene.

period

A unit of the geologic time scale; smaller than an era, larger than an epoch. We are currently in the Quaternary period.

permafrost

Soil and rock which is below freezing for long periods of time.

permeability

The ability for a fluid to travel between pores, or, how connected the pores are within a rock or sediment.

Permian

The last period of the Paleozoic, 299-252 million years ago.

Permian Mass Extinction

The largest mass extinction in history, where an estimated 83% of genera went extinct. Linked to the Siberian Trapps as a cause.

permineralization

Style of fossilization where materials are replaced by minerals in groundwater fluids.

petroleum

A liquid fossil fuel derived from shallow marine rocks (also known as crude oil).

petrology

The study of rocks, either macroscopically or microscopically. This study is typically divided into one of the three rock types (e.g. igneous petrology).

phaneritic

Large, easy-to-see crystals within an igneous rock. This is common in intrusive rocks.

Phanerozoic

Meaning 'visible life,' the most recent eon in Earth's history, starting at 541 million years ago and extending through the present. Known for the diversification and evolution of life, along with the formation of Pangea.

phase diagram

Chart that show the stability of different phases of a substance at different conditions.

phenocryst

A large crystal within an igneous rock. These can be seen within phaneritic and porphyritic rocks.

phosphate

Minerals that are bonded with the phosphate anion, PO_4^{+3}.

phyllite

A rock more metamorphosed than slate, to the point that microscopic (but larger) mica gives the rock a glow, called a sheen. Crenulation, or small bends/folds in the foliation can be present.

piercing point

An object that is cut by a fault which allows the amount of movement to be determined. This is useful for all faults, but more commonly used in strike-slip faults.

placer

Deposit of heavy ores in stream or beach sediments.

plane bed

A specific layer of rock formed by flowing fluid, either in the lowest part of the lower flow regime or lower part of the upper flow regime.

planet

A large astronomical body that is neither a star nor a stellar remnant.

planetary system

The generic term for a group of planets and other bodies circling a star.

planetesimal

A body that could or did come together with many others under gravitation to form a planet.

plate

A solid part of the lithosphere which moves as a unit, i.e. the entire plate generally moves the same direction at the same speed.

plate boundary

Location where two plates are in contact, allowing a relative motion between the two plates. These are the locations where most earthquakes and volcanoes are found.

Plate tectonics

The theory that the outer layer of the Earth (the lithosphere) is broken in several plates, and these plates move relative to one another, causing the major topographic features of Earth (e.g. mountains, oceans) and most earthquakes and volcanoes.

platform

Part of a craton that is covered, mainly by sedimentary rocks.

playa

A dry lake bed in a desert valley.

pluton

A coherent body of intrusive rock (which formed underground) which is now at (or near) the surface.

pluvial lake

Lakes that form via increased precipitation with glacial climate shifts.

point bar

Depositional portion of a meandering channel.

point source

Pollution that comes from one known source.

Polar Cell

Part of the global circulation pattern where air sinks at the poles (90° latitude) and rises at 60° latitude.

polar desert

Deserts formed by descending air at the poles.

polarity

A molecule (like water) which has a positive side and a negative side.

polymorph

Minerals with the same composition and different crystal structures.

polymorphism

A specific chemical composition that forms different minerals and different temperatures and pressures. Quartz has several different polymorphs, including coesite, tridimite, and stishovite.

pore

Empty space in a geologic material, either within sediments, or within rocks. Can be filled by air, water, or hydrocarbons.

porosity

Amount of empty space within a rock or sediment, including space between grains, fractures, or voids.

porphyritic

An igneous rock with two distinctive crystal sizes.

porphyry

Large metallic mineral deposit that forms near magma bodies like plutons. Commonly contains copper, lead, zinc, molybdenum, and gold.

positive feedback

A process that exacerbates the effects of an input, amplifying the output. That is, a one-directional loop that self-reinforces change.

potentiometric surface

The height of the water table, if no confining layers or other hindrances present.

Precambrian

A term for the collective time before the Phanerozoic (pre-541 million years ago), including the Hadean, Archean, and Proterozoic. Known for a lack of easy-to-find fossils.

precession

Wobbles in the Earth's axis.

precipitation

The act of a solid coming out of solution, typically resulting from a drop in temperature or a decrease of the dissolving material.

Principle of Cross-Cutting Relationships

A geologic object can not be altered until it exists, meaning, the change to the object must be younger than the object itself.

Principle of Faunal Succession

The fossils found at any time are unique, and the fossils in layers of different ages have progressed and changed as time has moved forward. Fossils found in layers that are not as old have organisms that more resemble organisms that are alive today.

Principle of Lateral Continuity

Layered rocks can be assumed to continue if interrupted within its area of deposition.

Principle of Original Horizontality

Layered rocks are generally laid down flat at their formation.

Principle of Superposition

In an undisturbed sequence of strata, the rocks on the bottom are older than the rocks on the top.

Principle of Uniformitarianism

Idea championed by James Hutton that the present is the key to the past, meaning the physical laws and processes that existed and operated in the past still exist and operate today.

proglacial lake

Lake that forms next to a glacier because of crustal loading.

prokaryote

A type of single-celled organism with no nucleus.

Proterozoic

Meaning "earlier life," the third eon of Earth's history, starting at 2.5 billion years ago and ending at 541 million years ago. Marked by increasing atmospheric oxygen and the supercontinent Rodinia.

protolith

The rocks that existed before the changes that lead to a metamorphic rock, i.e. what rock would exist if the metamorphism was reversed.

provenance

The study of the components of a rock, mainly sedimentary rocks, and the information that can be obtained by understanding the origin of the components.

proxy indicator

A measurement which can specify a change in another system. For example, changes in climate can change the amount of certain isotopes of oxygen and carbon in sea creatures.

Pseudoscience

A method of investigation that claims to be scientific, but does not hold up to full scientific scrutiny. Examples include astrology, paranormal studies, young-Earth creationism, and cryptozoology (i.e. the study of creatures like Bigfoot and the Loch Ness Monster).

pumice

Low density, highly vesiculated, usually white to tan volcanic rock. Typically arises from felsic volcanism.

pyroclastic

Rocks (or rock textures) that are formed from explosive volcanism.

pyroclastic flow

A collapsed part of the eruption column that travels down at very hot temperatures and very fast speeds. They are the most dangerous immediate volcanic hazard.

pyroxene

$XY(Al,Si)_2O_6$, in which X typically equals Na, Ca, Mg, or Fe and Y typically equals Mg, Fe, or Al. Typically black to dark green, blocky, with two cleavages at ~90°. Common in mafic igneous rocks and some metamorphic rocks. Structure is a single chain of silica tetrahedra.

qualitative

An observation which is based on non-numerical data. While these types of observations are not preferred, they can still be useful.

quantitative

An observation which is based on numerical data. These observations are preferred because they can be used in calculations.

quartz

SiO2. Transparent, but can be any color imaginable with impurities. No cleavage, hard, and commonly forms equant masses. Perfect crystals are hexagonal prisms topped with pyramidal shapes. One of the most common minerals, and is found in many different geologic settings, including the dominant component of sand on the surface of Earth. Structure is a three-dimensional network of silica tetrahedra, connected as much as possible to each other.

quartzite

A metamorphosed sandstone.

Quaternary

The most recent, and current, period within the Cenozoic era, starting 2.58 million years ago.

radial drainage

Drainage pattern emanating from a high point.

radioactive

The process of atoms breaking down randomly and spontaneously.

rain shadow desert

Deserts that form as air loses moisture traveling over mountains.

raindrop impression

Small circular pits formed by raindrops impacting soft sediments.

Rayleigh wave

Surface waves that have an up and down motion.

recessional moraine

A terminal moraine that forms as a glacier melts.

recharge

Area where water infiltrates into the ground and adds to the overall groundwater.

recrystallization

The process of changing a mineral without melting.

rectangular drainage

Drainage pattern in an area of low topography, dominated by bedding planes, joints, and fracture patterns.

recurrence

Average time between earthquakes calculated based on past earthquake records.

red shift

The increase in wavelength of light resulting from the fact that the source of the light is moving away from the observer.

redox

Reactions that are related to the availability of oxygen. Many minerals or ions change their solubility based on redox conditions.

redshift

A change in starlight that occurs as light moves away from a source.

reduced

Reduction involves a half-reaction in which a chemical species decreases its oxidation number, usually by gaining electrons.

reef

A topographic high found away from the beach in deeper water, but still on the continental shelf. Typically, these are formed in tropical areas by organisms such as corals.

refining

Removing trace elements from desired elements.

reflection

Waves that bounce off of a boundary between mediums of different properties.

refraction

Waves that change direction due to changing speeds, typically caused by a change in density of the medium.

regional metamorphism

Metamorphism that occurs with large-scale tectonic processes, like collision zones.

regolith

Loose material that is a mixture of soil components and weathered bedrock sediments.

regression

Sea level fall over time.

relative dating

Determining a qualitative age of a geologic item in relation to another geologic item.

remediation

The process of cleaning up a polluted site.

renewable

A resource which is replaced on human time scales.

reservoir

Rocks which allow petroleum resources to collect or move.

resonance

An amplification of earthquake waves due to a structure of buildings or structures.

reverse fault

A dip-slip fault that has the hanging wall moving up with respect to the foot wall.

revolve

To move in a circular or curving course or orbit. Not to be confused with rotate, when something spins on an axis.

rhizolith

Root systems preserved in rocks.

rhyolite

General name of a felsic rock that is extrusive. Generally has a white, tan, or pink groundmass color.

Richter scale

A magnitude scale using the amplitude of shaking via a seismograph.

rift

Area of extended continental lithosphere, forming a depression. Rifts can be narrow (focused in one place) or broad (spread out over a large area with many faults).

rip current

Currents that push seaward.

ripple

Ridges of sediment that form perpendicular to flow in the lower part of the lower flow regime.

rivers

Channels of water that flow downhill due to gravity.

rock cycle

The concept that any rock type (igneous, sedimentary, and metamorphic) can change into another rock type under the right conditions over geologic time.

rock fall

Detached, free-falling rocks from very steep slopes.

Rodinia

The supercontinent that existed before Pangea, about 1 billion years ago. North America was positioned in the center of the land mass.

root wedging

A process where plants and their roots wedge into cracks in bedrock, and widen them.

rotate

To spin on an axis. Not to be confused with revolve, when something moves in a circular or curving course or orbit.

rotational slide

Movement of regolith along a curved slip plane.

rounding

How smooth or rough the edges are within a sediment.

runoff

Water that flows over the surface.

s wave

Second-fastest seismic wave that has a shear motion.

saltation

Silt and sand that is lifted from the bed and transported for short distances.

sandstone

A rock primarily made of sand.

saturation

A solution that has the maximum allowed dissolved component, and is unable to dissolve more.

schist

Rock more metamorphosed than phyllite, to the point that mica grains are visible. Larger porphyroblasts are sometimes present.

schistosity

Term for coarse grained, visible, platy minerals in a planar fabric, typical of schists.

science denial

The act of purposely ignoring or dissenting from science for political or cultural gains.

scientific method

The idea in science that phenomena and ideas need to be scrutinized using hypothesizing, experimentation, and analysis. This can eventually result in a consensus or scientific theory.

seamount

An eroded island. Since wave and weather action does not extend deep into the ocean, the root of the island is preserved as a seamount. Reefs can grow around seamounts.

sediment

Pieces of rock that have been weathered and possibly eroded.

sediment-hosted copper

Diagenetic copper deposit within sedimentary rocks.

sediment-hosted disseminated gold

Low grade, broad deposits of microscopic gold found in sedimentary rocks with diagenetic alteration.

sedimentary

Relating to sediment, pieces of rock that have been weathered.

sedimentary basin

A local or regional depression which allows sediments to accumulate.

sedimentary rock

Rocks that are formed by sedimentary processes, including sediments lithifying and precipitation from solution.

sediments

Pieces of rock that have been weathered and possibly eroded.

seismic anomoly

Areas that have an unpredicted change in seismic data, indicating a change in properties.

seismic gap

Length of fault without earthquake activity, due to a locked segment of a fault.

seismic wave

Energy that radiates from fault movement via earthquakes.

seismograph

Instrument used to measure seismic energy.

semidiurnal tide

Location with two unequal tide cycles per tidal day.

sequence stratigraphy

The study of changes in the rock record caused by changing sea level over time.

serpentinite

Rock formed from hydrothermal alteration of basalt, made of serpentine.

shale

A very fine-grained rock with very thin layering (fissile).

shear

Stress within an object that causes a side-to-side movement within an internal fabric or weakness.

shear force

Component of the gravitational force which pushes material downslope.

shear strength

The relationship between shear force and normal force in a block of material on a slope. When shear force is greater than normal force, mass wasting can occur.

sheetwash

Planar flow of water over land surfaces.

shield

An exposed part of a craton.

shield volcano

Volcano with a gentle slope, formed from low viscosity, low volatile, mafic, basaltic lava.

shock metamorphism

Metamorphism caused by bolide impacts.

shoreface

Part of the coastal depositional environment, near the tidal zone but below. Lower shoreface is the part of the coastline which is only disturbed by storm waves, upper shoreface is disturbed by typical, daily wave action.

shoreline

The part of the coastline which is directly related to water-land interaction, specifically the tidal zone and the range of wave base.

Siberian Trapps

One of the largest volcanic eruptions on Earth, with over 3 million cubic kilometers of lava erupted, based on evidence found in Siberia.

silicate

Mineral group in which the silica tetrahedra, SiO_4-4, is the building block.

silicon-oxygen tetrahedra

A anion structure of one silicon bonded to four oxygens, in the shape of a tetrahedron, with the silicon in the center and four oxygens at the corners of the structure. It has a net charge of -4, and can bond to cations to form silicate minerals.

sill

A sheet-like igneous intrusion that has intruded parallel to bedding planes within the bedrock.

siltstone

A rock made of primarily silt.

Silurian

The third period of the Paleozoic, 444-420 million years ago.

sinistral

A strike-slip or transform motion in which the relative motion is to the left. As viewed across the fault, objects will move to the left.

skarn

Carbonate rock that reacts with hot magmatic fluids, creating concentrated ore deposits, which include copper, iron, zinc, and gold.

slab

Name given to the subducting plate, where volatiles are driven out at depth, causing volcanism.

slate

Metamorphic rock with a strong foliation but no visible minerals, derived from mudstones or shales.

slaty cleavage

A microscopic foliation in slate, in which flat slabs and planes of rock develop.

slickenside

A polished surface of rock from fault movement, covered with grooves.

sliding stone

Rocks that move along thin ice sheets with high winds.

smelting

A process which chemically separates desired element(s) from ore minerals.

Snowball Earth hypothesis

A hypothesis which states the entire ocean froze and continental glaciation covered the planet about 700 million years ago.

snowline

The line between the zone of accumulation and the zone of ablation.

soft sediment deformation

Weak, typically saturated sediments that deform and contort before lithification.

soil

A type of non-eroded sediment mixed with organic matter, used by plants. Many essential elements for life, like nitrogen, are delivered to organisms via the soil.

soil creep

Very slow movement of the soil downhill.

soil horizon

Specific layers within a soil profile with specific properties.

soil profile

A hypothetical or real section cut through soil, showing the different layers (horizons) that exist.

solar nebula

Rotating, flattened disk of gas and dust from which the solar system originated.

solar system

The generic term for a group of planets and other bodies circling a star is planetary system. Our planetary system is the only one officially called "solar system," because our Sun is sometimes called Sol.

sole mark

A series of sedimentary structures formed on the base of a flow, eroding into underlying sediment. Examples include scour marks, flute casts, groove casts, and tool marks.

solid solution

Two or more elements that can easily substitute for each other, due to similarities in ionic size and charge.

solution

The act of taking a solid and dissolving it into a liquid. This commonly occurs with salts and other minerals in water.

SONAR

An acronym for SOund Navigation And Ranging, sonar uses sound waves to navigate and map surfaces. Sound waves created by an observer reflect off of surfaces and return to the observer. The amount of time it takes for the sound to return is a function of the distance the surface is from the observer. Bats use sonar to navigate through the dark. Ships use sonar to map the ocean floor.

sorting

The range of sediment sizes within a sediment or sediment within sedimentary rocks. Well sorted means the sediment has the same sizes, poorly sorted means many different sizes are present.

source rock

A rock that contains material which can be turned into petroleum resources. Organic-rich muds form good source rocks.

specific gravity

Related to density; the ratio of the weight of a mineral vs. the weight of an equal volume of water.

spectroscopy

The study of the details of light, which can tell you the chemical makeup of light and even the movement of a light source.

spheroidal weathering

A type of exfoliation where homogenous rocks weather into round shapes.

spit

A ridge of sediment that occurs out into a body of water, formed via longshore currents.

spring

A place where pressurized groundwater flows onto the surface.

spring tide

Highest high tide of the month.

stack

Rock spire that is offshore and a remnant of a rock layer.

star dune

Dunes that form from many different wind directions.

stoping

The process of surrounding bedrock being broken off and passed through a magma.

storm wave base

The depth that waves can reach in large storms, such as hurricanes.

straight channel

Channels that form straight, typically near the headwaters.

strain

The deformation that results from application of a stress.

stratigraphic correlation

Matching disconnected rock strata over large distances.

stratigraphy

The study of rock layers and their relationships to each other within a specific area.

stratovolcano

Volcano with steep sides, made of a composite of many types of eruption styles, from low viscosity mafic magma, higher viscosity felsic lava, but most commonly, intermediate andesite lava.

streak

The color(s) that a mineral produces when powdered or rubbed against a hard surface, usually a porcelain tile.

stream

A channelled body of water.

streams

A channelled body of water.

stress

Force applied to an object, typically dealing with forces within the Earth.

strike

A measure of a geologic plane's orientation in 3-D space. Used for beds of rocks, faults, fold hinges, etc. Using the right hand rule, dip is perpendicular, and to the right 90° of the strike.

strike slip

Faulting that occurs with shear forces, typically on vertical fault plaines as two fault blocks slide past each other.

strip mine

Mining that occurs as entire layers of ore and gangue are removed.

stromatolite

A fossil that forms as algal mats grow and capture sediment into mounds.

structural basin

A basin formed structurally by symmetrical synclines.

sturzstrom

Large and mysterious landslides that travel for long distances.

subduction

A process where an oceanic plate descends bellow a less dense plate, causing the removal of the plate from the surface. Subduction causes the largest earthquakes possible, as the subducting plate can lock as it goes down. Volcanism is also caused as the plate releases volatiles into the mantle, causing melting.

subduction zone metamorphism

Metamorphism that occurs in subduction zones, typically lower temperature and higher pressure.

subhedral

A mineral which only shows some characteristics of its true crystal habit, and is not perfectly grown.

subjective

An observation which is influenced by the observer's personal bias.

submarine canyon

Canyon carved into a continental shelf.

submarine fan

Broad cone of coarse sediment deposited from a submarine flow or turbidity flow.

submergent coastline

Features of a coastline where relative sea level is rising.

subsidence

The act of the land surface down-warping, typically referred to when discussing sedimentation or with rapid ground-water removal.

subsoil

Lower layer of the soil (B) which is a mixture of weathered bedrock, leached materials, and organic material. Has two sublayers: the upper part, or regolith (with more organic materials), and the lower part, saprolite, which is only slightly weathered bedrock.

substratum

Lowest layer of the soil (C), which is mechanically weathered (not chemically weathered) bedrock.

sulfate

Minerals bonded via a sulfate ion, SO_4-2.

sulfide

Minerals bonded via a sulfur (S-2) atom.

summer berm

Lower, seaward berm that forms with lower wave energy in summer months.

supercontinent

An arrangement of many continental masses collided together into one larger mass. According to the Wilson Cycle, this occurs every half billion years or so.

superfund site

A federally-supported pollution clean-up effort.

supergene enrichment

Oxidation that occurs in sulfide deposits which can concentrate valuable elements like copper.

supernova

Large explosion when the largest stars end fusion; cause of the formation of heavy elements in the Universe, like gold and uranium.

surf zone

Shoreline area of breaking waves.

surface mine

Mining that occurs near the Earth's surface.

surface wave

Seismic waves that only move along the surface, mainly R waves and L waves.

suspended load

Bedload sediments that can be carried by higher-velocity flows.

syncline

A U-shaped, upward-facing fold with younger rocks in its core.

system

An interconnected set of parts that combine and make up a whole.

tafoni

Rounded cavities within rocks that form in various ways, including mineral growth, mainly salt.

talus

Loose blocks of rock that fall down from steep surfaces and cover slopes.

tar sand

Sands or sandstones that contain high-viscosity petroleum.

tectonic

Relating to the movement of plates of lithosphere.

Temperature

The measure of the vibrational (kinetic) energy of a substance.

tension

Stresses that pull objects apart into a larger surface area or volume; stretching forces.

tephra

General term for solid, but fragmented, material erupted from a volcano. Has three subcomponents: ash (<2mm), lapilli (2-64 mm), blocks and bombs (>64mm).

terminal moraine

Moraine that forms at the end of a glacier.

terrace

An elevated erosional surface caused by glacial or fluvial action.

terrane

A geological province which is added (accreted) to a continental mass via subduction and collision.

terrestrial

Depositional environments that are on land.

texture

Arrangement of minerals within a rock.

thalweg

Deepest part of a meandering channel.

theory

An accepted scientific idea that explains a process using the best available information.

thermohaline circulation

A connected global ocean circulation pattern that distributes water and heat around the globe.

thick-skinned

Faulting that is deep into the crust, and typically involves crystalline basement rocks.

thin-skinned

Faulting that is not deep into the crust, and typically only involves sedimentary cover, not basement rocks.

thrust fault

A low-angle reverse fault, common in mountain building.

tidal day

The amount of time that the moon takes to appear over the same location of Earth, slightly more than 24 hours.

tidal flat

Wide and flat area of land covered by ocean water during high tide, but exposed to air by low tide.

tide

Movements of water (rising and falling) due to the gravity of the moon and sun. This is most often seen in marine settings.

till

General term for very poorly sorted sediment that is of glacial origin.

tillite

Term for a rock made definitively of glacial till.

tombolo

Sand bar that connects a stack and the shore.

tomography

A process of using 3D seismic arrays to get subsurface images.

topsoil

Upper layer of soil, made mainly out of organic material.

trace fossil

Evidence of biologic activity that is preserved in the fossil record, but it not the organism itself. Examples include footprints and burrows. Ichnology is the study of trace fossils.

trade wind desert

Desert that forms near 30 degrees latitude due to atmospheric circulation.

trade winds

Wind patterns that move from east to west near the equator, due to global circulation patterns.

transform

Place where two plates slide past each other, creating strike slip faults.

transgression

Sea level rise over time.

translational slide

A landslide that moves along an internal plane of weakness.

transpression

A segment along a transform or strike-slip fault which has a compressional component, sometimes creating related thrust faulting and mountains.

transtension

A place along a transform or strike-slip fault with an extensional component, sometimes including normal faulting, basin formation, and volcanism.

trap

A geologic circumstance (such as a fold, fault, change in lithology, etc.) which allows petroleum resources to collect.

travertine

Porous, concentric, or layered variety of carbonate that forms with often heated water in springs and/or caves.

trellis drainage

A drainage pattern which forms between ridge lines in deformed (typically sedimentary) rocks.

trench

Deepest part of the ocean where a subducting plate dives below the overriding plate.

Triassic

The first period of the Mesozoic era, from 252-201 million years ago.

tributary

A natural stream that flows into a larger river or other body of water.

trigger

An event that causes a landslide event. Water is a common trigger.

triple junction

Place where three plate boundaries (typically divergent) extend from a single point at 120° angles.

truncated spur

An eroded arête that forms a triangular shape.

tsunami

A series of waves produced from a sudden movement of the floor of a ocean basin (or large lake), caused by events such as earthquakes, volcanic eruptions, landslides, and bolide impacts.

tufa

Porous variety of carbonate that form in relatively unheated water, sometimes as towers and spires.

tuff

Rocks made from pyroclastic tephra: either ash, lapilli, and/or bombs. Tephra type can be used as an adjective, i.e. ash-fall tuff. If deposited hot, where material can fuse together while hot, the rock is then called a welded tuff.

turbidite

Turbidite is the rock that forms from a turbidity flow, a relatively coarse and dense sediment transported to the abyssal plain.

turbidity current

Dense flow of sediment that goes down submarine canyons, forming submarine fans and turbidites.

ultramafic

An igneous rock with extremely low silica composition, being made of almost all olivine and pyroxene. Ultramafic rocks contain very low amount of silica and are common in the mantle. Primary ultramafic rocks are komatiite (extrusive) and peridotite (intrusive).

unconformity

Missing time in the rock record, either because of a lack of deposition and/or erosion.

underground mine

Mining that occurs within tunnels and shafts inside the Earth.

universal solvent

A chemical that can dissolve a wide range of other chemicals.

universe

All of space and time and their contents, including planets, stars, galaxies, and all other forms of matter and energy.

vadose zone

Place where pores are filled with some water and some air, above the water table.

valley glacier

An alpine glacier that fills a mountain valley.

varve

A type of lamination that is cyclical, perhaps seasonal or diurnal.

vent

Opening of a volcano where lava can erupt.

ventifact

Rock with abraded surfaces formed in deserts.

vertebrate

An animal that possesses a spinal column or backbone.

vesicular

An extrusive rock filled with small bubble structures, frozen in place as gases escaped from the cooling lava.

viscosity

The resistance of a fluid to flow, where a high value means a fluid which does not like to flow (like toothpaste), and a low value means a fluid which flows easily (like water).

volatiles

Components of magma which are dissolved until it reaches the surface, where they expand. Examples include water and carbon dioxide. Volatiles also cause flux melting in the mantle, causing volcanism.

volcanic arc

Place with a chain of mountain volcanism on a continent, from oceanic-continental subduction.

volcano

Place where lava is erupted at the surface.

volcanogenic massive sulfide

Metallic mineral deposit which forms near mid-ocean ridges.

Wadati-Benioff zone

A zone of earthquakes that descend into the Earth with the subducting slab. This is commonly used as evidence for plate tectonics.

water right

A purchase or claim to a legal allotment of a water source, obtained through the state government, such as a spring, stream, well, or lake.

water table

The depth of the groundwater system below which has pore space 100% filled with water.

wave base

The depth in which the movement of waves can be felt, specifically by sediments. This is approximately equal to 1/2 the wavelength. Wave base can change depending on fair weather verses stormy weather.

wave crest

Top of a wave.

wave cut platform

Flat erosional surface cut by wave action.

wave height

Twice the amplitude, or, the distance between the crest and trough of a wave.

wave notch

Erosional notch in bedrock cut by waves.

wave period

The time between like parts of a wave passing a fixed point.

wave train

A series of waves that form and move as a group.

wave trough

Bottommost part of a wave.

wave velocity

Speed at which a wave travels past a fixed point.

wavelength

The distance between any two repeating portions of a wave (e.g., two successive wave crests).

weather

Current conditions within the atmosphere.

weathering

Breaking down rocks into small pieces by chemical or mechanical means.

westerlies

Winds that move from west to east between 30° and 60° latitude due to global circulation patterns.

Wilson Cycle

The cycle of opening ocean basins with rifting and seafloor spreading, then closing the basin via subduction and collision, creating a supercontinent.

winter berm

Higher, landward berm that forms with higher wave energy in winter months.

xenolith

A piece of foreign rock that has been incorporated into a magma body. This can be a different type of magma, or a mantle xenolith, a rock from the mantle brought up near the surface.

yardang

Erosional rock face caused by sand abrasion.

yazoo stream

A tributary that runs parallel to a main stream within the floodplain.

yield point

An amount of strain where the substance has a maximum amount of elastic deformation and switches to ductile deformation.

zircon

ZrSiO4. Relatively chemically inert with a hardness of 8.5. Common accessory mineral in igneous and metamorphic rocks, as well as detrital sediments. Uranium can substitute for zirconium, making zircon a valuable mineral in radio-metric dating.

zone of accumulation

Part of a glacier in which there is a net gain over the course of a year.

zone of melting

Part of a glacier which has a net loss of material over the course of a year.

VERSION NOTES

Most chapters in this textbook match the organization of An Introduction to Geology (CC BY NC SA) with the exception of an added final chapter, chapter 17, which was adapted from parts of Chapter 22 of Physical Geology, 2nd edition (CC BY), and parts of Chapter 7 of OpenStax Astronomy, 2nd edition (CC BY).

Overall or Major Changes

- Chapters and data: Updated charts and data to the most recently available.
- Figure captions: Edited some existing figure captions for clarity.
- Existing tables: Adapted some existing tables to enhance quality and clarity.
- Embedded assessment questions: Updated embedded assessment questions to match the most recent data available.
- In-copyright figures: Identified and replaced in-copyright figures with openly-licensed or original work. Obtained permission to release figures by Peter Davis in Chapter 6 Metamorphic Rocks under CC BY NC SA 4.0 license.
- Glossary: Added/edited glossary terms and linked within chapter text: hydraulic, igneous, rivers, sedimentary, tectonic, vertebrate, solar system, revolve, asteroids, AU, comets, meteor, reduction, oxidation, differentiation, K (kelvin), solar nebula, moons, planetesimal, planet, dwarf planet, rotate, crater, mountain, astronomy, diameter, metal, atmosphere, electron, gamma ray, universe, exoplanetary, exoplanet, evaporate, Kuiper belt, Oort cloud, active margin, albedo, alluvium, alpine glacier, angular unconformity, dendritic drainage, epoch, fissile, groin, headwaters, landslide, Milankovitch Cycles, nonmetallic, oceanic crust, oxide, positive feedback, rock cycle, sill, tsunami, water table.
- Accessibility: Updates and expansion of alternative text (AltText) for figures are ongoing in the HTML/Pressbooks version of this book throughout its first year.

Specific Chapter-Level Changes

Chapters 1-3, 5-7, 9-16

- Edits as described in Overall or Major Changes

Chapter 4: Igneous Processes and Volcanoes

Source: An Introduction to Geology (CC BY NC SA)

- Removed some embedded videos and edited text accordingly: Alvin video Part 3; one out of two hotspot videos; and Japan's Mount Ontake volcano erupted/eruption video
- Removed text section describing carbonatites
- Removed sections 4.3.2 Decompression Melting, 4.3.3 Flux Melting, 4.3.4 Heat-Induced Melting, and 4.4 Partial Melting and Crystallization
- Removed Key concept "Explain partial melting and fractionation, and how they change magma compositions"

Chapter 8: Earth History

Source: An Introduction to Geology (CC BY NC SA)

- Removed objectives "Explain the big-bang theory and origin of the elements" and "Explain the solar system's origin and the consequences for Earth"
- Removed sections 8.1 Origin of the Universe and 8.2 Origin of the Solar System: The Nebular Hypothesis

Chapter 17: Origin of the Universe and our Solar System

Sources: Section 22.1 of Chapter 22 from Physical Geology, 2nd edition and sections 7.1, 7.2, 7.3, and 7.4 of Chapter 7 from OpenStax' Astronomy, 2nd edition

- Changed introduction section: adapted the Summary of Chapter 22 from Physical Geology
- Removed "LINK TO LEARNING" sections from OpenStax
- Removed "Voyagers in Astronomy" from OpenStax section 7.1
- Removed "Names in the Solar System" from OpenStax section 7.1
- Removed "There's No Place Like Home" from OpenStax section 7.2
- Removed "Geological Activity" from OpenStax section 7.2
- Added new Summary section adapted from https://openstax.org/books/astronomy/pages/7-summary (trimmed down for this book)
- Inserted new text block from An Introduction to Geloogy (specifically, from Section 8.2.1)
- Imported and linked glossary terms from OpenStax throughout chapter
- Removed Key Concepts: "Model the solar system with distances from everyday life to better comprehend distances in space," "Explain how astronomers can tell whether a planetary surface is geologically young or old," and "Describe the characteristics of planets that are used to create formation models of the solar system"
- Added new Key Concepts: "Explain the formation of the universe and how we observe it" and "Understand the origin of our solar system"

Made in the USA
Monee, IL
27 May 2024